Advances in Food Safety and Food Microbiology

New Trends in Marine and Freshwater Toxins

Food and Safety Concerns

ADVANCES IN FOOD SAFETY AND FOOD MICROBIOLOGY

Dr. Anderson de Souza Sant'Ana
and Dr. Bernadette D. G. M. Franko (Series Editors)
Department of Food and Experimental Nutrition,
Faculty of Pharmaceutical Sciences, University of Sao Paulo, San Paulo, Brazil

Food safety is a specific area of food science focused in the study of microbiological, chemical and physical hazards in foods and beverages. Several tools, systems and approaches have been developed last years to systematize the management of food production chain aiming at protecting public health. Food Microbiology is an interesting and exciting field of microbiology concerned with the study of foodborne microorganisms, their occurrence, interactions and responses to the environmental found in foods and beverages. Depending mainly on the intrinsic or extrinsic properties of foods and beverages, one or more of microorganisms such as protozoan, viruses, yeasts, moulds and bacteria or toxins will be of significance.

Food safety is a specific area of food science focused in the study of microbiological, chemical and physical hazards in foods and beverages. Several tools, systems and approaches have been developed in recent years to systematize the management of food production chain aiming at protecting public health.

Food Microbiology is an interesting and exciting field of microbiology concerned with the study of foodborne microorganisms, their occurrence, interactions and responses to the environmental found in foods and beverages. Depending mainly on the intrinsic or extrinsic properties of foods and beverages, one or more of microorganisms such as protozoan, viruses, yeasts, moulds and bacteria or toxins will be of significance.

In fact, Food Microbiology and Food Safety can be considered a "boiling and changing" field of microbiology and food science, respectively. These fields have risen to that status mainly in the last 30 years. Data from epidemiological studies of foodborne disease outbreaks or from investigations of spoilage episodes have highlighted relevant changes in the pillars supporting the microbial ecology of foods: consumer habits, practices and consumers health status, food preservation techniques, responses of microorganisms to preservation methods, microbial expression of virulence factors and food production chain systems.

In addition, a strong trend in food microbiology and safety is the application of computational, statistical, molecular biology and chemical approaches, resulting in a deeper approach to study and understand the microbial interactions with foods and among them. Trends indicate that the insertion of these four sciences into food safety and microbiology will make an important difference in the evolving of studies of microbial ecology and food safety. Based on this scenario, it is the main aim of the Series "Advances in Food Safety and Food Microbiology" to cover "hot topics" in the field of Food Microbiology and Food Safety. Coverage will be given to applied food microbiology and safety with a collection of texts focused on the study of spoilage, pathogens and industrial applications of foodborne microorganisms. Focus will be given to responses of microorganisms to preservation methods, consumer food safety practices, advances in microbiological methods and on the applications of statistical, molecular biology, computational and the "omics" in food microbiology and safety. Risk analysis, integration between epidemiology and food microbiology, chemical and physical contaminants of foods are also given consideration.

We hope that the books in this Series will be widely used by food microbiologists and those concerned with food safety in their studies or as references for new approaches to be considered in an effort to continuously evolve food microbiology and food safety from farm to fork, from tradition to technology.

Advances in Post-Harvest Treatments and Fruit Quality and Safety
Manuel Vázquez and José A. Ramírez de Leon
2011. ISBN: 978-1-61122-973-8 (Hardcover)
ISBN: 978-1-61470-700-4 (E-book)

Clostridium Botulinum: A Spore Forming Organism and a Challenge to Food Safety
Christine Rasetti-Escargueil and Susanne Surman-Lee (Editors)
2011. ISBN: 978-1-61470-575-8 (Hardcover)
ISBN: 978-1-61470-653-3 (E-book)

Enterococcus and Safety
Teresa Semedo-Lemsaddek, Maria Teresa Barreto-Crespo and Rogério Tenreiro (Editors)
2011. ISBN: 978-1-61470-569-7 (Hardcover)
ISBN: 978-1-61470-664-9 (E-book)

Probiotic and Prebiotic Foods: Technology, Stability and Benefits to Human Health
Nagendra P. Shah, Adriano Gomes da Cruz and Jose de Assis Fonseca Faria (Editors)
2011. ISBN: 978-1-61668-842-4 (Hardcover)
ISBN: 978-1-61728-825-8 (E-book)

Stress Response of Foodborne Microorganisms
Hin-Chung Wong (Editor)
2011. ISBN: 978-1-61122-810-6 (Hardcover)
ISBN: 978-1-61324-410-4 (E-book)

Bacteriophages in Dairy Processing
Andrea del Luján Quiberoni and Jorge Alberto Reinheimer (Editors)
2012. ISBN: 978-1-61324-517-0 (Hardcover)

Essential Oils as Natural Food Additives: Composition, Applications, Antioxidant and Antimicrobial Properties
Luca Valgimigli (Editor)
2012. ISBN: 978-1-62100-241-3 (Hardcover)
ISBN: 978-1-62100-282-6 (E-book)

Foodborne Protozoan Parasites
Lucy J. Robertson and Huw V. Smith (Editors)
2012. ISBN: 978-1-61470-008-1 (Hardcover)
ISBN: 978-1-61942-569-9 (E-book)

Molecular Typing Methods for Tracking Foodborne Micoorganisms
Steven L. Foley, Rajesh Nayak and Timothy J. Johnson (Editors)
2012. ISBN: 978-1-62100-643-5 (Hardcover)
ISBN: 978-1-62100-728-9 (E-book)

On-Farm Strategies to Control Foodborne Pathogens
Todd R. Callaway and Tom S. Edrington (Editors)
2012. ISBN: 978-1-62100-411-0 (Hardcover)
ISBN: 978-1-62100-480-6 (E-book)

Pathogenic Vibrios and Food Safety
Yi-Cheng Su (Editor)
2012. ISBN: 978-1-62100-866-8 (Hardcover)
ISBN: 978-1-62100-903-0 (E-book)

Predictive Mycology
Philippe Dantigny and Efstathios Z. Panagou (Editors)
2012. ISBN: 978-1-61942-675-7 (Hardcover)
ISBN: 978-1-61942-684-9 (E-book)

Progress on Quantitative Approaches of Thermal Food Processing
Vasilis P. Valdramidis and Jan F. M. Van Impe (Editors)
2012. ISBN: 978-1-62100-842-2 (Hardcover)
ISBN: 978-1-62100-899-6 (E-book)

New Trends in Marine and Freshwater Toxins: Food and Safety Concerns
Ana G. Cabado and Juan Manuel Vieites (Editors)
2012. ISBN: 978-1-61470-324-2 (Hardcover)
2013. ISBN: 978-1-62257-787-3 (Softcover)
ISBN: 978-1-61470-398-3 (E-book)

ADVANCES IN FOOD SAFETY AND FOOD MICROBIOLOGY

NEW TRENDS IN MARINE AND FRESHWATER TOXINS

FOOD AND SAFETY CONCERNS

ANA G. CABADO
AND
JUAN MANUEL VIEITES
EDITORS

New York

For permission to use material from this book please contact us:
Telephone 631-231-7269; Fax 631-231-8175
Web Site: http://www.novapublishers.com

LIBRARY OF CONGRESS CATALOGING-IN-PUBLICATION DATA

ISBN: 978-1-62257-787-3

Library of Congress Control Number: 2011946168

Published by Nova Science Publishers, Inc. † New York

Contents

PREFACE

This book, included in the Series "Advances in Food Safety and Food Microbiology" and edited by Dr. Anderson De Souza Sant'Ana, was a challenge of writing about a very exciting topic. We asked ourselves if another book of phycotoxins was needed. Other books in the same field have been published but we have undertaken this task to approach what we believed was missing. With the enormous growth of information and debate in phycotoxins in recent years the need for a book in this field has risen. This book covers topics in food safety related to marine and freshwater toxins, HABs and early warning systems. This project also deals with the existing methods to detect phycotoxins and toxin monitoring systems in different parts of the world. In addition there are selected bibliographies listing recent and the most significant references in the field. In our minds, such book would satisfy the needs of graduate students, scientists and authorities.

I was very fortunate that so many of the leading researchers in the field were willing to take the time to contribute with such high quality chapters. It was a pleasure to work with them and it was gratifying that authors were able to take time out of their busy schedules to write the chapters. We wish to express our deepest appreciation to them.

I would like to thank Nova Science Publishers, Inc and Dr. Anderson De Souza Sant'Ana for inviting me to edit this book. We are especially grateful to all professionals in ANFACO-CECOPESCA, in particular my colleagues in the Area of Microbiology and Biotoxins who are always there and helped in the preparation of this book.

To our families for their patience and understanding
The Editors

In: New Trends in Marine and Freshwater Toxins
Editors: A. G. Cabado and J. M. Vieites
ISBN: 978-1-62257-787-3

Chapter 1

Introduction to Natural Seafood and Freshwater Toxins: Geographic Distribution and Food-Borne Illness

Vitor Vasconcelos
Laboratory of Ecotoxicology, Genomics and Evolution,
Centro Interdisciplinar de Investigação Marinha e Ambiental (CIIMAR/CIMAR),
Departament of Biology, Faculty of Sciences, University of Porto, Portugal

Abstract

Organisms usually produce toxins to defend themselves from predators or from competitors or to attack other organisms and use them as food items. Nevertheless, there is a wide array of toxins, from fresh water or from marine origin that have no primary function well defined yet. Among those are toxins produced by cyanobacteria, diatoms, dinoflagelates and others present in many other marine organisms.

In freshwaters, the main toxins are produced by cyanobacteria, and may have peptide or alkaloid structure, being their effects mainly hepatotoxic, neurotoxic, cytotoxic or dermatotoxic. In marine and brackish waters, toxins are produced by a higher dioversity of taxa and probably much more diverse in terms of chemistry, from relatively simple molecules such as saxitoxins and analogues, to huge molecules such as palytoxins. Nevertheless, their effects are mostly neurotoxic, with few hepatotoxic exceptions.

Freshwater cyanotoxins are prevalent all over the world, from temperate reservoirs in Europe to volcanic lakes in North America, from hot springs to the Antarctica region. Marine toxins seem to be more diverse and with a higher incidence in tropical regions, but recently reports of the so called "warm water toxins" are being issue from temperate areas of the world.

The diversity of toxins from marine and freshwater origin increases every year with the description of new toxins or new variants of known toxins. This chapter describes the main toxins studied all over the world with a special emphasis on the freshwater toxins microcystins, cylindrospermopsin, anatoxin-a, the brackiswater toxin nodularin and the marine toxins tetrodotoxin, okadaic acid, domoic acid, palytoxin, ciguatoxin. β-N-methylamino-L-alanine (BMAA) and saxitoxins and analogues will also be referred, being examples of toxins that occur both in freshwater and in marine environments.

1.1. Freshwater Toxins Diversity and Occurrence

1.1.1. Microcystins (MC)

The most common and important hepatotoxins produced by cyanobacteria are the microcystins (MC). They include more than 70 structural variants (Zurawell et al., 2005) being MC-LR the most prevalent one (Dawson, 1998). MC are potent inhibitors of protein phosphatases 1A and 2A for animals and higher plants (Hastie et al., 2005). Those proteins are involved in several physiological and molecular processes (Sheen, 1993, Takeda et al., 1994).

1.1.2. Cylindrospermopsin (CYN)

CYN is an alkaloid, a sulfate ester of a tricyclic guanidine moiety (rings A, B & C), with a uracil ring (D) (1) and its zwitterionic nature makes it a highly water soluble molecule. CYN is a very stable compound showing no degradation in boiling temperature (100º C for 5 minutes) (Chiswell et al.,1999). When exposed to pH changes between 4 and 10 for a period of 8 weeks it only degrades 25% and in the solid state or in pure aqueous solutions, CYN does not show any degradation. *Cylindrospermospsis raciborskii* was the first species described as producer of CYN but other cyanobacterium species have been until present identified as CYN producers, namely *Umezakia natans* (Harada *et al*., 1994), *Aphanizomenon ovalisporum* (Banker *et al*., 1997), *Anabaena bergii* (Schembri *et al*., 2001), *Raphidiopsis curvata* (Li *et al*., 2001), *Aphanizomenon flos-aquae* (Preußel *et al*., 2006), *Anabaena lapponica* (Spoof *et al*., 2006) and more recently *Lyngbya wollei* (Seifert *et al*., 2007). Reports from Australia, Japan, China, Israel, Brazil, Spain, Germany, Finlad and USA among others shows that CYN has a worldwide distribution (Moreira et al., in pub).

The first report of human toxicity associated with CYN has been attributed to the Palm Island hepatoenteritis incident in 1979, where 148 people were hospitalized. During October of that year a cyanobacterial bloom occurred in the Solomon Dam, the sole source of the reticulated water supply on Palm Island, a tropical island about 28 Km off the northeast Australian coast. After it became dense, the bloom was treated with copper sulphate and five days after this treatment the first case of hepatoenteritis occurred. An epidemiological investigation conducted by Bourke *et al.* (1983) revealed that the people affected were of Aboriginal descent, mainly children, which had drunk the water from wells that are supplied by the reticulated system. Until now no other cases of human intoxication by CYN were ever reported.

CYN has been classified as a cytotoxic compound, although liver and kidney seem to be the main target organs. Exposure to CYN caused a depletion in rat hepatocytes GSH attributed to the inhibition of the final common pathway of GSH synthesis and, not to the increase in GSH efflux or GSH utilization (Runnegar *et al*., 1994). More recently, CYN was proved to inhibit the eukaryotic protein synthesis apparatus, and did not correlate with the ribosome content. These findings suggest that CYN may target another protein of the translation system but not the ribossomes (Froscio et al., 2008). CYN also increases the frequency of necrotic cells and necrotic cell death in a dose and time-dependent manner and has a very small impact on apoptosis (Lankoff et al., 2007).

1.1.3. Anatoxin-a

Anatoxin-a, is produced by several cyanobacterial genera: *Aphanizomenon, Anabaena, Microcystis, Planktothrix, Raphidiopsis, Arthrospira, Cylindrospermum, Phormidium* and *Oscillatoria* (Park, Watanabe et al. 1993; Bumke-Vogt, Mailahn et al. 1999; Namikoshi, Murakami et al. 2003; Viaggiu, Melchiorre et al. 2004; Araóz, Nghiem et al. 2005; Ballot, Krienitz et al. 2005; Gugger, Lenoir et al. 2005). Anatoxin-a has a 50% median lethal dose (LD_{50}) of 250 $\mu g.g^{-1}$ for the mouse after i.p. administration (Rogers, Hunter III et al. 2005). It is a very potent cholinergic agonist that competes with the acetylcholine for nicotinic and muscarinic receptors (Aronstam and Witkop 1981; Campos, Durán et al. 2006). Intoxication in terrestrial vertebrates, leads to symptoms such as loss of muscle coordination, difficulty in breathing, staggering, muscle fasciculation, gasping and convulsions (Smith 2008).

1.1.4. Nodularins (NOD)

Nodularin (NOD) is a cyclic pentapeptide produced by the cyanobacterium *Nodularia spumigena* that inhibits protein phosphatases and causes liver disruption that may end in a severe liver failure and death by hipovolemic shock (Honkanen et al., 1991). Both PP2A and PP1 were potently inhibited (IC50 = 0.026 and 1.8 nM, respectively) by nodularin, whereas PP2B was inhibited to a lesser extent (IC50 = 8.7 microM). Nodularin had no apparent effect on PP2C.

1.2. Marine Toxins Diversity and Occurrence

1.2.1. Tetrodotoxin (TTX)

Tetrodotoxin is a potent low molecular weight neurotoxin that was first isolated in 1909 from the ovaries of the puffer fish and named after the Tetraodontidae puffer fish family (Tahara and Hirata, 1909, Soong and Venkatesh, 2006). However, its chemical structure was only discovered later in 1964 (Tsuda *et al.*, 1964, Woodward 1964, Goto *et al.*,1965). TTX main occurrence is in the warm waters regions of the Indo-Pacific ocean. This geographical area is home to the main TTX-bearing species like fish belonging to the families Tetraodontidae (*Takifugu niphobles, T. pardalis, T. rubripes, Arothron firmamentum, Lagocephalus lunaris, Tetraodon nigroviridis*)(Kanoh, 1988, Fuchi et al., 1991, Khora et al., 1991), Diodontidae (*Diodon hystrix*) (Malpezzi et al., 1997), Canthigasteridae (*Canthigaster rivulata*) (Sugita et al., 1987), Gobiidae (*Yongeichthys criniger*) (Noguchi et al., 1973), gastropods (*Rapana rapiformis, R. venosa venosa, Nassarius glans, Charonia sauliae, Zeuxis siquijorensis, Tutufa lissostoma, Babylonia japonica*) (Narita et al., 1981, 1984, Noguchi et al., 1981, 1982, 1984, Hwang et al., 1991, 2005,), blue-ringed octopus (*Hapalochlaena maculosa*) (Sheumack and Howden, 1978, Yotsu-Yamashita et al., 2007), horseshoe crab (*Carcinoscorpius rotundicauda*) (Kungsuwan et al., 1987), xanthid crabs (*Demania toxica, Lophozozymus pictor, Atergatis floridus*) (Noguchi et al., 1983, Tsai et al., 2006,) , arthropods parasites belonging to the family Caligidae (*Taeniacanthus* sp., *Pseudocaligus fugu*)(Ikeda et al., 2006, Ito et al., 2006), toxic starfish, (*Astropecten scoparius, Astropecten latespinosus,* Astropecten polyacanthus) (Noguchi et al., 1982, Maruyama et al., 1984, 1985, Narita et al., 1987, Shi and Hwang, 2001) , ribbonworms (Lineus fuscoviridis, Tubulanus punctatus, Cephalothrix sp.) (Mioyazawa et al., 1988, Ali et al., 1990, Asakawa et al., 2003), flatworms (Planocera spp., Planocerid sp.) (Mioyazawa et al., 1986, 1987, Ritson-Williams et al., 2006).

1.2.2. Okadaic Acid (OA)

Okadaic acid (OA) inhibits protein phosphatases 1 and 2A but is a polyether produced by *Dynophysis* species and has a neurotoxic effect (Tanti et al., 1991). Okadaic acid (OA) is a specific inhibitor of the serine-threonine phosphatases 2A (PP2A) that are enzymes required in several processes during embryonic development. Ingestion of OA contaminated shellfish by humans produces the Diarrheic Shellfish Poisoning (DSP). (More information related to OA and analogues can be found in chapter 5 and other chapters of this book).

1.2.3. Domoic Acid (DA)

Domoic acid (DA) is synthesized by marine algae, such as the red algae *Chondria armata* and species of the diatom genus *Pseudo-nitzschia* (Takemoto and Daigo, 1958,Bates, 2000). DA is an analogue of kainic acid (KA), which shows excitotoxic activity. Both DA and KA are similar structurally to glutamate, which is the predominant neurotransmitter in the central nervous system. This structural similarity, allows DA and KA to bind the glutamate receptor [GluRs] family, inducing neuroexcitatory and neurotoxic effects (Hampson and Manalo, 1998, Clayden et al., 2005, Hald et al., 2007).

The first documented case of DA intoxication occurred in 1987, when at least four people died and over hundred others suffered severe neurological and gastrointestinal disorders (Bates et al., 1989). Since then, other DA intoxication events have been resulting in mass mortalities of sea birds and sea lions (Sierra et al., 1997, Lefebvre et al., 1999, Silvagni et al., 2005, Busse et al., 2006). DA has been recognized as a harmful food web-transferred phycotoxin to humans, through the consumption of mussels (Quilliam and Wright, 1989) and to marine mammals and birds, through planktivorous fish and krill (Bargu et al., 2002, Bargu and Silver, 2003).

DA-induced neurotoxicity in fish is poorly understood and the majority of the available information is related to DA-induced excitotoxicological signs that appeared following the intraperitoneal (i.p.) injection of DA. When juvenile leopard sharks (*Triakis semifasciata*) were injected intraperitoneally with doses that ranged from 9-27 mg DA kg^{-1} body weight (bw), no signs of toxicity were observed (Schaffer et al., 2006). In contrast, the i.p. injection of doses from 1-14 mg DA kg^{-1} bw induced excitotoxicological signs and death in anchovies (*E. mordax*) (Lefebfre et al., 2001). Further evidence of DA neurotoxicity in fish was also obtained after the i.p. injection of killifish (*Fundulus heteroclitus*) with 5 mg DA kg^{-1} bw, which led to behavioural alterations and increased the neuronal expression of c-Fos (Salierno et al., 2006). Similarly, the i.p injection of coho salmon (*Oncorhynchus kisutch*) with doses of 6.3 ± 0.6 mg DA kg^{-1} bw resulted in neurotoxic signs (Lefebvre et al., 2007). In addition, changes in metabolic activity in the brain of the Atlantic salmon (*Salmo salar*) were found after the i.p. injection of 6.0 mg DA kg^{-1} bw (Lefebvre et al., 1999).

1.2.4. Palytoxin (PlTX)

Palytoxin (PlTX) is a large, very complex molecule with a long polyhydroxylated and partially unsaturated aliphatic backbone, containing 64 chiral centers (Kan et al., 2001). This latter feature, coupled with the presence of eight double bonds able to exhibit cis/trans-isomerism means that PlTX can have more than 10^{21} stereoisomers (Katikou, 2007). PlTX was originally isolated in Hawaii from the tropical soft coral *Palythoa* sp., a zoanthid (Moore and Scheuer, 1981). In 1981, its unique chemical structure was elucidated independently by two groups (Moore and Bartolini, 1981, Uemura et al., 1981) while in 1994 it was fully synthesized for the first time (Suh and Kishi, 1994). Since then, chemical structures of some of PlTX analogues were also achieved, e.g. ostreocin (Usami et al., 1995, Ukena et al., 2001), mascarenotoxins (Lenoir et al., 2004, 2006) and ovatoxin (Ciminiello et al., 2008). Notably, all are produced by dinoflagellates from the genus *Ostreopsis*.

Although primarily found on *Palythoa* spp., PlTX was also detected in organisms living in close association with the colonial zoanthids (Gleibs and Mebs, 1999). Moreover, PlTX and analogues were extracted from many other marine organisms (see (Wu, 2009) for an extensive list) including primary producers such as the red alga *Chondria crispus* (Maeda et al., 1985) and the benthic dinoflagellates *Ostreopsis* spp. (Usami et al., 1995). In addition, bacteria associated with preceding organisms have also been studied as a possible source for the production of this non proteinaceous toxin. This is supported by the fact that PlTX hemolytic activity was detected in extracts of bacteria such as *Pseudomonas* (Carballeira et al., 1998), *Brevibacterium, Acinetobacter* and from the *Bacillus cereus* group (Seeman et al., 2009). It was also found that *Vibrio* sp. and *Aeromonas* sp. are able to produce compounds antigenically related to PlTX (Frolova et al., 2000). Thereby, the presence of PlTX and analogues in this myriad of marine organisms once can suggest a bacterial origin for the toxin production (Seeman et al., 2009, Piel, 2009).

1.2.5. Ciguatoxin (CTX)

Ciguatoxin (CTX) is a neurotoxin produced by dinoflagellates of the genus *Gambierdiscus*, among others *G toxicus,* and accumulates in the skin, head, viscera and roe of many species of reef fish. CTX are potent, lipophilic sodium channel activator toxins which bind to the voltage sensitive sodium channel on the cell membranes of all excitable tissues (Bagnis et al., 1980).

1.2.6. Brevetoxin (BTX)

Brevetoxins (BTX) are a group of cyclic polyether compounds produced by the dinoflagellate *Karenia brevis*. BTX are neurotoxins that bind to voltage-gated sodium channels in nerve cells, leading to disruption of normal neurological processes and causing the illness clinically described as Neurotoxic Shellfish Poisoning (NSP). There are several variants, being PbTX-3 one of them, a congener of type A brevetoxins (one of 10) (Colman et al., 2003).

1.3. Toxins from Freshwater and Marine Environments

1.3.1. β-N-methylamino-L-alanine (BMAA)

The methylated amino acid β-N-methylamino-L-alanine (BMAA) is biosynthesized by several strains of cyanobacteria in freshwater as well as in marine environments, in almost every continent (Vega and Bell, 1967, Rao et al, 2006, Li et al., 2010). It has been suggested that BMAA is linked to nerve degenerative diseases (Montine et al., 2005). BMAA has been recently detected in high concentrations in the brain tissues of patients with Amyotrophic Lateral Sclerosis-Parkinsonism Dementia Complex (ALS/PDC) in the South Pacific island of Guam (Murch et al., 2004; Papapetropoulos, 2007). BMAA for a long time was believed to occur only in cycads (Vega and Bell, 1967; Murch et al., 2004; Ince and Codd, 2005). But, Cox et al. (2003) reported that BMAA was produced by symbiotic cyanobacterial resident in specialized roots within the cycad tree. Chamorro traditional diets are rich in BMAA. People consume *Cycas* flour products made from the cycad seeds. They also usually eat flying foxes (*Pteropus mariannus*), which forage on the seeds of the cycads resulting in biomagnifications of BMAA (Cox and Sacks, 2002, Banack et al., 2006).

In 2005, Cox et al., found that culture collection strains of free-living freshwater and marine cyanobacteria contained BMAA and concluded that most, if not all, cyanobacteria produce BMAA. Banack, et al. (2007) report detection of BMAA in laboratory cultures of a free-living marine species of *Nostoc*. They detected BMAA in this marine species of *Nostoc* with five different analytical methods. They also mentioned that protein-associated BMAA can accumulate in increasing levels within food chains, and that biomagnification of BMAA could occur in marine and in terrestrial ecosystems, implying that marine cyanobacterial may be another route of human exposure to BMAA. Esterhuizen and Downing (2008) reported the presence of both free and Protein-associated BMAA in 96 % of cultures collected in Southern Africa.

1.3.2. Saxitoxin and Analogues (STXs)

Saxitoxins (STXs) represent a group of neurotoxic alkaloids produced both by dinoflagellates and freshwater cyanobacteria. They include pure saxitoxin (STX), neosaxitoxin (neoSTX), the gonyautoxins (GTX) and decarbamoylsaxitoxin (dcSTX). STX act on the voltage-gated sodium channels of nerve cells, preventing normal cellular function and leading to paralysis (Boczar et al., 1988). Ingestion of shellfish contaminated with STX produces the Paralytic Shellfish Poisoning (PSP) in humans. (More information regarding STX and analogues can be found in chapter 5 and other chapters of this book)

1.4. Food-Born Illness Produced by Marine and Freshwater Toxins

The major differences between the food-borne illnesses produced by marine and freshwater toxins are due to the different exposure routes. Food-borne illnesses due to exposure to marine toxins have been reviewed extensively for the past few years (Lanbdsberg et al., 2005, Stommel and Watters, 2007, Zaccaroni and Scaravelli, 2008). Most of the human intoxications are due to consumption of raw or cooked fish or shellfish contaminated with marine toxins.

Freshwater toxins pose other problems because they may cause food-borne illness due to the consumption of aquatic animals that co-habit with the toxin producing organisms, but also due to water consumption or vegetables. Nowadays most modern Water Treatment Plants (WTP) have technology that allows them to remove cyanotoxins. Nevertheless, in many regions of the world, such efficient WTP are not available, so the consumption of water contaminated with hazardous levels of cyanotoxins is possible. Recent evidences show that vegetables grown with water contaminated with cyanotoxins may uptake them and transfer the toxins along food chains.

Cyanotoxins may affect terrestrial edible plants if they are exposed to contaminated water. Cood et al (1999) showed that lettuce exposed through spray irrigation with water contaminated with MC could retain up to 2.5 µg MC/g in the central part of the leaves. Experiments done with broccoli and mustard seedlings exposed to MC at ecologically relevant concentrations showed that they could accumulate low levels of the toxin in the roots only (Järvenpää et al., 2007). Until now there are not yet enough studies on plants accumulation and transfer of cyanotoxins so further research is need, in special in what concerns with toxin translocation in the plant and potential accumulation in fruits.

Concerning aquatic organisms, bivalves are those that accumulate the highest amounts of cyanotoxins, being MC the most taken up (Vasconcelos, 1995, 1999, Amorim and Vasconcelos, 1999). Nevertheless, NOD, STX and CYL are also accumulated by many different bivalve species such as *Anodonta cygnea, Corbicula fluminea, Mytilus galloprovinciallis, M. edulis* (Pereira et al., 2004, Saker et al., 2004, Kankaanpää et al., 2007, Martins and Vasconcelos, 2009, Martins et al., 2009). Anatoxin-a seems to be the only toxin that is not readily accumulated by bivalves such as *Mytillus galloprovincialis*, at least under laboratory conditions (Osswald et al, 2008). We should take into account that although most of these toxins are from freshwater origin, the cyanobacteria that produce them may be found in estuaries and by that be accumulated by estuarine and costal organisms. MC are very stable and even during cooking procedures the toxins are not destroyed (Morais et al., 2008). The data obtained from laboratory experiments showed that bivalve molluscs are resistant to cyanobacteria toxins, being efficient toxin vectors.

Other freshwater organisms accumulate the toxins up to levels that may cause human health hazards. Among them we should point out crayfish such as *Procambarus clarkii* that may accumulate up to 2.7 μg MC/g (Vasconcelos, 1999, Vasconcelos et al., 2001) or *Cherax quadricarinatus* that accumulated up to 0.9 μg CYN/g (Saker and Eaglesham, 1999). Nevertheless, most of the toxin (99 %) accumulates in the gut and in the hepatopancreas (Vasconcelos et al., 2001). Bioaccumulation of hepatotoxins although in low quantities, was shown in prawns (*Penaeus monodon*) farmed in northern New South Wales (Kankaanpaa et al., 2005) and in the white shrimp *Litopenaeus vannamei* in a south-eastern USA shrimp facility (Zimba et al., 2006). These reports puts in evidence the need to understand the impact of cyanobacterial toxins in farming, as well as the potential consequences of bioaccumulation and possible effects in human health. If we consume only the abdominal part of the organism, removing the green intestine, the risk is almost none even if the crayfish inhabited water contaminated with cyanotoxins.

Fish such as *Cyprinus, Barbus* and *Lisa* species were found to accumulate up to 0.3 μg MC/g in the edible muscle in a natural situation of a Portuguese river (Vasconcelos, 1999). MCs can accumulate in muscle of different fish species, although the concentration of MCs in muscle is usually much lower than that in other tissues (Vasconcelos, 1999; Magalhães et al., 2001, 2003; Mohamed et al., 2003; Xie et al., 2005; Soares et al., 2004).

So, taken into account field studies and laboratory experiments we should consider that bivalves are those food items that may pose a higher risk to human health if they are harvested in areas were cyanotoxins occur. Depuration of cyanotoxins usually takes more than two weeks (Vasconcelos, 1995, Amorium and Vasconcelos, 1999, Pereira et al 2004), so even when the cyanobacteria bloom disappear; the risk is high for several weeks. Monitoring of the toxins in the organisms is essential to decrease the possibility of food-borne intoxications.

Conclusion

Although the monitoring of marine and freshwater toxins is legislated in many countries, the diversity of toxins and the different routes that they can have, makes difficult to prevent some food-borne illnesses. New toxins and new variants of existing toxins are being found every year. New chemical, biochemical and molecular techniques allow us to go further into the amounts of toxins detected and we can now predict if a bloom of cyanobacteria will be toxic by analysing the percentage of strains that have the genes responsible for the production of a toxin. New research teams working on new geographical regions identify the occurrence of toxins in new areas suggesting that some hazards are in many cases distributed around the globe. Toxins found usually in tropical regions start to be detected in temperate climates. Global warming might have also an impact on the redistribution of many toxins. So the challenges are many for scientists working on the field, being our job not just to produce the first report of a new toxin but also to contribute actively to the prevention of food-borne illnesses.

Acknowledgments

This work was partially funded by the Atlantic Area Programme (Interreg IVB Trans-national): 2008-1/003 (Atlantox) and 2009-1/117 Pharmatlantic.

References

Ali, A.E., Arakawa, O., Noguchi, T., Miyazawa, K., Shida, Y. & Hashimoto, K. (1999). Tetrodotoxin and related substances in a ribbon worm *Cephalothrix linearis* (Nemertean). *Toxicon, 28*, 1083-1093.

Amorim, A. & Vasconcelos, V.M. (1999). Dynamics of microcystins in the mussel *Mytilus galloprovincialis*. *Toxicon,* 37, 1041-1052.

Araóz, R., Nghiêm, H.-O., Rippka, R., Palibroda, N., Tandeau de Marsac, N. & Herdman, M. (2005). Neurotoxins in axenic oscillatorian cyanobacteria: coexistence of anatoxin-a and homoanatoxin-a determined by ligand-binding assay and GC/MS. *Microbiology,* 151, 1263-1273.

Asakawa, M., Toyoshima, T., Ito, K., Bessho, K., Yamaguchi, C., Tsunetsugu, S., Shida, Y., Kajihara, H, Mawatari, S. F., Noguchi, T., Miyazawa, K. (2003). Paralytic toxicity in the ribbon worm *Cephalothrix* species (Nemertea) in Hiroshima Bay, Hiroshima Prefecture, Japan and the isolation of tetrodotoxin as a main component of its toxins. *Toxicon*, 41, 747-753.

Bagnis. R., Chanteau, S., Chungue, E., Hurtel, J.M., Yasumoto, T. & Inoue, A. (1980). Origins of ciguatera fish poisoning: a new dinoflagellate, *Gambierdiscus toxicus* Adachi and Fukuyo, definitively involved as a causal agent. *Toxicon 18,* 199-208.

Ballot, B., Krienitz, L., Kotut, K., Wiegand, C. & Pflugmacher, S., 2005. Cyanobacteria and cyanobacterial toxins in the alkaline crater lakes Sonachi and Simbi, Kenya. *Harmful Algae,* 1, 139-150.

Banker, R., Carmeli, S., Hadas, O., Teltsch, B., Porat, R. & Sukenik, A. (1997). Identification of cylindrospermopsin in the cyanobacterium Aphanizomenon ovalisporum (Cyanophyceae) isolated from Lake Kinneret. *J Phycol,* 33, 613–616.

Banack, S. A., Murch, S. J. & Cox, P. A. (2006). Neurotoxic flying foxes as dietary items for the Chamorro people, Marianas Islands. *J. Ethnopharm.* 106, 97-104.

Banack, S.A., Johnson, H.E., Cheng, R. & Cox, P.A. (2007). Production of the neurotoxin BMAA by a marine cyanobacterium. *Mar. Drugs,* 5, 180–196.

Bargu, S., Powell, C., Coale, S., Busman, M., Doucette, G., Silver, M. (2002). Krill: a potential vector for domoic acid in marine food webs. *Mar. Ecol. Prog. Ser.,* 237, 209-216.

Bargu, S. & Silver, M. (2003). Field evidence of krill grazing on the toxic diatom genus *Pseudo-nitzschia* spp. in Monterey Bay, California. *Bull. Mar. Sci.,* 72, 629-638.

Bates, S.S. (2000). Domoic acid-producing diatoms: another genus added. *J. Phycol.,* 36, 978-983.

Bates, S.S., Bird, C.J., de Freitas, A.S.W., Foxall, R., Gilgan, M., Hanic, L.A., Johnson, G.R., McCulloch, A.W., Odense, P., Pocklington, R., Quilliam, M.A., Sim, P.G., Smith, J.C., Subba Rao, D.V., Todd, E.C.D., Walter, J.A. & Wright, J.L.C. (1989). Pennate diatom *Nitzschia pungens* as the primary source of domoic acid, a toxin in shellfish from Eastern Prince Edward Island, Canada. *Can. J. Fish. Aquat. Sci.*, 46, 1203-1215.

Boczar, B.A., Beitler M.K., Liston, J. & Sullivan, J., (1988). Cattolico, R.A. Paralytic shellfish toxins in *Protogonyaulax tamarensis* and *P. catenella* in axenic culture. *Plant Physiol.,* 88, 1285–1290.

Bourke, A.T.C., Hawes, R.B., Neilson, A. & Stallman, N.D. (1983). An outbreak of hepato-enteritis (the Palm Island mystery disease) possibly caused by algal intoxication. *Toxicon* Vol 21 Supplement, 3,45-48.

Bumke-Vogt, C., Mailahn, W. & Chorus, I., (1999). Anatoxin-a and neurotoxic cyanobacteria in German lakes and reservoirs. *Environmental Toxicology,* 14, 117-125.

Busse, L.B., Venrick, E.L., Antrobus, R., Miller, P.E., Vigilant, V., Silver, M.W., Mengelt, C., Mydlarz, L. & Prezelin, B.B. (2006). Domoic acid in phytoplankton and fish in San Diego. *Harmful Algae,* 5, 91-101.

Carballeira, N.M., Emiliano, A., Sostre, A., Restituyo, J.A., González, I.M., Colón, G. M., Tosteson, C.G. & Tosteson, T.R. (1998). Fatty acid composition of bacteria associated with the toxic dinoflagellate *Ostreopsis lenticularis* and with Caribbean *Palythoa* species. *Lipids,* 33, 627-632.

Clayden, J., Read, B. & Hebditch, K.R. (2005). Chemistry of domoic acid, isodomoic acids, and their analogues. *Tetrahedron,* 61, 5713-5724.

Chiswell, R.K., Shaw, G.R., Eaglesham, G.K., Smith, M.J., Norris, R.L., Seawright, A.A. & Moore, M.R., (1999). Stability of cylindrospermopsin,the toxin from the cyanobacterium *Cylindrospermopsis raciborskii*: effect of pH, temperature, and sunlight on decomposition. Environ.*Toxicol.,* 14 (1), 155–165.

Ciminiello, P., Dell'Aversano, C., Fattorusso, E., Forino, M., Tartaglione, L., Grillo, C. & Melchiorre, N. (2008). Putative palytoxin and its new analogue, ovatoxin-a, in *Ostreopsis ovata* collected along the Ligurian coasts during the 2006 toxic outbreak. *J. Am. Soc. Mass Spectrom.,* 19, 111–120.

Codd, G.A., Metcalf, J.S. & Beattie, K.A. (1999). Retention of *Microcystis aeruginosa* and microcystin by salad lettuce (*Lactuca sativa*) after spray irrigation with water containing cyanobacteria. *Toxicon,* 37, 1181–1185.

Colman, J.R. & Ramsdell, J.S. (2003). The type B Brevetoxin (PbTx-3) adversely affects development, cardiovascular function, and survival in Medaka (*Oryzias latipes*) embryos. *Environmental Health Perspectives,* 111, 1920-1925.

Cox, P.A. & Sacks, O.W. (2002). Cycad neurotoxins, consumption of flying foxes, and ALS–PDC disease in Guam. *Neurology,* 58, 956–959.

Cox, P. A., Banack, S. A. & Murch, S. J. (2003). *Proc. Natl. Acad. Sci.* 100, 13380-13383.

Cox, P.A., Banack, S.A., Murch, S.J., Rasmussen, U., Tien, G., Bidigare, R.R., Metcalf, J.S., Morrison, L.F., Codd, G.A. & Bergman, B. (2005). Diverse taxa of cyanobacteria produce beta-N-methylamino-L-alanine, a neurotoxic amino acid. *Proc. Natl. Acad. Sci.,* 102, 5074–5078.

Esterhuizen,M and Downing, T.G. b-N-methylamino-L-alanine (BMAA) in novel South African cyanobacterial isolates. *Ecotoxicology and Environmental Safety* 71 (2008) 309–313.

Fisher, W. J., & Dietrich, D. R. (2000). Pathological and biochemical characterization of Microcystin-induced hepatopancreas and kidney damage in carp (*Cyprinus carpio*). *Toxicol. Appl. Pharmacol.,* 164, 73–81.

Frolova, G.M., Kuznetsova, T.A., Mikhailov, V.V. & Elyakov, G.B. (2000). An enzyme linked immunosorbent assay for detecting palytoxin-producing bacteria. *Russ. J. Bioorg. Chem., 26,* 285-289.

Froscio, S.M., Humpage, A.R., Wickramasinghe, W., Shaw, G. & Falconer, I.R. (2008). Interaction of the cyanobacterial toxin cylindrospermopsin with the eukaryotic protein synthesis system. *Toxicon,* 51 (2), 191–198.

Fuchi, Y., Narimatsu, H., Nakama, S., Kotobuki, H., Hirakawa, H., Torishima, Y., Noguchi, T. & Ohtomo, N. (1991). Tissue distribution of toxicity in a pufferfish, Arothron firmamentum ("hoshifugu"). *J. Food Hyg. Soc.* Japan, 32, 520-524.

Gleibs, S. & Mebs, D. (1999). Distribution and sequestration of palytoxin in coral reef animals. *Toxicon,* 37, 1521-1527.

Goto, T., Kishi, Y., Takashi, S. & Hirata, Y. (1965). Tetrodotoxin. *Tetrahedron,* 21, 2059-2088.

Gugger, M., Lenoir, S., Berger, C., Ledreux, A., Druart, J.-C., Humbert, J.-F., Guette, C. & Bernard, C., (2005). First report in a river in France of the benthic cyanobacterium *Phormidium favosum* producing anatoxin-a associated with dog neurotoxicosis. *Toxicon,* 45, 919-928.

Hald, H., Naur, P., Pickering, D. S., Sprogoe, D., Madsen, U., Timmermann, D. B., Ahring, P. K., Liljefors, T., Schousboe, A., Egebjerg, J., Gajhede, M. & Kastrup, J. S (2007). Partial agonism and antagonism of the ionotropic glutamate receptor iGLuR5: structures of the ligand-binding core in complex with domoic acid and 2-amino-3-[5-tert-butyl-3-(phosphonomethoxy)-4-isoxazolyl]propionic acid. *J. Biol. Chem.,* 282 (35), 25726-25736.

Hampson, D.R. & Manalo, J.L. (1998). The activation of Glutamate receptors by kainic acid and domoic acid. *J. Nat. Toxins* 53-158.

Harada, K., Ohtani, I., Iwamoto, K., Suzuki, M., Wantanabe, M.F., Wantanabe, M. & Terao, K. (1994). Isolation of cylindrospermopsin from a cyanobacterium Umezakia natans and its screening method. *Toxicon,* 32, 73–84.

Honkanen, R. E., Dukelow, M., Zwiller, J., Moore, R.E., Khatra, B.S. & Boynton, A. L. (1991). Cyanobacterial nodularin is a potent inhibitor of type 1 and type 2A protein phosphatases. *Molecular Pharmacology,* 40, 577-583.

Hwang, D.F., Lu, S.C. & Jeng, S.S. (1991). Occurrence of tetrodotoxin in the gastropods *Rapana rapiformis* and *R. venosa venosa*. *Mar. Biol.,* 111, 65–69.

Hwang, P.A., Tsai, Y.H., Deng, J.F., Cheng, C.A., Ho, P.H. & Hwang, D.F. (2005). Identification of tetrodotoxin in a marine gastropod (*Nassarius glans*) responsible for human morbidity and mortality in Taiwan. *J. Food Prot.,* 68:8, 1696–1701.

Ince, P. G. & Codd, G. A. (2005). Return of the cycad hypothesis - does the amyotrophic lateral sclerosis/parkinsonism dementia complex (ALS/PDC) of Guam have new implications for global health? Neuropathol. *Appl. Neurobiol.,* 31, 345-353.

Ikeda, K., Venmathi Maran, B.A., Honda, S., Ohtsuka, S., Arakawa, O., Takatani, T., Asakawa, M. & Boxshall, G.A. (2006). Accumulation of tetrodotoxin (TTX) in *Pseudocaligus fugu*,a parasitic copepod from panther puffer *Takifugu pardalis*, but without vertical transmission—Using an immunoenzymatic technique. *Toxicon,* 48, 116–122.

Ito, K., Okabe, S., Asakawa, M., Bessho, K., Taniyama, S., Shida, Y. & Ohtsuka, S. (2006). Detection of tetrodotoxin (TTX) from two copepods infecting the grass puffer *Takifugu niphobles*: TTX attracting the parasites? *Toxicon,* 48, 620–626.

Järvenpää, S., Lundberg-Niinistö C., Spoof, L., Sjövall, O., Tyystjärvi, E. & Meriluoto J, (2007). Effects of microcystins on broccoli and mustard, and analysis of accumulated toxin by liquid chromatography-mass spectrometry.*Toxicon, 49*(6), 865-74.

Kan, Y., Uemura, D., Hirata, Y., Ishiguro, M. & Iwashita, T. (2001). Complete NMR signal assignment of palytoxin and N-acetylpalytoxin. *Tetrahedron Lett.,* 42, 3197-3202.

Kankaanpää, H. T., Holliday, J., Schroder, H., Goddard, T.J., Fister, R., & Carmichael, W. W. (2005). Cyanobacteria and prawn farming in northern New South Wales, Australia—a case study on cyanobacteria diversity and hepatotoxin bioaccumulation. *Toxicol. Appl. Pharmacol.,* 203, 243– 256.

Kankaanpää, H., Leiniö, S., Olin, M., Sjövall, O., Meriluoto, J. & Lehtonen, K. K. (2007). Accumulation and depuration of cyanobacterial toxin nodularin and biomarker responses in the mussel *Mytilus edulis. Chemosphere,* 68(7), 1210-7.

Kanoh, S. (1988). Distribution of tetrodotoxin in vertebrates. In *Recent Advances in Tetrodotoxin Research*; Hashimoto, K., Eds.; Koseisha-Koseikaku: Tokyo, Japan, 32-44.

Katikou, P. (2007). Chemistry of Palytoxins and Ostreocins. In *Phycotoxins, Chemistry and Biochemistry*; Botana, L.M., Ed.; Blackwell Publishing: Ames, IA, USA, pp.75-93.

Khora, S.S., Isa, J. & Yasumoto, T. (1991). Toxicity of puffers from Okinawa, Japan. *Nippon Suisan Gakk.,* 57, 163-167.

Kungsuwan, A., Nagashima, Y., Noguchi, T., Shida, Y., Suvapeepan, S., Suwansakornkul, P. & Hashimoto, K. (1987). Tetrodotoxin in the horseshoe crab *Carcinoscorpius rotundicauda* inhabiting Thailand. *Nippon Suisan Gakk.,* 53, 261-266

Landsberg, J., Dolah, F. V. & Doucette, G. (2005). Marine and Estuarine Harmful Algal Blooms: Impacts on Human and Animal Health. In "*Oceans and Health: Pathogens in the Marine Environment".* Ed. Bekin and Colwell, Springer, New York, , 165-215

Lankoff, A., Wojcik, A., Lisowska, H., Bialczyk, J., Dziga, D. & Carmichael, W.W. (2007). No induction of structural chromossomal aberrations in cylindrospermopsin-treated CHO-K1 cells without and with metabolic activation. *Toxicon,* 50, 1105-1115.

Lefebvre, K.A., Dovel, S.L. & Silver, M.W. (2001). Tissue distribution and neurotoxic effects of domoic acid in a prominent vector species, the northern anchovy *Engraulis mordax. Mar. Biol.,* 138, 693-700.

Lefebvre, K.A., Noren, D.P., Schultz, I.R., Bogard, S.M., Wilson, J. & Eberhart, B.T.L. (2007). Uptake, tissue distribution and excretion of domoic acid after oral exposure in coho salmon (*Oncorhynchus kisutch*). *Aquat. Toxicol*., 81, 266-274.

Lefebvre, K.A., Powell, C.L., Busman, M., Doucette, G.J., Moeller, P.D.R., Silver, J.B., Miller, P., Hughes, M.P., Singaram, S., Silver, M.W. & Tjeerdema, R.S. (1999). Detection of domoic acid in northern anchovies and California sea lions associated with an unusual mortality event. *J. Nat. Toxins,* 7, 85-92.

Lenoir, S., Ten-Hage, L., Turquet, J., Quod, J. P., Bernard, C. & Hennion, M. C. (2004). First evidence of palytoxin analogues from an *Ostreopsis mascarenensis* (Dinophyceae) benthic bloom in Southwestern Indian Ocean. *J. Phycol.,* 40, 1042-1051.

Lenoir, S., Ten-Hage, L., Turquet, J., Quod, J. P. & Hennion, M. C. (2006). Characterisation of new analogues of palytoxin isolated from an *Ostreopsis mascarenensis* bloom in the south-western Indian Ocean. *Afr. J. Mar. Sci.,* 28, 389-391.

Li, A., Tian, Z., Li, J., Yu, R., Banack, S. & Wang, Z. (2010). Detection of the neurotoxin BMAA within cyanobacteria isolated from freshwater in China. *Toxicon,* 55(5), 947-53.

Li, R., Carmichael, W.W., Brittain, S., Eaglesham, G.K., Shaw, G.R., Liu, Y. & Watanabe, M.M. (2001). First report of the cyanotoxins cylindrospermopsin and deoxycylindrospermopsin from Raphidiopsis curvata (cyanobacteria). *J. Phycol.,* 37, 1121–1126.

Maeda, M., Kodama, R., Tanaka, T., Yohizumi, H., Nomyoto, K., Takemoto, T. & Fujita, M. (1981). Structures of insecticidal substances isolated from a red alga, *Chondria armata.* In *Proceedings of the 27th Symposium on the Chemistry of Natural Products.* Symposium Organizing Committee: Hiroshima, Japan, 1985; p. 616.Moore, R.E.; Bartolini, G. Structure of palytoxin. *J. Am. Chem. Soc.,* 103, 2491-2494.

Magalhães, V. F., Moraes Soares, R. & Azevedo, S., (2001). Microcystin contamination in fish from the Jacarepaguá Lagoon (Rio de Janeiro, Brazil): ecological implication and human health risk. *Toxicon,* 39, 1077–1085.

Magalhães, V. F., Marinho, M. M., Domingos; P., Oliveira, A. C., Costa, S. M., Azevedo, L. O. & Azevedo, S.M.F.O. (2003). Microcystins (cyanobacteria hepatotoxins) bioaccumulation in fish and crustaceans from Sepetiba Bay (Brasil, RJ). *Toxicon,* 42, 289–295.

Malpezzi, E.L.A., Freitas, J.C. & Rantin, F.T. (1997). Occurrence of toxins, other than paralyzing type, in the skin of tetraodontiformes fish. *Toxicon,* 35, 57-65.

Martins, J.C. & Vasconcelos, V.M., (2009). Microcystin distribution and dynamics in aquatic organisms – a review. J. Toxicology and Environmental Health. *Part B. Critical reviews,* 12, 1-18.

Martins, J.C., Leão, P. & Vasconcelos, V.M., (2009). Differential protein expression in *Corbicula fluminea* upon exposure to a *Microcystis aeruginosa* toxic strain. *Toxicon.,* 53, 409-416.

Maruyama, J., Noguchi, T., Jeon, J.K., Harada, T. & Hashimoto, K. (1984). Occurrence of tetrodotoxin in the starfish *Astropecten latespinosus. Experientia,* 40, 1395-1396.

Maruyama, J., Noguchi, T., Narita, H., Nara, M., Jeon, J.K., Otsuka, M. & Hashimoto, K. (1985). Occurrence of tetrodotoxin in a starfish, *Astropecten scoparius. Agric. Biol. Chem.*, 49, 3069-3070.

Miyazawa, K., Jeon, J.K., Maruyama, J., Noguchi, T., Ito, K. & Hashimot, K. (1986). Occurrence of tetrodotoxin in the flatworm *Planocera multitentaculata. Toxicon,* 24, 645-650.

Miyazawa, K., Jeon, J.K., Noguchi, T., Ito, K., Hashimoto, K. (1987). Distribution of tetrodotoxin in the tissues of the flatworm *Planocera multitentaculata* (Platyhelminthes). *Toxicon,* 25, 975-980.

Miyazawa, K., Higashiyama, M., Ito, K., Noguchi, T., Arakawa, O., Shida, Y. & Hashimoto, K. (1988). Tetrodotoxin in two species of ribbon worm (nemertini), *Lineus fuscoviridis* and *Tubulanus punctatus. Toxicon,* 26, 864–874.

Mohamed, Z., Carmichael, W., & Hussein, A. (2003). Estimation of microcystins in the freshwater fish *Oreochromis niloticus* in an Egyptian fish farm containing a *Microcystis* bloom. *Environ. Toxicol.,* 18, 137–141.

Moore, R.E. & Scheuer, P.J. (1971). Palytoxin - new marine toxin from a coelenterate. *Science,* 172, 495-498.

Montine, T. J., Li, K., Perl, D. P. & Galasko, D. (2005). *Neurology,* 65(5), 768.

Morais, J., Augusto, M., Carvalho, A.P., Vale, M. & Vasconcelos, V.M. (2008). Microcystins - cyanobacteria hepatotoxins- bioavailability in contaminated mussels exposed to different environmental conditions. *European Food Research & Technology,* 227,949-952.

Moreira, C., Martins, A., Azevedo' J., Freitas, M., Vale M., Regueiras, A., Antunes, A. & Vasconcelos, V. (2011). Application of real-time PCR in monitoring *Cylindrospermopsis raciborskii* in a Portuguese freshwater system: abundance and toxicological evaluation. *Applied Microbiology and Biotechnology*, (in press)

Murch, S.J., Cox, P.A., Banack, S.A., Steele, J.C. & Sacks, O.W. (2004). Occurrence of *β*-methylamino-L-alanine (BMAA) in ALS/PDC patients from Guam. *Acta Neurol. Scand.,* 110, 267–269.

Namikoshi, M., Murakami, T., Watanabe, M.F., Oda, T., Yamada, J., Tsujimura, S., Nagai, H. & Oishi, S., (2003). Simultaneous production of homoanatoxin-a, anatoxin-a, and a new non-toxic 4-hydroxyhomoanatoxin-a by the cyanobacterium *Raphidiopsis mediterranea* Skuja. *Toxicon,* 42 (5), 533-538.

Narita, H., Noguchi, T., Maruyama, J., Ueda, Y., Hashimoto, K., Watanabe, Y. & Hida, K. (1981). Occurrence of tetrodotoxin in a trumpet shell, "boshubora" *Charonia sauliae.* Bull. *Japan. Soc. Sci. Fish.,* 47, 935-941.

Narita, H., Noguchi, T., Maruyama, J. & Hashimoto, K. (1984). Occurrence of a tetrodotoxin-associated substance in a gastropod, "hanamushirogai" *Zeuxis siquijorensis. Bull. Japan. Soc. Sci. Fish.*, 50, 85-88.

Narita, H., Sorokuro, M., Miwa, N., Akahane, S., Murakami, M., Goto, T., Nara, M., Noguchi, T., Saito, T., Shida, Y. & Hashimoto, K. (1987). *Vibrio alginolyticus*, a TTX-producing bacterium isolated from the starfish *Astropecten polyacanthus. Nippon Suisan Gakk.*, 53, 617-621.

Noguchi, T., Narita, H., Maruyama, J. & Hashimoto, K. (1982). Tetrodotoxin in the starfish *Astropecten polyacanthus* in Association with Toxification of a Trumpet Shell, "Boshubora" *Charonia sauliae. Bull. Japan. Soc. Sci. Fish.*, 48:8, 1173-1177.

Noguchi, T., Maruyama, J., Ueda, Y., Hashimoto, K., Harada, T. (1981).Occurrence of tetrodotoxin in the Japanese ivory shell *Babylonia japonica. Bull. Japan. Soc. Sci. Fish.,* 47, 901-913.

Noguchi, T., Maruyama, J., Narita, H.; & Hashimoto, K. (1984). Occurrence of tetrodotoxin in the gastropod mollusk *Tutufa lissostoma* (frog shell). *Toxicon,* 22, 219-226.

Noguchi, T. & Hashimoto, Y. (1973). Isolation of tetrodotoxin from a goby *Gobius criniger. Toxicon,* 11, 305-307.

Noguchi, T., Uzu, A., Koyama, K. & Hashimoto, K. (1983). Occurrence of tetrodotoxin as the major toxin in xanthid crab *Atergatis floridus. Bull. Japan. Soc. Sci. Fish.,* 49, 1887-1892.

Osswald, J., Rellan, S., Gago, A. & Vasconcelos V.M., (2008). Uptake and depuration of anatoxin-a by *Mytillus galloprovincialis* under laboratory conditions. *Chemosphere*, 72, 1235-1241

Otero, P., Pérez, S., Alfonso, A., Vale, C., Rodríguez, P., Gouveia, N.N., Gouveia, N., Delgado, J., Vale, P., Hirama, M., Ishihara, Y., Molgo, J. & Botana, L.M.. (2010). First toxin profile of ciguateric fish in Madeira Arquipelago (Europe*). Analytical Chemistry,* 82, 6032-6039.

Papapetropoulos, S. (2007). Is there a role for naturally occurring cyanobacterial toxins in neurodegeneration? The beta-N-methylamino-L-alanine (BMAA) paradigm. *Neurochem. Int.,* 50, 998-1003.

Park, H.D., Watanabe, M.F., Harda, K., Nagai, H., Suzuki, M., Watanabe, M. & Hayashi, H., (1993). Hepatotoxin (microcystin) and neurotoxin (anatoxin-a) contained in natural blooms and strains of cyanobacteria from Japanese freshwaters. *Natural Toxins,* 1(6), 353-360.

Pereira, P., Dias, E., Franca, S., Pereira E., Carolino M. & Vasconcelos, V. (2004). Accumulation and depuration of cyanobacterial paralytic shellfish toxins by the freshwater mussel *Anodonta cygnea*. *Aquatic Toxicology,* 68, 339-350.

Piel, J. (2009). Metabolites from symbiotic bacteria. *Nat. Prod. Rep.,* 26, 338-362.

Preußel, K., Stüken, A., Wiedner, C., Chorus, I. & Fastner, J. (2006). First report on cylindrospermopsin producing Aphanizomenon flos-aquae (Cyanobacteria) isolated from two German lakes. *Toxicon,* 47, 156–162.

Quilliam M.A. & Wright J.L.C. (1989). The amnesic shellfish poisoning mystery. *Anal. Chem.,* 61, 1053A-1060A.

Rao, S., Banack, S.A., Cox, P.A. & Weiss, J.H. (2006). BMAA selectively injures motor neurons via AMPA/kainate receptor activation. *Experimental Neurology,* 201. 244-252.

Rogers, E.H., Hunter, III, E.S., Moser, V.C., Phillips, P.M., Herkovitz, J., Muñoz, L., Hall, L.L. & Chernoff, N., (2005). Potential developmental toxicity of anatoxin-a, a cyanobacterial toxin. *Journal of Applied Toxicology*, 25, 527-534.

Ritson-Williams, R., Yotsu-Yamashita, M. & Paul, V. J. (2006). Ecological functions of tetrodotoxin in a deadly polyclad flatworm. *Proc. Nat. Acad. Sci. U.S.A.,* 103:9, 3176-3179.

Rodriguez, P., Alfonso, A., Vale, C., Alfonso, C., Vale, P., Tellez, A. & Botana, L.M. (2008). First toxicity report of tetrodotoxin and 5,6,11-trideoxyTTX in the trumpet shell Charonia lampas lampas in Europe *Analytical Chemistry,* 80, 5622-5629.

Runnegar, M.T., Kong, S.M., Zhong, Y.Z., Ge, J.L. & Lu, S.C. (1994). The role of glutathione in the toxicity of a novel cyanobacterial alkaloid cylindrospermopsin in cultured rat hepatocytes. *Biochem Biophys Res Commun,* 201, 235–241.

Saker, M.L. & Eaglesham, G.K., (1999). The accumulation of cylindrospermopsin from the cyanobacterium *Cylindrospermopsis raciborskii* in tissues of the Redclaw crayfish *Cherax quadricarinatus*. *Toxicon,* 37, 1065-1077.

Saker, M.L., Metcalf, J.S., Codd, G.A. & Vasconcelos, V.M. (2004). Accumulation and depuration of the cyanobacterial toxin cylindrospermopsin in the freshwater mussel *Anodonta cygnea*. *Toxicon,* 43, 185-194.

Salierno, J.D., Snyder, N.S., Murphy, A.Z., Poli, M., Hall, S, Baden, D. & Kane, A.S. (2006). Harmful algal bloom toxins alter c-Fos protein expression in the brain of killifish, *Fundulus heteroclitus*. *Aquat. Toxicol.,* 78, 350-357.

Schaffer, P., Reeves, C., Casper, D.R. & Davis, C.R. (2006). Absence of neurotoxic effects in leopard sharks, *Triakis semifasciata*, following domoic acid exposure. *Toxicon,* 47, 747-752.

Schembri, M.A., Neilan, B.A. & Saint, C.P. (2001). Identification of genes implicated in toxin production in the cyanobacterium *Cylindrospermopsis raciborskii*. *Environ Toxicol,* 16, 413-421.

Seemann, P., Gernert, C., Schmitt, S., Mebs, D. & Hentschel, U. (2009). Detection of hemolytic bacteria from *Palythoa caribaeorum* (Cnidaria, Zoantharia) using a novel palytoxin-screening assay. Anton. Leeuw. Int. J. G., doi:10.1007/s10482-009-9353-4

Seifert, M., McGregor, G., Eaglesham, G., Wickramasinghe, W., Shaw, G. (2007). First evidence for the production of cylindrospermopsin and deoxy-cylindrospermopsin by the freshwater benthic cyanobacterium, *Lyngbya wollei* (Farlowex Gomont) Speziale and Dyck. *Harmful Algae,* 6, 73–80.

Sheumack, D.D. & Howden, M.E.H. (1978). Maculotoxin: a neurotoxin from the venom glands of the octopus *Hapalochlaena maculosa* identified as tetrodotoxin. *Science,* 199, 188-189.

Shi, J. L., & Hwang, D.F. (2001). Possible source of tetrodotoxin in the starfish *Astropecten scoparius*. *Toxicon,* 39, 573-579.

Sierra B., A., Palafox-Uribe, M., Grajales-Montiel, J., Cruz-Villacorta, A. & Ochoa J.L. (1997). Sea bird mortality at Cabo San Lucas, Mexico: evidence that toxic diatom blooms are spreading. *Toxicon,* 35, 447-453.

Silvagni, P.A., Lowestine, L.J., Spraker, T., Lipscomb, T.P. & Gulland, F.M.D. (2005). Pathology of domoic acid toxicity in california sea lions (*Zalophus californianus*). *Vet. Pathol.,* 42, 184-191.

Sivonen, K., & Jones, G. (1999). Toxic cyanobacterial toxins. In *Toxic cyanobacteria in water: A guide to their public health consequences,monitoring and management*, eds. *I. Chorus and J. Bartram,* pp. 41–111. London: E & FN Spon.

Soong, T. W. & Venkatesh, B. (2006). Adaptive evolution of tetrodotoxin resistance in animals. *Trends Genet.,* 22:11, 621-626.

Spoof, L., Berg, K. A., Rapala, J., Lahti, K., Lepistö, L., Metcalf, J.S., Codd, G.A. & Meriluoto, J. (2006). First observation of cylindrospermopsin in *Anabaena lapponica* isolated from the Boreal Environment (Finland). *Environ Toxicol,* 21, 552–560.

Stommel E.W. & Watters M., (2007). Marine neurotoxins: Ingestible toxins. *Current Treatment Options in Neurology,* 6, 105-114,

Sugita, H., Noguchi, T., Hwang, D.F., Furuta, M., Motokane, T., Sonoda, T., Hashimoto, K. & Degushi, Y. (1987). Intestinal Microflora of Coastal Puffer Fishes. *Nippon Suisan Gakk,.* 53:12, 2201-2207.

Suh, E.M. & Kishi, Y. (1994). Synthesis of palytoxin from palytoxin carboxylic acid. *J. Am. Chem. Soc.,* 116, 11205-11206.

Tahara, Y. & Hirata, Y. (1909). Studies on the puffer fish toxin. *J. Pharm. Soc. Jpn.*, 29, 587-625.

Takemoto, T. & Daigo, K. (1958). Constituents of *Chondria armata. Chem. Pharm. Bull.,* 6, 578-580.

Tanti, J. F., Grémeaux, T., Van Obberghen, E. & Le Marchand-Brustel, Y. (1991). Effects of okadaic acid, an inhibitor of protein phosphatases-1 and -2A, on glucose transport and metabolism in skeletal muscle. *The Journal of Biological Chemistry,* 266, 2099-2103.

Tsai, Y.H., Ho, P.H., Hwang, C.C., Hwang, P.A., Cheng, C.A. & Hwang D.F. (2006). Tetrodotoxin in several species of xanthid crabs in southern Taiwan. *Food Chem.,* 95, 205–212.

Tsuda, K., Ikuma, S., Kawamura, M., Tachikawa, R., Sakai, K., Tamura, C. & Amakasu, O. (1964). Tetrodotoxin. VII. On the Structures of Tetrodotoxin and its derivatives. Chem. Pharm. *Bull. Japan,* 12, 1357-1374.

Uemura, D., Ueda, K., Hirata, Y., Naoki, H., & Iwashita, T. (1981). Further-studies on palytoxin .II. Structure of palytoxin. *Tetrahedron Lett.,* 22, 2781-2784.

Ukena, T., Satake, M., Usami, M., Oshima, Y., Naoki, H., Fujita, T., Kan, Y. & Yasumoto, T. (2001). Structure elucidation of ostreocin D, a palytoxin analog isolated from the dinoflagellate *Ostreopsis siamensis. Biosci. Biotech. Bioch., 65*, 2585-2588.

Usami, M., Satake, M., Ishida, S., Inoue, A., Kan, Y. & Yasumoto, T. (1995). Palytoxin analogs from the dinoflagellate *Ostreopsis siamensis. J. Am. Chem. Soc.,* 117, 5389-5390.

Vasconcelos, V.M., (1995). Uptake and depuration of the peptide toxin microcystin-LR in the mussel *Mytilus galloprovinciallis. Aquatic Toxicology,* 32, 227-237.

Vasconcelos, V.M., 1999. Cyanobacteria toxins in Portugal: effects on aquatic animals and risk for human health. *Brazilian Journal of Medical and Biological Research,* 32, 249-254.

Vasconcelos, V.M., Oliveira, S. & L.F. Oliva Teles. (2001). Impact of a toxic and non toxic strain of cyanobacterium *Microcystis aeruginosa* on the crayfish *Procambarus clarkii* . *Toxicon,* 39, 1461-1470.

Vega, A. & Bell, E.A. (1967). a-Amino-ß-methylaminopropionic acid, a new amino acid from seeds of Cycas circinalis. *Phytochemistry,* 6, 759–762.

Viaggiu, E., Melchiorre, S., Volpi, F., Di Corcia, A., Mancini, R., Garibaldi, L., Crichigno, G. & Bruno, M., (2004). Anatoxin a toxin in the cyanobacterium *Planktothrix rubescens* from a fishing pond in northern Italy. *Environmental Toxicology*, 19(3), 191-197.

Xie, L., Xie, P., Guo, L., Li, L., Miyabara, Y. & Park, H.-D. (2005). Organ distribution and bioaccumulation of microcystins in freshwater fish at different trophic levels from the eutrophic Lake Chaohu, China. *Environ. Toxicol*., 20, 293–300.

Woodward, R.B. (1964). The structure of tetrodotoxin. *Pure Appl. Chem.,* 9, 49-74.

Wu, C.H. (2009). Palytoxin: Membrane mechanisms of action. *Toxicon,* 54, 1183-1189.

Yotsu-Yamashita, M., Mebs, D. & Flachsenberger, W. (2007). Distribution of tetrodotoxin in the body of the blue-ringed octopus (*Hapalochlaena maculosa*). *Toxicon,* 49, 410-412.

Zaccaroni, A. & Scaravelli, D. (2008). Toxicity of Sea Algal Toxins to Humans and Animals . Algal Toxins: nature, occurrence effect and detection. NATO Science for Peace and Security Series A: *Chemistry and Biology,* 91-158

Zimba, P. V., Camus, A., Allen, E. H. & Burkholder, J. M. (2006). Co-occurrence of white shrimp, *Litopenaeus vannamei*, mortalities and microcystin toxin in a southeastern USA shrimp facility. *Aquaculture,* 261, 1048–1055.

In: New Trends in Marine and Freshwater Toxins
Editors: A. G. Cabado and J. M. Vieites
ISBN: 978-1-62257-787-3

Chapter 2

TOXIC HARMFUL ALGAL BLOOMS: NATURAL AND ANTHROPOGENIC CAUSES

Larry E. Brand
Marine Biology and Fisheries,
Rosenstiel School of Marine and Atmospheric Science,
University of Miami, Miami, FL, US

ABSTRACT

A small percentage of algae produce toxins that can greatly harm human health and even kill. The functional role of these toxins in the algae is not known, but it is clear that these complex molecules are secondary metabolites with a function and not simply metabolic byproducts. Some of these compounds just happen to bind strongly to certain active sites in humans and cause health problems-the main ones being gastrointestinal distress, liver toxicity, and neurotoxicity.

Increased anthropogenic nutrient fluxes into aquatic ecosystems have led to a significant increase in algal blooms throughout the world. These blooms (usually referred to as Harmful Algal Blooms, HABs) often alter and have negative impacts on the ecosystem. When the bloom is formed by an algal species that produces toxins, many animals in the food web can die as a result. Lipid soluble toxins tend to biomagnify up the food chain. Depending on food chain pathways, some of these toxins can end up in seafood and harm or kill humans. Globally, such events are increasing.

1. INTRODUCTION

There is widespread agreement that there has been a significant increase in Harmful Algal Blooms (HABs) around the world – their size, duration, and geographic distribution (Smayda, 1990, 2008; Hallegraeff, 1993; Chorus and Bartram, 1999; Glibert et al., 2005a, b; Anderson et al., 2002, 2008; Glibert and Burkholder, 2006; Paerl and Fulton, 2006; Heisler et al., 2008). Some of this can be attributed to better and more widespread monitoring by better trained researchers and better detection techniques. Some can be attributed to more people interacting with water bodies for recreational or commercial purposes and thus reporting more observations. Some can be attributed to increased environmental concerns and thus more reports of problems. Despite these biases, there is general agreement that there has been an overall increase in HABs throughout the world.

To be addressed here are the environmental factors that help generate toxic HABs throughout the world and how human activities may affect them. Most HABs are quite sporadic in time and space, thus sampling them and the associated environmental factors is usually sparse and inadequate. What sampling is conducted is often carried out during and after a bloom, not before. As a result, drawing firm conclusions about the cause of a particular HAB in a certain location at a particular time is usually difficult. Nevertheless, if we take a global view, certain patterns can be discerned. Every algal species that forms HABs has its own specific biochemistry, physiology, ecology, and evolutionary strategy. This review will take the approach of making broad generalizations.

1.1. Definition of Harmul Algal Blooms

First we must define HABs, as this term has been used in many different ways. This review will focus primarily on HABs that affect humans directly, although data from other HABs will be used to gain insight into their general ecology. We need to define Harmful Algal Blooms, each word separately.

1.1.1. Harmful

What do we mean by "harmful"? Harmful to what or whom? Naturally eutrophic waters such as upwelling regions and polar regions have blooms of algae, mostly diatoms, that are considered beneficial, as they provide more organic carbon for the food web, including seafood.

Ecosystem Harm

In normally oligotrophic waters, such as many areas of the tropics, algal blooms can alter the structure of the community and disrupt the normal food web (Kemp et al., 1983; Orth and Moore, 1983; Pasterak and Bilyard, 1985; Birkeland, 1988; Bell, 1992; Wittenberg and Hunte, 1992; Smith et al., 1981; Maragos et al., 1985; Cambridge and McComb, 1986; Silberstein et al., 1986; Tomasko and Lapointe, 1991; Lapointe et al, 1994; Worm and Sommer, 2000; Greening and Janicki, 2006; Bauman et al., 2010). Despite the higher primary productivity by algal blooms, lower seafood production is often the result. For example, benthic plants in tropical lakes and coastal waters, and coral reefs depend on low abundance of algae in the water column above so they can obtain sufficient light. Persistent algal blooms and algal epiphytes can lead to the decline or death of these benthic ecosystems and their associated food webs.

In stratified waters, a large increase in sinking organic carbon as a result of algal blooms can lead to hypoxic or anoxic bottom water (Conley et al., 2009), and thus the death of virtually all eukaryotic organisms. Even if the bottom waters do not go anaerobic, the sediments can become more anaerobic, affecting sea grasses and the benthic community in general.

Algal species such as *Phaeocystis* form blooms that create large amounts of mucilage that greatly increase the viscosity of the water. This reduces diffusion rates and alters chemistry, thus probably changing the relationships of virtually all the aquatic organisms in the ecosystem. The mucilage also clogs the filtering devices of animals, gills of fish, and nets of fishers. This also reduces the aesthetic and recreational value. Mucilage blooms have developed in the last few decades in the North Sea and northern Adriatic Sea as a result of river runoff from western and eastern Europe respectively (Riegman et al., 1991). Dense blooms of filamentous or spiny algae can form mats that also clog filter feeders.

A large literature exists on the ecosystem effects of HABs and will not be reviewed further here. See Pasterak and Bilyard (1985), Birkeland (1988), Wittenberg and Hunte (1992) Smith et al. (1981), Maragos et al. (1985), Cambridge and McComb (1986), Silberstein et al. (1986), Tomasko and Lapointe (1991) Lapointe et al. (1994), Worm and Sommer (2000), Greening and Janicki (2006); and Bauman et al. (2010) for more details.

Human Harm

Of greater interest here are toxins produced by HABs that harm humans in a more direct way. Less than 5% of algal species produce toxins that are known to be able to harm humans (Steidinger and Garcés, 2006; Burkholder et al, 2006, Graneli and Turner, 2006; Zaccaroni and Scaravelli, 2008b), but more are being discovered all the time. See Table 1 for a list of the major toxins and their characteristics. As some HABs produce toxins that harm other animals but not yet known to harm humans (Landsberg, 2002), they will be considered here, as they have the potential to harm humans in the future. Species that produce toxins that can harm or potentially harm humans will be referred to as toxic HABs.

Algal toxins can have a wide array of effects on humans, from a mere nuisance to deadly. Some individual toxins can produce a wide array of effects. Many species produce more than one toxin type and/or a range of congeners. Similarly there is a wide range of time scales upon which they have an effect and the degree to which humans can be aware of their exposure. Some are detected immediately. Others go undetected but can generate serious health problems years or decades later. There are also a number of mechanisms and pathways by which humans are exposed to these toxins.

At the more innocuous end of the spectrum are compounds that make drinking water smell or taste bad (Chorus and Bartram, 1999; Watson, 2003). Where freshwater lakes and rivers are used as a drinking water supply, blooms of cyanobacteria can cause problems and even result in the cutoff of drinking water. Such compounds can also propagate through the food chain to seafood, making it unpalatable (Tucker, 2000). Blooms of cyanobacteria and other algae that form noxious compounds in freshwater or coastal waters also reduce their aesthetic and recreational value. As far as we know, most of these compounds are not toxic and are simply a nuisance. Furthermore, they are immediately detected by smell or taste, thus humans are alerted to their presence and can avoid them.

Perhaps slightly worse than the compounds that smell or taste bad, are compounds such as lyngbyatoxin and aplysiatoxin that cause varying degrees of skin irritation (Sellner, 1997; Codd and Poon, 1988; Codd et al., 1999; Osborne et al., 2001; Funari and Testai, 2008; Sivonen and Borner, 2008).

Table 1. Toxin characteristics; origin, biochemical action, health effects

Microcystin Type: hydrophilic cyclic peptide Source: freshwater and brackish-*Anabaena*, *Aphanocapsa*, *Anabaenopsis*, *Hapalosiphon*, *Microcystis*, *Nostoc*, *Oscillatoria* Main pathway: water, aerosol Biochemical action: protein phosphatase inhibitor Health effects: gastrointestinal distress, liver toxin, tumor promotor
Nodularin Type: hydrophilic cyclic peptide Source: freshwater and brackish-*Nodularia* Main pathway: water Biochemical action: protein phosphatase inhibitor Health effects: gastrointestinal distress, liver toxin, tumor promotor
Cylindrospermopsin Type: hydrophilic polycyclic uracil derivative Source: freshwater and brackish-*Aphanizomenon*, *Cylindrospermopsis*, *Anabaena* Main pathway: water Biochemical action: protein phosphatase inhibitor Health effects: gastrointestinal distress, liver toxin
Anatoxin Type: hydrophilic bicyclic amine alkaloid Source: freshwater and brackish-*Anabaena*, *Aphanizomenon*, *Oscillatoria* Main pathway: water Biochemical action: cholinergic and anticholinesterase Health effects: gastrointestinal distress, tingling, numbness, respiratory paralysis, convulsions, death
Lyngbyatoxin Type: hydrophilic cyclic bipeptide Source: marine-*Lyngbya* Main pathway: surface contact Biochemical action: Health effects: skin irritation, tumor promoter
Aplysiatoxin Type: phenolic bislactone Source: marine-*Lyngbya*, *Oscillatoria*, *Schizothrix* Main pathway: skin contact Biochemical action: Health effects: skin irritation, tumor promoter
Saxitoxin Type: hydrophilic alkaloid Source: marine,-*Alexandrium spp. Gymnodinium catenatum*, *Gonyaulax acatenella*, *Pyrodinium bahamense*, *P. phoneus* freshwater-*Anabaena*, *Aphanizomenon*, *Lyngbya*, *Cylindrospermopsis* Main pathway: shellfish Biochemical action: Na^+ channel blocker Health effects: gastrointestinal distress, respiratory distress, death
Brevetoxin Type: lipophilic cyclic polyether

Source: marine-*Karenia brevis*, (possibly also *K. bicuneiformis, K. brevisulcata, K. concordia, K. cristata, K. papilionacea*, and *K. selliformis*), *Chattonella antigua, Heterosigma akashiwo, Fibrocapsa japonica*
Main pathway: shellfish, aerosol Biochemical action: Na^+ channel activator Health effects: gastrointestinal distress, tingling, numbness, temperature sensory reversal, respiratory distress, eye irritation
Ciguatoxin Type: lipophilic ladder polyether Source: marine-*Gambierdiscus spp*, possibly *Prorocentrum sps., Ostreopsis spp. Coolia spp.* Main pathway: tropical reef fish Biochemical action: Na^+ channel activator Health effects: gastrointestinal distress, tingling, numbness, temperature sensory reversal, respiratory distress, low bloom pressure and heart rate, death
Maitotoxin Type: hydrophilic cyclic polyether Source: marine *Gambierdiscus spp* Main pathway: tropical reef fish Biochemical action: Ca^{++} channel activator Health effects: unknown
Gambierol Type: polycyclic ether Source: *Gambierdiscus spp* Main pathway: tropical reef fish Biochemical action: K^+ channel blocker Health effects: neurotoxin
Okadaic acid Type: Lipophilic polyether Source: marine-*Dinophysis spp. Prorocentrum lima* Main pathway: shellfish Biochemical action: phosphorylase phosphatase inhibitor Health effects: gastrointestinal distress, chills, fever, tumor promotor
Domoic acid Type: hydrophilic amino acid Source: marine-*Pseudonitzschia spp.* Main pathway: shellfish Biochemical action: glutamate receptor activator Health effects: gastrointestinal distress, disorientation, hallucinations, memory loss, seizures, coma, death
Azaspiracids Type: lipophilic polycyclic polyether Source: marine-*Azadinium spinosum* Main pathway: shellfish Biochemical action: unknown Health effects: gastrointestinal distress, paralysis
Yessotoxin Type: lipohilic disulfated ladder polyether Source: marine *Protoceratium reticulatum, Lingulodinium polyedrum, Gonyaulax spinifera* Main pathway: shellfish Biochemical action: unknown Health effects: unknown

Brevetoxin is a neurotoxin produced by the delicate dinoflagellate *Karenia brevis*. The cells are destroyed and the brevetoxin is released at the air-sea surface if there is significant turbulence at the sea surface or along the beach. As a result, the brevetoxin ends up in an aerosol that can be inhaled by humans, resulting in eye, nose, and throat irritation, and respiratory distress (Backer et al., 2003; Fleming et al., 2005a, b; Pierce et al., 2005). Inhalation of brevetoxin appears to be much more toxic to humans than ingestion (Steidinger et al., 2008). The same appears to be true for inhaled microcystin from cyanobacteria (Stewart and Falconer, (2008).

The "good" aspect of all these compounds is that humans are immediately aware that they are exposed to the compounds and can proceed to reduce or eliminate their exposure.

More serious are compounds such as microcystin, nodularin, cylindrospermopsin, anatoxin, saxitoxin, brevetoxin, ciguatoxin, okadaic acid, domoic acid, and azaspiracids that produce gastrointestinal disorders (Gorham and Carmichael, 1988; Van Dolah, 2000; Daranas et al., 2001; Van Dolah et al., 2001; Humpage, 2008; Falconer, 2008; Funari and Testai, 2008; Zaccarini and Scaravelli, 2008a, b). Most of these algal toxins are odorless and tasteless, and are not destroyed by normal cooking procedures, so humans are not initially aware of their exposure and can eat large quantities of contaminated seafood. Vomiting and diarrhea usually develops within hours however and can last for days. At least the body is getting rid of the toxin. In some cases, however, if a large amount of the toxin has been consumed, death can result.

Some of the toxins produced by cyanobacteria such as microcystin, nodularin, and cylindrospermopsin are not detected upon exposure, but can produce long-term liver damage (Gorham and Carmichael, 1988; Carmichael, 1996; Codd, 1999; Daranas et al., 2001;Van Dolah et al, 2001; Humpage, 2008; Falconer, 2008; Funari and Testai, 2008; Zaccarini and Scaravelli, 2008a; Sivonen and Borner, 2008).

More damaging are the neurotoxins such as anatoxin, saxitoxin, brevetoxin, ciguatoxin, domoic acid, and azaspiracids that cause neurological disorders for weeks, months, years, or even permanent damage (Codd, 1999, 2000; Van Dolah, 2000; Daranas et al., 2001; Van Dolah et al., 2001; Sivonen and Borner, 2008; Humpage, 2008; Falconer, 2008; Funari and Testai, 2008; Zaccarini and Scaravelli, 2008a, b). The human body does not detect these neurotoxins until they are already embedded in the tissues. Unlike the gastrointestinal toxins, the neurotoxins are usually not expelled readily.

Perhaps most insidious are the toxins that produce no obvious symptoms initially and accumulate in the body, eventually causing cancer or neurodegenerative diseases that develop many years or even decades later. A number of algal toxins such as microcystin, nodularin, lyngbyatoxin, aplysiatoxin, and okadaic acid have been shown to be tumor promotors (Falconer, 1993; Fujiki and Suganuma, 1993; Fujiki et al, 1988, 1996; Falconer and Humpage, 1996; Landsberg et al., 1999; Van Dolah et al., 2001; Steidinger et al, 2008; Zaccarini and Scaravelli, 2008a; Sivonen and Borner, 2008; Falconer, 2008; Humpage, 2008; Codd 2000). Beta-N-methylamino-L-alanine (BMAA), a neurotoxic amino acid that may be produced by all known groups of cyanobacteria has been hypothesized to be involved in the development of neurodegenerative diseases such as Alzheimer's Disease, Parkinson's Disease, and Amyotrophic Lateral Sclerosis (ALS) (Humpage, 2008; Bradley and Mash, 2009, Metcalf and Codd, 2009).

Toxin Pathways

Toxins produced by algae can reach humans by a number of pathways. Some toxins end up in the water as a result of algal excretion, cell leakage, or cell lysis, and can cause skin irritation. If the water is swallowed, it can taste bad or cause gastrointestinal or neurological problems, and potentially cancer Falconer, 1999, 2005; Falconer and Humpage, 2005, 2006). Some toxins end up in the air and either smell bad or cause respiratory distress (Chorus and Bartram, 1999; Pierce et al., 2005).

Many toxins enter the food chain and then potentially humans through seafood (Shumway, 1995; Matsunaga et al, 1999; Tester et al., 2000; Van Dolah et al, 2001; Van Buynder et al., 2001; Landsberg, 2002; Xie et al., 2005; Doucette et al., 2006: Ibelings and Chorus, 2007; Stewart et al, 2008; Funari and Testai, 2008; Zaccaroni and Scaravelli, 2008a, b; Miller et al., 2010). One of the more lethal pathways for humans is shellfish, primarily for two reasons-their filtering abilities and the fact that humans eat the entire animal. Filter feeding bivalve molluscs such as clams and mussels are very efficient at filtering algae out of the water. They subsequently store lipophilic toxins in their fatty tissues, primarily in the liver and other organs, and not so much in the muscle. Unlike most seafood like fish, in which only the muscle tissue is eaten, the entire shellfish is eaten. Therefore, humans tend to get large doses of lipophilic toxins in shellfish. As a result, some of the more widespread and best known HABs that harm humans are known as Paralytic Shellfish Poisoning (PSP), Neurotoxic Shellfish Poisoning (NSP), Diarrhetic Shellfish Poisoning (DSP), and Amnesic Shellfish Poisoning (ASP).

Historically, much of the research on toxic HABS began when large numbers of humans got sick or died from eating shellfish contaminated with algal toxins. Ironically, because filter feeding molluscan shellfish do not need dense blooms of toxic algae to eventually accumulate amounts of toxin deadly to humans, many of the most serious algal human health hazards are not necessarily associated with dense, obvious blooms. For example, *Karenia brevis* blooms typically kill fish and become obvious at concentrations above 100,000 cells/l, but shellfish can become extremely toxic at concentrations of only around 5,000 cells/l (Tester et al, 1998; Steidinger, 2009). Therefore, medically oriented researchers tend to refer to Shellfish Poisoning and ecologically oriented researchers tend to refer to Harmful Algal Blooms.

In seafood such as fish, smaller concentrations of many algal toxins end up in the muscle tissue, which is the only part that humans eat in most cases. As a result, humans get a smaller dose of algal toxins in non-shellfish seafood, resulting in fewer and weaker harmful effects. When people have died from algal toxins from seafood other than shellfish, it is usually the result of eating the liver or other organs, and not just the muscle tissue.

Ciguatera poisoning is also not associated with any obvious bloom. In a somewhat similar fashion to shellfish, herbivorous fish eat the epibenthic dinoflagellate *Gambierdiscus toxicus* containing ciguatoxin, maitotoxin, and gambierol and end up retaining the lipophilic toxins. The toxins then biomagnify up the food chain to top carnivores, which can make humans extremely sick if eaten (Bagnis, 1993; Tindall and Morton, 1998; Bienfang, et al., 2008; Litaker et al., 2010). Unlike some of the other toxins, high concentrations of ciguatoxin do end up in the muscle tissue. Ciguatera is by far the most common type of algal toxin poisoning in the world. Five times as many cases are reported as Paralytic Shellfish Poisoning and Neurotoxic Shellfish Poisoning combined. Approximately 50,000 ciguatera cases are reported each year, but it is suspected that more like 500,000 cases actually occur because of underreporting (Bienfang et al., 2008).

Fish and other organisms that do not eat algae directly may still acquire some of these toxins from the water across the surfaces of their gills. These animals may then be eaten or the toxin may biomagnify up the food chain to seafood (Landsberg, 2002). The same is true for low sublethal concentrations of toxic algae. These low concentrations can be taken up and biomagnified in the food web, leading to high concentrations in top carnivores (Flewelling et al., 2005; Brand and Compton, 2007; Naar et al., 2007).

Some algal species that form dense blooms and produce toxins that may kill large numbers of fish or other biota (Landsberg, 2002), but apparently do not affect humans (as far as we know at the present time). Some of this may be due to specific biochemistry, but much is due to exposure and food web pathways. Therefore, algal toxins that affect animals but not humans now, may affect humans in the future as aquatic community structure changes. Therefore, we will not focus exclusively on human toxins.

Sublethal Effects

Consumption of shellfish and ciguatoxic fish is by far the major way in which humans are obviously harmed by algal toxins. Many other toxin pathways lead to the death of aquatic animals but not necessarily humans. Humans may be getting substantial sublethal doses however. Given the wide range of algal concentrations, toxicities, and pathways, one would expect a range of doses and effects from lethal to sublethal to impossible to detect. There is also a range of time scales for the effects to develop from immediate to decades later. Most research has focussed on HABs that cause dramatic human health problems rather quickly that can be documented relatively easily. Very little is known about the long term and/or sublethal effects. When one considers the biochemical effect of the algal toxins, sublethal effects certainly seem plausible.

Cyanobacteria produce a wide variety of secondary metabolites (Moore, 1996; Welker and von Dohren, 2006; Sivonen and Borner, 2008; Humpage, 2008). Dinoflagellates do the same (Smayda, 1997b; Hansen, 1998; Tindell and Morton, 1998; Cembella, 2003; Legrand et al., 2003; Graneli and Hansen, 2006; Graneli et al., 2008). We do not know the bioactivity of most of these compounds, but some are certainly known to be toxic. The vast majority have not been investigated, but it seems likely that some produce sublethal or long term effects that are not easily detected epidemiologically.

1.1.2. Algal

What do we mean by "algal"? Technically, algae are photosynthetic protists, but other groups of organisms with toxins are included here. Phylogenetically, cyanobacteria are eubacteria, but ecologically are similar to the eukaryotic algae (hence their earlier name blue-green algae). Cyanobacteria grow in the same habitats as eukaryotic algae using light and nutrients, and can form toxic blooms just like eukaryotic algae.

Many species of the phylogenetic groups called algae are not completely photosynthetic, having varying capabilities for heterotrophy (Stoecker, 1999; Jones, 2000). Many are capable of mixotrophy and appear to bloom in waters that have unusually high ratios of organic to inorganic nutrients (Glibert and Legrand, 2006). Many dinoflagellates and haptophytes that are photosynthetic can also be phagotrophic (Graneli and Carlsson, 1998; Stoecker et al., 2006). While most still need light, the consumption of prey by phagotrophy allows many of them to grow faster (Stoecker et al., 2006).

Some toxic species are completely heterotrophic and thus are protozoa, not algae (Glibert and Legrand, 2006). Approximately half of all dinoflagellates are heterotrophic (Gaines and Elbrachter, 1987; Smayda, 1997; Glibert and Legrand, 2006). Heterotrophic species that form toxic blooms will be considered here because of their similarities to Harmful Algal Blooms.

As a broad generalization, species of cyanobacteria are the dominant toxin-producing algae that form HABs in freshwater (Shapiro, 1973; Paerl and Ustach, 1982; Paerl, 1988; Riegman, 1998; Chorus and Bartram, 1999; Paerl and Fulton, 2006; Vasconcelos, 2006; Zaccaroni and Scaravelli, 2008a). In seawater, dinoflagellates are the dominant toxin-producing species of algae (Paerl 1988; Smayda, 1997; Cembella, 1998; Riegman, 1998; Cembella and John, 2006; Zaccaroni and Scaravelli, 2008b). In addition, some species of raphidophytes, haptophytes, and diatoms also produce toxins (Edvardsen and Imai. 2006; Bates et al., 1998; Bates and Trainer, 2006; Trainer et al., 2008).

Role of Bacteria

While it is usually assumed that the algae are the ones that produce the toxin, there has been considerable discussion over the years that bacteria associated or symbiotic with the algae may be the actual producers of the toxins (Doucette, et al., 1998; Bates et al., 2004; Kodama et al, 2006). This hypothesis has been stimulated by observations of loss of toxin production capabilities in algal cultures over time, and that some bacteria are known to produce some of the same toxins as algae. In some cases, the genes involved in toxin synthesis have been found in the algae (Snyder et al, 2005; Tillett et al., 2000; Mihali et al., 2008). In other cases, these questions remain unresolved and the idea remains a hypothesis. At a minimum, if bacteria are the ultimate producers, they still tend to associate strongly with just certain species of algae. This topic will be analyzed in more detail in chapter 5.

1.1.3. Blooms

What do we mean by “blooms”? First we need to distinguish between the use of the word “bloom” as a verb and as a noun. HABs generally refer to high concentrations of harmful algae. It is sometime implied or assumed that “bloom” is a verb and that the algae are growing rapidly. Certainly many algae can grow fast, referred to as “blooming”, but most of the species that form toxic HABs are slow growers (Cembella and John, 2006; Stolte and Garcés, 2006; Paerl and Fulton, 2006). Dinoflagellates are incapable of dividing more than once a day (Steidinger et al., 2008). While some cyanobacteria can grow fast, the ones that form most toxic HABs tend to be slow growers (Paerl and Fulton, 2006). Dense biomass of toxic species is often the result of physical aggregation or slow accumulation, not rapid growth.

A “bloom” cannot be easily defined, but the extreme cases are obvious. Concentrations of algae high enough to discolor the water and be easily visible to the naked eye are generally regarded as blooms. Algae that aggregate at the surface to form scums can be regarded as blooms. In general, concentrations orders of magnitude higher than the concentrations the species usually exist at is generally regarded as a bloom. Blooms often but not always have very distinct boundaries, as if occupying a particular water mass (Paerl, 1988).

Bloom species are often but not always the dominant species in the algal community and often essentially monospecific, suggesting that they have somehow excluded other species (Steidinger and Vargo, 1988; Anderson, 1998). Naturally eutrophic ecosystems tend to have low diversity but still have a number of co-occurring species within the phytoplankton community. Many HABs tend to be monospecific or have low concentrations of relatively few other species in the community. This could reflect unusual physical, chemical, or biological conditions that allow this one species to outcompete the others, but it could also indicate that the toxin is being used for allelopathic purposes. It has also been speculated that phagotrophy of their competitors may allow them to form monospecific blooms (Stoecker et al., 2006).

While dense aggregations of toxic species are more likely to cause harm, low concentrations of toxic algae can also be deadly to humans. Because many filter feeding bivalve molluscs such as clams and mussels are so efficient at filtering algae out of the water, they can accumulate the toxins from algal populations that are not particularly dense. No obvious high densities of toxic algae need to be apparent for some filter feeding molluscs to accumulate enough toxin in them to kill humans. These situations are included in the term "HAB", so focussing on concentrations of the algae is not particularly useful. That said, certainly higher concentrations of toxic algae are more likely to cause harm.

1.2. Environmental Factors that Influence HABs

HABs are sporadic in both time and space, so there are often few data collected in the early days of a bloom, and often no data collected before the bloom (Burkholder et al., 2006). Furthermore, there are very few long-term data sets in ecosystems where HABs occur. Therefore, views on the factors that help cause HABs tend to be somewhat speculative with relatively few data to back them up. While the species that form toxic HABs are quite diverse, some patterns in the type of species and environmental factors associated with HABs can be discerned.

In general, the environmental requirements of toxic HAB species are the same as other algal species. Therefore the question is: How do slow growing toxic HAB species outcompete the large number of other algal species with which they are competing? Is there something different in the ecology of toxic algal species relative to non-toxic species? A related question is the "purpose" of the toxins produced.

It is fairly easy to predict and explain blooms of algae. It remains difficult to predict and explain blooms of a particular species of algae because we understand so little of the mechanisms of competition among algal species.

Ecological Role of Toxins

Stolte and Garcés (2006) have compared toxic and non-toxic species and concluded that on average toxic species grow slower than non-toxic species within phylogenetic groupings, apparently the result of the energetic costs of toxin production. Clearly, the toxin must provide a benefit to the algae. For species that end up forming monospecific blooms, are their toxins used to inhibit or kill their competitors or predators?

Current evidence suggests that many algal species use allelochemicals to inhibit or kill their competitors, but these chemicals may be different from the toxins that harm humans (Smayda, 1997; Hansen, 1998; Tindell and Morton, 1998; Cembella, 2003; Legrand et al., 2003; Graneli and Hansen, 2006; Graneli et al., 2008). Saxitoxin, brevetoxin, okadaic acid, and domoic acid have not been found to have significant allelochemical properties (Graneli and Hansen, 2006). Many HAB species are phagotrophic and use toxins to attack their prey (Smayda, 1997b; Carlsson and Graneli, 1998; Hansen, 1998; Stoecker et al., 2006) but it is not clear that these toxins are the same ones that harm humans.

While it is becoming clear that many algae use allelochemicals against their competitors and toxins as an aid in phagotrophy, it is not yet clear if there is a difference between bloom formers and non-bloom formers in these characteristics. We do not know at the present time if these characteristics are necessary for bloom development.

Residence Time and Hydrography

Because of their slow growth rates, blooms of cyanobacteria and dinoflagellates cannot develop if they are flushed out of a water body too quickly or if they are dispersed too quickly by mixing. In general, enclosed water bodies with long residence times such as lakes and lagoons are more susceptible to HAB development (Chorus and Bartram, 1999; Paerl, 2008). Because of low physical losses, it is easier to establish a strong connection between environmental factors and HAB development in such water bodies. Open waters with short local residence times such as open coastlines are less susceptible to HABs. Furthermore, it is harder to determine causal relationships in such waters when HABs do develop in them.

A fast running mountain stream with a small watershed will have very little accumulation of algae. A water body that has a large watershed but an outflow dominated by evaporation and groundwater sinking rather than surface outflow will tend to accumulate nutrients and algae, thus be more susceptible to HABs. Lakes show a wide range of characteristics, depending on their watershed and the residence time of the water, which is roughly the lake volume divided by the flow rate in and out.

Estuaries tend to accumulate nutrients because estuarine circulation tends to export nutrient poor surface water and import nutrient rich bottom water. Because of the increasing ionic strength at the freshwater-seawater interface, many dissolved substances in the incoming freshwater coagulate and precipitate. The net flux to the bottom of these substances, along with fecal pellets and dead organisms result in estuaries acting as nutrient traps. As a result, high biomass of algae is often observed in estuaries.

Coastal waters exhibit a wide range of hydrographic regimes, including upwellings, downwellings, convergence zones, and gyres. It is thought that the development of HABs in these more open waters is the result of the swimming behavior of dinoflagellates and floating behavior of cyanobacteria in interaction with these complex physical regimes (Seliger et al., 1970; Tyler and Seliger, 1978; Fraga et al., 1988; Donaghay and Osborne, 1997; Cullen and MacIntyre, 1998; Figueiras et al., 2006). A new way of analyzing mixing regimes in the ocean (Lagrangian Coherent Structures) is revealing parcels of relatively unmixed water within what has traditionally been seen as mixed waters (Olascoaga et al., 2006). These relatively unmixed parcels of water may serve as incubators for developing HABs, as the slow algal growth rate must be faster than the mixing rate that disperses the bloom.

The construction of dams on rivers increases the residence time and new HABs often form in the reservoirs created behind dams (Paerl, 2008). The construction of marinas and breakwaters to protect docked boats tends to also create embayments with long residence times. HABs often develop in these harbors.

Role of Flotation and Swimming

Because most toxic HAB species are slow growers (Riegman 1998; Cembella and John, 2006; Stolte and Garcés 2006; Paerl and Fulton, 2006), physical aggregation is probably a major factor in the development of many HABs. This is particularly true for cyanobacteria and dinoflagellates because they can move through the water quickly using their gas vesicles and swimming abilities respectively. Indeed they move through the water more quickly than other types of algae.

Many cyanobacteria generate more gas vesicles during the spring, forming scums at the surface during the summer (Klemer and Konopka, 1980; Walsby, 1987; Oliver, 1994; Carmichael, 1996; Paerl et al., 2006). These surface blooms then shade their competitors in the water column below. Winds can then concentrate these surface scums in convergence zones and along the shoreline, resulting in very high concentrations of cyanobacteria and toxins (Chorus and Bartram, 1999).

Typical dinoflagellate swimming speeds are around one meter per hour (Steidinger et al, 2008). This allows them to swim many meters in the water column and also against upwelling or downwelling waters. In stratified waters, many dinoflagellates undergo diel vertical migration, photosynthesizing in nutrient-poor surface waters during the daytime and taking up nutrients in nutrient-rich water below the pycnocline during the night (Cullen and MacIntyre, 1998). This allows them to outcompete other algae that cannot take advantage of the spatial separation of light and nutrients in the water column.

In downwelling waters, dinoflagellates carried into deep waters with little light will become positively phototactic and swim upwards. If the downwelling speed is less than the swimming speed, high concentrations of dinoflagellates can accumulate as more water approaches the shoreline or convergence zone and downwells, leaving the dinoflagellates in the photic zone. In this way, low concentrations of dinoflagellates can end up forming a dense bloom along the shoreline or in a convergence zone without any significant active growth. The same can occur in an upwelling situation if the algae become negatively phototactic to avoid photoinhibition at the surface.

As humans cause increasing turbidity in aquatic habitats, thus reducing sunlight to the deeper water column, cyanobacteria and dinoflagellates that can move quickly to the surface will have an advantage over diatoms and slower swimming algae.

Temperature and Water Column Stratification

Most cyanobacteria and dinoflagellates tend to dominate phytoplankton communities in the summer when the water column is stratified (Paerl, 1988; Carmichael, 1996; Smayda, 1997b; Hallegraeff and Fraga, 1998; Chorus and Bartram, 1999; Erdner et al., 2008; Alvarez-Salgado et al, 2011). Their floating and swimming abilities give them an advantage in stratified waters (Klemer and Konopka, 1980; Paerl and Ustach, 1982; Smayda, 1997b; Donaghay and Osborne, 1997; Paerl and Fulton, 2006). Cyanobacteria are particularly well adapted for high temperatures. Some can tolerate temperatures as high as 74^{0}C (Stewart and Falconer, 2008). *Gambierdiscus toxicus* appears to also be more abundant during the summer, and in at least one case, was particularly high during a coral bleaching event as a result of high temperatures. (Bienfang et al, 2008). By contrast, diatoms tend to be more prevalent in colder upwelling or polar waters.

As the earth warms and waters become more stratified, it seems likely that cyanobacteria and dinoflagellates will have an advantage over other species of algae (Moore et al., 2008; Paul, 2008; Paerl and Huisman, 2009). Crouch et al. (2003) documented a large increase in dinoflagellates at the Paleocene-Eocene boundary when a large increase in atmospheric CO_2 increased the temperature of the earth. A warmer earth will allow cyanobacteria and dinoflagellates to expand their biogeographic ranges poleward, so toxic HABS can be expected to develop in areas where they have not occurred before (Moore et al., 2008).

Benthic Ecology

Most research treats HABs as a water column phenomenon because that is the part that is visible, but many HAB species have benthic stages. Many cyanobacteria have benthic akinetes that overwinter and then surface in the spring to form summertime blooms (Paerl, 1988; Tsujimura and Okubo, 2003). Many dinoflagellate species have benthic cysts in their life cycle (Anderson 1998; Anderson et al., 2005; Steidinger and Garcés, 2006). Some species have distinct seasonal cycles – cysts that overwinter and then germinate in the spring to develop into water column blooms in the summer.

The environmental factors involved in cyst germination rates are not well known but temperature, photoperiod, and endogenous clocks are known to be involved (Anderson 1998; Anderson et al., 2005). *Alexandrium* species spend most of the year as a cyst in the sediments and only a few weeks in the water column actually photosynthesizing (Wyatt and Jenkinson, 1997). Given their sporadic nature, many HAB species probably spend more time in the sediments than in the water column. While outside the expertise of most HAB researchers, more understanding of benthic ecology and sedimentology would probably give us more insight into the factors that govern this part of the life cycle of HABs that have a benthic stage.

Because aquatic ecosystems are net depositional environments as a result of the net particulate flux from the surface water to the deep water and sediments, even moderate nutrient additions could lead to more significant accumulation of nutrients in the sediments than in the water column. Therefore eutrophication could affect the benthic phase of HAB life cycles long before any effects are seen in the water column. This aspect of the role of eutrophication in HAB development has not been explored.

Benthic trawling and dredging resuspend sediments. It is not known how this affects the development of HAB species that depend on benthic cysts.

Immigration

One possible explanation for a new development of a local toxic HAB is transport from another location (Hallegraeff and Gollesch, 2006; Smayda, 2007). More natural causes could be changes in currents or environmental conditions that allow a species to move into a new area (Dale et al., 2006). Transport could be as cysts attached to migratory birds or other animals. Local habitat changes could explain why the species had not successfully invaded earlier but now have established themselves. In reviewing the evidence, Hallegraeff and Gollesch (2006) and Bolch and Salas, (2007) found good evidence that *Gymnodinium catenatum* was transported by ship ballast water to Tasmanian coastal waters for the first time in the 1970s, leading the first severe bloom in 1980. In some cases in which toxic HABs appeared recently for the first time and were suspected to have been introduced, instead had simply existed in low, non-noticed concentrations for a long time previously, as shown in sediment records (Dale, 2001; Dale et al., 2006). In most cases of new HABs, the cause is not clear. Anthropogenic transport mechanisms include ballast water in ships, transport of aquacultural products, and attachment to ship hulls.

Grazing

An increase in HAB biomass is the result of growth and/or physical aggregation, but it is opposed by grazing losses. A HAB cannot develop if grazing is high enough to counteract the increase. The idea that toxins reduce grazing and thus allow HABs to develop has been an attractive idea for a long time. The data to support this hypothesis has been equivocal however (Turner et al, 1998; Turner, 2006; Turner and Graneli, 2006; Buskey, 2008). While some studies indicate that algal toxins can deter grazers, others show no effects.

From a theoretical point of view, there is no selective advantage for an algal genotype to kill a metazoan grazer if the toxin has an effect on the grazer only after the algal cell has been eaten and its genes lost to future generations. For this reason, it is suspected that the function of the toxin is not to kill grazers but something else that is not yet understood. Indeed, in the case of many of the toxins, the filter feeding molluscs are not killed, but rather humans and others that eat the molluscs. The same is true for ciguatoxic dinoflagellates that are eaten (thus losing their genetic future) but do not kill herbivorous fish. The ciguatoxin can then biomagnify up the food chain to top carnivorous fish, which can then be extremely toxic to humans (Bienfang et al., 2008). Toxins from one algal cell are not likely to deter a metazoan grazer that feeds on many cells at a time. Furthermore, any antigrazing effects help both toxic and non-toxic genotypes as well as other algal species (competitors), giving the toxic cell no selective advantage.

Algal toxins in a dense monospecific bloom could deter grazing and help prolong a bloom, but it is hard to see how these toxins could have a significant impact before a bloom develops when the toxic species is sparse and only one species among many in a diverse community (Buskey, 2008). The fact that so many animals end up with these toxins in their bodies and sometimes dead (Shumway, 1995; Matsunaga et al, 1999; Van Dolah et al, 2001; Landsberg, 2002: Ibelings and Chorus, 2007; Stewart et al., 2008) suggests that any antigrazing activity against metazoans is not very effective.

Antigrazing properties of toxins could be selected if they diffuse from the algal cells and inhibit unicellular predators such as protozoa, bacteria, or viruses that interact with one algal cell at a time. In this case, the toxic cell itself survives the attack and perpetuates its genotype. Such allelopathy appears widespread among toxic HAB species (Smayda, 1997; Hansen, 1998; Tindell and Morton, 1998; Cembella, 2003; Legrand et al., 2003; Graneli and Hansen, 2006; Graneli et al., 2008). Metazoans eat both algae and protozoa, so metazoan grazing can be either positive or negative for algae, depending on the grazing impact on protozoans that eat algae (Stoecker et al., 2008). It is more likely that other factors in the ecosystem may on occasion cause metazoan grazing pressure to decline, allowing blooms to then develop.

Eutrophication and fishing have altered aquatic community structures throughout the world. Community structure is, in most cases, not well enough understood to be able to say if the changes have resulted in more or less grazing on HAB species.

Nutrients

Obviously, more algal biomass needs more nutrients. This does not necessarily imply high nutrient concentrations are necessary however, as in some cases, sparse algal biomass in low nutrient waters can be physically aggregated into a dense bloom. Additionally, high concentrations of algal toxins can accumulate in filter feeding molluscs in water with relatively low concentrations of toxic algae. Overall however, many HABs are the result of high biomass found in eutrophic waters.

N:P ratios well below the Redfield ratio of 16:1 give nitrogen-fixing cyanobacteria an advantage over eukaryotic algae (Smith, 1983, 1990). In addition to inorganic nutrients, it appears that most HAB species are capable of using organic nutrients and/or phagotrophy (Prakesh, 1987; Paerl, 1988; Smayda, 1997; Carlsson and Graneli, 1998; Graneli and Carlsson, 1998; Hansen, 1998; Glibert and Legrand, 2006; Stoecker et al., 2006; Burkholder et al., 2008). Burkholder et al. (2008) provides long lists of species with these capabilities.

Many species of cyanobacteria and dinoflagellates tend to be K-selected, occurring at the end of successional sequences after natural turbulent events. Events that inject nutrients into the photic zones usually lead to initial blooms of fast growing diatoms and other groups of algae that grow rapidly. The diatom bloom generates eutrophic conditions high in organics and microbial biomass that many cyanobacteria and dinoflagellates seem to prefer (Glibert and Legrand, 2006). Slow growing cyanobacteria and dinoflagellates then tend to increase in abundance later in the successional sequence as the water column stratifies in the summer (Carmichael, 1996; Smayda, 1997; Paerl, 1988; Hallegraeff and Fraga, 1998; Chorus and Bartram, 1999; Erdner et al., 2008; Alvarez-Salgado et al, 2011).

It is well established that humans are injecting increasing amounts of nutrients into various water bodies around the world (Nixon, 1995; Galloway et al., 1995, 2003; Vitousek et al., 1997; Howarth et al., 2000; Pinckney et al., 2001; Cloern, 2001; Rabalei et al., 2002; Seitzinger et al., 2002, 2005; Kemp et al., 2005; Glibert et al, 2006; de Jonge et al., 2002; Turner et al., 2006; Schindler, 2006; Smith 2006; Smith et al., 2006; Burkholder et al., 2006; Norring and Jorgensen, 2009; Clarke et al., 2006; Fisher et al., 2006; Paerl et al., 2006; Howarth, 2008; Bricker et al., 2008). Over the past 50 years, the global human population has increased 5-fold (Glibert and Burkholder. 2006). Associated with this is an increase in habitat disturbance, fertilizer use, and release of sewage and animal wastes. This has resulted in a 20-fold increase of nitrogen runoff and 4-fold increase in phosphorus runoff from land (Glibert and Burkholder, 2006).

In most cases, an increase in nutrients will lead to an increase in algae. As algae are the base of most aquatic food chains, a small increase could be considered as positive, especially if there is no change in species composition, leading to greater fish yield, etc. A large increase in algal abundance however usually has a number of negative impacts, which have been documented throughout the world (Ryther and Dunstan, 1971; Larsson et al., 1985; Cooper and Brush, 1991; Bodeanu, 1992; Mee, 1992; Nehring, 1992; Vollenweider, 1992; Turner and Rabalais, 1994; Cociasu et al., 1996; Havens et al., 1996; Richardson and Jorgensen, 1996; Harding and Perry, 1997; Havens, 2008). The increased algal biomass in the water column leads to less light reaching benthic plants such as seagrasses and the zooxanthellae of corals, reducing their productivity or even causing their death and the demise of the community dependent on them (Kemp et al., 1983; Orth and Moore, 1983; Pasterak and Bilyard, 1985; Birkeland, 1988; Bell, 1992; Wittenberg and Hunte, 1992; Smith et al., 1981; Maragos et al., 1985; Cambridge and McComb, 1986; Silberstein et al., 1986; Tomasko and Lapointe, 1991; Lapointe et al, 1994; Worm et al., 2000; Greening and Janicki, 2006; Bauman et al., 2010). High algal biomass can lead to hypoxia or anoxia, particularly in stratified waters. There has been a significant increase in these anoxic "dead zones", where essentially all eukaryotic organisms are dead, throughout the world, and anthropogenic eutrophication is generally seen as the major cause (Diaz and Rosenberg, 2008).

This has also resulted in a decrease in Si/N and Si/P ratios, giving dinoflagellates and cyanobacteria a competitive advantage over diatoms in eutrophic waters (Officer and Ryther, 1980; Radach et al., 1990; Smayda, 1990, 1997, 2004; Justic et al., 1995). As a broad generalization, diatoms tend to dominate naturally eutrophic waters. Land runoff of fertilizer, animal wastes and sewage sources has lower Si/N and Si/P ratios, giving dinoflagellates and cyanobacteria a competitive edge over diatoms in these eutrophic waters, thus making toxic HABs more likely.

Increased fertilizer, animal wastes and sewage also changes the quality of the nutrients. Natural upwelling nitrogen is in the form of nitrate, whereas anthropogenic nitrogen tends to be in the form of ammonia, urea, and organic nitrogen. Again, as a broad generalization, dinoflagellates and cyanobacteria tend to use these more reduced forms of nitrogen better than diatoms (Riegman, 1998; Glibert et al, 2006).

There is general agreement that this increased nutrient loading to water bodies is a major factor behind the overall increase in HABs worldwide (Paerl and Ustach, 1982; Paerl, 1988, 1997, 2008; Smayda, 1989, 1990, 2008; Hallegraeff, 1993; Carmichael, 1996; Riegman, 1998; Chorus and Bartram,1999; Glibert et al, 2005a, b; Anderson et al., 2002, 2008; Giani et al., 2005; Burkholder et al., 2006; Vasconcelos, 2006; Bricker et al., 2008; Heisler et al., 2008; Glibert and Burkholder, 2006).

That anthropogenic eutrophication can lead to increases in algal biomass and the harmful effects of benthic shading and hypoxia or anoxia has been known for a long time and is well established. These effects are the result of overall algal biomass and do not depend significantly upon which species dominate. Some species are harmful at high concentrations but not low because of their physical effects. Species that produce large amount of mucilage can clog the feeding mechanisms of filter feeders and the gills of fish and other animals. Diatoms with large spines can form aggregations that can also have clogging effects. Such species have harmful effects on the ecosystem at high concentrations but are innocuous at low concentrations. Thus anthropogenic eutrophication can lead to harmful impacts on the ecosystem.

All these harmful effects impact the ecosystem, but not human health in any direct way. While it is well established that anthropogenic eutrophication causes HABs that harm the ecosystem, it is less clear that it causes toxic HABs that impact human health. The primary way in which algae harm human health is through the production of toxins. Only a small fraction of algal species produces toxins harmful to humans. Most of these are cyanobacteria or dinoflagellates, but toxic species of raphidophytes, haptophytes, and diatoms are also known (Edvardsen and Imai. 2006; Bates et al., 1998; Bates and Trainer, 2006; Trainer et al., 2008). Predictions of eutrophication effects on particular species are more difficult than general algal biomass.

While the hypothesis that an increased input of nutrients into a water body will likely lead to an increase in algal abundance is obvious, data supporting the hypothesis that it will increase a particular species are often sparse. One of the effects of eutrophication is a change in species competition and it is not easily predictable. Many HABs are quite sporadic, both spatially and temporally, making adequate sampling difficult. Most of the time, sampling and research are not initiated until after a problem HAB has been identified. As a result, there are no earlier data for determining if the HAB has increased over time or was simply undetected before. Most water bodies do not have long term data on algal species abundance and nutrient inputs and concentrations to rigorously test the hypothesis that eutrophication has led to an increase of a particular HAB species.

While toxic HABs are more likely in eutrophic waters, they can be found in oligotrophic waters. Because filter feeding molluscs such as clams and mussels can filter large volumes of water for algal food, they can end up with deadly concentrations of algal toxins even if the toxic algae are quite sparse. High nutrients are not needed to generate these types of toxic HABs. Humans were aware of toxic water, shellfish, and fish hundreds and even thousands of years ago, before land runoff of nutrients had increased significantly (Paerl, 1988; Kusek et al., 1999).

Similarly, eutrophic conditions do not necessarily lead to toxic HABs. Most eutrophic habitats do not have toxic HABs, so clearly toxic HABs are not an inevitable consequence of nutrient enrichment. Both naturally and anthropogenically eutrophic waters are often dominated by diatoms, which are mostly non-toxic. Once a toxic HAB species has outcompeted other species however, higher nutrients do allow for the production of more of the toxic algae and its toxin.

That a connection between increased nutrients and increased HABs is not necessarily obvious or inevitable can be seen with a simplistic hypothetical situation. Overall, many more cyanobacteria and dinoflagellate species produce toxins than diatoms. In general, diatoms grow faster than dinoflagellates (Brand and Guillard, 1981; Cembella and John, 2006; Stolte and Garcés 2006; Steidinger et al., 2008). Overall, diatoms are adapted to higher nutrient regimes and dominate in phytoplankton communities in nutrient rich waters, while cyanobacteria and dinoflagellates are better adapted for stratified nutrient poor waters (Smayda, 1997; Chorus and Bartram, 1999; Erdner et al., 2008). In this generalized scenario, non-toxic diatoms are more likely to increase and displace toxic dinoflagellates if nutrient loading to a water body increases if there are no other competitive processes occurring. With respect to the hypothesized link between anthropogenic eutrophication and toxic HABs, four scenarios can be envisioned.

One scenario is that increasing nutrients increases all algal species equally, with no change in species composition. The biomass of both toxic and non-toxic species would increase equally and there would be an increase in toxins (perhaps from undetectable to detectable levels). This could be considered a neutral hypothesis, given no information on the details of competition between species.

A second scenario is one in which increasing nutrients alter the algal community is a way that causes the reduction or local extinction of a toxic species as a result of a change in algal species composition. Total algal biomass could increase at the same time as a particular toxic algal species actually decreases in abundance. A typical scenario would be upwelling that leads to a fast growing diatom bloom that outcompetes toxic dinoflagellates.

A third scenario is that increased nutrients help an indigenous toxic species compete against other species. An increase in total algal biomass as well as a higher fraction of it being toxic would then increase toxin exposure to humans.

A fourth scenario is that anthropogenic eutrophication could alter community structure in a way that a toxic species that could not previously survive in the water body as a result of low nutrients, competition or predation is now able to survive there and proliferate to toxic concentrations. It could enter the new water body by natural dispersal processes or human transport mechanisms (such as ballast water).

Anthropogenic eutrophication could lead to either a decrease or increase in toxic HABs, depending on the characteristics of the water body, its community structure, and the particular toxic algal species and its adaptations. Each situation has to be evaluated on its own.

The one way in which eutrophication could lead to more toxic HABs is if high nutrients, organics, or biomass somehow selected for toxic species. Allelopathy and phagotrophy both use toxins and rely upon cell to cell interactions. Such interactions are more frequent in waters with higher concentrations of cells. Therefore, it is plausible that allelochemical and phagotrophic toxins are more advantageous in more eutrophic waters. A good test of this hypothesis would be to compare estuarine, coastal and oceanic species of similar phylogenetic lineage. It appears that most toxic HABs are capable of mixotrophy and phagotrophy (Glibert and Legrand, 2006; Stoecker et al., 2006). This suggests that they will grow better in ecosystems with higher concentrations of organic compounds and other microbes – thus eutrophic conditions.

The original definition of eutrophication refers to natural ecosystems that are rich in organic compounds, often the result of long term accumulation of nutrients (Nixon, 2009). Over time, the term cultural eutrophication has come to mean overenrichment of nutrients as a result of human activities. Upon reflection of this change in meaning, Smayda (2008) has recommended the use of the term nutrification for the process of nutrient enrichment. While much focus has been on the role of nutrification in HAB development, the old term eutrophic may in fact be relevant to many HABs, given their heterotrophic capabilities.

Some species at high concentrations do kill many animals and also can form monospecific blooms. It is thus plausible that the toxins at high concentrations may help certain species maintain their dominance in the community and acquire more nutrients from dying animals. Some other mechanism would be required to generate the high concentrations from an initially low concentration. In this way, anthropogenic eutrophication may not help generate toxic blooms, but could help sustain them.

1.3. Environmental Generalizations

While exceptions clearly exist, certain broad patterns appear with respect to which types of algae tend to form toxic HABs and under which ecological conditions. There are really two questions to be addressed. What conditions give cyanobacteria and dinoflagellates an advantage over other types of algae and what conditions give toxic species an advantage over non-toxic species? Freshwater toxic HABs are mostly cyanobacteria. Marine toxic HABs are mostly dinoflagellates, or occasionally other types of flagellates. Prokaryotic cyanobacteria are the most primitive algae. Dinoflagellates are also considered quite primitive because of their unusual genetic and nuclear organisation, leading some to call them mesokaryotes, a rather primitive type of eukaryote (Hackett et al., 2004). These two phylogenetic groups are well known for producing a great many secondary products, including most of the algal toxins known to harm invertebrates, vertebrates, mammals, and humans (Moore, 1996; Welker and von Dohren, 2006; Sivonen and Borner, 2008; Humpage, 2008; Smayda, 1997b; Hansen, 1998; Tindell and Morton, 1998; Cembella, 2003; Legrand et al., 2003; Graneli and Hansen, 2006; Graneli et al., 2008). They also appear to utilize allelopathy and phagotrophy more than other algae (Smayda, 1997; Hansen, 1998; Tindell and Morton, 1998; Cembella, 2003; Legrand et al., 2003; Graneli and Hansen, 2006; Graneli et al., 2008; Carlssen and Graneli, 1998; Hansen, 1998; Stoecker et al., 2006). Allelopathy and phagotropy may be more advantageous when cells are crowded together and interact more frequently (eutrophic conditions). At the opposite end of the ecological and evolutionary spectrum, diatoms, the last major phylogenetic group of algae to evolve, appear to produce few toxins.

In moving through the water column quickly, cyanobacteria and dinoflagellates have the fastest mechanisms, using gas vesicles buoyancy and swimming, respectively This give them an advantage in stratified waters over other types of algae, particularly diatoms. Spatially, diatoms often dominate in upwelling and polar regions, cyanobacteria and dinoflagellates tend to dominate tropical stratified waters. Seasonally, diatoms tend to dominate the colder times of the year when there is little or no stratification, while cyanobacteria and dinoflagellates tend to dominate the warmer times of the year when there is strong stratification.

1.4. Unknown Toxins and Effects

Historically, research has focussed on toxic HABs that have the most obvious, immediate and serious health effects. The research starts with the effects on humans and works back to the causative toxin. These known toxins however are only a very small fraction of the secondary compounds known to be produced by cyanobacteria and dinoflagellates. The biochemical, physiological, and medical effects of most of these have not been investigated.

Almost certainly these secondary compounds are entering the food web and seafood. There is huge potential for sublethal and long-term effects that has not been explored. This requires research proceeding from the other direction – from compound to effects.

1.5. Anthropogenic Effects

How could human activities lead to increased toxic HABs? The building of dams, breakwaters, and harbors increases the residence time of the water, allowing the slower growing cyanobacteria and dinoflagellates a better chance to build up biomass. These structures will often help allow stronger stratification of the water column, which gives cyanobacteria and dinoflagellates an advantage over other types of algae because of their ability to move faster up and down in the water column. Global warming will also increase vertical stratification in many cases. Increased turbidity from watershed disturbance and dredging reduces light to the deeper part of the water column, which again gives cyanobacteria and dinoflagellates an advantage over the other algae. Dredging and trawling resuspends sediments and associated cysts of toxic HAB species. Fishing and other activities alter the food web and ecosystem, changing the grazing pressure on algal species in ways that are not easily predicted. Altered food webs mean altered pathways by which toxins can make it to seafood. Altered watersheds and increased runoff of fertilizer, sewage and animal wastes result in higher nutrients which usually lead to increased algal biomass and often altered species composition. This runoff is usually higher in organic nutrients than upwelled water, giving an advantage to cyanobacteria and dinoflagellates. This runoff also has lower Si/N and Si/P ratios than upwelled water, giving an advantage to cyanobacteria and dinoflagellates over diatoms. The resulting higher biomass in these eutrophic conditions leads to more cell-cell interactions, which may give an advantage to species capable of allelopathy and/or phagotrophy, and thus capable of producing toxins. There are many mechanisms by which human activities could lead to more toxic HABs. In many cases, an increase in toxic HABs can be documented, but there are often not enough data to determine the cause.

More information related to the environmental factors that affect toxin production is gathered in Chapter 3.

1.6. Specific Cases

Giani et al., (2005), Kotak and Zurawell (2007), and Bigham et al (2009) have shown that microcystin is more frequently encountered in more eutrophic lakes. Jaworski (1990) found that increases in *Microcystis* corresponded to nutrients in the Potomac estuary. Finni et al. (2000) and Poutanen and Nikkila (2000) have documented the increase in cyanobacterial blooms in the Baltic Sea during the 20th century. While there is no evidence that anthropogenic nutrients influence populations of *Alexandrium fundyense* in the open Gulf of Maine, they do appear to enhance blooms in estuaries and embayments along its coastline (Anderson et al., (2008). Trainer et al. (2003) found increases in paralytic shellfish toxins in Puget Sound shellfish over the past half century to correspond with the increase in the human population in the Puget Sound watershed. Heil et al. (2005) and Glibert et al. (2008) have concluded that *Prorocentrum minimum* is a species that tends to bloom in eutrophic waters and it has increased worldwide as a result of eutrophication. Heil et al. (2001) and Glibert et al. (2002) reported a fish kill by a bloom of *Karenia selliformis* that is thought to be a result of excess sewage discharge into Kuwait Bay. Hattenrath et al. (2010) concluded that sewage was responsible for larger blooms of *Alexandrium fundyense* in estuaries on Long Island, New York. Brand and Compton (2007) found a large increase in the overall abundance of *Karenia brevis* from the 1954-1963 time period to 1994-2002 time period along the west coast of Florida, which parallels a dramatic increase in the human population and land runoff along that coast and documented increases in nutrients during that time. Parsons et al. (2002) documented a large increase in *Pseudonitzschia* in five sediment cores near the mouth of the Mississippi River since the 1950s. Lam and Ho (1989) have documented an increase in HABs in Hong Kong harbor, which parallels the increase in anthropogenic eutrophication. Imai et al. (2006) showed that a number of HAB species increased in the Seto Inland Sea of Japan during the 1960s and 1970s during a period of increased nutrient input. Those HABs then declined during the 1980s after new regulations reduced the nutrient input.

CONCLUSION

There is strong evidence that toxic HABs have increased globally over the past century. Toxic HABs are dominated by the primitive cyanobacteria in freshwater and primitive dinoflagellates in seawater. Current data suggest that these two groups of primitive algae produce many more allelochemicals than the other groups of algae that evolved later. It is hypothesized that these allelochemicals target unicellular competitors and predators, not metazoan predators. Because of biochemical similarities, a small fraction of these secondary metabolites do affect multicellular organisms, including humans. The primary human health effects are gastrointestinal distress, liver damage, and neurotoxicity, occasionally leading to death.

Because of the sporadic nature of HABs, data supporting any particular hypothesis on the cause of a particular HAB is usually weak, but taking a broad perspective, it appears that the anthropogenic increase in nutrients in aquatic ecosystems has led to an increase in toxic HABs that are a hazard to human health.

Cyanobacteria and dinoflagellates are the slowest growing algae, so physical factors are important in the development of high concentrations of these toxic species. Using gas vesicles and swimming, respectively, cyanobacteria and dinoflagellates move the fastest through the seawater compared to other groups of algae. This gives them a competitive advantage in certain physical regimes, particularly strongly stratified water columns. Cyanobacteria and dinoflagellates tend to predominate over other groups of algae in warm stratified water, as found the temperate summer and in the tropics. It appears likely that global warming will give cyanobacteria and dinoflagellates a competitive advantage over other groups of algae. It is predicted that increased eutrophication and global warming will lead to a further increase in toxic HABs.

References

Àlvarez-Salgado, X.A. et al. (2011). Control of lipophilic shellfish poisoning outbreaks by seasonal upwelling and continental runoff. *Harmful Algae, 10*, 121-129.

Anderson, D.M. (1998). Physiology and bloom dynamics of toxic *Alexandrium* species, with emphasis on life cycle transitions. In D.M. Anderson, A.D. Cembella, A.D. & G.M. Hallegraeff (Eds.), *Physiological Ecology of Harmful Algal Blooms (*NATO ASI Series, Vol G 41, pp. 29-48). Berlin: Springer Verlag.

Anderson, D.M., Glibert, P.M., & Burkholder, J.M. (2002). Harmful algal blooms and eutrophication: nutrient sources, composition and consequences. *Estuaries, 25,* 704-726.

Anderson, D.M. et al. (2005). *Alexandrium fundyense* cyst dynamics in the Gulf of Maine. *Deep Sea Research, 52*, 2522-2542.

Anderson, D.M. et al. (2008). Harmful algal blooms and eutrophication: Examining linkages from selected coastal regions of the United States. *Harmful Algae, 8*, 39-53.

Backer, L. et al. (2003). Recreational exposure to aerosolized brevetoxins during Florida red tide events. *Harmful Algae, 2*, 19-28.

Bagnis, R. (1993). Ciguatera fish poisoning. In: I.R. Falconer (Ed.), *Algal Toxins in Seafood and Drinking Water* (pp. 105-115). New York: Academic Press.

Bates, S.S., Garrison, D.L., & Horner, R.A. (1998). Bloom dynamics and ecophysiology of domoic-acid-producing *Pseudo-nitzchia* species. In D.M. Anderson, A.D. Cembella, & G.M. Hallegraeff, (Eds.), *Physiological Ecology of Harmful Algal Blooms* (pp. 267-292). Heidelberg: Springer-Verlag.

Bates, S.S., Gaudet, J., Kaczmarska, I., and Ehrman, J.M. (2004). Interaction between bacteria and the domoic acid producing diatom Pseudo-nitzschia multiseries (Hasle) Hasle: can bacteria produce domoic acid autonomously? *Harmful Algae, 3*, 11-20.

Bates, S.S. & Trainer, V.L. (2006). The ecology of harmful diatoms. In: E. Granéli, & J.T. Turner (Eds.), *Ecology of Harmful Algae* (81-94). Berlin: Springer.

Bauman, A. G., et al. (2010). Tropical harmful algal blooms: An emerging threat to coral reef communities? *Marine Pollution Bulletin. 60*, 2117-2122.

Bell, P.R.F. (1992). Eutrophication and coral reefs-Some examples in the Great Barrier Reef lagoon. *Wat. Res., 26*, 553-568.

Bienfang, P.K. et al. (2008). Ciguatera fish poisoning: a synopsis from ecology to toxicity. In: Walsh, P.J. et al. (Eds.), *Oceans and Human Health: Risks and Remedies from the Sea* (pp. 257-270). Elsevier.

Bigham, D.L., Hoyer, M.V., & Canfield, D.E. (2009). Survey of toxic algal (microcystin) distribution in Florida lakes. *Lake and Reservoir Management, 25,* 264-275.

Birkeland, C. (1988). Second-order ecological effects of nutrient input into coral communities. *Galaxea, 7*, 91-100.

Bodeanu, N. (1992). Algal blooms and the development of the main phytoplanktonic species at the Romanian Black Sea littoral in conditions of intensification of the eutrophication process. In R.A. Vollenweider, R. Marchetti, and R. Viviani (Eds.), *Marine Coastal Eutrophication* (pp. 891-906). Amsterdam: Elsevier.

Bolch, C.J.S. & de Salas, M.F. (2007). A review of the molecular evidence for ballast water introduction of the toxic dinoflagellates *Gymnodinium catenatum* and the *Alexandrium* "tamarensis complex" to Australasia. *Harmful Algae, 6,* 465-485.

Bradley, W.G. & Mash, D.C. (2009). Beyond Guam: The cyanobacteria/BMAA hypothesis of the cause of ALS and other neurodegenerative diseases. *Amy. Lat. Scler., 10 (suppl. 2),* 7-20.

Brand, L.E. & Guillard, R.R.L. (1981). The effects of continuous light and light intensity on the reproduction rate of twenty-two species of marine phytoplankton. *J. Exp. Mar. Biol. Ecol., 50,* 119-132.

Brand, L.E. & Compton, A. (2007). Long-term increase in *Karenia brevis* abundance along the southwest Florida coast. *Harmful Algae, 7,* 232-252.

Bricker, S.B., et al. (2008). Effects of nutrient enrichment in the nation's estuaries: A decade of change. *Harmful Algae, 8,* 21-32.

Burkholder, J.M. et al. (2006). Comprehensive trend analysis of nutrients and related variables in a large eutrophic estuary: a decadal study of anthropogenic and climatic influences. *Limnology and Oceanography, 51,* 463-487.

Burkholder, J.M., Glibert, P.M., and Skelton, H.M. (2008). Mixotrophy, a major mode of nutrition for harmful algal species in eutrophic waters. *Harmful Algae, 8,* 77-93.

Buskey, E.J. (2008). How does eutrophication affect the role of grazers in harmful algal bloom dynamics? *Harmful Algae, 8,* 152-157.

Cambridge, J.L. and McComb, A.J. (1986). The loss of seagrass in Cockburn Sound, Western Australia. II. Possible causes of seagrass decline. *Aquat. Bot, 24,* 269-285.

Carlsson P. & Granéli, E. (1998). Utilization of Dissolved Organic Matter (DOM) by phytoplankton, including harmful species. In D.M. Anderson, A.D. Cembella, A.D. & G.M. Hallegraeff (Eds.), *Physiological Ecology of Harmful Algal Blooms* (NATO ASI Series, Vol G 41, pp.509-524). Berlin, Heidelberg: Springer Verlag.

Carmichael, W.W. (1996). Toxic Microcystis and the environment., In: M.F. Watanabe et al (Eds.), *Toxic Microcystis* (pp. 1-11). CRC Press.

Carmichael W.W. (2001). Health effects of toxin producing cyanobacteria: The cyanoHABS. *Human Ecol Risk Assess, 7,* 1393-1407.

Cembella, A.D. (1998). Ecopysiological processes and mechanisms: towards common paradigms for harmful algal blooms microalgae. In D.M. Anderson, A.D. Cembella, A.D. & G.M. Hallegraeff (Eds.), *Physiological Ecology of Harmful Algal Blooms* (NATO ASI Series, Vol G 41, 649-652). Berlin, Heidelberg: Springer Verlag.

Cembella, A.D. (2003). Chemical ecology of eukaryotic microalgae in marine ecosystems. *Phycologia, 42 (4),* 420-447.

Cembella, A.D., & John, U. (2006). Molecular physiology of toxin production and growth regulation in harmful algae. In E. Granéli & J.T. Turner (Eds.), *Ecology of Harmful Algae* (pp. 215-228). Berlin: Springer.

Chorus, I. & Bartram J. (1999). *Toxic Cyanobacteria in Water: A Guide to Their Public Heath Consequences, Monitoring and Management.* London: E&FN Spoon.

Clarke, A.L. et al. (2006). Long-term trends in eutrophication and nutrients in the coastal zone. *Limnol. Oceanogr., 51,* 385-397.

Cloern, J. E. (2001). Our evolving conceptual model of the coastal eutrophication problem. *Marine Ecology Progress Series, 210,* 223-253.

Cociasu, A., Dorogan, L., Humborg, C., & Popa, L. (1996). Long-term ecological changes in Romanian coastal waters of the Black Sea. *Mar. Poll. Bull., 32,* 32-38.

Codd, G.A, & Poon, G.K. (1988). Cyanobacterial toxins. In L.J. Rogers, & J.R. Gallon (Eds.), *Biochemistry of the Algae and Cyanobacteria* (pp. 283-296). Oxford Science Publications.

Codd, G.A. (1999). Cyanobacterial toxins: Their occurrence in aquatic environments and significance to health. In L. Charpy, & A.W.D. Larkum (Eds.), *Marine Cyanobacteria; Bulletin de L'Institut Oceanographique, Special Paper 19* (pp.483-500). Monaco Musee Oceanographique.

Codd, G.A. (2000). Cyanobacterial toxins, the perception of water quality, and the prioritisation of eutrophication control. *Ecol. Eng., 16*, 51-60.

Codd, G.A. et al. (1999). Cyanobacterial toxins, exposure routes and human health. *Eur. J. Phycol., 34,* 405-415.

Conley, D.J. et al. (2009). Ecosystem thresholds with hypoxia. *Hydrobiol., 629*, 21-29.

Cooper, S.R. & Brush, G.S. (1991). Long-term history of Chesapeake Bay anoxia. *Science, 254*, 992-995.

Crouch, E.M. et al. (2003). The *Apectodinium* acme and terrestrial discharge during the Paleocene-Eocene thermal maximum: New palynological, geochemical and calcareous nannoplankton observations in Tawanui, New Zealand. *Palaeogeography, Palaeoclimatology, Palaeoecology, 194(4)*, 387-403.

Cullen, J.J. & MacIntyre, J.G. (1998). Behavior, physiology and the niche of depth-regulating phytoplankton. In D.M. Anderson, A.D. Cembella, A.D. & G.M. Hallegraeff (Eds.), *Physiological Ecology of Harmful Algal Blooms* (NATO ASI Series, Vol G 41, pp. 559-580). Berlin, Heidelberg: Springer Verlag.

Dale, B. (2001). Marine dinoflagellate cysts as indicators of eutrophication and industrial pollution: a discussion. *Science of the Total Environment, 264,* 235-240.

Dale, B., Edwards, M., & Reid, P.C. (2006). Climate change and harmful algal blooms. In E. Granéli & J.T. Turner (Eds.), *Ecology of Harmful Algae* (pp. 367-378). Berlin: Springer.

Daranas, A.H., Norte, M. & Fernandez, J.J. (2001). Toxic marine microalgae. *Toxicon, 39,* 1101-1132

de Jonge, V.N., Elliott, M., & Orive, E. (2002). Causes, historical development, effects and future challenges of a common environmental problem: eutrophication. *Hydrobiologia, 475/476,* 1-19.

Diaz, R.J. & Rosenberg, R. (2008). Spreading dead zones and consequences for marine ecosystems. *Science, 321,* 926-929.

Donaghay, P.L. & Osborn, T.R. (1997). Toward a theory of biological-physical control of harmful algal bloom dynamics and impacts. *Limnology and Oceanography, 42 (5, part 2),* 1283-1296.

Doucette, G.J., Kodama, M., Franca, S., & Gallacher, S. (1998). Bacterial interactions with harmful algal bloom species: Bloom ecology, toxigenesis and cytology. In D.M. Anderson, A.D. Cembella, A.D. & G.M. Hallegraeff (Eds.), *Physiological Ecology of Harmful Algal Blooms* (NATO ASI Series, Vol G 41, 619-648). Berlin, Heidelberg: Springer Verlag.

Doucette, G.J., Maneiro, I., Riveiro, I., & Svensen, C. (2006). Phycotoxin pathways in aquatic food webs: transfer, accumulation and degradation. In E. Granéli & J.T. Turner (Eds.), *Ecology of Harmful Algae* (pp. 283-297). Berlin: Springer.

Edvardsen, B. & Imai, I. (2006). The ecology of harmful flagellates within Prymnesiophyceae and Raphidophyceae. , In E. Granéli & J.T. Turner (Eds.), *Ecology of Harmful Algae* (pp. 67-79). Berlin Heidelber, Germany: Springer-Verlag.

Erdner, D.L. et al. (2008). Centers for Oceans and Human Health: a unified approach to the challenge of harmful algal blooms. *Environmental Health , 7:S2.*

Falconer, I.R. (1993). Mechanism of toxicity of cyclic peptide toxins from blue-green algae. In I.R. Falconer (Ed.), *Algal Toxins in Seafood and Drinking Water* (pp. 177-186). Academic Press.

Falconer, I.R. (2008). Health effects associated with controlled exposures to cyanobacterial toxins. In H.K. Hudnell (Ed.), *Cyanobacterial Harmful Algal Blooms: State of the Science and Research Needs* (pp. 607-612). Springer.

Falconer, I.R. (1999). An overview of problems caused by toxic blue green algae (cyanobacteria) in drinking and recreational water. *Env. Toxicol.,14,* 5-12.

Falconer, I.R. (2005). *Cyanobacterial Toxins of Drinking Water Supplies*. CRC Press.

Falconer, I.R. & Humpage, A.R. (2005). Health risk assessment of cyanobacterial (blue-green algal) toxins in drinking water. *Int J Env Res Pub Health, 2,* 43-50.

Falconer, I.R. & Humpage, A.R. (2006). Cyanobacterial (blue-green algal) toxins in water supplies: Cylindrospermopsins. *Env. Tox.,21*, 299-304.

Falconer, I.R. & Humpage, A.R. (1996). Tumor promotion by cyanobacterial toxins. *Phycol., 35*, 74-79.

Figueiras, F.G., Pitcher, G.C., & Estrada, M. (2006). Harmful algal bloom dynamics in relation to physical processes. In E. Granéli & J.T. Turner (Eds.), *Ecology of Harmful Algae* (pp. 127-138). Berlin: Springer.

Finni, T. et al. (2000). The history of cyanobacterial blooms in the Baltic Sea. *Ambio., 30,* 172-178.

Fisher, T.R., Hagy, J.D., Boynton, W.R., & Williams, M.R. (2006). Cultural eutrophication in the Choptank and Patuxent estuaries of Chesapeake Bay. *Limnol. Oceanogr., 51*, 435-447.

Fleming, L.E., Backer, L.C., & Baden, D.G. (2005a). Overview of aerosolized Florida red tide toxins: Exposures and effects. *Env. Health Persp., 113,* 618-620.

Fleming, L.E. et al., (2005b). Initial evaluation of the effects of aerosolized Florida red tide toxins (brevetoxins) in persons with asthma. *Env. Health Persp., 113,* 650-657.

Flewelling, L.J. et al. (2005). Red tides and marine mammal mortalities. *Nature, 435*, 755-756.

Fraga, S. et al. (1988). Influence of upwelling relation on dinoflagellates and shellfish toxicity in Ria de Vigo Spain. *Est. Coastal Shelf Sci., 27*, 349-361.

Fujiki, H. & Suganuma, M. (1993). Tumor promotion by inhibitors of protein phosphatase-1 and phosphatase-2a - the okadaic acid class of compounds. *Adv. Cancer Res., 61,* 143-194.

Fujiki, H. et al. (1988). Diarrhetic shellfish toxin, dinophysistoxin-1, is a potent tumor promotor on mouse skin. *Jpn. J. Cancer Res., l79*, 1089-1093.

Fujiki, H., Sueoka, E., & Suganuma, M. (1996). Carcinogenesis of microcystins. In J.F. Watanabe, K. Harada, W.W. Carmichael, & H. Fujiki (Eds.), *Toxic Microcystis* (pp. 203-231). CRC Press.

Funari, E. & Testai, E. (2008). Human health risk assessment related to cyanotoxins exposure. *Crit. Rev. Tox., 38,* 97-125.

Gaines, G. & Elbrachter, M. (1987). Heterotrophic nutrition. In F.J.R. Taylor (Ed.), *The Biology of Dinoflagellates* (pp. 225-268). Oxford, England: Blackwell Scientific.

Galloway, J.N. et. al. (1995). Nitrogen fixation: anthropogenic enhancement-environmental response. *Global Biogeochem. Cycles, 9,* 235-252.

Galloway, J.N. et al. (2003). The nitrogen cascade. *Bioscience, 53,* 341-356.

Giani, A., Bird, D.F., Prairie, Y.T., & Lawrence, J.F. (2005). Empirical study of cyanobacterial toxicity along a trophic gradient of lakes. *Canadian Journal of Fisheries and Aquatic Science, 62,* 2100-2109.

Glibert, P.M. et al. (2005a). The global, complex phenomena of harmful algal blooms. *Oceanography, 18(2),* 136-147.

Glibert, P.M. et al. (2005b). The role of eutrophication in the global proliferation of harmful algal blooms. *Oceanogr., 18 (2),* 198-209.

Glibert, P.M. & Burkholder, J.M. (2006). The complex relationships between increases in fertilization of the Earth, coastal eutrophication and proliferation of harmful algal blooms. In E. Granéli & J.T. Turner (Eds.), *Ecology of Harmful Algae* (pp. 341-354). Berlin: Springer.

Glibert, P.M., Harrison, J., Heil, C., & Seitzinger, S. (2006). Escalating worldwide use of urea-a global change contributing to coastal eutrophication. *Biogeochemistry, 77,* 441-463.

Glibert, P.M., et al. (2002). A fish kill of massive proportion in Kuwait Bay, Arabian Gulf, 2001: the roles of bacterial disease, harmful algae, and eutrophication. *Harmful Algae, 1,* 215-231.

Glibert, P.M. & Legrand, C. (2006). The diverse nutrient strategies of HABs: focus on osmotrophy. In E. Granéli & J.T. Turner (Eds.), *Ecology of Harmful Algae* (pp. 163-176). Berlin: Springer.

Glibert, P.M., Mayorga, E., & Seitzinger, S. (2008). *Prorocentrum minimum* tracks anthropogenic nitrogen and phosphorus inputs on a global basis: Application of spatially explicit nutrient export models. *Harmful Algae*, *8*, 33-38.

Gorham, P.R. & Carmichael, W.W. (1988). Hazards of freshwater blue-green algae (cyanobacteria). In C.A. Lembi and J. R. Waaland (Eds.), *Algae and Human Affairs* (pp. 403-431). Cambridge University Press.

Granéli, E., & Carlsson, P. (1998). The ecological significance of phagotrophy in photosynthetic flagellates. In D.M. Anderson, A.D. Cembella, A.D. & G.M. Hallegraeff (Eds.), *Physiological Ecology of Harmful Algal Blooms (*NATO ASI Series, Vol G 41, pp. 539-558). Berlin: Springer Verlag.

Granéli, E., & Hansen, P.J. 2006. Allelopathy in harmful algae: a mechanism to compete for resources? In E. Granéli & J.T. Turner (Eds.), *Ecology of Harmful Algae* (pp. 189-201). Berlin: Springer.

Graneli, E. & Turner, J.T. (2006). An introduction to harmful algae. In E. Granéli & J.T. Turner (Eds.), *Ecology of Harmful Algae* (pp. 3-7). Berlin: Springer.

Granéli, E., Weberg, M., & Salomon, P.S. (2008). Harmful algal blooms of allelopathic microalgal species: The role of eutrophication. *Harmful Algae, 8*, 94-102.

Greening, H. & Janicki, A. (2006). Toward reversal of eutrophic conditions in a subtropical estuary: Water quality and seagrass response nitrogen loading reductions in Tampa Bay, Florida USA. *Env. Manag., 38,* 163-178.

Hackett, J.D., Anderson, D.M., Erdner, D.L., & Bhattacharya, D. (2004). Dinoflagellates: A remarkable evolutionary experiment. *American Journal of Botany, 91(10),* 1523-1534.

Hallengraeff, G.M. (1993). A review of harmful algal blooms and their apparent global increase. *Phycologia, 32,* 79-99.

Hallegraeff, G.M, & Fraga, S. (1998). Bloom dynamics of the toxic dinoflagellate *Gymnodinium catenatum*, with emphasis on Tasmanian and Spanish coastal waters. In D.M. Anderson, A.D. Cembella, A.D. & G.M. Hallegraeff (Eds.), *Physiological Ecology of Harmful Algal Blooms (*NATO ASI Series, Vol G 41, pp. 59-80). Berlin: Springer Verlag.

Hallengraeff, G. & Gollasch, S. (2006). Anthropogenic introductions of microalgae. In E. Granéli & J.T. Turner (Eds.), *Ecology of Harmful Algae* (pp. 379-390). Berlin: Springer.

Hansen, P.J. (1998). Phagotrophic mechanisms and prey selection in mixotrophic phytoflagellates. In D.M. Anderson, A.D. Cembella, A.D. & G.M. Hallegraeff (Eds.), *Physiological Ecology of Harmful Algal Blooms (*NATO ASI Series, Vol G 41, pp. 525-538). Berlin: Springer Verlag.

Harding, L.W. & Perry, E.S. (1997). Long-term increase of phytoplankton biomass in Chesapeake Bay, 1950-1994. *Mar. Ecol. Prog. Ser., 157*, 39-52.

Hattenrath, T.K., Anderson, D.M., & Gobler, C.J. (2010). The influence of anthropogenic nitrogen loading and meteorological conditions on the dynamics and toxicity of *Alexandium fundyense* blooms in a New York (USA) estuary. *Harmful Algae, 8,* 402-412.

Havens, K.E. (2008). Cyanobacteria blooms: Effects on aquatic ecosystems. In H.K. Hudnell (Ed.), *Cyanobacterial Harmful Algal Blooms: State of the Science and Research Needs* (pp. 733-747). Berlin: Springer.

Havens, K.E., Aumen, N.G., James, R.T., and Smith, V.H. (1996). Rapid ecological changes in a large subtropical lake undergoing cultural eutrophication. *Ambio., 25,* 150-155.

Heil, C.A. et al. (2001). First record of a fish-killing *Gymnodinium* sp. bloom in Kuwait Bay Arabian Sea: Chronology and potential causes. *Mar. Ecol. Prog. Ser., 214,* 15-23.

Heil, C.A., Glibert, P.M., & Fan, C. (2005). *Prorocentrum minimum* (Pavillard) Schiller—a review of a harmful algal bloom species of growing worldwide importance. *Harmful Algae, 4,* 449-470.

Heisler, J. et al. (2008). Eutrophication and harmful algal blooms: A scientific consensus. *Harmful Algae, 8,* 3-13.

Howarth, R.W. (2008). Coastal nitrogen pollution: A review of sources and trends globally and regionally. *Harmful Algae, 8*, 14-20.

Howarth, R.W. et al. (2000). Nutrient pollution of coastal rivers, bays, and seas. *Issues Ecol., 7,* 1-15.

Humpage, A. (2008). Toxin types, toxicokinetics and toxicodynamics. In H.K. Hudnell (Ed.), *Cyanobacterial Harmful Algal Blooms: State of the Science and Research Needs* (pp. 383-415). Springer.

Ibelings, Bas W. & Chorus, Ingrid. (2007). Accumulation of cyanobacterial toxins in freshwater "seafood" and its consequences for public health: A review. *Environmental Pollution, 150*, 177-192.

Imai, Ichiro, et al. (2006). Eutrophication and occurrences of harmful algal blooms in the Seto Island Sea, Japan. *Plankton Benthos Research, 1(2)*, 71-84.

Jaworski, N. (1990). Retrospective of the water quality issues of the upper Potomac estuary. *Aquatic Science, 3,* 11-40.

Jones, R.I. (2000). Mixotrophy in planktonic protists: An overview. *Fresh. Biol.*, *45*, 219-226.

Justic, D. et al. (1995). Changes in nutrient structure of river-dominated coastal waters: Stoichiometric nutrient balance and its consequences. *Est. Coast. Shelf Sci.*, *40*, 339-356.

Kemp, W.M. et al. (1983). The decline of submerged vascular plants in upper Chesapeake Bay: Summary of results concerning possible causes. *Mar. Tech. Soc. J.*, *17*, 78-89.

Kemp, W.M. et al. (2005). Eutrophication of Chesapeake Bay: historical trends and ecological interactions. *Marine Ecology Progress Series*, *303,* 1-29.

Klemer, A.R. & Konopka, A.E. (1980). Causes and consequences of blue-green algal (cyanobacterial) blooms. *Lake Res. Mang.*, *5*, 9-19.

Kodama, M. et al. (2006). Relationships between bacteria and harmful algae. In: In E. Granéli & J.T. Turner (Eds.), *Ecology of Harmful Algae* (pp. 243-257). Berlin: Springer.

Kotak, B.G. & Zurawell, R.W. (2007). Cyanobacterial toxins in Canadian freshwaters: A review. *Lake Res. Manag.*, *23*, 109-122.

Kusek, K.M., Vargo, G., & Steidinger, K. (1999). *Gymnodinium breve* in the field, in the lab, and in the newspaper-a scientific and journalistic analysis of Florida red tides. *Contri. Mar. Sci., 34*, 1-229.

Lam, C.W.Y. & Ho, K.C. (1989). Red tides in Tolo Harbour Hong Kong. In T. Okaichi, D.M. Anderson, & T. Nemoto (Eds.), *Red Tides Biology, Environmental Science and Toxicology* (pp. 49-52). Elsevier.

Landsberg, J.H. (2002). The effects of harmful algal blooms on aquatic organisms. *Rev. Fish. Sci.*, *10*, 113-390.

Landsberg, J.H., et al. (1999). The potential role of natural tumor promotors in marine turtle fibropapillomatosis. *Journal of Aquatic Animal Health, 11,* 199-210.

Lapointe, B.E., Tomasko, D.A., and Matzie, W.R. (1994). Eutrophication and trophic state classification of seagrass communities in the Florida Keys. *Bull. Mar. Sci.*, *54*, 696-717.

Larsson, U., Elmgren, R., & Wulff, F. (1985). Eutrophication and the Baltic Sea: Causes and Consequences. *Ambio, 14*, 9-14.

Legrand, C., Rengefors, K., Fistarol, G.O., & Granéli, E. (2003). Allelopathy in phytoplankton – biochemical, ecological and evolutionary aspects. *Phycologia, 42 (4),* 406-419.

Litaker, R.W. et al. (2010). Global distribution of ciguatera causing dinoflagellates in the genus Gambierdiscus. *Toxicon, 56*, 711-730.

Maragos, J.E., Evans, C., & Holtbus, P. (1985). Reef corals in Kaneohe Bay six years before and after termination of sewage discharges. *Proc. Fifth Int. Coral Reef Congress, 4*, 189-194.

Matsunaga, H. et al. (1999). Possible cause of unnatural mass death of wild birds in a pond in Nishinomiya, Japan: Sudden appearance of toxic cyanobacteria. *Nat. Toxins, 7*, 81-88.

Mee, L.D. (1992). The Black Sea in crisis: A need for concerted international action. *Ambio, 21*, 278-285.

Metcalf, J.S. & Codd, G.A. (2009). Cyanobacteria, neurotoxins and water resources: Are there implications for human neurodegenerative disease? *Amy. Lat. Scler., 10 (suppl. 2)*, 74-78.

Mihali, T.K. et al. (2008). Characterization of the gene cluster responsible for Cylindrospermopsin biosynthesis.

Miller, M.A. et al. (2010). Evidence for a novel marine harmful algal bloom: cyanotoxin (microcystin) transfer from land to sea otters. *PLOS One, 5(9),* e12576.

Moore RE. (1996). Cyclic peptides and depsipeptides from cyanobacteria: A review. *J. Ind. Microbiol,, 16*, 134-143.

Moore, S.K. et al. (2008). Impacts of climate variability and future climate change on harmful algal blooms and human health. *Environmental Health, 7 (Suppl 2),* S4.

Naar, J.P. et al. (2007). Brevetoxins, like ciquatoxins, are potent ichthyotoxic neurotoxins that accumulate in fish. *Toxicon, 50*, 707-723.

Nehring, D. (1992). Eutrophication in the Baltic Sea. In R.A. Vollenweider, R. Marchetti, & R. Viviani (Eds.), *Marine Coastal Eutrophication* (pp. 673-682). Elsevier.

Nixon, S.W. (1995). Coastal marine eutrophication: a definition, social causes and future concerns. *Ophelia, 41,* 199-219.

Nixon, Scott W. (2009). Eutrophication and the macroscope. *Hydrobiologia, 629,* 5-19.

Norring, Niels Peter, & Jorgensen, Erik. (2009). Eutrophication and agriculture in Denmark: 20 years of experience and prospects for the future. *Hydrobiologia, 629,* 65-70.

Officer, C.B. & Ryther, J.H. (1980). The possible importance of silicon in marine eutrophication. *Marine Ecology – Progress Series*, 3, 83-91.

Olascoaga, M.J. et al. (2006). Persistent transport barrier on the west Florida shelf. *Geophys. Res. Let.*, 33, L22603,doi:10.1029/2006GL027800.

Oliver. R.L. (1994). Floating and sinking in gas-vacuolate cyanobacteria. *J. Phycol., 30*, 161-173.

Orth, R.J. and K.A. Moore. 1983. Chesapeake Bay: An unprecedented decline in submerged aquatic vegetation. *Science, 222*, 51-53.

Osborne N.J.T., Webb P.M., & Shaw G.R. (2001). The toxins of *Lyngbya majuscula* and their human and ecological effects. *Env. Int., 27*, 381-392.

Paerl, H.W. (1988). Nuisance phytoplankton blooms in coastal, estuarine, and inland waters. *Limnology and Oceanography, 33 (4, part 2),* 823-847.

Paerl, H.W. (1997). Coastal eutrophication and harmful algal blooms: Importance of atmospheric deposition and groundwater as "new" nitrogen and other nutrient sources. *Limnol. Oceanogr., 42*, 1154-1165.

Paerl, H. (2008). Nutrient and other environmental controls of harmful cyanobacterial blooms along the freshwater-marine continuum. In H.K. Hudnell (Ed.), *Cyanobacterial Harmful Algal Blooms: State of the Science and Research Needs* (pp. 217-237). Springer.

Paerl, H.W. & Fulton III, R.S. (2006). Ecology of harmful cyanobacteria. In E. Granéli & J.T. Turner (Eds.), *Ecology of Harmful Algae* (pp. 95-110). Springer.

Paerl, H.W. & Ustach, J.F. (1982). Blue-green algal scums: An explanation for their occurrence during freshwater blooms. *Limnology and Oceanography, 27*, 212-217.

Paerl, H. et al. (2006). Anthropogenic and climatic influences on the eutrophication of large estuarine systems. *Limnology and Oceanography, 51*, 448-462,

Paerl, H.W. & Huisman, J. (2009). Climate change: A catalyst for global expansion of harmful cyanobacterial blooms. *Env. Microbiol. Rep., 1*, 27-37.

Parsons, M.L., Dortch, Q., & Turner, R.E. (2002). Sedimentological evidence of an increase in *Pseudo-nitzschia* (Bacillariophyceae) abundance in response to coastal eutrophication. *Limnology and Oceanography, 47(2),* 551-558.

Pastorak, R.A. & Bilyard, G.R. (1985). Effects of sewage pollution on coral-reef communities. *Mar. Biol. Prog. Ser., 21*, 175-189.

Paul, V.J. (2008). Global warming and cyanobacterial harmful algal blooms. In H.K. Hudnell, (Ed.), *Cyanobacterial Harmful Algal Blooms: State of the Science and Research Needs* (pp. 239-257). Springer.

Pierce, R.H. et al. (2005). Brevetoxin composition in water and marine aerosol along a Florida beach: Assessing potential human exposure to marine biotoxins. *Harmful Algae, 4*, 965-972.

Pinckney, J.L., Paerl, H.W., Tester, P., & Richardson, T.L. (2001). The role of nutrient loading and eutrophication in estuarine ecology. *Env. Health Persp., 109 (suppl. 5)*, 699-706.

Poutanen, E.L. & Nikkila, K. (2000). Carotenoid pigments as tracers of cyanobacterial blooms in recent and post-glacial sediments of the Baltic Sea. *Ambio, 30*, 179-183.

Prakash, A. (1987). Coastal organic pollution as a contributing factor to red-tide development. *Rapp. P-v. Reun. Cons. Int. Explor. Mer, 187,* 61-65.

Rabalais, N.N. et al. (2002). Nutrient-enhanced productivity in the northern Gulf of Mexico: Past, present and future. *Hydrobiol., 475/476*, 39-63.

Radach, G., Berg, J., & Hagmeier, E. (1990). Long-term changes of the annual cycles of meteorological, hydrographic, nutrient and phytoplankton time series at Helgoland and at LV ELBE 1 in the German Bight. *Cont. Shelf Res., 10*, 305-328.

Richardson, K. & Jorgensen, B.B. (1996). Eutrophication: Definition, history and effects. In *Eutrophication in Coastal Marine Ecosystems. Coastal and Estuarine Studies* (Vol. 52, pp. 1-19). American Geophysical Union.

Riegman, R. (1998). Species composition of harmful algal blooms in relation to macronutrient dynamics. In D.M. Anderson, A.D. Cembella, A.D. & G.M. Hallegraeff (Eds.), *Physiological Ecology of Harmful Algal Blooms* (NATO ASI Series, Vol G 41, pp. 474-488). Berlin: Springer Verlag.

Riegman, R., Row, A., Noordeloos, A.A.M., & Cadee, G.C. (1991). Evidence for eutrophication induced *Phaeocystis* sp. Blooms in the Marsdiep area (The Netherlands). In T.J. Smayda & Y. Shimizu (Eds.). *Toxic Phytoplankton Blooms in the Sea* (pp. 799-805). Elsevier Science.

Ryther, J.H. & Dunstan, W.M. (1971). Nitrogen, phosphorus and eutrophication in the coastal marine environment. *Science, 171*, 1008-1013.

Schindler, D.W. (2006). Recent advances in the understanding and management of eutrophication. *Limnology and Oceanography, 51,* 356-363.

Seitzinger, S.P. et al. (2005). Sources and delivery of carbon, nitrogen and phosphorus to the coastal zone: an overview of global nutrient export from watersheds (NEWS) models and their application. *Global Biogeochemistry Cycles, 19*, GB4S09.

Seitzinger, S.P. et al. (2002). Global patterns of dissolved inorganic and particulate nitrogen inputs to coastal systems: Recent conditions and future projections. *Estuaries, 25 (4b)*, 640-655.

Seliger, H.H, Carpenter, J.H., Loftus, M., & McElroy, W.D. (1970). Mechanisms for the accumulation of high concentration of dinoflagellates in a bioluminescent bay. *Limnol. Oceanogr., 15*, 234-245.

Sellner, Kevin G. (1997). Physiology, ecology, and toxic properties of marine cyanobacteria blooms. *Limnol. Oceanog., 42 (5, part 2)*, 1089-1104.

Shapiro, J. (1973). Blue-green algae: Why they become dominant. *Science, 179,* 382-384.

Shumway, Sandra E. (1995). Phycotoxin-related shellfish poisoning: bivalve molluscs are not the only vectors. *Reviews in Fisheries Science, 3(1),* 1-31.

Silberstein, K., Chiffings, A.W., & McComb, A.J. (1986). The loss of seagrass in Cockburn Sound, Western Australia. III. The effect of epiphytes on productivity of *Posidonia australis* Hook. F. *Aquat. Bot., 24,* 355-371.

Sivonen K, & Borner, T. (2008). Bioactive compounds produced by cyanobacteria. In A. Herrero & E. Flores (Eds.), *The Cyanobacteria: Molecular Biology, Genomics and Evolution* (pp. 159-197). Caister Academic Press.

Smayda, T.J. (1989). Primary production and the global epidemic of phytoplankton blooms in the sea: a linkage? In E.M. Cosper, E.J. Carpenter, & M. Bricelj (Eds.), *Novel Phytoplankton Blooms: Causes and Impacts of Recurrent Brown Tide and Other Unusual Blooms* (pp. 213-228). Springer-Verlag.

Smayda, T.J. (1990). Novel and nuisance phytoplankton blooms in the sea: Evidence for a global epidemic. In *Toxic marine phytoplankton: Proceedings from the 4th Internation Conference.* Elsevier.

Smayda, T.J. (1997b). Harmful algal blooms: Their ecophysiology and general relevance to phytoplankton blooms in the sea. *Limnol. Oceanogr., 42 (5, part 2),* 1137-1153.

Smayda, T.J. (2007). Reflections on the ballast water dispersal-harmful algal bloom paradigm. *Harmful Algae, 6*, 601-622.

Smayda, T.J. (2008). Complexity in the eutrophication-harmful algal bloom relationship, with comment on the importance of grazing. *Harmful Algae, 8*, 140-151.

Smith, S. V. et al.(1981). Kaneohe Bay sewage diversion experiment: perspective on ecosystem responses to nutritional perturbation. *Pac. Sci., 35*, 279-385.

Smith, V.H. (1983). Low nitrogen to phosphorus ratios favor dominance by blue-green algae in lake phytoplankton. *Science, 221,* 669-671.

Smith, V.H. (1990). Nitrogen, phosphorus, and nitrogen fixation in lacustrine and estuarine ecosystems. *Limnol. Oceanogr., 35*, 1852-1859.

Smith, V.H. (2006). Responses of estuarine and coastal marine phytoplankton to nitrogen and phosphorus enrichment. *Limnol Oceanogr., 51*, 377-384.

Smith, V. H., Joye, S.B., & Howarth, R.W. (2006). Eutrophication of freshwater and marine ecosystems. *Limnol. Oceanogr., 51 (1, part 2),* 351-355.

Snyder, R.V. et al. (2005). Localization of polyketide synthase encoding genes to the toxic dinoflagellate *Karenia brevis. Phytochem., 66*, 1767-1780.

Steidinger, K.A. (2009). Historical perspective on *Karenia brevis* red tide research in the Gulf of Mexico. *Harmful Algae, 8*, 549-561.

Steidinger, K.A & Garcés, E. (2006). Importance of life cycles in the ecology of harmful microalgae. In E. Graneli & J.T. Turner (Eds.), *Ecology of Harmful Algae* (pp. 37-49). Springer-Verlag.

Steidinger, K.A& Vargo, G.A. (1988). Marine dinoflagellate blooms: dynamics and impacts. In C. Lembi & J.R. Waaland (Eds.), *Algae and Human Affairs* (pp. 373-401). Cambridge.

Steidinger, K.A., Landsberg, J.H., Flewelling, L.J. & Kirkpatrick, B.A. (2008). Toxic dinoflagellates. In P. J. Walsh et al. (Eds.), *Oceans and Human Health: Risks and Remedies from the Sea* (pp. 239-256). Elsevier.

Stewart, I. & Falconer, I.R. (2008). Cyanobacteria and cyanobacterial toxins. In P. J. Walsh et al. (Eds.), *Oceans and Human Health: Risks and Remedies from the Sea* (pp. 271-296). Elsevier.

Stewart, I., Seawright, A.A., & Shaw G.R. (2008). Cyanobacterial poisoning in livestock, wild mammals and birds – an overview. In H.K. Hudnell (Ed.), *Cyanobacterial Harmful Algal Blooms: State of the Science and Research Needs* (pp. 613-637). Springer.

Stoecker, D.K. (1999). Mixotrophy among dinoflagellates. *Journal of Eukaryotic Microbiology*, *46,* 397-401.

Stoecker, D.K, Thessen, A.E., & Gustafson, D.E. (2008). "Windows of opportunity" for dinoflagellate blooms: Reduced microzooplankton net growth coupled to eutrophication. *Harmful Algae*, *8*, 158-166.

Stoecker, D., Tillmann, U., & Granéli, E. (2006.) Phagotrophy in harmful algae. In E. Granéli and J.T. Turner (Eds.), *Ecology of Harmful Algae* (pp. 177-188). Springer.

Stolte, W. & Garcés, E. (2006). Ecological aspects of harmful algal in situ population growth rates. In E. Granéli & J.T Turner (Eds.), *Ecology of Harmful Algae* (pp. 139-152). Springer.

Tester, P.A., Stumpf, R.P., & Steidinger, K.A. (1998). Ocean color imagery: What is the minimum detection level for *Gymnodinium breve* blooms. In B. Reguera, J. Blanco, M.L. Fernandez, & T. Wyatt (Eds.), *Harmful Algae* (pp. 149-151).

Tester, P.A., Turner, J.T., & Shea D. (2000). Vectorial transport of toxins from the dinoflagellate *Gymnodinium breve* through copepods to fish. *Journal of Plankton Research, 22*, 47-62.

Tillett, D. et al. (2000). Structural organization of microcystin biosynthesis in Microcystis aeruginosa PCC 7806: An integrated peptide-polyketide synthetase system. *Chem. Biol., 7i*, 753-764.

Tindall, D. R. & Morton, S.L. (1998). Community dynamics and physiology of epiphytic/benthic dinoflagellates associated with ciguatera. In D.M. Anderson, A.D. Cembella, & G.M. Hallegraeff (Eds.), *Physiological Ecology of Harmful Algal Blooms* (NATO ASI Series, Vol G 41, pp. 293-314). Berlin: Springer Verlag.

Tomasko, D.A. & Lapointe, B.E. (1991). Productivity and biomass of *Thalassia testudinum* as related to water column nutrient availability and epiphyte levels: field observations and experimental studies. *Mar. Ecol. Prog. Ser., 75*, 9-17.

Trainer, V.L., Hickey, B.M, & Bates, S.S. (2008). Toxic diatoms. In P.J. Walsh et al. (Eds.), *Oceans and Human Health: Risks and Remedies from the Sea* (pp. 219-238). Elsevier.

Trainer, V.L. et al. (2003). Paralytic shellfish toxins in Puget Sound, Washington State. *Journal of Shellfish Research, 22 (1)*, 213-223.

Tsujimura, S. and T. Okubo. (2003). Development of Anabaena blooms in a small reservoir with dense sediment akinete population, with special reference to temperature and irradiance. J. *Plankton Res., 25*, 1059-1067.

Tucker, C.S. (2000). Off-flavor problems in aquaculture. *Rev. Fish. Sci., 8*, 45-88.

Turner, J.T. (2006). Harmful algae interactions with marine planktonic grazers. In E. Granéli & J.T. Turner (Eds.), *Ecology of Harmful Algae* (pp. 259-270). Springer.

Turner, J.T., & Granéli, E. (2006). "Top-down" predation control on marine harmful algae. In E. Granéli & J.T. Turner (Eds.), *Ecology of Harmful Algae* (pp. 355-366). Springer.

Turner, J.T., Tester, P.A., & Hansen, P.J. (1998). Interactions between toxic marine phytoplankton and metazoan and protistan grazers. In D.M. Anderson, A.D. Cembella, & G.M. Hallegraeff (Eds.), *Physiological Ecology of Harmful Algal Blooms* (NATO ASI Series, Vol G 41, pp. 453-474). Berlin: Springer Verlag.

Turner, R.E. & Rabalais, N.N. (1994). Coastal eutrophication near the Mississippi river delta. *Nature, 368*, 619-621.

Turner, R.E. et al. (2006). Paleo-indicators and water quality change in the Charlotte Harbor estuary (Florida). *Limnol. Oceanogr., 51*, 518-533.

Tyler, M.A. & Seliger, H.H. (1978). Annual subsurface transport of a red tide dinoflagellate to its bloom area: water circulation patterns and organism distributions in the Chesapeake Bay. *Limnol. Oceanogr., 23,* 227-246.

Van Buynder, P.G. et al. (2001). Nodularin uptake by seafood during a cyanobacterial bloom *Env. Tox., 16*, 468-471.

Van Dolah, F.M. (2000). Marine algal toxins: origins, health effects and their increased occurrence. *Environmental Health Perspectives, 108 (S1),* 133-141.

Van Dolah, F.M., Roelke, D. & Greene, R.M. (2001). Health and ecological impacts of Harmful Algal Blooms: Risk assessment needs. *Human and Ecological Risk Assessment, 7(5)*, 1329-1345.

Vasconcelos, V. (2006). Eutrophication, toxic cyanobacteria and cyanotoxins: When ecosystems cry for help. *Limnetica, 25*, 425-432.

Vitousek, P.M. et al. (1997). Human alteration of the global nitrogen cycle: sources and consequences. *Ecol. Appl., 7*, 737-750.

Vollenweider, R.A. (1992). Coastal marine eutrophication: Principles and control. In R.A. Vollenweider, R. Marchetti, & R. Viviani (Eds.), *Marine Coastal Eutrophication* (pp. 1-20). Elsevier.

Walsby, A.E. (1987). Mechanisms of buoyancy regulation by planktonic cyanobacteria with gas vesicles. In P. Fay & C. Van Baalen (Eds.), *The Cyanobacteria* (pp. 377-392). Elsevier.

Watson, S. B. (2003). Cyanobacterial and eukaryotic algal odour compounds: Signals or by-products? A review of their biological activity. *Phycologia, 42 (4)*, 332-350.

Welker, M. & von Döhren, H. (2006). Cyanobacterial peptides – nature's own combinatorial biosynthesis. *FEMS Microbiological Review, 30*, 530-563.

Wittenberg. M. & Hunte, W. (1992). Effects of eutrophication and sedimentation on juvenile corals. I. Abundance, mortality and community structure. *Mar. Biol., 112*, 131-138.

Worm, B. and B. Sommer. (2000). Rapid direct and indirect effects of a single nutrient pulse in a seaweed-epiphyte-grazer system. *Mar. Ecol. Prog. Ser., 202*, 283-288.

Wyatt, T. & Jenkinson, I.R. (1997). Notes on *Alexandrium* population dynamics. *J. Plankton Res., 19*, 551-575.

Xie, L. et al. (2005). Organ distribution and bioaccumulation of microcystins in freshwater fish at different trophic levels from the eutrophic Lake Chaohu, China. *Env. Tox., 20,* 293-300.

Zaccaroni, A. & Scaravelli, D. (2008a). Toxicity of fresh water algal toxins to humans and animals. In V. Evangelisa et al. (Eds.), *Algal Toxins: Nature, Occurrence, Effect and Detection* (pp. 45-89). Springer Science + Business Media B.V.

Zaccaroni, A. & Scaravelli, D. (2008b). Toxicity of sea algal toxins to humans and animals. In V. Evangelisa et al. (Eds.), *Algal Toxins: Nature, Occurrence, Effect and Detection* (pp. 91-158). Springer Science + Business Media B.V.

In: New Trends in Marine and Freshwater Toxins
Editors: A. G. Cabado and J. M. Vieites
ISBN: 978-1-62257-787-3

Chapter 3

Harmful Algae Blooms and Food Safety: Physiological and Environmental Factors Affecting Toxin Production and Their Accumulation in Shellfish

Beatriz Reguera[1], Francisco Rodríguez[1] and Juan Blanco[2]
[1]IEO, Centro Oceanográfico de Vigo, Cabo Estay, Canido, Aptdo, Vigo, Spain
[2]CIMA, Pedras de Corón s/n. Apdo., Vilanova de Arousa, Pontevedra, Spain

Abstract

Harmful Algal Blooms (HAB) of toxin producing microalgae that, even at low cell densities, cause accumulation of unsafe levels of toxins in filter-feeders, pose a major natural hazard for public health and the sustainable exploitation of shellfish resources. Key questions are why blooms of the same species have chronic devastating effects in some parts of the world and hardly cause any damage elsewhere, and what are the suite of environmental conditions triggering the expression of toxin synthesis and its seasonal and interannual variability. Answers to these questions require understanding of the complex processes involved in toxin production by microalgae and their availability to filter and deposit feeders, and the species-specific behaviour of molluscan shellfish organisms. This chapter revises current knowledge on intrinsic (genetically determined) and environmental factors affecting toxin production by microalgae, their accumulation in dense populations (controlled by hydroclimatic processes) and the species-specific responses of filter feeders to the uptake, biotransformation and elimination of each group of toxins. New avenues opened by advanced molecular biology applications to address these questions are discussed.

1. Introduction

Practically every coastal area in the world suffers from some kind of harmful algal blooms (HABs), a term that designates any proliferation of microalgae perceived as a nuisance from an anthropogenic perspective, regardless of their cell concentration (Anderson et al., 2010). There are many kinds of HABs caused by different species and associated with various negative impacts on public health, commercial fisheries, recreation and tourism, that in the European Union alone accounts for more than 800 million US$ annual losses (Hoagland & Scatasta, 2006). Some high biomass HABs are formed by exotoxin-producing planktonic microalgae that excrete hemolythic and oxidative substances that above a threshold concentration kill fish and benthic fauna. There are toxigenic HAB species that, even at low cell densities (low biomass HABs) produce endotoxins that are transferred through the food web, mainly through filter-feeding bivalves, and render shellfish unsuitable for human consumption. These kinds of HABs, herein called toxic algae events, constitute a major threat to public health and shellfish exploitations, and will be the main focus of this chapter. In coastal areas with no seafood safety control, toxic syndromes such as paralytic shellfish poisoning (PSP), diarrheic shellfish poisoning (DSP) and amnesic shellfish poisoning (ASP) are responsible for chronic human intoxications that include fatal cases. Conversely, in regions where costly monitoring programs on potentially toxic phytoplankton and their phycotoxins have been established, public health is protected, but different kinds of toxic algae events lead to lengthy shellfish harvesting closures enforced by health and fisheries authorities every time toxin content in shellfish exceeds regulatory levels (Van Egmond et al., 1993; FAO, 2004).

It is frequently stated that "harmful algae blooms are increasing in frequency, intensity and geographic extension" and they are often associated with coastal pollution. Considering that more than 300 taxa from different microalgae groups, with species—or even strain-specific physiological requirements—have been associated with harmful effects, this is an oversimplified statement. In the case of toxic algae events, about 100 species from different algal groups have been confirmed to produce toxins (Moestrup et al., 2009) that belong to different families of toxic compounds with their characteristic molecular structure and biological activity (Table 1). Further, some of the most intense shellfish poisoning events have occurred in pristine coastal areas in Patagonian waters (Sar et al., 2002) and on the northwest coast of North America (Taylor & Trainer, 2002). Certainly, increased aquaculture exploitations—that have acted as the canary in the mine and put into evidence already existing phenomena—and increased monitoring efforts and seafood safety regulations have led to an exponential growth of the number of reports on phycotoxin-contaminated shellfish.

Toxins are secondary metabolites. Their metabolic and ecological roles, and in recent years the identification of genes controlling their production, are topics under intense research. Most intriguing is why blooms of the same species have devastating effects on shellfish contamination in some parts of the world but are practically harmless elsewhere. Moreover, blooms of a given species in the same location exhibit large interannual variability in their toxigenic capability. These differences are largely due to variability in the toxin composition—toxin profiles—and cellular toxin content—toxin per cell (Q_t)—of the causative agents (Cembella & John, 2006; Granéli & Flynn, 2006). Information on toxin profiles and content of the main toxin-producing species have been obtained from laboratory cultures, but also from field populations, in particular from species difficult to establish in culture (i.e. *Dinophysis* spp.). At least in the case of PSP toxin producers, toxin profiles in strains of the same species are relatively constant, and support the view that they are genetically determined (Ishida et al., 1998). The situation is more complex when the same species can produce more than one group of toxins subject to different metabolic regulators. In any case, the production and accumulation of toxins in microalgal cells will be affected by intrinsic, genetically-controlled factors, and by the responses of different strains to environmental (physical, chemical and biological) factors.

Table 1. List of species that have been associated with shellfish poisoning events

ASP	DSP	PSP	AZP
Pseudo-nitzschia australis	*Dinophysis acuminata*	*Alexandrium acatenella*	*Azadinium spinosum*
P. calliantha	*D. acuta*	*A. andersonii*	
P. cuspidata	*D. caudata*	*A. catenella*	NSP
P. delicatissima	*D. fortii*	*A. fundyense*	*Karenia brevis*
P. fraudulenta	*D. infundibulus*	*A. minutum*	
P. galaxiae	*D. miles*	*A. ostenfeldii*	YTX
P. multiseries	*D. norvegica*	*A. peruvianum*	*Gonyaulax spinifera*
P. multistriata	*D. ovum*	*A. tamarense*	*Lingulodinium polyedrum*
P. pungens	*D. rapa*	*A. tamiyavanichii*	*Protoceratium reticulatum*
P. seriata	*D. rotundata*	*Gymnodinium catenatum*	
P. turgidula	*D. sacculus*	*Pyrodinium bahamense*	
	D. tripos		
	Phalacroma mitra		

Notes: ASP = Amnesic Shellfish Poisoning; DSP = Diarrhetic Shellfish Poisoning; PSP = Paralytic Shellfish Poisoning; AZP = Azaspiracid Shellfish Poisoning; NSP = Neurotoxic Shellfish Poisoning; YTX = Yessotoxins. (Information extracted from Moestrup et al.. 2009; Reguera & Pizarro 2008).

Toxin composition and content of the toxigenic cells and their density (cells $\cdot L^{-1}$) are not the only factors. Toxic algae events are reported following the accumulation of toxins in filter-feeders, and different species exposed to the same noxious bloom in a given locality may respond in very different ways. For example, during an intense PSP outbreak in the Seto Inland Sea in 1976, record values of 91,200 µg STX equiv. kg^{-1} were found in the purple clam (*Saxidomus purpuratus*), whereas a maximum of 3,400 µg STX-equiv. kg^{-1} was reached in the littleneck clam (*Tapes japonica*) growing in the same location (Ono et al., 1996). The next steps to consider are the feeding behaviour of different shellfish species in response to toxigenic and other accompanying cells and suspended matter in the plankton, and their capability to accumulate and biotransform the toxins in their tissues.

This chapter revises current knowledge on different intrinsic (genetically determined) and external (environmental) factors affecting production and accumulation of phycotoxins in toxigenic microalgal populations—both from field and laboratory cultures—and their accumulation by shellfish. The main focus will be on microalgal species causing major shellfish poisoning events—PSP, DSP and other lipophilic toxins, and ASP—that affect commercially exploited shellfish resources.

1.1. Intrinsic Factors that Affect Toxin Production by Microalgae

The toxigenic capability of a microalgal bloom, i.e., its potential to render shellfish unsafe for human consumption, is dependent on its cell concentration, but most importantly on the toxin profile and cell toxin quota (Q_t) of the particular species/strain.

1.1.1. Differences in Toxin Profile and Cell Toxin Content (Q_t)

The toxin profile is defined here as the relative contribution (percentage) on a per mol basis of different toxins to the overall toxin content (Q_t) of a species or a strain. Q_t is a quantitative term that refers to the total amount of toxin (moles or weight) accumulated per cell, i.e. the balance between gains (production) and losses (release in the water). It is well known that a significant proportion of the produced toxins can be released in the water—in particular when cells are senescent—and be tracked with adsorbing resins deployed *in situ* (MacKenzie et al., 2004) (see chapter 4). Nevertheless, it is assumed here that these "dissolved" (<0.22μm) toxin fractions are not taken up by shellfish; but efforts should be made to measure them in any reliable study of toxin production dynamics.

It is generally accepted that the toxin profile or "fingerprint" of a given species/strain is genetically determined and inherited in a Mendelian manner, and changes within certain ranges of this profile should be expected in response to environmental conditions (Ishida et al., 1998). Moreover, each species may exhibit considerable genetic variability, each strain being able to develop subpopulations distributed in geographic locations that may extend from restricted embayments to larger mesoscale distributions. Therefore, within the same genus, important interspecific differences in toxin profile and maximum toxin content may be found. Furthermore, intraspecific variability (different strains of the same species) can be at least as large as interspecific differences.

1.1.1.1. Paralytic Shellfish Toxin (PST) Producers

PST are produced by different dinoflagellates species of the genus *Alexandrium*, plus *Gymnodinium catenatum* and *Pyrodinium bahamense*. Since the structure of the parent compound saxitoxin (STX) was first described, about 57 analogues have been described (Wiese et al., 2010) that on the basis of their chemical structure and toxigenic potential can be classified as: i) High potency carbamoyl toxins (STX, neoSTX, GTX 1-4); ii) Low potency N-sulfocarbamoyl toxins (GTX5, GTX6, C1-C4); iii) High potency decarbamoyl derivatives (dc-GTX1-4, dc-neoSTX, dc-STX). Differences over 1-2 orders of magnitude exist between the toxicity of one mol of the parent toxin, STX, and some of the mild sulfocarbamoyl analogues (Oshima, 1995). Thus, given the same Q_t, PST producers with a profile dominated by carbamoyl and decarbamoyl toxins are expected to be considerably more toxic than those with a profile dominated by N-sulfocarbamoyl toxins. Nevertheless, some bivalve species are able to transform sulfocarbamoyl toxins into the more toxic forms (see section 3.5.4, biotransformations and chapter 5 for further information on toxicity equivalences). The most toxic PSP events recorded are those associated with *Alexandrium catenella* blooms in southern Chile and northwestern North America (Benavides et al., 1995; Taylor & Trainer, 2002), and with *A. tamarense* and *A. fundyense* in the Argentine Patagonia and northwestern

US respectively. In contrast, *A. catenella* has been reported in the southern Benguela Current system (South Africa) as a low to moderately toxic species with a high proportion of the less toxic sulfocarbamoyl toxins, but its blooms can reach concentrations of several million cells L^{-1} and lead to toxin concentrations of tens of thousands μg STX-equiv. kg^{-1} shellfish according to standard mouse bioassay (MBA) (Pitcher et al., 2001). Nevertheless, there are non-toxic strains of *A. tamarense* in other parts of the world, and on a genetic basis the three "morphospecies" are grouped nowadays in the "*Alexandrium tamarense/fundyense/catenella* species complex" (John et al., 2005) that includes toxic and non-toxic clades. Two extremes of toxigenic potential can be illustrated by comparing blooms of *A. tamarense* in South America, where an estimate of 325 pg STX-equiv. $cell^{-1}$ and cell concentrations up to $82 \cdot 10^4$ cell L^{-1} were reported (Benavides et al., 1995), and PSP events in Chinhae Bay (Korea), with Q_t estimates in culture of 10 pg $cell^{-1}$ and a bloom maximum of $9 \cdot 10^3$ cell L^{-1} (Kim et al., 1996).

The large intraspecific variability in toxin profile and content of PST producers has been demonstrated in laboratory cultures of different strains from the same species in a given region. In Korean coastal waters, *G. catenatum* strains were classified into three groups: Group I from the Yellow Sea, with predominance of carbamoyl toxins, Group II from the South Sea with predominance of N-sulfocarbamoyl toxins, and Group III with characteristics intermediate between the two previous groups (Park et al., 2004). *Alexandium minutum*, associated with mild to moderate PSP outbreaks in warm and temperate seas, has strains with profiles dominated by GTX1 and GT4 in Iberian waters (Franco et al., 1994), whereas the less toxic GTX2 and GTX3 dominate the profiles of strains from Brittany (France) and the Adriatic Sea (Honsell et al., 1996), and strains with the two kinds of profiles co-occur in coastal waters of Taiwan (Chou et al., 2004). This species develops very dense blooms (10^7-10^8 cell L^{-1}) but its Q_t can be quite low (< 1 fmol GTX $cell^{-1}$). In northwestern Iberia, cultured strains isolated in the same location showed a similar toxin profile but their maximum Q_t was quite variable under identical culture conditions (1-24 fmol $cell^{-1}$), and led to their classification into toxic, moderately toxic and very mild strains (Franco et al., 1994). Isolates of the *A. tamarense* species complex from the western Pacific have shown a 100-fold variation in total toxicity among clones from the same area (Ogata et al., 1987) and 20-fold differences among sub-clones isolated from the same clonal culture.

1.1.1.2. Diarrheic Shellfish Toxin (DST) Producers

The main DST producers are planktonic dinoflagellates of the genus *Dinophysis*. These species produce two groups of polyether toxins with different biological activity and molecular structure: i) Okadaates—or okadaic acid (OA) and its analogues the Dinophysistoxins (DTX)— and ii) Pectenotoxins (PTX). Okadaates are the only toxins from the "old DSP complex" with diarrheic effects. Further, there is controversy as to whether pectenotoxins pose a real threat to public health because oral administration to mice does not have the same toxic effect as intraperitoneal administration (Miles et al., 2004). Benthic dinoflagellates from the genus *Prorocentrum* also produce DSP toxins. Nevertheless benthic *Prorocentrum* spp. have rarely been unequivocally associated with DSP outbreaks—except in some shallow coastal lagoons, or under very high turbulence and/or storms that detached them from their substrate and made them available to filter-feeders—but they contribute to the complex toxin profiles associated with Ciguatera Fish Poisoning (CFP) as part of the benthic HAB communities in tropical and subtropical regions (Reguera & Pizarro, 2008).

DSP outbreaks with the highest impact on public health and shellfish contamination are those where the causative *Dinophysis* has a high Q_t dominated by okadaates. Pectenotoxins also count as they contribute to the global response obtained in conventional mouse bioassays (MBA), but they are not regulated worldwide. Moreover, some shellfish species, including mussels and oysters, very rapidly transform PTX2 (dominant PTX form in several strains of *D. acuminata*, *D. acuta* and *D. forti,* the main DST producers around the world) into PTX2SA (Suzuki et al., 2001), an analogue that is not toxic to mice following standard MBA. The predominance of PTX2 in *D. fortii* and *D. acuminata* strains from the western Japanese coasts would explain why DSP events there are so mild in contrast with those in the north and northeast, where DTX1 is an important component of the toxin profile of *Dinophysis* spp. (Suzuki & Mitsuya, 2001). Further, scallops, that are not as efficient as mussels and oysters to convert PTX2 into its non-diarrheogenic seco-acid derivative, are the main commercial bivalve on the northeast coast. A predominance of PTX2 and only traces of okadaates were found in the profile of cultured strains of *D. acuminata* from northeast USA (Hackett et al., 2009)—a region where *Dinophysis* spp. have been traditionally considered harmless—and oysters are the main shellfish resource commercialized in Chesapeake Bay where exceptional blooms of OA-producing *D. acuminata* were recorded in 2001 (Tango et al., 2004). PTX2 was the only toxin detected in *D. acuminata* from northern Chile, an area with no "tradition" of DSP outbreaks (Blanco et al., 2007). Chronic autumn DSP outbreaks in the Galician Rías (NW Spain) are caused by *D. acuminata*—with a profile dominated by OA—and/or *D. acuta*—with OA, DTX2 and PTX2—. DSP events were not detected in the autumn of 2009 when the predominant *Dinophysis* sp. in the plankton was *D. tripos*, a species that in culture showed a toxin profile dominated by PTX2 (Rodríguez et al., submitted).

1.1.1.3. Amnesic Shellfish Toxin (AST) Producers

Domoic (DA) and isodomoic acid, that cause amnesic shellfish poisoning, are produced by diatoms of the genus *Pseudo-nitzschia*. The species with highest impacts on food safety are *P. australis, P. multiseries and P. seriata. Pseudo-nitzschia multiseries* was identified as the source of neurotoxins that poisoned humans in Canada in 1987. *P. australis,* due to its larger size, is probably the main source of domoic acid in the upwelling systems in northwest Iberia, California and South Africa. *P. australis* has caused great damage to king scallop (*Pecten maximus*) aquaculture in Europe, because these scallops take months (if ever) to depurate domoic acid from their flesh. Curiously *P. australis* has caused mass mortalities of birds and marine mammals with anchovies acting as vector of the toxins, but levels of DA in shellfish were often below regulatory limits in California (Bates & Trainer 2006; Trainer et al., 2010). A possible explanation is that *P. australis* blooms are mainly in offshore areas. Accurate taxonomic identification of *Pseudo-nitzschia* species requires laborious observations with the electron microscope. Blooms of toxic and non-toxic species, impossible to distinguish with the light microscope, co-occur. Blooms of the same species may have undetectable levels of DA until the population is at the stationary phase. Thus, it is difficult in most cases to ascribe toxic impacts to a single *Pseudo-nitzschia* species. Laboratory experiments reveal a high inter- and intraspecific variability in toxin content that ranges from less than 0.1 pg to more than 50 pg $cell^{-1}$ (Bates, 1998).

1.1.2. Genetic Analyses of Toxin Producing Microalgae

Genetic variability in dinoflagellates constitutes an active field of research in modern taxonomy given its potential to discriminate among cryptic species (Montresor et al., 2003), biogeographical clades (Richlen et al., 2008; McCauley et al., 2009), and within species complexes sharing similar morphologies (Lilly et al., 2005, 2007). The widespread use of molecular methods has revealed higher taxonomical diversity than previously thought before the advent of these techniques. As a consequence, there exist a growing number of molecular clades that await confirmation of their taxonomic identity by means of recognizable phenotypical characters. Some representative cases of this are, for example, the dinoflagellate genera *Alexandrium* and *Gambierdiscus* (Kuno et al., 2010; Lilly et al., 2007; Litaker et al., 2010). Unambiguous taxonomic characterization at species level is relevant not only to increase our knowledge of biodiversity but, regarding toxic and non-toxic dinoflagellate species, to provide detection tools in the field. Most often, molecular taxonomic studies involve sequencing of ribosomal genes and internal transcribed spacer regions (Edvardsen et al., 2003; Litaker et al., 2007) or mitochondrial genes (Zhang et al., 2005; Kamikawa et al., 2008) to distinguish dinoflagellate species.

Characterization of genetic structure in field population studies is also useful to unveil the temporal trends of cryptic species, intraspecific variability and environmental conditions triggering growth of toxic proliferations (John et al., 2004; Nagai et al., 2007; Alpermann et al., 2009, 2010; Masseret et al., 2009; Touzet et al., 2010). These studies at population level usually employ non-sequencing techniques such as amplified fragment length polymorphism (AFLP), microsatellites, or fluorescence *in situ* hybridization (FISH) methods. However, the latter technique takes advantage of the accumulated body of knowledge from DNA sequences. FISH rDNA molecular probes have been successfully applied in many cases to discern similar morphospecies. One representative case is the *Alexandrium tamarense* complex, which includes toxic and non-toxic strains separated into geographical clades, which can be distinguished and quantified using ribosomal molecular probes (Sako et al., 2004; Hosoi-Tanabe & Sako, 2005; John et al., 2005).

DNA barcodes are intended also to provide a fast and effective means of species identification through the amplification of short genetic sequences. At present, universal DNA barcodes have already been identified for animals (cytochrome oxidase I; Hebert et al., 2003) and land plants (concatenated chloroplast genes; Hollingsworth et al., 2009). Regarding protists, in particular dinoflagellate species, such DNA regions with universal resolution have not yet been reported. Despite this, there are some advances, and different DNA markers have been tested with positive results in a variety of cultures and field samples (Litaker et al., 2007; Lin et al., 2009; Stern et al., 2010). But for instance, in the case of the genus *Dinophysis*, the lack of specific resolving gene markers precludes a robust molecular taxonomy at species level (Raho et al., 2008). Therefore, the distinct genetic variability among species in different dinoflagellate genera could represent an obstacle towards defining universal specific rules in these organisms. To conclude, molecular approaches will progressively increase in importance due to their continuous development and application in future taxonomic studies. However, the correct identification of protist species should not be reduced solely to DNA sequences but supported (and enriched) by their linkage with morphology, phenotypic variation and, whenever possible, the distribution and ecology of the organisms under study (McManus & Katz, 2009).

1.1.3. Genetic Analyses of Toxin Production by Microalgae

Understanding the genetic bases for the synthesis and regulation of toxic compounds in harmful algae is highly challenging due to the complexity of their genome and the paucity of suitable model organisms, particularly in the case of dinoflagellates. For these reasons there is still limited knowledge about the biosynthetic pathways, metabolic roles and the genes ultimately associated with the production of marine toxins.

Using labelled precursors to PST-producing organisms a biosynthetic pathway for saxitoxin (STX) was proposed from arginine, acetate, and methionine methyl groups (Shimizu et al., 1984). However, the identity of the multiple enzymes, and the number of steps involved in this route have not been totally characterized and the physiological regulation of STX also remains unknown.

Apart from marine dinoflagellates, the only STX producers are freshwater cyanobacteria of the genera *Cylindrospermopsis, Anabaena, Lyngbya and Aphanizomenon*. The standing hypothesis has been that the STX pathway is the same in all these organisms despite their distant phylogenetic origins. Have dinoflagellates acquired the STX-genes separately or by horizontal gene transfer from prokaryotes? To address this question would require sequencing of STX-genes in cyanobacteria and dinoflagellates. Recently, Kellmann et al. (2008), based on previous knowledge of STX biosynthesis, identified an operon in *Cylindrospermopsis raciborskii* incorporating all predicted genes which were only present in STX-producing strains. Based on these results, a complete biosynthetic route was also proposed (Mihali et al., 2008). The majority of STX genes were homologous with those of other cyanobacteria, suggesting that STX biosynthesis may have evolved in an ancestral cyanobacterium that acquired the remaining genes by horizontal gene transfer from other bacteria. Similar genes to those of cyanobacteria have not been found in expressed sequence tag (EST) libraries of *A. minutum* (Yang et al., 2010), but 192 genes were differentially expressed between toxic and non-toxic strains. This result suggests STX genes in dinoflagellates may be more divergent than would be expected in the case of a recent gene transfer from cyanobacteria.

Brevetoxins, okadaic acid (OA) and dinophysistoxins (DTX) are polyketides, probably derived from dicarboxylic acids (Shimizu et al., 1990). The polyketide nature of brevetoxins was established in the 80's from stable isotope incorporation experiments (Nakanishi, 1985; Chou & Shimizu, 1987; Shimizu et al., 1990). Polyketides are structurally diverse and constituted by a carbon chain synthesized by a family of enzymes called polyketide synthases (PKS). PKS genes were first cloned and sequenced in bacteria and fungi, and later identified and associated with the dinoflagellate *Karenia brevis* by Snyder et al. (2003, 2005).

Finally, the biochemical pathway for domoic acid, a potent neurotoxin synthesized by some species of the diatom genus *Pseudo-nitzschia*, is unknown and the genes expressed during toxic episodes in the field or in laboratory cultures have not been identified.

Toxin composition and production (at least in saxitoxin- and brevetoxin-producing organisms) are variable, and this could be useful to discover the genes responsible of biosynthetic routes for toxin production. To study the genetic profiles and physiological conditions determining toxin production in harmful algae requires the establishment of clonal cultures to examine their variations in toxin content and composition, and how these are affected by environmental changes. However, the search for genetic traits specifically related to toxin production represents a formidable challenge as it goes a step beyond the molecular characterization of dinoflagellate species.

Observations on the large intraspecific variability of Q_t in *Alexandrium* spp. (Cembella et al., 1987; Orlova et al., 2007) led to the suggestion that toxin production might not be inherited in a Mendelian way, but caused by mutations and/or epi-genetic factors. However, the toxin profile has been shown to be a relatively stable characteristic of a given species (*A. tamarense*, Cembella, 1998; *Karenia brevis*, Errera et al., 2010), sometimes slightly variable (Boczar et al., 1988; Errera et al., 2010) or rather constant (Wang et al., 2006), depending on culture conditions and the species under study.

Ideally, the establishment of non-toxic subclones derived from toxic clonal strains would be an excellent source to reveal any genes related to toxin production, since non-toxic and toxic subclones would display identical genomes. However, at present, available comparative genomic analyses of this nature have not revealed correlations between any gene alleles and toxicity levels (Cho et al., 2008). It is already known that in some dinoflagellates such as *Alexandrium*, clonal toxic strains exhibit changes in Q_t with time and become heterogeneous in terms of their toxicity, or even become non toxic altogether (Martins et al., 2004). The genetic bases explaining these differences, when other mechanisms such as recombination from sexual reproduction are not involved, could be related to spontaneous mutation, unavoidable in any organism as a consequence of DNA replication. These mutations can be fixed in the original toxic clonal culture if they are advantageous to the cells. In a hypothetical scenario where cellular resources are directed to growth instead of toxin production, non-toxic mutants would eventually out-compete toxin producers, changing the phenotype of the culture (Lakeman et al., 2009).

The regulation of gene expression related to toxin production, as well as DNA sequences, must be studied to identify proteins and metabolites potentially involved in toxin biosynthesis. Thus, a different approach to avoid the complexities associated with examining dinoflagellate genomes is represented by transcriptomic studies. Library and microarray analysis of expressed sequence tags (EST; Hackett et al., 2005; Yang et al., 2010) and proteome analysis by 2-D gel electrophoresis (Chan et al., 2004; Martins, 2007) are some of the techniques applied. Dinoflagellate proteomics is still a new field and a relatively small number of expressed proteins have been described (Chan et al., 2002, 2004). These studies allow comparative proteomic analyses, known also as functional proteomics, to identify different protein expression levels between toxic and non-toxic clonal strains. These proteins differentially expressed could later be identified through mass spectrometry, and eventually related to known metabolic roles in the cells (Chan et al., 2006). Available comparative proteomic studies aim to reveal any "biomarker of toxicity" in *Alexandrium* species (Chan et al., 2005, 2006; Martins, 2007; Yang et al., 2010). Some differentially expressed proteins of unknown function have been identified in toxic strains (Chan et al., 2006), or associated with other biosynthetic routes such as the Calvin cycle (Martins, 2007). These unknown proteins, detected in toxic *versus* non-toxic *A. tamarense* strains, may potentially be used in immunoassay studies to detect toxic algae.

Therefore, the large size and particular characteristics of dinoflagellate genomes has hindered identification of the biosynthesis pathways for toxins. Nevertheless, the implementation of next-generation sequencing techniques, such as the 454 long read technology on cDNA (Metzker, 2010), will probably provide a wealth of knowledge about toxin-producing dinoflagellates in coming years. Not only genomic approaches, but the parallel development of proteomics will be crucial to unravel the mechanisms responsible of phenotypic changes in toxin production among clonal isolates.

1.1.4. Influence of Cell Cycle Stages, Life Cycle and Behaviour

Toxin production is not a constant process in a given strain but is subject to fluctuations in connection with cellular rhythms which range from diurnal (cell cycle, vertical migration) to seasonal (life cycle, stages of population growth) time scales.

1.1.4.1. The Cell Cycle

The cell cycle, or cell-division cycle, is the series of events in a cell leading to its division and duplication. In eukaryotic cells, the cell cycle is divided in two periods: interphase—during which the cell grows (G_1 or Gap 1 phase), accumulates nutrients needed for DNA synthesis (S phase) and resumes growing (G_2 or Gap2 phase)—and the mitosis (M) phase, during which cell growth stops and energy is focused on two tightly coupled processes: nuclear division (the cell's chromosomes are divided between the two daughter cells) and cytokinesis (the cell's cytoplasm divides in half forming two distinct cells). Activation of each phase is dependent on the proper progression and completion of the previous one. Some cells may temporarily stop division and enter a state of quiescence called G_0 phase.

Many microalgal species, particularly dinoflagellates, exhibit a phased cell division, i.e. all dividing cells in a population undergo mitosis during a narrow time-window of the day, usually coupled to the photocycle. It would be reasonable to expect that newly divided cells, following distribution of the parent Q_t between the two offspring, have much lower toxin content than larger cells that have gone through the interphase and are about to divide (megacytic cells). Further, quiescent cells that stop division but continue toxin production might reach extremely high Q_t values. Johansen (2008) analyzed toxins in single cell isolates of recently divided cells of *Dinophysis* and observed that toxin content was almost half of that from fully grown vegetative cells. A good correlation between cell volume ($10\text{-}64 \cdot 10^3$ μm^3) and toxin content (20-140 fmol $cell^{-1}$) was found in cultures of the spirolide producer *Alexandrium ostenfeldii* (Quilliam et al., 2008). Therefore, for comparable results, Q_t should be estimated during the same cell cycle phase (e.g. at the end of the G1 phase) and related to cell size.

Field and laboratory studies on the relation of toxin production with the cell cycle—or with the supply of precursor molecules of the toxins, also coupled to the diurnal cycle—are rare. Cultures, for this purpose, need to be "synchronized", which is achieved by a blockage of light or nutrient supply and release after about 3 light/dark cycles or by using metabolic inhibitors. Often the induced stress prevents full recovery of the culture. Kim et al. (1993) found that in cultures of the PST producers *A. tamarense* and *A. catenella,* Q_t increased gradually from the latter half of the light period through the dark period until the M phase. Taroncher-Oldenburg et al. (1997) observed that in cultures of *Alexandrium fundyense*, toxin synthesis was triggered by light in the G_1 phase and ceased by the end of this phase; thus it appeared to be decoupled from DNA synthesis (S phase). In cultures of *Alexandium ostenfeldii* (where complete synchronization was not achieved), the amount of spirolides (SPX) per cell increased (over 50%) at the beginning of the dark phase and decreased in the light, suggesting that SPX synthesis is affected by light-dependent metabolic processes coupled to photoperiod and the cell cycle (John et al., 2001). In contrast, in cultures of *P.lima* increased toxin per cell was observed in the light, but extended through several phases of the cell cycle: production of DTX4 derivatives was initiated in the G1 phase and continued into the S phase, whereas OA and DTX1 were produced later, in the S and G2 phases. Toxin levels did not increase when the dinoflagellates were maintained in the dark, with production starting when cultures were returned to the regular light-dark cycle (Pan et al., 1999).

Pizarro et al. (2008) followed the variability in toxin profile and content in a natural late season population of *Dinophysis acuta* in which division rate was practically zero. Strains of this species in Galicia produce okadaates and pectenotoxins, both lipohilic toxins, but with different chemical structures and biological activity. The OA:DTX2 ratio was constant throughout the cycle. The proportion of PTX2 varied throughout the study; thus production of the two groups of toxins exhibited different temporal patterns. Similar results were found by Fux et al. (2008) in field populations of the same species in southwest Ireland, where maximum OA and toxin content per cell was observed in the middle of the dark period (0100 h). These two field studies showed that in cells containing different groups of toxins, the toxin profile (and the predominant toxin) varied with sampling hour. From all the above, we can conclude that in studies planned to describe the seasonal variability of toxin content of a toxigenic species in a given location, samplings must be performed at similar hours of the day and phase of the cell cycle.

1.1.4.2. Life Cycle

The life cycle comprises all different stages of a species which succeed each other by means of asexual or sexual reproduction. Dinoflagellates and diatoms exhibit complex polymorphic life cycles including haplont and diplont vegetative, sexual and resting stages. To distinguish all the different life cycle stages of a determined species is a difficult task that only few specialists are able to do. Most important, species that produce large numbers of resting cysts that finally accumulate in the sediments extend their contamination potential from epifaunal suspension feeders to infaunal siphonate deposit feeders. Preliminary work suggested that cysts could have toxin contents up to one order of magnitude higher than those of vegetative cells, but more recent studies have shown them to have the same or even lower Q_t (Bravo et al., 1998; Cembella, 1998). In *G. catenatum*, the toxin content of cysts produced in the laboratory (12-32 pg $\cdot$ cyst^{-1}) was lower than that of vegetative cells (40-45 pg $\cdot$ cell^{-1}), and there were important differences in the toxin profile, dominated by C1 and GTX5 in cysts and by C2 and GTX6 in vegetative cells, but sulfocarbamoyl toxins were predominant in both stages (Bravo et al., 1998). Epimerization associated with different life cycle stages has also been reported in natural populations of *Alexandrium* (Cembella et al., 1990; Oshima et al., 1992), and this conversion has been considered a natural step in the change from vegetative cells to cysts with decreased metabolic activity.

Temporary (pellicle) and resting (sexual) cysts of *Alexandrium* could represent a stage to escape digestion by shellfish, because temporary and resting cysts may survive gut passage better than vegetative cells as revealed from viable material found in shellfish fecal pellets (Laabir et al., 2007). Conversely, the ingestion of *G. catenatum* cysts—that were abundant in the digestive tract—was suggested to explain toxicity in the siphonate deposit feeder peppery furrow shell (*Scrobicularia plana*), long after vegetative cells were no longer observed in the water column (Artigas et al., 2008).

1.1.4.3. Behaviour

Dinoflagellates may perform daily vertical migration (DVM) of more than 10m or aggregate in density discontinuity layers of the water column; pennate diatoms, such as *Pseudo-nitzschia* spp, may form high density layers (see section 1.2.1.4). In any case, the vertical distribution of HAB species in the water column will have a great influence on the distribution patterns of toxin accumulation in filters feeders, in particular on those cultivated on long lines and floating rafts. Whether a microalgae population exhibits DVM or aggregates around physical gradients is part of the (genetically determined) species characteristics, and this behaviour may vary during different phases of population growth.

1.1.5. Phases of Population Growth

Steidinger (1975) established an analogy between microalgae population growth in laboratory cultures and in the field, and summarized different phases in the development of harmful algae populations into initiation, growth, maintenance (plateau) and transport. Laboratory results support the view that toxin synthesis is not a constitutive component of algal metabolism. Instead, both the rate of toxin production and the extent to which toxins are accumulated by microalgae (Q_t), are strongly influenced by environmental growth conditions. Changes in toxin accumulation are generally associated with disturbed physiology, i.e. uncoupling between cellular division and toxin production: a fast division rate has a dilution effect on Q_t, whereas stationary growth will favour accumulation until the onset of conditions limiting toxin production, or when leakage due to cell senescence overcomes accumulation.

Within PST producers, laboratory experiments in batch cultures have shown that toxin production peaks during early and mid-exponential growth phase, and accumulation (Q_t) takes place during late exponential phase (Anderson et al., 1990; Flynn et al., 1994, 1996).

Observations on DST producers (*Dinophysis* spp.) were mainly focused on field populations—due to difficulties to establish these species in culture—or in laboratory cultures using benthic dinoflagellates of the genus *Prorocentrum* as a model organism. Differences of over one order of magnitude were observed in Q_t in field populations in western Iberia, and maximum Q_t was observed during the stationary phase, particularly in senescent populations, with no active division, transported to the coast (Pizarro et al., 2009). Recent experiments with *Dinophysis* cultures also found maximum Q_t during the stationary phase, although growth and therefore net toxin production rate per unit of culture volume were higher at higher temperatures during the exponential growth phase (Kamiyama et al., 2010). Tong et al. (in press) found that OA, DTX1, PTX2, and OA-D8 cell toxin quota remained low during middle and late exponential phases, but significantly increased by mid-plateau phase. The general pattern of variability was different for the two groups of toxins, okadaic acid analogues and pectenotoxins, produced by *Dinophysis* both in natural populations (Pizarro et al., 2008, 2009) and in cultures (Kamiyama et al., 2010). In cultures of *Prorocentrum hoffmannianum*, toxin production rates (μmol toxin $cell^{-1}$ d^{-1}) are similar to growth rates (div d^{-1}), with maximum values observed during the exponential phase (Aikman et al., 1993, Rausch de Traubenberg & Morlaix, 1995), but production continues in the stationary phase after division has ceased, leading to higher Q_t (accumulation) (Jackson et al., 1993, Holmes et al., 2001)

Domoic acid production by *Pseudo-nitzschia* spp. is undetectable or extremely low during exponential growth in laboratory batch cultures, and increases during the stationary phase due to imbalance between division and toxin production rates; if cells in stationary phase (with high Q_t) are re-inoculated into fresh media, Q_t decreases as division rate increases in the exponential phase (Bates, 1998). Pan et al. (1998) interpreted this in terms of competition for reduction of energy (ATP), which would be preferentially used on biosynthesis of primary metabolites during the exponential growth phase.

1.2. Environmental Factors that Affect Toxin Production by Microalgae

Toxins are secondary metabolites. Rates of toxin synthesis depend on the metabolism of the primary metabolites which are their precursors. In turn, synthesis of primary metabolites varies with extrinsic abiotic factors—light, salinity, temperature, nutrient concentration, ratio and supply rates, turbulence—that affect growth, and the accumulation of allelochemical and autoinhibitory metabolites (Flynn, 1998; Cembella & John, 2006). Therefore, any environmental factor affecting growth—by causing stress or imbalance—will directly or indirectly affect toxin production and accumulation.

1.2.1. Abiotic Factors

Microalgae need light and nutrients to grow and each species in a region needs a particular set of environmental conditions that defines its "optimum environmental window".

1.2.1.1 Light

Light harvesting efficiency and phosynthetic capacity are linked to nutrient status and it is quite difficult to separate their respective effects. Light dependence of toxin production is likely to be related to the photoassimilation of toxin precursors. Growth and toxin production were found to stop under extremely low light conditions in cultures of *Alexandrium tamarense* (Ogata, 1996). In cultures of *Prorocentrum* spp. toxin production stops in the dark (Pan et al., 1999), and OA is mainly localized in the peripheral chloroplasts (Zhou & Fritz, 1994). All known toxin producing species are phototrophic. Even in mixotrophic species of *Dinophysis*, growth and toxin production require light as well as live prey (Kim et al., 2008). It has been questioned whether the heterotrophic *D. rotundata* produces toxin *de novo* (González-Gil et al., 2010), and it was recently found that the heterotroph *Protoperidinium crassipes* is only a vector of the azaspiracid toxins produced by its prey, *Azadinium spinosum* (Tillmann et al., 2009).

Each toxigenic species exhibits a characteristic photosynthesis *versus* light intensity (P-I) curve that will in turn influence the capability to synthesize toxins, and obviously, each species is adapted to the natural environment where it blooms: depth-regulators in particular occupy the appropriate depth in the water column where there is a trade-off between optimum harvesting of light and nutritional sources (Cullen & MacIntyre, 1998).

Light intensity and quality (wavelength) are important, and so too is photoperiod, i.e. the number of dark and light hours during a daily cycle. Some species are very seasonal and grow only in spring or summer, whereas others seem to extend their growth season throughout most of the year provided light is not limiting. The onset of many *Alexandrium* species, after cyst germination, follows a solar calendar, the annual population maximum appearing within a narrow seasonal time window (Wyatt & Jenkinson, 1997). Photoperiod differentially affects growth of the two dominant *Pseudo-nitzschia* spp in Scotland, *P. delicatissima* and *P. seriata*, and toxin production of the latter (Fehling et al., 2005). Photoperiod also affects sexual reproduction in *Pseudo-nitzschia* spp. which in turn affects growth and toxin production (Bates, 1998).

1.2.1.2. Temperature and Salinity

Temperature has a strong influence on cellular growth, and each species' strain exhibits an optimum "temperature window for growth". Growth will proceed between a minimum temperature—or threshold that may act as a "trigger" for cyst germination and/or vegetative growth—and a maximum—beyond which growth is halted—with a value in between at which maximum growth occur. Each species may have strains adapted to diverse ecosystems with quite different conditions of temperature and salinity. Eurythermic/eurohaline species are able to tolerate a wide range of temperature/salinity in the environment and are well adapted to coastal and estuarine ecosystems with fluctuating conditions, in contrast to stenothermic/stenohaline species, which tolerate only narrow ranges of these parameters. A good example is the pair *D. acuminata*/*D. acuta* in western Iberia. The coastal *D. acuminata* is observed under wide ranges of salinity and temperature, and persists from spring to autumn; in contrast *D. acuta,* from shelf waters, is very seasonal (late summer-autumn) and is associated with a narrow margin of warmer and saltier waters (Reguera et al., 1993; Palma et al., 1998). Decreased salinity often has a beneficial effect on the growth of microalgae cultures. Flynn et al. (1996) suggested that decreased salinity would influence positively toxin synthesis by affecting the distribution of metabolites used for osmotic regulation, but they did not find changes in toxin content caused by variations in salinity in cultures of *Gymnodinium catenatum.*

It has long been believed that strains from colder boreal waters are more toxic than strains of the same species with more meridional distributions. A careful review of results from toxin profiles and Q_t from field populations and laboratory experiments does not provide any solid scientific evidence. Higher Q_t in colder seas could result from the uncoupling between toxin production (less sensitive to temperature) and growth (considerably decreased at lower temperatures), leading to increased toxin accumulation without a significant increase in production (toxin·cell d^{-1}). This would explain lower Q_t in laboratory cultures of *Alexandrium* spp. under excess nutrients and higher temperature conditions than those from field populations (Cembella, 1998).

1.2.1.3. Nutritional Sources

All toxin producing microalgae are dependent on the uptake and assimilation of dissolved or particulate nutritional resources for growth and toxin biosynthesis, but differences are observed according to the chemical composition of the toxin produced, the composition of their cell wall—diatoms require Si—and to the autotrophic or mixotrophic feeding behaviour of the species. Absolute concentrations of nutrients are important as well as their relative abundance.

P-stress has often been invoked as a trigger for toxin accumulation in PST producers in short-term batch cultures, and it has been hypothesized—but not proved—as a promoter of toxin production in unbalanced environments with a high N:P ratio; on longer time scales, P limitation finally affects diverse metabolic functions that require reducing power (ATP), and will limit growth and toxin production (reviewed in Cembella, 1998; Graneli & Flynn et al., 2006). The main changes in Q_t in cultures have been observed during refeeding on N-starved cells (up-shock) or in cells getting into the stationary phase (down-shock). There is no information available for field populations, but a situation of up-shock could be expected during transient conditions when nutrients become again available to N-limited cells through upwelling pulses, runoff, etc. Reduced N-forms (NH_4^+) enhance synthesis of arginine, a precursor of hydropurines, and lead to higher toxin content in *Alexandrium* spp., the diatom

Pseudo-nitzschia multiseries and *G. catenatum*. Therefore, we may expect cells to become more toxic in field populations when regenerated production (ammonia) predominates over new production (nitrates).

Si is essential for diatoms as a constituent of their frustules. Both P and Si limitation negatively affects growth of *Pseudo-nitzschia* species and promote synthesis of domoic acid (Bates, 1998). More recent studies have shown a positive correlation between DA production and Fe and/or Cu, which was interpreted on the basis of a possible role of DA as a chelator, thus enhancing Fe bioavailability and reducing the toxicity of free Cu to the cells (reviewed in Bates & Trainer, 2006).

Among mixotrophic DSP toxin-producing *Dinophysis* spp., no clear correlation between inorganic nutrients and growth/toxin production has been found (Maestrini, 1998). Cultures of *Dinophysis* feeding on the ciliate *Myrionecta rubra* (itself fed with cryptophytes) have shown that growth and toxin production are maximal when there is no limitation of their live ciliate prey (Kamiyama & Suzuki, 2009; Kamiyama et al., 2010).

1.2.1.4. Turbulence versus Physical Stability

Fast dividing diatoms have a competitive advantage in turbulent and nutrient rich waters, whereas many dinoflagellates thrive in stable and stratified water columns. Stratification can be thermally driven—a situation typically found during summer heating—or salinity driven—after heavy rain—or a combination of both. In any case, it will often be difficult to determine what triggers bloom formation when several factors are involved. Thus, the effect of increased temperature often co-occurs with increased water column stability and haline stratification after heavy rain provides a combination of decreased salinity, physical stability and a heavy input of macro- and micronutrients and organic material from terrestrial runoff. A combination of exceptionally calm weather following nutrient inputs (from upwelling, runoff and other sources) is often reported in association with exceptional harmful algal blooms. Further, stratified conditions favour the formation of thin layers of harmful algae, i.e., aggregation of high densities of microalgae in narrow layers in the water column (GEOHAB, 2008). *Pseudo-nitzschia* spp. and *Dinophysis* spp. have often been observed forming thin layers that may act as "toxic carpets" transferring high levels of toxins to shellfish exposed to them. Formation of thin layers of *Pseudo-nitzschia* spp. during late bloom stages, and displacement of these layers to deeper waters during downwelling events, could explain why benthic shellfish resources often become toxic before mussels in floating rafts (Velo-Suárez et al., 2008).

1.2.2. Biotic Factors

Harmful algae are embedded in microbial communities in which complex biological interactions, such as secretion of growth promoting and/or allelopathic substances, take place. Nevertheless, most information on the effects of biotic factors on toxin production comes from laboratory cultures that may not accurately reflect field conditions (more information in chapter 8).

There is a long standing debate about the bacterial origin of the toxins associated with dinoflagellates and with *Pseudo-nitzschia*. Several studies have demonstrated that removal of bacteria from *Alexandrium* cultures does not affect toxin production (Uribe & Espejo, 2003; Lu et al., 2000). Moreover, the inheritance of toxin profiles follows a Mendelian segregation

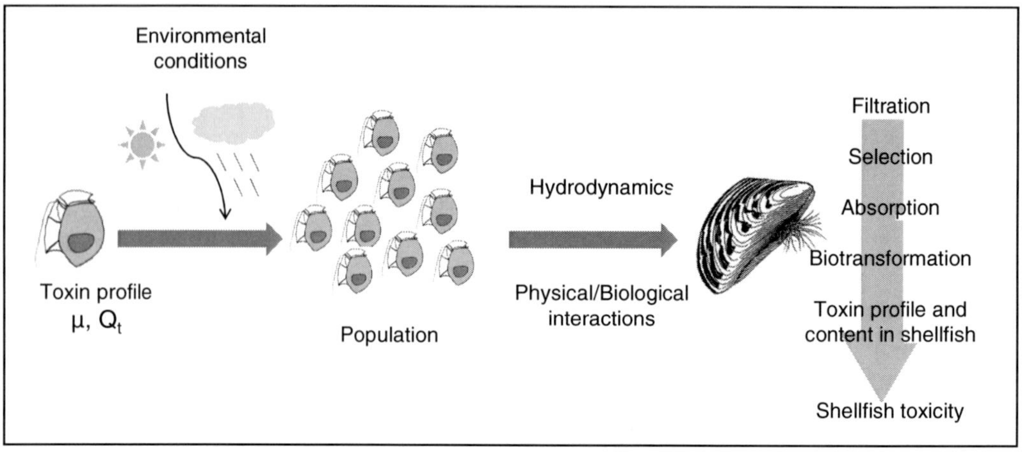

Figure 1. Simplified diagram of different factors and processes affecting the accumulation of toxins from toxigenic microalgae to filter-feeding bivalves.

pattern (Ishida et al., 1998), indicating that at least the interconverting enzymes are encoded by dinoflagellate nuclear genes. In contrast, some bacterial strains enhance toxin synthesis in *Pseudo-nitzschia multiseries* cultures 100-fold, despite the fact that there is no conclusive evidence of autonomous production of DA by bacteria (Bates et al., 1995). Here again, the potential role of DA as a chelator is invoked to explain these findings: the bacteria produce chelators that remove essential trace metals, and to counter this DA is produced by *Pseudo-nitzschia* spp (reviewed in Bates & Trainer, 2006).

Other external factors, such as the presence of potential grazers like copepods, have been shown to substantially promote PST production in *A. minutum* cultures (Selander et al., 2006); whether here PST production was a direct chemotactic response to the presence of copepods, or simply a beneficial indirect effect due to the excretion of reduced N-compounds remains an open question.

1.3. Hydrodynamic Conditions, Eutrophication and Shellfish Contamination

Once a toxin producing microalgae population with a good level of toxin per cell has reached an appropriate density (cell L^{-1}), contamination of shellfish resources requires that toxin-producing cells become available to filter and/or deposit feeders (Figure 1). There are many examples in the literature of HAB populations developed offshore, far from shellfish grounds, which are eventually advected to the coast and become accessible to filter feeders. In other cases, HAB populations aggregate around density gradients (thermoclines, haloclines) and reach very high densities at certain depths. Further, some physical structures, such as anticyclonic eddies, may constitute retention areas or "incubators" of noxious species.

Hydrodynamics and the mechanisms through which HAB populations are transported from source or "incubator" to their final destinations or "sinks" are site- and species-specific, and tightly coupled to the general circulation and meteorological influences. Case by case studies is therefore necessary, since the same species may exploit different physical conditions in different regions. Thus, autumn PSP and DSP outbreaks in northwest Iberia caused by *G. catenatum* and *Dinophysis* spp. at the end of the upwelling season are coupled to the onset of southerly winds that promote advection of shelf waters into the Galician Rías (Fraga et al., 1988; Reguera et al., 1995). PSP outbreaks in the Gulf of Maine caused by *A. fundyense* are triggered by wind-promoted transport to the coast of a buoyant (lower salinity) plume in which the dinoflagellate population accumulates (Keafer et al., 2005a,b). The intensity of *A. tamarense* blooms off Río de La Plata, in Uruguayan waters, correlates well with the intensity of the estuarine front determined by the magnitude of river outflow (rainfall) (Méndez & Ferrari, 2002). Toxic blooms of *Pseudo-nitzschia* are found in seasonally developed offshore retentive sites, such as the Juan de Fuca eddy region in the northeast Pacific. Intensification of ASP events there are linked to wind promoted transport of these eddies to the coast (Trainer et al., 2001). Nevertheless, it is important to remember that optimum environmental conditions for inoculation and growth of HAB populations are required before those enhancing accumulation and transport linked to the annual toxicity maximum. These examples illustrate how important climate variability must be for interanual variations in the intensity, extension and duration of toxic algae events.

Anthropogenic nutrient enrichment has been often invoked to account for the occurrence of HABs in coastal waters, and increased numbers of harmful algae events have been used as indicators of eutrophic conditions and counter-indicators of ecosystem health. Analyses of time series data from monitoring programs in different North Atlantic regions provides no convincing evidence—with the possible exception of high biomass blooms of the non toxic *Phaeocystis* spp. in Belgian coastal waters—that harmful algal blooms, either in their intensity or bloom-species selection, are linked to eutrophication processes, to elevated nutrient concentrations, or to altered nutrient ratios at the evaluated locations (ICES, 2006). Most available data sets are too short to evaluate the effects of climate change on phytoplankton dynamics, i.e., it is not possible to state whether observed changes are due to "global warming" or to climate variability. Nevertheless, this datasets permit to establish simple relationships between climate variability and HAB events that contribute to the development of predictive models.

1.4. Toxin Accumulation in Shellfish: A Balance between Uptake and Elimination

Shellfish, and especially bivalves, are the main vectors of toxins from phytoplankton to humans. They acquire the toxins mostly by feeding toxic phytoplankton, and accumulate them with an efficiency that depends on the balance between toxin absorption/uptake and elimination. In general, absorption of toxins and organic matter by bivalves involves several processes (Figure 2). Water, with cells in suspension, is pumped through the bivalve gill, where some particles are retained (filtered) and channelled to the mouth. In this process, some particles are rejected and others are ingested. After ingestion, the particles pass to the stomach, where the initial digestion (extracellular) takes place, and where some additional particle selection is carried out (post-ingestive selection; Owen, 1955). Nevertheless, shellfish toxicity depends not only on the amounts of toxin accumulated, but also on the biomass of the filtering organisms and on biotransformations that could change the potency of the microalgal toxins.

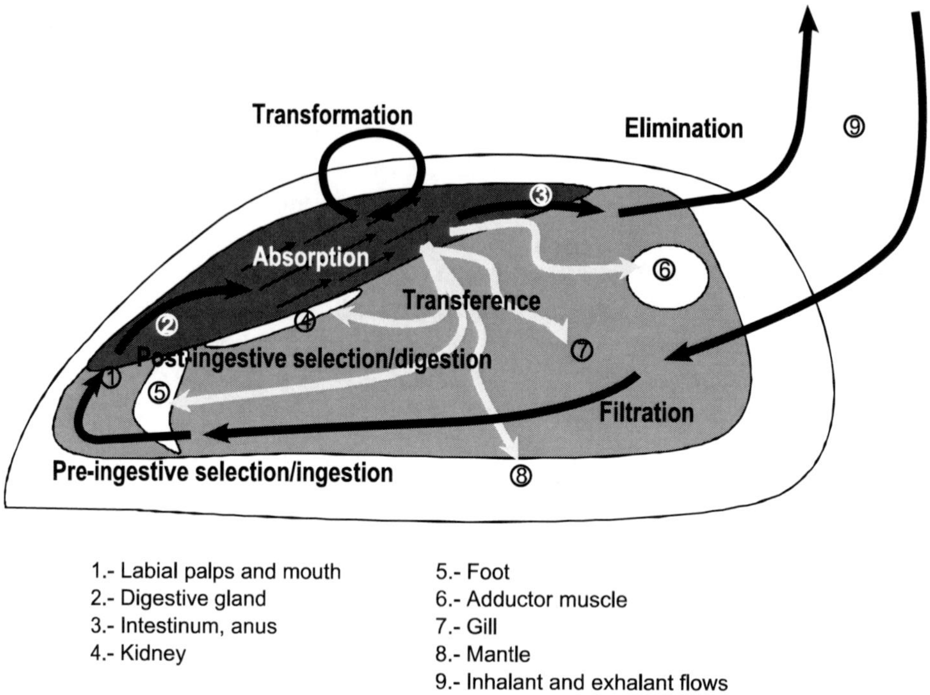

Figure 2. Main processes and organs involved in the accumulation of toxins in bivalves.

1.4.1. Filtration

The first step of the ingestion process—pumping/filtration—depends on the bivalve species and on the environmental conditions (including species/strain-specific characteristics of the phytoplankton populations). Bivalve species differ substantially in their maximum filtration rates. The filtration capability of a 5-year old flat oyster *Ostrea edulis*, for example, does not exceed 1.7 $L \cdot h^{-1}$ (computed from Rodhouse, 1978), while the blue mussel *Mytilus galloprovincialis* and the king scallop *Pecten maximus* can reach 7 $L \cdot h^{-1}$ (Filgueira et al., 2008) and nearly 30 $L \cdot h^{-1}$ (Acosta & Blanco, 2003), respectively.

For a particular species, filtration rate is affected by environmental conditions and size. Environmental conditions usually do not have important effects on filtration so long as they are within the tolerance range of the species (Riisgard & Seerup, 2003), perhaps due in some cases—as suggested by Bayne et al. (1976)—to acclimation to different conditions. Nevertheless, sudden changes in environmental conditions usually lead to reduced filtration rates for a period of time while acclimation proceeds (Bayne et al., 1976). Further, some interactions between environmental factors may occur as in the case of the effect of temperature and salinity on *Pecten maximus* (Strand et al., 1993). These interactions can also significantly affect the ingestion of toxic cells when their populations are associated with non-average environmental conditions. For example, many *Alexandrium* blooms are associated with lower salinity buoyant plumes (Anderson et al., 2005; Keafer et al., 2005a, b; Fauchot et al., 2008; Bravo et al., 2010), which means that when bivalves are exposed to these toxic populations, they are simultaneously exposed to reduced salinities, which in turn leads to reduced filtration rates and consequently to reduced toxin absorption. This kind of scenario would explain the situation during an *A. minutum* in the Ría de Pontevedra (Spain) in which the estimated incorporation of toxins by cultured mussels was substantially lower than expected (J. Blanco, unpubl. data).

Seston—particulate matter, including plankton, suspended in the water—concentration also has an important effect on filtration rates (Iglesias et al., 1996; Figure 3.3). If seston concentration is very low, the filtration rate is substantially reduced to maintain the energetic cost. Filtration is also reduced if seston concentration is very high, because the processing capability of the digestive system is exceeded. With intermediate concentrations, filtration is optimized. The optimal concentration depends on the shellfish species, but also on the characteristics of the seston particles. When seston has a high organic content, filtration usually peaks at lower concentrations. It can therefore be expected that, in both situations, low seston and high organic matter concentration, toxin absorption will be lower than average bivalve feeding behaviour. This could be the case when *Dinophysis* populations start to develop without accompanying species, or when high biomass (red tides) blooms of toxigenic organisms develop.

Ingested phycotoxins may also affect bivalve filtration because they have serious effects on their health. The toxin-induced alterations are species (or even strain)- and toxin-specific. Although some information can be inferred from the compilations of Landsberg (2002), Blanco-Pérez (2003) and others, this aspect is best illustrated with the finding of one mutant strain of the clam *Mya arenaria* by Bricelj et al. (2005). Normal clams are sensitive to PST, and when they feed on PST-producing dinoflagellates, their filtering and burying capacities are hindered, which result in the accumulation of lower quantities of toxins. The mutant strain is much more resistant to PST, hence it is able to accumulate much higher levels of toxins from PST producers.

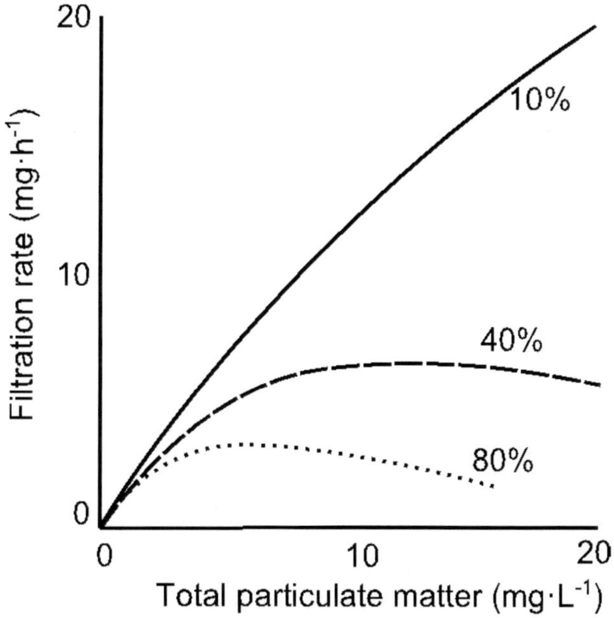

Figure 3. Effect of seston characteristics on filtration rate of the cockle *Cerastoderma edule* (partially redrawn from Iglesias et al., 1996).

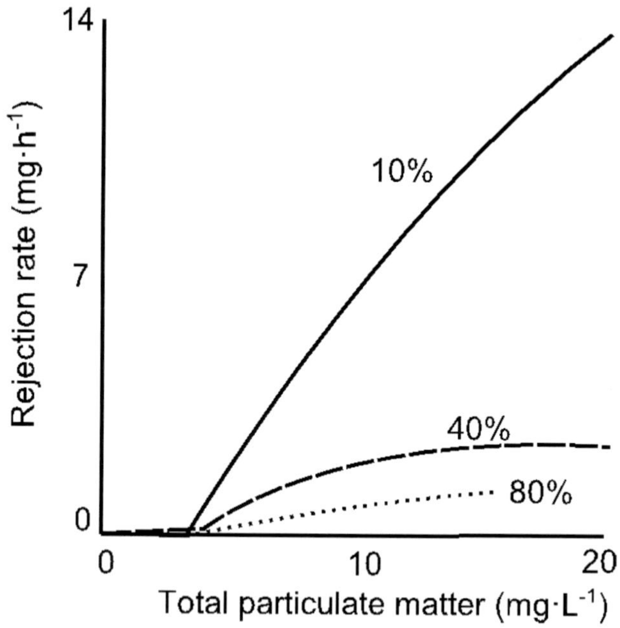

Figure 4. Effect of seston characteristics on the selection efficiency of the cockle *Cerastoderma edule* (partially redrawn from Iglesias et al., 1996).

1.4.2. Pre-Ingestive Selection and Rejection

Not all filtered particles are finally ingested. Bivalves can reject a large proportion of filtered material when the particle concentration is too high (Figure 4), but they are also able to select particles according to their characteristics (Figure 5). Some of the initially retained particles can be rejected by different mechanisms, the most widely known being the production of pseudofeces, i.e. mucous threads in which the rejected particles are embedded, but other mechanisms, such as accumulation in the pallial cavity followed by expulsion mixed with intervalvar water, have also been observed (Ward & Shumway, 2004; Bricelj et al., 1998).

Increased particles rejection at high seston concentrations, together with reduced filtration, leads to maximum particle ingestion at intermediate concentrations (Figure 6). Consequently, during very dense blooms, less toxin is transferred to bivalves than during moderate ones. Single phytoplankton species or strains can be rejected or a very small proportion of them retained by some bivalve species. Such is the case with the filtration of the toxic dinoflagellate *Alexandrium* by *Ostrea edulis* (Bricelj et al., 1998) and of the toxic diatom *Pseudo-nitzschia pungens* by the oyster *Crassostrea virginica* (Mafra et al., 2010). In contrast to mussels (*Mytilus edulis*), oysters are able to select and reject *Pseudo-nitzschia* cells, and especially its short-celled morphotypes. Its large selection capability, that is temperature-dependent, was the basis of the difference found in the accumulation of DA by two shellfish species, in particular at low temperature: *M. edulis* accumulated over 200μg DA g^{-1} while *C. virginica* did not exceed 0.9 $\mu g{\cdot}g^{-1}$ (Mafra et al., 2010).

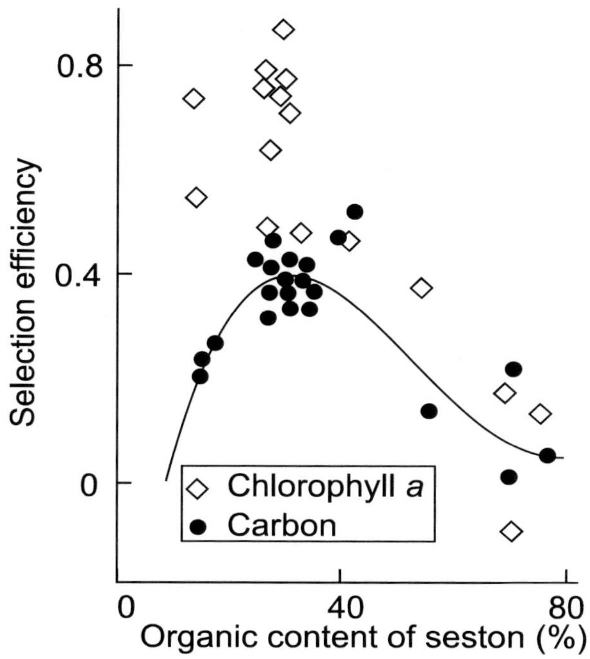

Figure 5. Selection efficiency of the cockle *Cerastoderma edule* for carbon-containing and chlorophyll-containing particles, as a function of the organic content of seston (partially redrawn from Iglesias et al., 1996).

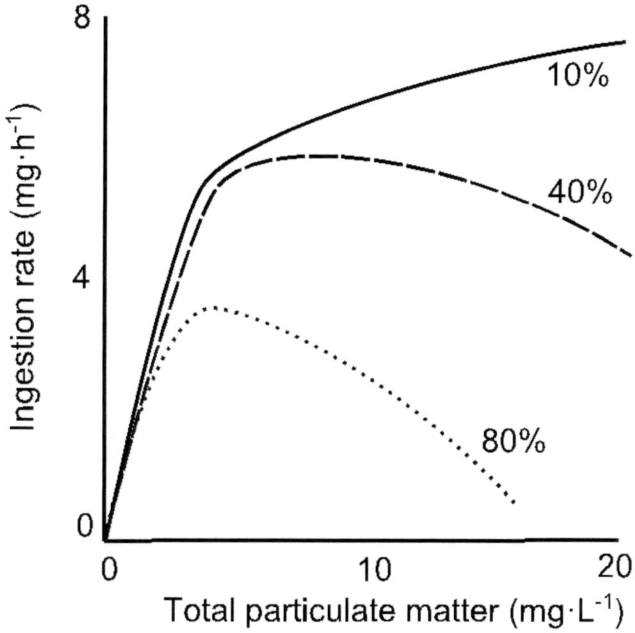

Figure 6. Effect of seston characteristics on the ingestion rate of the cockle *Cerastoderma edule* (partially redrawn from Iglesias et al., 1996).

1.4.3. Post-Ingestive Selection and Absorption

Once the selected cells and particles are ingested they pass to the stomach, where they are subjected to the initial phases of extracellular digestion and suffer from additional selection. Some particles are directly channeled to the intestine and expelled within the feces. The positively selected particles and digestive fluids pass to the digestive tubules where extracellular digestion is completed, and the resulting materials are absorbed by the bivalve digestive cells to proceed to intracellular digestion. The efficiency of this process—efficiency of absorption (of organic matter or toxins) —is known to be dependent on the residence time of the organic (or toxin-containing) particles in the digestive system, which in turn depends on the ratio 'organic matter (or toxin)/total volume of seston', usually known as "food quality" when it refers to organic matter. As digestibility differs substantially between phytoplankton species, it is clear that the whole digestion process is also species-specific. Concerning digestibility and post-ingestive selection, Bricelj & Shumway (1998) found that intact and viable *Alexandrium* cells were massively present in fecal ribbons of the quahog *Mercenaria mercenaria*. As for the modification of the efficiency of absorption by the ratio of toxin to seston volume, Moroño et al. (2001) demonstrated that in *Mytilus galloprovincialis*, PST are affected in the same way as organic matter, and consequently a higher proportion are absorbed when the seston volume is smaller. Reduced accumulation of DST from natural populations of *Dinophysis* was noted when the accompanying species were abundant (Sampayo et al. 1990).

1.4.4. Enzymatic Biotransformation

Once toxins are inside the bivalve cells, they undergo a series of transformations which in many cases result in a substantial modification of toxicity. Among the PSTs, the low-potency sulfocarbamate toxins (C1-C4, GTX5-6) are frequently desulfo-decarbamoylated by bivalves (Shimizu & Yoshioka, 1981; Fast et al., 2006), yielding products that are one or two orders of magnitude more toxic; other toxins can undergo reductions and epimerizations that also involve changes (increases or decreases) in toxicity of lesser magnitude (Blanco et al., 2003). Within the OA-related toxins, diol-esters (DTX4, DTX5) present in the dinoflagellate cells are mildly or not toxic because their carboxylic moeities are esterified. Nevertheless, they seem to be readily hydrolysed by bivalves to okadaic acid, thus producing an important increase in shellfish toxicity in relation to the toxicity of the source cells. These transformations are responsible for increases in bivalve toxicity for some days, even when the toxic cells are no longer available in the water column (reviewed in Blanco et al., 2005). Conversely, rapid transformation of PTX2 to PTX2SA in mussels considerably reduces the toxicity estimates by standard MBA (Suzuki et al., 2001). Other transformations among DSTs, such as the esterification of OA, DTX1 and DTX2 to give DTX3-type compounds, do not produce substantial changes in toxicity, but are required to eliminate the toxins from the bivalve body (Rossignoli et al., 2011).

1.4.5. Elimination (Depuration)

The elimination of toxins from bivalve tissues proceeds at different rates, mostly depending on the combination of bivalve species and phycotoxins. As depuration is usually proportional to toxin concentration in bivalves, large toxin accumulations are not usually found in rapidly depurating organisms. In contrast, slowly depurating organisms can attain large toxin concentrations and maintain them for long periods of time, thus constituting a serious risk for consumer safety. As far as we know, environmental conditions have little effect on toxin elimination, and consequently it can be assumed to be relatively constant for any species/toxin combination (reviewed in Fernández et al., 2003).

1.4.6. Allometric Effects

Bivalves toxicity depends on toxin concentration (μg toxins per unit of shellfish meat), which in turn depends on the amount of toxin accumulated and on the biomass in which it is diluted. In many cases bivalves undergo substantial and sudden changes in biomass of all or some of their organs that lead to changes in toxin concentration, and hence of toxicity, even without changes in the amount of toxin accumulated. During periods of intense growth, for example, toxins are diluted in the increasing bivalve biomass, and toxin concentration decreases independently of the toxin elimination processes. In those cases, an "apparent" depuration takes place, which can be substantially faster than the actual depuration (Bricelj & Cembella, 1995; Blanco et al., 2002). But increases of toxicity due to loss of shellfish biomass, mostly due to sudden biomass losses during spawning, are much more frequent and intense than decreased toxicity due to increased biomass. The extent to which toxin concentrations can increase due to biomass losses depends on the location of the accumulated toxins and on the reduction of biomass. For example, DSTs are accumulated in the mussel (*Mytilus galloprovincialis*) digestive gland (Blanco et al., 2007). When mussels spawn, most of the gonad and mantle biomass is lost in the released spermatozoids or oocytes, but hardly any toxin is lost from the digestive gland. As a consequence, the overall weight of the organism is reduced while the amount of toxin is maintained, and therefore the toxin concentration is increased. A similar situation occurs concerning accumulation of domoic acid in the king scallop *Pecten maximus*. In this species, most of the toxins are located in the digestive gland and gonad—toxins are associated with the intestinal loop, a part of the digestive system that runs through the gonads—. In other species, such as *Saxidomus giganteus*, that accumulate PSTs, mostly in the siphon, or the razor clams *Siliqua patula,* that strongly retains domoic acid, these increases in toxin concentration should also take place frequently.

CONCLUSION

The capacity of microalgae to produce certain toxins (toxin profile), and the inter-and intra-specific variability observed in the amount of intracellular toxin produced (toxin quota, Q_t) is genetically determined, but the actual production of toxins (gene expression) is controlled by intrinsic factors—cell cycle, life cycle, behaviour, population growth—which are also genetically determined, and modulated by biotic and abiotic environmental factors. Thus, the toxin profile of a strain and its cellular content (toxin cell quota) may vary, within certain ranges, depending on the time of the day, the proportion of different life cycle stages, and the phase the population growth when the sample is collected. Microalgae behaviour—daily vertical migration (DVM), aggregation in patches or in thin layers—will influence the distribution patterns of toxicity in shellfish beds and in floating rafts and long-lines. Species producing large amounts of resting stages which accumulate in the sediments expand their contamination potential from epifaunal suspension feeders to infaunal siphonate deposit feeders.

Toxins are secondary metabolites. Toxin production rates and their accumulation in the cells (cell toxin quota) are tightly coupled to general growth processes and to the generation of primary metabolites that are precursors of the toxins. Therefore, any biotic and abiotic environmental factor affecting growth, will also indirectly affect toxin production. More toxic cells (higher toxin quota) are often the result of imbalance between growth and toxin production rates. Much information on toxin production comes from laboratory experiments which, if carried out under standard procedures, provide knowledge of the microalgae physiology. Nevertheless, extrapolation of these results to multispecies field populations, subject to complex physical and biological interactions, should be made with caution.

The application of molecular approaches to microalgae taxonomy is developing rapidly. Discrimination between morphologically similar toxic and non-toxic strains of the same genus or species is essential to relate inter- and intraspecific variability and environmental conditions triggering growth and toxin production of different strains of toxigenic microalgae. Meanwhile, it is not possible to determine if unusually intense blooms are caused by the predominance of more toxic strains of the same species, or by optimum environmental conditions that promote maximum toxin production of one or several strains.

The large size and particular characteristics of the dinoflagellates genome have hindered the identification of biosynthetic pathways for toxins. Hopefully, the implementation of next-generation genomic approaches and the parallel development of proteomics will unveil the mechanisms responsible for phenotypical changes in toxin production among clonal isolates.

The environmental conditions promoting germination, growth, accumulation, and transport of toxic HAB populations from their source areas to shellfish grounds are site- and species-specific, and tightly coupled to meteorologically-driven macro- and microscale physical processes. It is an oversimplification to state that "HABs are increasing in intensity, duration and geographic extension". Some HAB species are, and some are not, as their distributions exhibit expansions and contractions coupled to climate variability. There is no convincing scientific evidence to relate toxin-producing HABs with eutrophication, although there are a few documented cases of high biomass non-toxic HABs linked to increased levels of nutrient inputs from anthropogenic sources. HAB time series available from research and monitoring program are providing evidence of relations between climate and interannual variability, but still these time series are not long enough to soundly relate changes in HAB populations to climate change.

Accumulation of toxins in shellfish is not a simple process that can be predicted from the concentration (cell counts) of toxic cells to which shellfish feeders are exposed. Most important factors include the cell toxin profile and content, and the proportion of toxic cells in the whole plankton community. The use of "trigger levels" of toxic cells as a monitoring tool before testing toxins in shellfish poses a risk to public health and the safe exploitation of resources.

The accumulation of toxins in shellfish is determined by the feeding behavior of each species and by their depuration capabilities. The rate at which organisms can ingest toxic algae is a key point in the toxin acquisition process. The actual rate depends on the filtration, rejection and selection capabilities, that are partly genetically determined and partly regulated by the environmental conditions, the toxic phytoplankton species (or strain) and the toxins involved, in ways specific to each bivalve species. Very little is known about depuration, but it is clear that it is also a toxin- and shellfish species- specific process.

Shellfish toxicity depends on toxin concentration, i.e. the amount of toxin (moles, µg) per unit of shellfish flesh, but also on the potency of the accumulated toxins. As toxin concentration is the ratio between toxin amount and shellfish biomass, any change in biomass will affect the toxicity of the organism. As the toxic power of each toxin from the same family can differ substantially, most toxin biotransformations carried out by bivalves, with no gains or losses of toxic compounds, will have important effects on toxicity (increased or decreased potency).

Knowledge of the seasonality of each toxin producing species, its response to environmental conditions, and of species-specific accumulation and detoxification responses allows the development of predictive capabilities about high risk conditions for shellfish contamination. The worst scenario would result from a combination of: i) environmental conditions favouring the development of relevant densities of populations of a very toxic species that reaches a mature phase of no growth and high toxin accumulation per cell ; ii) hydrodynamic transport and/or aggregation of this toxic population that then becomes available to filter feeders; this includes processes, such as downwelling, that select swimming dinoflagellates and increase the proportion of toxic cells in the plankton community available to filter-feeders; iii) effective filtration, assimilation and biotransformation by shellfish, that will be even more harmful if the exploited resource is a slowly depurating organism.

ACKNOWLEDGMENTS

Thanks are given to Dr Toshiyuki Suzuki for fruitful discussions about biotransformations of *Dinophysis* toxins in shellfish. This work was partially funded by project *DIGEDINO* (CTM2009-12988-C02-01) from the Spanish Ministry of Science and Innovation, the Strategic Action *EPITOX* (CEI, Xunta de Galicia) and by the EU project ASIMUTH (FP7-SPACE-2010-1, contract # 261860).

REFERENCES

Acosta, C. P., Blanco, J. (2003). Clearance rate of *Pecten maximus* II: Optimal food concentration. 14th International Pectinid Workshop. St.Petersburg, Florida, USA. http://www.cimacoron.org/EpisodiosToxicos/Documentos/PFLORIDA2.pdf.

Aikman, K. E., Tindall, D. R. & Morton, S. L. 1993. Physiology and potency of the dinoflagellate *Prorocentrum hoffmannianum* (Faust) during one complete growth cycle. In T. J. Smayda & Y. Shimizu (Eds.), *Toxic phytoplankton blooms in the sea* (pp. 463-468). New York: Elsevier Science.

Alpermann, T. J, Beszteri, B., John, U., Tillmann, U. & Cembella, A. D. (2009). Implications of life-history transitions on the population genetic structure of the toxigenic marine dinoflagellate *Alexandrium tamarense*. *Mol. Ecol*., 18, 2122–2133.

Alperman, T. J; Tillmann, U., Beszteri, B., Cembella, A. D. & John, U. (2010). Phenotypic variation and genotypic diversity in a planktonic population of the toxigenic marine dinoflagellate *Alexandrium tamarense* (Dinophyceae). *J. Phycol*., 46, 18-32.

Anderson, D. M., Keafer, B. A., Geyer, W. R., Signell, R. P. & Loder, T. C. (2005). Toxic *Alexandrium* blooms in the western Gulf of Maine: The plume advection hypothesis revisited. *Limnol. Oceanogr*., 50, 328–345.

Anderson, D. M., Kulis, D. M., Sullivan, J. J., Hall, S. & Lee, C. (1990). Dynamics and physiology of saxitoxin production by dinoflagellates *Alexandrium* spp. *Mar. Biol.,* 104, 511-524.

Anderson, D. M., Reguera, B., Pitcher, G. C. & Enevoldsen, H. O. (2010). The IOC International Harmful Algal Bloom program: History and science impacts. *Oceanography*, 23, 72–85.

Artigas, M. L., Amorim, A., Vale, P., Gomes, S. S., Botelho, M. J., Rodrigues, S. M.. 2008. Prolonged toxicity of *Scrobicularia plana* after a PSP event and its relation to *Gymnodinium catenatum* cyst consumption and toxin depuration. In Ø. Moestrup (chief Ed.), *Proceedings of the 12th International Conference on Harmful Algae* (pp.273-275). Copenhagen: International Society for the Study of Harmful Algae and Intergovernmental Oceanographic Commission of UNESCO.

Bates, S. S. (1998). Ecophysiology and metabolism of ASP toxin production. In D. M. Anderson, A.D. Cembella, & G. M. Hallegraeff (Eds.). *Physiological Ecology of Harmful Algal Blooms. NATO ASI series, Series G, Ecological Sciences 41* (pp. 405-426). Berlin Heidelberg: Springer-Verlag.

Bates, S. S., Douglas, D. J., Doucette, G. J. & Leger, C. (1995). Enhancement of domoic acid production by reintroducing bacteria to axenic cultures of the diatom *Pseudo-nitzschia multiseries*. *Nat. Toxins,* 3, 428-435.

Bates, S. S. & Trainer, V. L. (2006). The ecology of harmful diatoms. In E. Granéli, & J. Turner (Eds), *Ecology of Harmful Algae*, Ecological Studies 189 (pp. 81-93). Berlin, Heidelberg: Springer Verlag.

Bayne, B.L., Thompson, R. J. & Widdows, J. (1976). Physiology: I. In B. L. Bayne (Ed.) *Marine Mussels: their Ecology and Physiology* (pp 121–206). Cambrige, U.K.: Cambridge University Press.

Benavides, H., Prado, L., Díaz, S. & Carreto, J.I. (1995). An excepcional bloom of *Alexandrium catenella* in the Beagle Channel, Argentina. In P. Lassus, G.Arzul, E. Erard-Le Denn, P. Gentien, & C. Marcaillou-LeBaut (Eds.), *Harmful Marine Algal Blooms* (113-119). Paris, France: Lavoisier Intercept Ltd.

Blanco, J., Acosta, C. P. & Bermúdez de la Puente, M., Salgado, C. (2002). Depuration and anatomical distribution of the amnesic shellfish poisoning (ASP) toxin domoic acid in the king scallop *Pecten maximus*. *Aquat. Toxicol*. 60, 111–121.

Blanco, J., Álvarez, G. & Uribe, E. (2007). Identication of pectenotoxins in plankton, filter feeders, and isolated cells of a *Dinophysis acuminata* with an atypical toxin profile, from Chile. *Toxicon*, 49, 710-716.

Blanco, J., Mariño, C., Martín, H. & Acosta, C. P. (2007). Anatomical distribution of Diarrhetic Shellfish Poisoning (DSP) toxins in the mussel *Mytilus galloprovincialis*. *Toxicon*, 50, 1011–1018.

Blanco, J., Moroño, A. & Fernández, M. L. (2005). Toxic episodes in shellfish produced by lipophilic phycotoxins: An overview. *Revista Galega dos Recursos Mariños (Monog.)*, 1, 1-70.

Blanco, J., Reyero, M. I. & Franco, J.M. (2003). Kinetics of accumulation and transformation of paralytic shellfish toxins in the blue mussel *Mytilus galloprovincialis*. *Toxicon*, 42, 777–784.

Blanco-Pérez, J. (2001). Episodios nocivos por fitoplancton. In A. N. Maeda-Martínez (Ed.) *Los Moluscos Pectínidos de Iberoamérica: Ciencia y Acuicultura* (pp 285–324). Mexico, Editorial Limusa.

Boczar, B. A., Beitler, M. K., Liston, J., Sullivan, J. J. & Cattolico, R. A. (1988). Paralytic shellfish toxins in *Protogonyaulax tamarensis* and *Protogonyaulax catenella* in axenic culture. *Plant Physiol.*, 88, 1285–90.

Bravo, I., Fraga, S., Figueroa, R. I., Pazos, Y., Massanet, A. & Ramilo, I. (2010). Bloom dynamics and life cycle strategies of two toxic dinoflagellates in a coastal upwelling system (NW Iberian Peninsula). *Deep-Sea Res. II*, 57, 222–234.

Bravo, I., Franco, J. M. & Reyero, M. I. (1998). PSP toxin composition of three life cycle stages of *Gymnodinium catenatum*. In B. Reguera, J. Blanco, M. L. Fernández, & T. Wyatt (Eds.), *Harmful Algae* (pp. 356-358). Santiago de Compostela, Spain: Xunta de Galicia and Intergovernmental Oceanographic Commission of UNESCO.

Bricelj, V. M. & Cembella, A. D. (1995) Fate of gonyautoxins in surfclams, *Spisula solidissima*, grazing upon toxigenic *Alexandrium*. In P. Lassus, G.Arzul, E. Erard-Le Denn, P. Gentien, & C. Marcaillou-LeBaut (Eds.), *Harmful Marine Algal Blooms* (413-418). Paris, France: Lavoisier, Intercept Ltd.

Bricelj, V. M., Connell, L., Konoki, K., Macquarrie, S. P., Scheuer, T., Catterall, W. A. & Trainer, V. L. (2005). Sodium channel mutation leading to saxitoxin resistance in clams increases risk of PSP. *Nature* 434, 763–767.

Bricelj, V. M. & Shumway, S. E. (1998). Paralytic shellfish toxins in bivalve molluscs: occurrence transfer kinetics and biotransformation. *Rev. Fish. Sci.* 6, 315–383.

Bricelj, V. M., Ward, J. E., Cembella, A. D & MacDonald, B.A. (1998). Application of video-endoscopy to the study of bivalve feeding on toxic dinoflagellates. In B. Reguera, J. Blanco, M. L. Fernández, & T. Wyatt (Eds.), *Harmful Algae* (pp. 453-456). Santiago de Compostela, Spain: Xunta de Galicia and Intergovernmental Oceanographic Commission of UNESCO.

Cembella, A.D. 1998. Ecophysiology and metabolism of paralytic shellfish toxins in marine microalgae. In D. M. Anderson, A. Cembella, G. M. Hallegraeff (Eds.). *Physiological Ecology of Harmful Algal Blooms. NATO ASI series, Series G, Ecological Sciences 41* (pp. 381-403). Berlin, Heidelberg: Springer-Verlag.

Cembella, A. D., Destombe, C. & Turgeon, J. (1990). Toxin composition of alternative life history stages of *Alexandrium*, as determined by high-performance liquid chromatography. In E. Graneli, B. E., Sundtröm, L. Edler, & D. M. Anderson (Eds.), *Toxic Marine Phytoplankton. Proceedings of the 4thInternational Conference on Toxic Marine Phytoplankton* (pp. 333-338). New York: Elsevier

Cembella, A.D & John, U. (2006). Molecular physiology of toxin production and growth regulation in harmful algae. In E. Granéli, & J. Turner (Eds), *Ecology of Harmful Algae*, Ecological Studies 189 (pp. 215-227). Berlin, Heidelberg: Springer Verlag.

Cembella, A. D., Sullivan, J. J., Boyer, G. L., Taylor, F. J. R., & Andersen, R. J. (1987). Variation in paralytic shellfish toxin composition within the *Protogonyaulax tamarensis/ catenella* species complex; red tide dinoflagellates. *Biochem. Systems Ecol.*, 15, 171-186.

Chan, L. L., Hodgkiss, I. J., Lu, S. & Lo, S. C. (2004). Use of two-dimensional gel electrophoresis proteome reference maps of dinoflagellates for species recognition of causative agents of harmful algal blooms. *Proteomics* 4: 180-192.

Chan, L. L., Hodgkiss, I. J., Lam, P. K.-S., Wan, J. M.-F., Chou, H.-N., Lum, J. H.-K., Lo, M. G.-Y., Mak, A. S.-C., Sit, W.-H. & Lo, S. C.-L. (2005). Use of two-dimensional gel electrophoresis to differentiate morphospecies of *Alexandrium minutum*, a paralytic shellfish poisoning toxin-producing dinoflagellate of harmful algal blooms. *Proteomics* 5: 1580–1593.

Chan, L. L., Lo, S. C. & Hodgkiss, I. J. (2002). Proteomic study of a model causative agent of harmful red tide, *Prorocentrum triestinum* I: Optimization of sample preparation methodologies for analyzing with two-dimensional electrophoresis. *Proteomics*, 2, 1169-1186.

Chan, L. L., Sit, W.-H., Lam, P. K.-S., Hsieh, D. P. H., Hodgkiss, I. J., Wan, J. M.-F., Ho, A. Y.-T., Choi, N. M.-C., Wang, D.-Z. & Dudgeon, D. (2006). Identification and characterization of a "biomarker of toxicity" from the proteome of the paralytic shellfish toxin-producing dinoflagellate *Alexandrium tamarense* (Dinophyceae). *Proteomics*, 6: 654–666.

Cho, Y., Hiramatsu, K., Ogawa, M., Omura, T., Ishimaru, T. & Oshima, Y. (2008). Non-toxic and toxic subclones obtained from a toxic clonal culture of *Alexandrium tamarense* (Dinophyceae): Toxicity and molecular biological features. *Harmful Algae,* 7: 740-751.

Chou, H. N., Chen, Y. M. & Chen, C.Y. (2004). Variety of PSP toxins in four culture strains of *Alexandrium minutum* collected from southern Taiwan. *Toxicon,* 43, 337-340.

Chou, H.N. & Shimizu, Y. (1987). Biosynthesis of brevetoxins. Evidence for the mixed origin of the backbone carbon chain and possible involvement of dicarboxylic acids. *J. Am. Chem. Soc.*, 109, 2184–2185.

Cullen, J. J. & MacIntyre, J. G. (1998). Behavior, physiology and the niche of depth-regulating phytoplankton. In D. M. Anderson, A. Cembella, G. M. Hallegraeff (Eds.). *Physiological Ecology of Harmful Algal Blooms. NATO ASI series, Series G, Ecological Sciences 41* (pp. 559-579). Berlin, Heidelberg: Springer-Verlag.

Edvardsen, B., Shalchian-Tabrizi, K., Jakobsen, K.S., Medlin, L.K., Dahl, E., Brubak, S. & Paasche, E. (2003). Genetic variability and molecular phylogeny of *Dinophysis* species (Dinophyceae) from Norwegian waters inferred from single cell analyses of rDNA. *J. Phycol.,* 39, 395–408.

Errera, R.M., Bourdelais, A., Drennan, M.A., Dodd, E.B., Henrichs, D.W. & Campbell L. (2010). Variation in brevetoxin and brevenal content among clonal cultures of *Karenia brevis* may influence bloom toxicity. *Toxicon*, 55: 195-203.

FAO (2004). *Marine Biotoxins*. FAO Food and Nutrition Paper, 80. Rome: Food and Agriculture Organization of the United Nations.

Fast, M. D., Cembella, A. D & Ross, N. W. (2006). In vitro transformation of paralytic shellfish toxins in the clams *Mya arenaria* and *Protothaca staminea*. *Harmful Algae*, 5, 79–90.

Fauchot, J., Saucier, F. J., Levasseur, M., Roy, S. & Zakardjian, B. (2008). Wind-driven river plume dynamics and toxic *Alexandrium tamarense* blooms in the St. Lawrence estuary (Canada): A modeling study. *Harmful Algae,* 7, 214–227.

Fehling, J., Davidson, K., Bates, S. S. (2005). Growth dynamics of non-toxic *Pseudo-nitzschia delicatissima* and toxic *P. seriata* (bacillariophyceae) under simulated spring and summer photoperiods. *Harmful Algae* 4, 763-769.

Fernández, M. L., Shumway, S. E. & Blanco, J (2003). Management of shellfish resources. In G. M. Hallegraeff, D. M. Anderson & A. D. Cembella (Eds), *Manual on Harmful Marine Microalgae. Monographs on Oceanographic Methodology 11* (pp 657–692). Paris: UNESCO Publishing.

Filgueira, R., Labarta, U. & Fernández-Reiriz, M. J. (2008). Efecto del índice de condición en las relaciones alométricas de la tasa de aclaramiento en *Mytilus galloprovincialis* Lamarck, 1819. *Rev. Biol. Mar. Oceanogr* 43, 391-398.

Flynn, K. J. (1998). Physiology of toxic microalgae with special emphasis on toxin production: construction of dynamic models. In B. Reguera, J. Blanco, M. L. Fernández, & T. Wyatt (Eds.), *Harmful Algae* (pp. 315-320). Santiago de Compostela, Spain: Xunta de Galicia and Intergovernmental Oceanographic Commission of UNESCO.

Flynn, K. J., Flynn, K., John, E. H., Reguera, B., Reyero, M. I. & Franco, J. M. (1996). Changes in toxins, intracellular and dissolved free aminoacids of the toxic dinoflagellate *Gymnodinium catenatum* in response to changes in inorganic nutrients and salinity. *J. Plankton Res.* 18, 2093-2011.

Flynn, K., Franco, J. M., Fernández, P., Reguera, B., Zapata, M., Wood, G. & Flynn, K.J. (1994). Changes in toxin content, biomass and pigments of the dinoflagellate *Alexandrium minutum* during nitrogen refeeding and growth into nitrogen and phosphorous stress. *Mar. Biol. Prog. Ser.* 111: 99-109.

Fraga, S., Anderson, D. M., Bravo, I., Reguera, B., Steidinger, K. A., Yentsch, C. M.(1988). Influence of upwelling relaxation on dinoflagellates and shellfish toxicity in Ría de Vigo, Spain. *Estuar. Coast. Shelf Sci.,* 27, 349-361.

Franco, J. M., Fernández, P. & Reguera, B. (1994). The toxin profiles of natural populations and cultures of *Alexandrium minutum* Halim from Galician coastal waters. *J. Appl. Phycol.*, 6, 275-279.

Fux, E.,González-Gil, S., Lunven, M., Gentien, P. & Hess, P. (2010). Production of diarrhetic shellfish poisoning toxins and pectenotoxins at depths within and below the euphotic zone. *Toxicon* 56, 1487–1496.

GEOHAB (2008). *Global Ecology and Oceanography of Harmful Algal Blooms, GEOHAB Core Research Project: HABs in Stratified Systems.* P. Gentien, B. Reguera, H. Yamazaki, L. Fernand, E. Berdalet, & R. Raine (Eds.). Paris, France and Newark, Delaware, USA: IOC and SCOR.

González-Gil, S., Pizarro, G., Paz, B., Velo-Suárez, L. & Reguera, B. 2010. Is Dinophysis *rotundata* a toxigenic species? In K. C. Ho, M. J. Zhou & Y. Z. Qi (Eds.). *Harmful Algae 2008* (pp. 128-131). Copenhagen, Denmark: International Society for the Study of Harmful Algae and Environmental Publications House Hong Kong.

Granéli, E. & Flynn, K. J. (2006). Chemical and physical factors influencing toxin content. In E. Granéli, & J. Turner (Eds), *Ecology of Harmful Algae*, Ecological Studies 189 (pp. 229-241). Berlin, Heidelberg: Springer Verlag.

Hackett, J. D., Scheetz, T. E., Yoon, H. S., Soares, M. B., Bonaldo, M. F., Casavant, T. L. & Bhattacharya, D. (2005). Insights into a dinoflagellate genome through expressed sequence tag analysis. *BMC Genomics,* 6, 80.

Hackett, J. D., Tong, M., Kulis, D.M., Fux, E., Hess, P., Bire, R. & Anderson, D.M. (2009). DSP toxin production de novo in cultures of *Dinophysis acuminata* (Dinophyceae) from North America. *Harmful Algae,* 8, 873-879.

Hebert, P. D. N., Cywinska, A., Ball, S. L. & deWaard J. R. (2003). Biological identifications through DNA barcodes. 270, 313-321

Hoagland, P. & Scatasta, S. (2006). The economic effects of harmful algal blooms. In E. Granéli, & J. Turner (Eds), *Ecology of Harmful Algae*, Ecological Studies 189 (pp. 391-402). Berlin, Heidelberg: Springer Verlag.

Hollingsworth, P. M. et. al. (CBOL Plant Working Group). (2009). A DNA barcode for land plants. *Proc. Natl. Acad. Sci. USA*, 106, 12794-12797.

Holmes, M. J., Chin Lee, F., Woo Khoo, H. & Ming Teo, S. L. (2001). Production of 7-deoxyokadaic acid by a new Caledonian strain of *Prorocentrum lima* (Dinophyceae). *J. Phycol.*, 37, 280-288.

Honsell, G., Poletti, R., Pompei, M., Sidari, L., Milandri, A., Casadei, C. & Viviani, R. (1996). *Alexandrium minutum* Halim and PSP contamination in the northern Adriatic Sea. In T. Yasumoto, Y. Oshima, Y., & Y. Fukuyo (Eds.), *Harmful and Toxic Algal Blooms* (pp. 77-80). Sendai, Japan, Intergovernmental Oceanographic Commission of UNESCO.

Hosoi-Tanabe, S. & Sako, Y. (2005). Rapid detection of natural cells of *Alexandrium tamarense* and *A. catenella* (Dinophyceae) by fluorescence in situ hybridization. *Harmful Algae*, 4, 319-328.

ICES, 2006. Report of the ICES Workshop on Time Series Data relevant to Eutrophication Ecological Quality Objectives (WKEUT). *International Council for the Exploration of the Sea*, ICES CM 2006/ACE:07, REF. MHC, OCC.

Iglesias, J. I. P., Urrutia, M. B., Navarro, E., Alvarez-Jorna, P., Larretxea, X., Bougrier, S. & Heral, M. (1996). Variability of feeding processes in the cockle *Cerastoderma edule* (L.) in response to changes in seston concentration and composition. *J. Exp. Mar. Biol. Ecol.,* 197, 121–143.

Ishida, Y., Uchida, A. & Sako, Y. (1998). Genetic and biochemical approaches to PSP toxin production of toxic dinoflagellates. In D. M. Anderson, A. Cembella, & G. M. Hallegraeff (Eds.). *Physiological Ecology of Harmful Algal Blooms. NATO ASI series, Series G, Ecological Sciences 41* (pp. 49-58). Berlin Heidelberg: Springer-Verlag.

Jackson, A. E., Marr, J. C. & McLachlan, J. L. (1993). The production of diarrhetic shellfish toxins by an isolate of *Prorocentrum lima* from Nova Scotia, Canada. In T. J. Smayda, & Y. Shimizu.(Eds.), *Toxic Phytoplankton Blooms in the Sea* (pp. 513-518). Amsterdam: Elsevier.

Johanssen, M. (2008). On *Dinophysis* : Occurrence and Toxin Content. Ph D thesis, University of Göteborg and Kristineberg Marine Biology Station, Sweden.

John, U., Groben, R., Beszteri, B. & Medlin, L. (2004). Utility of Amplified Fragment Length Polymorphisms (AFLP) to analyse genetic structures within the *Alexandrium tamarense* species complex. *Protist*, 155: 169-179.

John, U., Medlin, L. K. & Groben, R. (2005). Development of specific rRNA probes to distinguish between geographic clades of the *Alexandrium tamarense* species complex. *J. Plankton Res.*, 27, 199-204.

John, U., Quilliam, M. A., Medlin, L. & Cembella, A.D. (2001). Spirolide production and photoperiod-dependent growth of the marine dinoflagellate *Alexandrium ostenfeldii*. In G. M. Hallegraeff, S. I. Blackburn, C. J. Bolch, & R. J. Lewis (Eds). *Harmful Algal*

Blooms 2000 (299-302). Paris, Intergovernmental Oceanographic Commission of UNESCO.

Kamikawa, R., Hosoi-Tanabe, S., Yoshimatsu, S., Oyama, K., Masuda, I. & Sako, Y. (2008). Development of a novel molecular marker on the mitochondrial genome of a toxic dinoflagellate, *Alexandrium* spp., and its application in single-cell PCR. *J. Appl. Phycol.*, 20: 153-159.

Kamiyama, T., Nagai, S., Suzuki, T., Miyamura, K. (2010). Effect of temperature on production of okadaic acid, dinophysistoxin-1, and pectenotoxin-2 by *Dinophysis acuminata* in culture experiments. *Aquat. Microb. Ecol.*, 60, 193-202.

Kamiyama, T. & Suzuki, T. (2009). Production of Dinophysistoxin-1 and Pectenotoxin-2 by a culture of *Dinophysis acuminata* (Dinophyceae). *Harmful Algae*, 8, 312-317.

Keafer, B. A., Churchill, J. H. & Anderson, D. M. (2005a). Blooms of the toxic dinoflagellate, *Alexandrium fundyense* in the Casco Bay region of the western Gulf of Maine: Advection from offshore source populations and interactions with the Kennebec River plume. *Deep-Sea Res. II*, 52, 2631–2655.

Keafer, B. A., Churchill, J. H., Mcgillicuddy, D. J. & Anderson, D. M. (2005b). Bloom development and transport of toxic *Alexandrium fundyense* populations within a coastal plume in the Gulf of Maine. *Deep-Sea Res. II,* 52, 2674–2697.

Kellmann, R., Stüken, A., Orr, R. J. S., Svendsen, H.M. & Jakobsen, K. S. (2010). Biosynthesis and molecular genetics of polyketides in marine dinoflagellates. *Mar. Drugs*, 8, 1011–1048.

Kim, C. H., Sako, Y. & Ishida, Y. (1993). Comparison of toxin composition between populations of *Alexandrium* spp. from geographical distant areas. *Nipp. Suis. Gakk.*, 59, 641-646.

Kim, H. G., Lee, S. G., Lim, W. A., Lee, J. S. & Kim, J. H. (1996). Environmental physiology of *Alexandrium tamarense* isolate from Chinhae Bay, the South Sea of Korea. In T. Yasumoto, Y. Oshima, & Y. Fukuyo (Eds.), *Harmful and Toxic Algal Blooms* (pp. 57-60) Sendai, Japan, Intergovernmental Oceanographic Commission of UNESCO.

Kim, S., Kang, Y. G., Kim, H. S., Yih, W., Coats, D. W. & Park, M. G. (2008). Growth and grazing responses of the mixotrophic dinoflagellate *Dinophysis acuminata* as functions of light intensity and prey concentration. Aquat. Microb. Ecol., 51, 301-310.

Kuno, S., Kamikawa, R., Yoshimatsu, S., Sagara, T., Nishio, S. & Sako, Y. (2010). Genetic diversity of *Gambierdiscus* spp. (Gonyaulacales, Dinophyceae) in Japanese coastal areas. *Phycol. Res.*, 58: 44–52.

Laabir, M., Amzil, Z., Lassus, P., Masseret, E., Tapilatu, Y., De Vargas, R., Grzebyk, D. (2007). Viability, growth and toxicity of *Alexandrium catenella* and *Alexandrium minutum* (Dinophyceae) following ingestion and gut passage in the oyster *Crassostrea gigas. Aquat. Living Resour.*, 20, 51–57.

Lakeman, M. B., von Dassow, P. & Cattolico, R. A. (2009) The strain concept in phytoplankton ecology. *Harmful Algae*, 8, 746–758.

Landsberg, J. H. (2002). The effects of harmful algal blooms on aquatic organisms. *Rev. Fish. Sci.*, 10, 113–390.

Lilly, E. L., Halanych, K. M., Anderson, D. M. (2005). Phylogeny, biogeography, and species boundaries within the Alexandrium minutum group. *Harmful Algae*, 4, 1004-1020.

Lilly, E. L., Halanych, K. M. & Anderson, D. M. (2007). Species boundaries and global biogeography of the *Alexandrium tamarense* complex (Dinophyceae). *J. Phycol.*, 43, 1329–38.

Lin, S., Zhang, H., Hou, Y., Zhuang, Y. & Miranda, L. (2009). High-Level Diversity of Dinoflagellates in the natural environment, revealed by assessment of mitochondrial cox1 and cob genes for dinoflagellate DNA barcoding. *Appl. Environ. Microbiol., 5, 1279-*1290.

Litaker, R. W., Vandersea, M. W., Faust, M. A., Kibler, S. R., Nau, A. W., Holland, W. C., Chinain, M., Holmes, M. J. & Tester, P. A. (2010). Global Distribution of ciguatera causing dinoflagellates in the genus *Gambierdiscus*. *Toxicon* 56, 711-730.

Litaker, R. W., Vandersea, M. W., Kibler, S. R., Reece, K. S., Stokes, N. A., Lutzoni, F. M., Yonish, B. A., West, M. A., Black, M. N. D. & Tester, P. A. (2007). Recognizing dinoflagellates species using ITS rDNA sequences. *J. Phycol.*, 43, 344–355.

Lu, Y.-H., Chai, T.-J., Hwang, D.-F. (2000). Isolation of bacteria from dinoflagellate *Alexandrium minutum* and their effects on algae toxicity. *J. Nat. Toxins*, 9, 409–417.

MacKenzie, L., Beuzenberg, V., Holland, P., McNabb, P. & Selwood, A. 2004. Solid phase adsorption toxin tracking (SPATT): a new monitoring tool that simulates the biotoxin contamination of filter feeding bivalves. *Toxicon,* 44, 901-918.

Maestrini, S. Y. (1998). Bloom dynamics and ecophysiology of *Dinophysis* spp. In D. M. Anderson, A.D. Cembella, & G. M. Hallegraeff (Eds.). *Physiological Ecology of Harmful Algal Blooms. NATO ASI series, Series G, Ecological Sciences 41* (pp. 243-266). Berlin Heidelberg: Springer-Verlag.

Mafra, L. L., Bricelj, V. M., Ouellette, C. & Bates, S. S. (2010). Feeding mechanics as the basis for differential uptake of the neurotoxin domoic acid by oysters, *Crassostrea virginica*, and mussels, *Mytilus edulis*. *Aquat. Toxicol.*, 97, 160–171.

Martins, C. A. (2007). Functional genomics of a non-toxic *Alexandrium lusitanicum* culture. PhD Thesis. Massachusetts Institute of Technology, Dept. of Biology, and the Woods Hole Oceanographic Institution, 163 pp.

Martins, C. A., Kulis, D., Franca, S. & Anderson, D. M. (2004). The loss of PSP toxin production in a formerly toxic *Alexandrium lusitanicum* clone. *Toxicon* 43, 195-205.

Masseret, E., Grzebyk, D., Nagai, S., Genovesi, B., Lasserre, B., Laabir, M., Collos, Y., Vaquer, A. & Berrebi, P. (2009). Unexpected genetic diversity among and within populations of the toxic dinoflagellate *Alexandrium catenella* as revealed by nuclear microsatellite markers. *Appl. Environ. Microbiol.*, 7, 2037–2045.

McCauley, L. A. R., Erdner, D. L., Nagai, S., Richlen, M. L. & Anderson, D. M. (2009). Biogeographic analysis of the globally distributed harmful algal bloom species *Alexandrium minutum* (Dinophyceae) based on rRNA gene sequences and microsatellite markers. *J. Phycol.* 45, 454–463.

McManus, G. B. & Katz L. A. (2009). Molecular and morphological methods for identifying plankton: what makes a successful marriage? *J.Plankton Res.*, 31, 1119-1129.

Metzker, M.L. (2010). Sequencing technologies—the next generation. *Nat. Rev. Genet.*, 11, 31-46.

Méndez, S. & Ferrari, G. (2002). Floraciones algales nocivas en Uruguay: antecedents, proyectos en curso y revisión de resultados. In E. A. Sar, M. E. Ferrario & B. Reguera (2002), *Floraciones algales nocivas en el Cono Sur Americano* (pp. 269-288). Madrid: Instituto Español de Oceanografía.

Mihali, T. K., Kellmann, R., Muenchhoff, J., Barrow, K. D. & Neilan, B. A. (2008). Characterization of the gene cluster responsible for *Cylindrospermopsin* biosynthesis. *Appl. Environ. Microbiol.*, 74, 716–722.

Miles, C. O., Wilkins, A. L., Munday, R., Dines, M., Hawkes, A. D., Briggs, L. R., Sandvik, M., Jensen, D. J., Cooney, J. M., Holland, P.T., Quilliam, M. A., Mackenzie, L., Beutzenberg, V. & Towers, N. R. (2004). Isolation of pectenotoxin-2 from *Dinophysis acuta* and its conversion to pectenotoxin-2 seco acid, and preliminary assessment of their acute toxicities. *Toxicon,* 43, 1–9.

Moestrup, Ø., Akselman, R., Cronberg, G., Elbraechter, M., Fraga, S., Halim, Y., Hansen, G., Hoppenrath, M., Larsen, J., Lundholm, N., Nguyen, L. N., & Zingone, A. (Eds) (2009 onwards). IOC-UNESCO Taxonomic Reference List of Harmful Micro Algae. http://www.marinespecies.org/HAB.

Montresor M., Sgrosso S., Procaccini G. & Kooistra W. H. C. F. (2003). Intraspecific diversity in *Scrippsiella trochoidea* (Dinophyceae): evidence for cryptic species. *Phycologia*, 42: 56-70.

Moroño, A., Franco, J. M., Miranda, M., Reyero, M. I. & Blanco, J. (2001). The effect of mussel size, temperature, seston volume, food quality and volume-specific toxin concentration on the uptake rate of PSP toxins by mussels (*Mytilus galloprovincialis* LmK). *J. Exp. Mar. Biol. Ecol.,* 257, 117–132.

Nagai, S., Lian, C., Yamaguchi, S., Hamaguchi, M., Matsuyama, Y., Itakura, S., Shimada, H., Kaga, S., Yamauchi, H., Sonda, Y., Nishikawa, T., Kim, C.-H. & Hogetsu, T. (2007). Microsatellite markers reveal population genetic structure of the toxic dinoflagellates *Alexandrium tamarense* (Dinophyceae) in Japanese coastal waters. *J. Phycol.*, 43, 43–54.

Nakanishi, K. (1985). The chemistry of brevetoxins - a review. *Toxicon*, 23: 473–479.

Ogata, T., Kodama, M. & Ishimaru, T. (1987). Toxin production in the dinoflagellate *Protogonyaulax tamarensis. Toxicon* 25, 923–928.

Ogata, T., Koike, K., Nomura, S. & Kodama, M. (1996). Utilization of organic substances for growth and toxin production by *Alexandrium tamarense*. In T. Yasumoto, Y. Oshima, Y., & Y. Fukuyo (Eds.), *Harmful and Toxic Algal Blooms* (pp. 343-346). Sendai, Japan: Intergovernmental Oceanographic Commission of UNESCO.

Ono, C., Yoshimatsu, S. & Matsuoka, S. (1996). Monitoring system of harmful and toxic phytoplankton in Kagawa prefecture, Japan. In T. Yasumoto, Y. Oshima, Y., & Y. Fukuyo (Eds.), *Harmful and Toxic Algal Blooms* (pp. 23-26). Sendai, Japan: Intergovernmental Oceanographic Commission of UNESCO.

Orlova, T.Y., Selina, M.S., Lilly, E.L., Kulis, D.M. & Anderson, D.M. (2007). Morphogenetic and toxin composition variability of *Alexandrium tamarense* (Dinophyceae) from the east coast of Russia. *Phycologia*, 46, 534-548.

Oshima, T. 1995. Post-column derivatization HPLC methods for paralytic shellfish toxins. In: G. M. Hallegraeff, D. M. Anderson, & A. D. Cembella (Eds.), *Manual on Harmful Marine Microalgae. IOC Manual and Guides 33* (pp. 81-94). Paris, France: Intergovernmental Oceanographic Commission of UNESCO.

Oshima, Y., Bolch, C. J. S. & Hallegraeff, G. M. (1992). Toxin profiles of resting cysts of *Alexandrium tamarense* (Dinophyceae). *Toxicon* 30: 1539-1544.

Owen, G. (1955). Observations on the stomach and digestive diverticula of the Lamellibranchia: I. The Anisomyaria and Eulamellibranchia. *Q. J. Microsc. Sci.*, 96, 516-537.

Palma, A. S.,Vilarinho, M. G. & Moita, M. T.(1998). Interannual trends in the long-shore distribution of *Dinophysis* off the Portuguese coast. In B. Reguera, J. Blanco, M. L. Fernández, & T. Wyatt (Eds.), *Harmful Algae* (pp. 124-127). Santiago de Compostela,

Spain: Xunta de Galicia and Intergovernmental Oceanographic Commission of UNESCO.

Pan, Y., Bates, S. S. & Cembella, A. D. (1998). Environmental stress and domoic acid production by *Pseudo-nitzschia*: a physiological perspective. *Nat Toxins*, 6,127–136.

Pan, Y., Cembella, A. D. & Quilliam, M. A. (1999). Cell cycle and toxin production in the benthic dinoflagellate *Prorocentrum lima. Mar. Biol.*,134, 541-549.

Park, T.G, Kim, C. H. & Oshima, Y. 2004. Paralytic shellfish toxin profiles of different geographic populations of *Gymnodinium catenatum* (Dinophyceae) in Korean coastal waters. *Phycol. Res.*, 52, 300–305.

Pitcher, G.C., Franco, J. M., Doucette, G. J., Powell, C. L. & Mouton, A. (2001). Paralytic shellfish poisoning in the abalone *Haliotis midae* on the West Coast of South Africa. *J. Shellfish Res.*, 20, 895-904.

Pizarro, G., Escalera, L., González-Gil, S., Franco, J. M. & Reguera, B. (2008). Growth, behaviour and cell toxin quota of *Dinophysis acuta* Ehrenberg during a daily cycle. *Mar. Ecol. Prog. Ser.*, 353, 89-105.

Pizarro, G., Paz, B., González-Gil, S., Franco, J. M. & Reguera, B. (2009). Seasonal variability of lipophilic toxins during a *Dinophysis acuta* bloom in Western Iberia: Differences between picked cells and plankton concentrates *Harmful Algae*, 8, 926-937.

Quilliam, M. A., Lewis, N. I., Aasen, J. & Hardstaaf, W. (2008). Analysis of phycotoxins in hand-picked plankton cells by micro-column liquid chromatography-tandem mass spectrometry. In Ø. Moestrup (chief Ed.), *Proceedings of the 12th International Conference on Harmful Algae* (pp.377-380). Copenhagen: International Society for the Study of Harmful Algae and Intergovernmental Oceanographic Commission of UNESCO.

Raho, N., Pizarro, G., Escalera, L., Reguera, B. & Marín, I. (2008). Morphology, toxin composition and molecular analysis of *Dinophysis ovum* Schütt, a dinoflagellate of the "*Dinophysis acuminata* complex". *Harmful Algae*, 7: 839-848.

Rausch de Traubenberg, C. & Morlaix, M. 1995. Evidence of okadaic acid release into extracelular medium in cultures of *Prorocentrum lima* (Erhenberg) Dodge In P. Lassus, G.Arzul, E. Erard-Le Denn, P. Gentien, & C. Marcaillou-LeBaut (Eds.), *Harmful Marine Algal Blooms* (pp. 493-498). Paris, Lavoisier Intercept Ltd.

Reguera, B., Bravo, I. & Fraga, S. 1995. Autoecology and some life history stages of *Dinophysis acuta* Ehrenberg. *J. Plankton Res.*, 17, 999-1015.

Reguera, B., Mariño, J., Campos, M. J., Bravo, I., Fraga, S.& Carbonell, A.(1993).Trends in the occurrence of *Dinophysis* spp. in Galician waters. In T. J. Smayda, & Y. Shimizu.(Eds.), *Toxic Phytoplankton Blooms in the Sea* (pp. 559-564), Amsterdam, Holland: Elsevier.

Reguera, B. & Pizarro, G. (2008). Planktonic dinoflagellates that contain polyether toxins of the old "DSP Complex". In L. M. Botana (Ed.), *Seafood and freshwater toxins: pharmacology, physiology, and detection* (2nd edition; 257-284). Boca Raton, London, New York: CRC Press, Taylor and Francis Group.

Richlen, M.L., Morton, S.L., Barber, P.H. & Lobel, P.S. (2008). Phylogeography, morphological variation and taxonomy of the toxic dinoflagellate *Gambierdiscus toxicus* (Dinophyceae). *Harmful Algae*, 7: 614-629.

Riisgard, H. U. & Seerup, D. F. (2003). Filtration rates in the soft clam *Mya arenaria*: effects of temperature and body size. *Sarsia* 88, 415–428.

Rodhouse, P. G. (1978). Energy transformations by the oyster *Ostrea edulis* L. in a temperate estuary. *J. Exp. Mar. Biol. Ecol.,* 34, 1–22.

Rossignoli, A. E., Fernández, D., Regueiro, J., Mariño, C. & Blanco, J. (2011). Esterification of okadaic acid in the mussel *Mytilus galloprovincialis*. *Toxicon*, 57, 712-720.

Sako, Y., Hosoi-Tanabe, S. & Uchida, A. (2004) Fluorescence in situ hybridization using rRNA-targeted probes for simple and rapid identification of the toxic dinoflagellates *Alexandrium tamarense* and *Alexandrium catenella*. *J. Phycol.,* 40: 598–605.

Sampayo, M. A., Alvito, P., Franca, S. & Sousa, I. (1990). *Dinophysis* spp. toxicity and relation to accompanying species. In E. Granéli, B. Sundström, L. Edler, & D. M. Anderson (Eds), *Toxic Marine Phytoplankton* (pp 215–220). New York, USA: Elsevier.

Sar, E. A., Ferrario, M.E. & Reguera, B. 2002. *Floraciones algales nocivas en el Cono Sur Americano*. Madrid: Instituto Español de Oceanografía.

Selander, E., Thor, P., Toth, G., Pavia, H. (2006). Copepods induce paralytic shellfish toxin production in marine dinoflagellates. *Proc. R. Soc. Lond. Ser. B Biol. Sci.*, 273, 1673-1680.

Shimizu, Y., Norte, M., Hori, A., Almoursi, G. & Kobayashi, M. (1984). Biosynthesis of saxitoxin analogues: the unexpected pathway. *J. Am. Chem. Soc.*, 106: 6433-6434.

Shimizu, Y., Gupta, S. & Prasad, A. V. K. (1990). Biosynthesis of dinoflagellate toxins. In E. Graneli, B. E., Sundtröm, L. Edler, & D. M. Anderson (Eds.), *Toxic Marine Phytoplankton. Proceedings of the 4th International Conference on Toxic Marine Phytoplankton* (pp. 62-73). New York, U.S.: Elsevier.

Shimizu, Y. & Yoshioka, M. (1981) Transformation of paralytic shellfish toxins as demonstrated in scallop homogenates. *Science* 212, 547–549.

Snyder, R.V., Gibbs, P.D.L., Palacios, A., Abiy, L., Dickey, R., Lopez, J.V. & Rein, K.S. (2003). Polyketide Synthase Genes from Marine Dinoflagellates. *Mar.Biotech.,* 5, 1–12.

Snyder, R.V., Guerrero, M., Sinigalliano, C., Winshell, J., Perez, R., Lopez, J.V. & Rein, K.S. (2005). Localization of polyketide synthase encoding genes to the toxic dinoflagellate *Karenia brevis*. *Phytochemistry*, 66:1767-1780

Steidinger, K. A. (1975). Basic factors influencing red tides. In V. R. LoCicero (Ed.), *Proceedings of the First International Conference on Toxic Dinoflagellate Blooms* (pp. 153-162). Wakefield, Massachusetts, USA: Massachusetts Science and Technology Foundation.

Stern, R. F., Horak, A., Andrew, R. L., Coffroth, M.-A. & Andersen, R. A., Küpper, F. C., Jameson, I., Hoppenrath, M., Véron, B., Kasai, F., Brand, J., James, E.R. & Keelling, P. J., (2010). Environmental Barcoding Reveals Massive Dinoflagellate Diversity in Marine Environments. *PLoS ONE 5*, e13991. doi:10.1371/journal.pone.0013991.

Strand, Ø., Solberg, P. T., Andersen, K. K. & Magnesen, T. (1993). Salinity tolerance of juvenile scallops (*Pecten maximus* L.) at low temperature. *Aquaculture*, 115, 169–179.

Suzuki, T., MacKenzie, L., Stirling, D. & Adamson, J. (2001). Pectenotoxin-2 seco acid: a toxin converted from pectenotoxin-2 by the New Zealand Greenshell mussels, *Perna canaliculus. Toxicon,* 39, 507-514.

Suzuki, T. & Mitsuya, T. (2001). Comparison of dinophysistoxin-1 and esterified dinophysistoxin-1 (dinophysistoxin-3) contents in the scallop *Patinopecten yessoensis* and the mussel *Mytilus galloprovincialis*. *Toxicon*, 39, 905-908.

Suzuki, T., Mitsuya,T., Matsubara, H. & Yamasaki, M. (1998). Determination of pectenotoxin-2 after solid-phase extraction from seawater and from the dinoflagellate *Dinophysis fortii* by liquid chromatography with electrospray mass spectrometry and ultraviolet detection. Evidence of oxidation of pectenotoxin-2 to pectenotoxin-6 in scallop. *J. Chromatog. A*, 815, 155-160.

Tango, P., W. Butler, R. Lacouture, D. Gorshorn, R. Marnien, B. Michael, S. Hall, K. Browhawn, R. Wittman & W. Beatty. 2004. An unprecedented bloom of *Dinophysis acuminata* in Chesapeake Bay. In: K. A. Steidinger, J. H. Landsberg, C. R. Tomas, & G. A. Vargo (Eds.), *Harmful Algae 2002. Proceedings of the 10th International Conference on Harmful Algae* (pp. 358-360). Paris, France: Florida Fish and Wildlife Conservation Commission, Florida Institute of Oceanography and Intergovernmental Oceanographic Commission of UNESCO.

Taroncher-Oldenburg, G., Kulis, D. M. & Anderson, D. M. (1997). Toxin variability during the cell cycle of the dinoflagellate *Alexandrium fundyense. Limnol. Oceanogr.,* 42,1178-88.

Taylor, F.J.R. & Trainer, V.L. (2002). Harmful Algal Blooms in the PICES Region of the North Pacific. *PICES Sci. Rep.,* 23, 152 pp.

Tillmann, U., Elbrächter, M., Krock, B., John. U. & Cembella, A. (2009). *Azadinium spinosum* gen. et sp. nov. (Dinophyceae) identified as a primary producer of azaspiracid toxins. *Eur. J. Phycol.* 44, 63-79.

Tong, M., Kulis, D. M., Fux, E., Smith, J. L., Hess, P., Zhou, Q. & Anderson, D. M. (in press). The effects of growth phase and light intensity on toxin production by *Dinophysis acuminata* from the northeastern United States. *Harmful Algae,* doi:10.1016/ j.hal.2010.10.005

Touzet, N., Davidson, K., Pete, R., Flanagan, K., McCoy, G.,R., Amzil, Z., Maher, M., Chapelle, A. & Raine, R. (2010). Co-occurrence of the West European (Gr. III) and North American (Gr. I) ribotypes of *Alexandrium tamarense* (Dinophyceae) in Shetland, Scotland. *Protist*, 161: 370-384.

Trainer, V. L., Adams, N. G. & Wekell, J. C. (2001). Domoic acid-producing *Pseudo-nitzschia* species off the U.S. west coast associated with toxification events In G. M. Hallegraeff, S. I. Blackburn, C. J. Bolch, & R. J. Lewis (Eds). *Harmful Algal Blooms 2000* (46-49). Paris: Intergovernmental Oceanographic Commission of UNESCO.

Trainer, V. L., Pitcher, G. C., Reguera, B., Smayda, T. J. (2010). The distribution and impacts of harmful algal bloom species in eastern boundary upwelling systems. *Prog. Oceanogr.* 85: 33-52.

Uribe, P. & Espejo, R. T. (2003). Effect of associated bacteria on the growth and toxicity of *Alexandrium catenella. Appl. Environ. Microbiol.*, 69, 659-662.

Van Egmond, H. P., Aune, T., Lassus, P., Speijers, G. J. A. & Waldock, M. (1993). Paralytic and diarrhoeic shellfish poisons: occurrence in Europe, toxicity, analysis and regulation. *J. Nat. Toxins*, 2, 41-82.

Velo-Suárez, L., González-Gil, S., Gentien, P., Lunven, M., Bechemin, C., Fernand, L., Raine, R. & Reguera, B. (2008). Thin layers of *Pseudo-nitzschia* spp and the fate of *Dinophysis acuminata* during an upwelling-downwelling cycle in a Galician Ría. *Limnol. Oceanogr.,* 53, 1816-1834.

Wang, D.-Z., Zhang, S.-G., Gu, H.-F., Chan, L. L. & Hong, H.-S. (2006). Paralytic shellfish toxin profiles and toxin variability of the genus *Alexandrium* (Dinophyceae) isolated from the Southeast China Sea. *Toxicon*, 48: 138-151.

Ward, J. E. & Shumway, S. E. (2004). Separating the grain from the chaff: particle selection in suspension- and deposit-feeding bivalves. *J. Exp. Mar. Biol. Ecol.,* 300, 83–130.

Wiese, M., D'Agostino, P. M., Mihali, T. K., Moffitt, M. C., Neilan, B. A. (2010). Neurotoxic Alkaloids: Saxitoxin and Its Analogs. *Mar. Drugs*, 8, 2185-2211.

Wright, J. L. C. & Cembella, A. D. (1998). Ecophysiology and biosythesis of polyether marine biotoxins. In D. M. Anderson, A. Cembella, & G. M. Hallegraeff (Eds.). *Physiological Ecology of Harmful Algal Blooms. NATO ASI series, Series G, Ecological Sciences 41* (pp. 427-451). Berlin, Heidelberg: Springer-Verlag.

Wyatt, T. & Jenkinson, I. (1997). Notes on *Alexandrium* population dynamics. *J. Plankton Res.*, 19, 551-575.

Yang, I., John, U., Beszteri, S., Glöckner, G., Krock, B., Goessman, A. & Cembella, A. D. (2010). Comparative gene expression in toxic versus non-toxic strains of the marine dinoflagellate *Alexandrium minutum. BMC Genomics,* 11, 248.

Zhang, H., Bhattacharya, D., Lin, S. (2005). Phylogeny of dinoflagellates based on mitochondrial cytochrome b and nuclear small subunit rDNA sequence comparisons. *J. Phycol.*, 41, 411–420.

Zhou, J. & Fritz, L. (1994). Okadaic acid antibody localzes to chloroplasts in the DSP-toxin producing dinoflagellates *Prorocentrum lima* and *Prorocentrum maculosum. Phycologia*, 33, 455-461.

In: New Trends in Marine and Freshwater Toxins
Editors: A. G. Cabado and J. M. Vieites
ISBN: 978-1-62257-787-3

Chapter 4

SOLID-PHASE ADSORPTION PASSIVE SAMPLING: REVIEW OF A MONITORING TOOL TRACKING *IN SITU* MARINE AND FRESHWATER TOXINS

Jean-Pierre Lacaze
Marine Scotland Science, Aberdeen, Scotland

ABSTRACT

Solid-phase adsorption toxin tracking (SPATT) is a relatively recent passive sampling technique founded on the observation that when low levels of toxic algae are present in the water column, significant amounts of phycotoxins are released in the water. Coupled to sensitive detection techniques such as liquid chromatography-mass spectrometry (LC-MS) and enzyme-linked immunosorbent assay (ELISA), SPATT proves to be a reliable, sensitive, time-integrated sampling tool which monitors the occurrence of toxic algal bloom events by mimicking biotoxin contamination of shellfish. Since 2004 the SPATT technique has been mainly embraced by scientists working on a wide spectrum of marine biotoxins, especially the lipophilic shellfish toxins. However, within the last couple of years, work on monitoring freshwater toxins using this novel technique has emerged and offers interesting perspectives. SPATT is still in its infancy and more development work needs to be done, in conjunction with hydrographic studies, in order to validate its ability to provide an effective early warning tool tracking the evolution of toxicity of harmful algal blooms.

1. INTRODUCTION

In the past few decades, proliferation of marine and freshwater harmful algal blooms (HABs) has been observed on a global scale as well as increases in frequency, severity and duration (Hallegraeff et al. 2003). Although the parameters responsible for the HABs are still to be fully understood, a rise in occurrence has been linked to a general alteration of the water quality of the coastal zone and freshwater environment triggered by human activities and climate change (Harvell et al. 1999, Paerl et al. 2008). Of the 5000 known and referenced algae, which include unicellular marine and freshwater photosynthetic organisms, less than 2% have been found to cause harm by producing toxins that can kill other organisms including marine mammals, birds and humans (Hallegraeff et al. 2003).

Throughout the world, shellfish are successfully farmed and shellfish culture represents a legitimate use of the marine environment for sustainable food production (Shumway 2003). Shellfish filter feeding on harmful algae have the potential to accumulate large concentrations of phycotoxins mainly in their digestive organs. Consumption of such contaminated shellfish would raise food safety concerns by causing a significant threat to human health. The shellfish industry can also suffer severe economic losses when confronted with the closure of aquaculture production during HAB events.

To protect consumers from locally produced or imported toxin contaminated shellfish and reduce economic impacts, strict food safety regulations (e.g. EU regulation 854/2004) have been implemented, requiring the creation of national marine biotoxin monitoring programs. These programs involve, generally on a weekly basis, the routine monitoring of shellfish ensuring that when toxins are detected, they comply with the limits prescribed. Similarly, identification of planktonic species present in samples representative of the water column is carried out in order to provide information on the presence of toxin producing microalgae.

The results of the phytoplankton monitoring are used to provide an early warning of the presence of biotoxin producing species in shellfish harvesting areas. Although phytoplankton monitoring has demonstrated its effectiveness as an early warning method, it is labour intensive and skills demanding, and only provides a brief snapshot of the composition of the phytoplankton at one time and place. Furthermore, the use of light microscopy to identify phytoplankton does not always permit scientists to distinguish between those species that are toxin producers and those that are harmless, thus leading to unnecessary phytoplankton alerts.

Different studies have reported that significant amounts of toxins were found to be naturally released in seawater during harmful algal bloom events and in culture media during growth experiments of toxin producing microalgae (Paz et al. 2004).

Based on these observations, scientists in New Zealand (MacKenzie et al. 2004) were the first to understand the potential behind the ability to detect the presence of dissolved toxins in seawater during the lifespan of an HAB event. Mimicking the bioaccumulation of toxins by shellfish, they created a technique based on the passive adsorption of toxins onto a synthetic resin. They gave this technique the acronym of SPATT (Solid Phase Adsorption Toxin Tracking). By coupling this technique to sensitive analytical technologies such as liquid chromatography-mass spectrometry (LC-MS) and enzyme-linked immunosorbent assay (ELISA) they added a potentially new HAB monitoring tool to the existing shellfish testing and phytoplankton monitoring.

This chapter reviews the development of the SPATT technique since 2004 which has so far been mainly embraced by scientists working on a wide spectrum of marine biotoxins. However, within the last couple of years, work on monitoring freshwater toxins using this novel technique has emerged and offers interesting perspectives.

1.1. Passive Sampling Technique and Genesis of SPATT

The SPATT technique is based on the free flow of analyte molecules from the sample matrix to the receiving phase. The different concentration, pressure, temperature and electromotive force gradients of the analytes between the two media results in the capture and enrichment of the analytes into the receiving phase (Gorecki et al. 2002, Zabiegala et al. 2010). This concept was developed in the 1990s with a view to monitor the presence of organic pollutants in different environments (water, soil and air). Of all the passive samplers, semi-permeable membrane devices (SPMDs) have proven to be highly effective in their capacity to accumulate lipophilic organic contaminants from the water column (Huckins et al. 1990, Huckins et al. 1993). SPMDs consists of a low density polyethylene (LDPE) semi-permeable membrane (Figure 1A) containing triolein (1 mL), a high molecular weight non-polar lipid found in aquatic organisms that passively adsorb hydrophobic molecules. These molecules can be concentrated in the SPMD to levels comparable to or even greater than the values of their octanol/water partition coefficients or bioconcentration factors in aquatic organisms (Namiesnik et al. 2005). In effect, the LDPE material mimics a biological membrane in its ability to allow selective diffusion of hydrophobic contaminants. Different types of adsorption materials are used in SPATT depending on the phycotoxin group targeted, but on a general basis, the resins used are less selective than the triolein used in SPMDs. As a consequence, a wider range of compounds with varying degrees of polarity and hydrophibicity have the potential to be adsorbed by the SPATT sampler (MacKenzie et al. 2004). Furthermore, the mesh material generally used in SPATT does not constitute a rate limiting barrier for the transport of solutes to the adsorbent material, whereas the semi-permeable membrane for SPMDs represents a limiting transport barrier.

Among the other passive sampling methods available, the Polar Organic Chemical Integrative Sampler (POCIS) has been designed to sequester and concentrate polar organic contaminants in freshwater. It consists of a microporous polyethersulfone membrane enveloping various solid phase sorbents and/or mixtures of sorbents (Figure 1B). Its versatility allows for the sequestering medium and membranes to be tailored to specific applications. The POCIS device is calibrated with pure compounds to enable estimation of the time weighted averaged (TWA) of organic pollutants in the water column. Oasis® HLB resin has been one of the main adsorbents used in POCIS and several studies have shown its efficiency for measuring quantitatively pharmaceutical and personal care product residues in aquatic environments (Comerton et al. 2009, Li et al. 2010). So far, no studies have reported the use of POCIS to track algal biotoxins.

1.2. SPATT Sampler Designs: Bags and Discs

The first SPATT bags (MacKenzie et al. 2004) were made from 95 μm polyester mesh (Polymon PES 95/27) sewn with polyester thread and had dimensions of about 60 x 60 mm (Figure 2A). Into one seam a length of polyester string was sewn to fix the bags to the submerged mooring frame (Figure 2D).

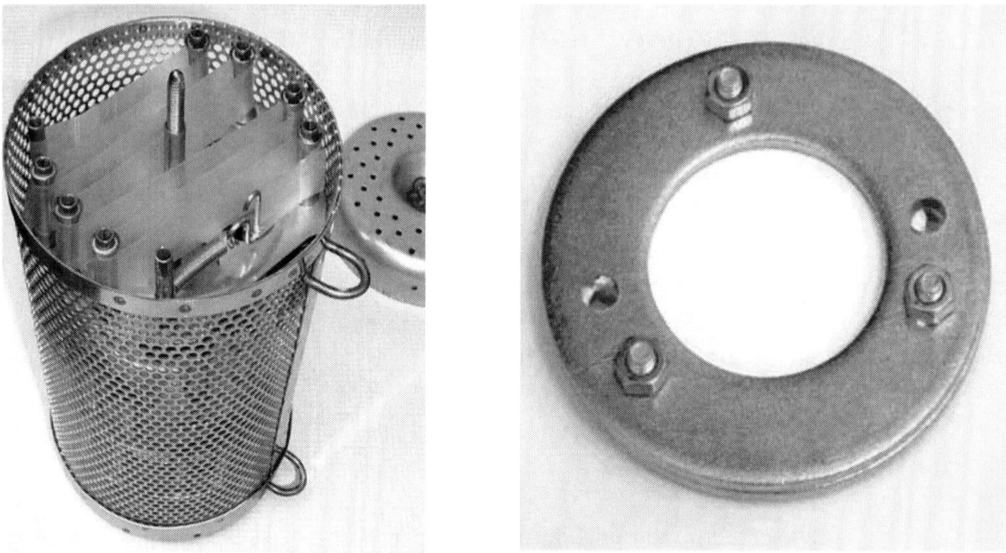

Figure 1. SPMD in its carrier (A) and POCIS (B).

During spring 2007 as part of EU project SPIES-DETOX, two different designs of SPATT bags were assessed during a field trial on the west coast of Scotland (SPIES-DETOX 2007). Both styles of bags partially filled with SP700 resin beads (ca. 12g) were two thirds closed with a non-releasable cable tie. A Midi clip (DTP Supplies, UK) was used to seal the top which served at the same time as an anchor point. The first bag was made of a folded 95 μm polyester mesh rectangle (175 x 90 mm, Bissett and Ross, UK) with a curved bottom which was sewn following a "J" shape around the opened edges (Figure 2C). The second bag made of a folded 100 μm nylon mesh rectangle (185 x 60 mm, Carbis Filtration Ltd. USA) was sewn along the length of each seam following an "L" shape using a piece of binding material (Figure 2B). After deployment for one week, neither type of bag was hampered by biofouling organisms and similar quantities of toxins were extracted from the resin they contained. The cost of the polyester mesh used for assembling a SPATT bag was ca. €12, while preparation of a SPATT bag using nylon mesh cost ca. €2.50. As performance of both containment materials appeared to be suitable for field deployment, the less expensive nylon mesh material was selected and used to prepare all the SPATT bags for the project.

Another team in Norway (Rundberget et al. 2009) decided to improve the practicability of the first described SPATT bags (MacKenzie et al. 2004). Their passive samplers were made from 100 μm nylon mesh (Sefar AG, Heiden, Switzerland) folded in half. The resin (HP20, 3 g) was placed between the two layers of nylon mesh and a 75 mm plastic embroidery circular frame was used to clamp tightly the passive sampler together, ensuring the formation of a thin layer of resin between the mesh material. A fishing swivel (No. 2, Mustad, Gjøvik, Norway) was attached to the outer ring of the embroidery frame to provide an attachment point during deployment. The authors described their design as "simple, cheap, more easily assembled and dissembled than the sewn SPATT bags" which would be "well suited to high throughput processing of samples". The frames and mesh material could be washed and reused and the frames held the resin in a reproducible thin layer increasing its exposure to the phycotoxins in the water column.

For the detection of cyanobacteria produced anatoxins (see section 1.4) in the Waipoua River, Lower North Island, New Zealand (Wood et al. 2007), a robust method of supporting small bags containing SPATT materials in river flows was developed. SPATT bags were made from 5 μm polyester mesh with one gram of adsorbent. The bags were attached with a hose clip to a holding tube made from aluminium alloy (Figure 2E).

A description of different SPATT designs and adsorbents used since 2004 to monitor the presence of phycotoxins in marine and freshwater environments has been gathered in Table 1.

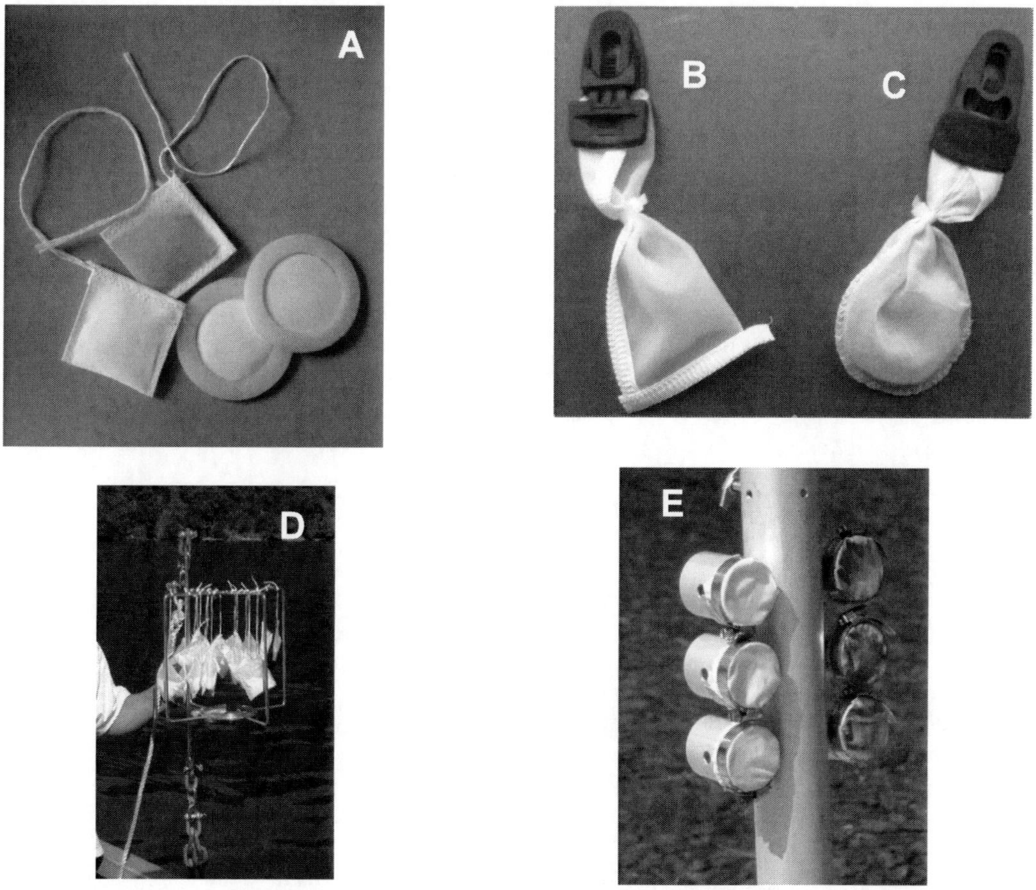

Figure 2. Different SPATT designs used in New Zealand (MacKenzie) (A), 2 different designs of SPATT bags tested in Scotland (SPIES-DETOX) (B, C), SPATT bags on a frame before deployment in New Zealand (MacKenzie) (D), SPATT bags, holding and mounting tubes. Strata-X was contained in the SPATT bags on the left while the three on the right contained Powdered Activated Carbon (Wood) (E).

Table 1. Review of different SPATT designs and adsorbents used since 2004 to monitor the presence of phycotoxins in marine and freshwater environments

SPATT material	Dimen-sions	Adsor-bents	Conditioning/ Washing	Weight (g)	Toxins adsorbed	Country	References
95 µm polyester mesh (Polymon PES 95/27) sewn bag	60 mm x 60 mm	HP20 SP207 HP2MG	Methanol Water	3	OA, DTX1, PTX2, YTX	New Zealand	MacKenzie 2004
100 µm nylon mesh (Carbis Filtration Ltd. USA)	185 mm x 60 mm	SP700	N.A.	12	OA, DTX1, DTX2, PTX2, YTX, AZA1, DA	Scotland	Spies-Detox 2007
95 µm polyester mesh (Polymon PES 95/27) sewn bag	60 mm x 60 mm	HP20	Methanol/ Water	3	OA, PTX2, GD	Australia	Takahashi 2007
95 µm nylon mesh (John Staniar & Co., UK) sewn bag (1) or disc (2)	Sachet dimensions unknown Disc diameter: 88 mm	SP850 SP825L XAD4 L-493 HP20	Methanol/ Water	3	OA, DTX1, DTX2, AZA1, AZA2, AZA3, PTX2, SPX-13-desMeC	Ireland	Fux 2008 Fux 2009
77µm plankton mesh cylindrical PVC frame	95 mm length 25 mm diameter	HP20	Methanol/ Water	3	OA, DTX2, PTX2, PTX2sa	Spain	Pizarro 2008
100 µm nylon mesh (Sefar AG, Switzlerland)	75 mm diameter	HP20	Methanol/ Water	3	OA/DTXs, PTX2, PT2sa, AZA1, AZA2,20-MeSPX-G	Norway	Rundberget 2009
100 µm Nitex bolting cloth (Wildlife Supply Company, ref. 24-C34) heated sealed	55 mm x 55 mm	HP20 SP700 SP207 SP207SS	Methanol/ Water	3	DA, PSP toxins	USA	Lane 2010
5 µm polyester mesh bag heated sealed	40 mm x 60 mm	PAC G-60 Strata-X AG 50W-X4 Amberlite IRP-64	Methanol/ Water	1	ATX, HTX	New Zealand	Wood 2007
100 µm Nitex bolting cloth with plastic embroidery hoops	unknown	HP20	Methanol/ Water	3	MCY-LR	USA	Miller 2010

Notes: Bold text indicates the best resin identified in the study.

1.3. SPATT AND ADSORPTION OF HYDROPHILIC ALGAL TOXINS

To date, the vast majority of published studies have reported the application of SPATT in laboratory and field trials for the detection of various lipophilic toxins in water. However, less work has been done on the detection of the more polar biotoxin groups consisting of the saxitoxin (STX) and related hydrophilic paralytic shellfish toxins (PSTs) and the neurotoxin domoic acid (DA), although it is known that PST-producing *Alexandrium* spp. (Lefebvre et al. 2008) and DA-producing diatoms from the genus *Pseudo-nitzschia* (Bates et al. 1998) release soluble toxins into the water during blooms.

The removal of a variety of PSTs from water using three different activated carbons (ActCs) from Norit Americas Inc. was investigated during laboratory scale experiments (Robertson et al. 2006). Powdered activated carbon HYDRODARCO® C (PAC) and granular activated carbon NORIT® GAC 1240 Plus (GAC 1) and HYDRODARCO® 3000 (GAC 2) were evaluated. The results showed the three ActCs, previously pre-treated in methanol, were able to effectively remove the main STX analogues from water, within approximately thirty minutes for GAC 1 and GAC 2. Several extraction procedures using different solvent and acid combinations were then tested to remove the PSTs from the ActCs. One ActC proved to be promising for subsequent experiments after combining toxin capture, concentration and clean-up in one step with consistent recovery of PST analogues (80-85%). Desorption of the PSTs was carried out by boiling the ActC for ten minutes in a solution of 80% ethanol with 1% hydrochloric acid at pH 2.

Likewise in the same study, water spiked with shellfish extracts contaminated by DA resulted in complete adsorption of the neurotoxin to GAC 2. However, sub-optimal recoveries using different solvent and acid combinations were obtained for DA. The use of ActC as a possible SPATT adsorbent candidate for DA was then abandoned and a variety of alternative adsorbents were further assessed. Although promising, since 2006 no further work on PSTs and DA has been reported by this group.

In 2008, the use of computationally designed polymers (CDPs) to remove two PSP toxins (STX and neoSTX) from seawater in laboratory experiments was reported (Turrell 2008). It was found that ethylene glycol metacrylate phosphate (EGMP) base polymer was able to efficiently adsorb both PSP toxins. Several extraction procedures were evaluated and a protocol enabling the recovery of 97% of neoSTX and 92% of STX from the CDP was established. In another experiment, a different CDP and other types of adsorbents (activated glass beads, resins and zeolites) were assessed for the adsorption and the recovery of DA from seawater. Amberlite® XAD761 was found to be the best adsorbent, binding nearly 100% of available DA after 96 hours exposure. However, it was not possible to efficiently recover the adsorbed DA.

Additionally, as part of the SPIES-DETOX project (data not published yet), an hydrophobic porous polymeric resin (SEPABEADS® SP700) already used in SPATT bags for the adsorption of lipophilic toxins (SPIES-DETOX, 2007) was tested for the adsorption of domoic acid. The aim was to find out if the SP700 resin could be used as a universal sorbent capable of tracking the presence of both lipophilic and hydrophilic marine biotoxins in the environment. Laboratory experiments where seawater was spiked with known amounts of DA were carried out to assess the SP700 resin ability to adsorb DA. It was found that nearly 90% of the spiked DA had been adsorbed by the resin after 7 days exposure. Desorption of the DA from the SP700 resin was investigated using 50% methanol to mimic the routine extraction of DA from shellfish (Quilliam et al. 1995). The SP700 resin (ca. 5 g) which had previously adsorbed DA was washed using deionised water (100 mL) to remove salts and was then soaked in 50% methanol (20 mL) for 5 minutes at room temperature. The methanolic fraction was collected and the procedure was repeated four times. In total, five fractions were collected which were subsequently evaporated to dryness using a TurboVap LV concentration evaporator workstation (water bath set at 45°C) before re-suspension in 50% methanol (500 μL) and analysis by LC-MS. Only 9% of the DA adsorbed by the resin was successfully recovered, mainly in the first fraction.

A bloom of *Pseudo-nitzschia* had developed in late summer 2008 at Loch Ewe, on the west coast of Scotland, where SP700 filled SPATT bags (ca. 11 g) were deployed and retrieved on a weekly basis along with mussels (*Mytilus edulis*) in an on-going program monitoring the presence of lipophilic toxins. This was a good opportunity to test the potential of SP700 resin to monitor DA in the water column during a toxic event. Each SPATT bag deployed during the September-October 2008 period was opened and the resin was aliquoted into two equal portions (ca. 5 g). The first portion was extracted for lipophilic toxins while the second one was extracted for DA. The resin was washed with deionised water (100 mL) and extraction of DA was carried out using only 20 mL of 50% methanol. In the same way as described in the initial laboratory experiment, the collected unique fraction was evaporated to dryness and re-suspended in 50% methanol before ELISA testing (Biosense, Norway). Mussels collected at the same time as the SPATT bags were also extracted using 50% methanol (Quilliam et al. 1995). Results in Figure 3 showed that low levels of DA (ca. 3-4 ng $DA.g^{-1}$ of resin) were detected in SPATT bags during the bloom of *Pseudo-nitzschia spp.* which reached a maximum of 300000 $cells.L^{-1}$. Although the laboratory trial showed only 9% of DA could be recovered from the SP700 resin, this proved to be sufficient to detect the variation of DA released by the phytoplankton during the bloom. The SPATT results were matched by low levels of DA being also detected in the mussels following almost an identical toxicity pattern.

In a very recent study (Lane et al. 2010), the use of SPATT for the detection of DA in seawater was assessed in both laboratory trials and coastal field deployments in Monterey Bay, California, USA. Four candidate resins (DIAION® HP20 and SEPABEADS® SP700, SP207 and SP207SS) were selected and their DA-adsorption and extraction efficiencies under controlled conditions were evaluated. Adsorption characteristics for the different SPATT resins were assessed after the SPATT were incubated for several days in both deionised and filtered seawater fortified with DA. Adsorption profiles followed an exponential curve for the three SEPABEADS® resins while it fitted a linear regression for the HP20 in both media. After 7 days incubation, HP20 showed the slowest adsorption rate (19%) while faster adsorption was observed for SP700 (non-activated), SP700 (activated), SP207 and SP207SS (64%, 70%, 97% and 93% respectively). Full adsorption of DA by the respective resins was achieved after 36, 22, 17, 13 and 10 days for HP20, SP700 (non-activated), SP700 (activated), SP207 and SP207SS. Extraction of DA followed an adapted unpublished protocol developed at Marine Scotland Science (Aberdeen, UK). The SPATT bags were washed with deionised water (3 x 200 mL), inserted into a 1.5 x 12 cm polypropylene chromatography column with a porous 30 μm polyethylene bed (Bio-Rad, cat No. 732-1010) before addition of 50% methanol (10 mL). The contents were vortex mixed for 1 minute and the SPATT extract was eluted on a manifold into a scintillation vial. The extraction sequence was repeated three times using 10 mL ammonium acetate (1 M) in 50% methanol. It was found that DA was much more easily recovered from the HP20 resin than from the SEPABEADS® although for HP20, some DA was found in the wash. The authors decided to use for field trials the HP20 resin and to consider the wash step as an extraction step. The 17 months monitoring period included two significant *Pseudo-nitzschia* bloom events (10^5 to 10^6 cells.L^{-1}) as well as a small bloom of *Alexandrium catenella* (ca. 400 cells.L^{-1}). DA amounts adsorbed to the SPATT resin increased significantly in March 2009. This was followed by detection of toxigenic *Pseudo-nitzschia* in late April, 6 weeks after the beginning of the increase of DA in SPATTs. Shellfish toxicity was detected in May, 7-8 weeks after the increase of DA in SPATTs. A similar toxic event with a very similar chronology of events occurred from late September through to November. PSP toxins were also detected in the HP20 filled SPATT bags as an additional quick trial and the exhibited patterns were similar to those for DA. Extraction of the PSP toxins from the HP20 resin followed the same procedure as that described for DA but the final step was the 50% methanol (10 mL) elution. Concentrations of PSP toxins in HP20 extracts analysed by ELISA reflected both the presence of *Alexandrium catenella* and the toxin levels in the mussel samples.

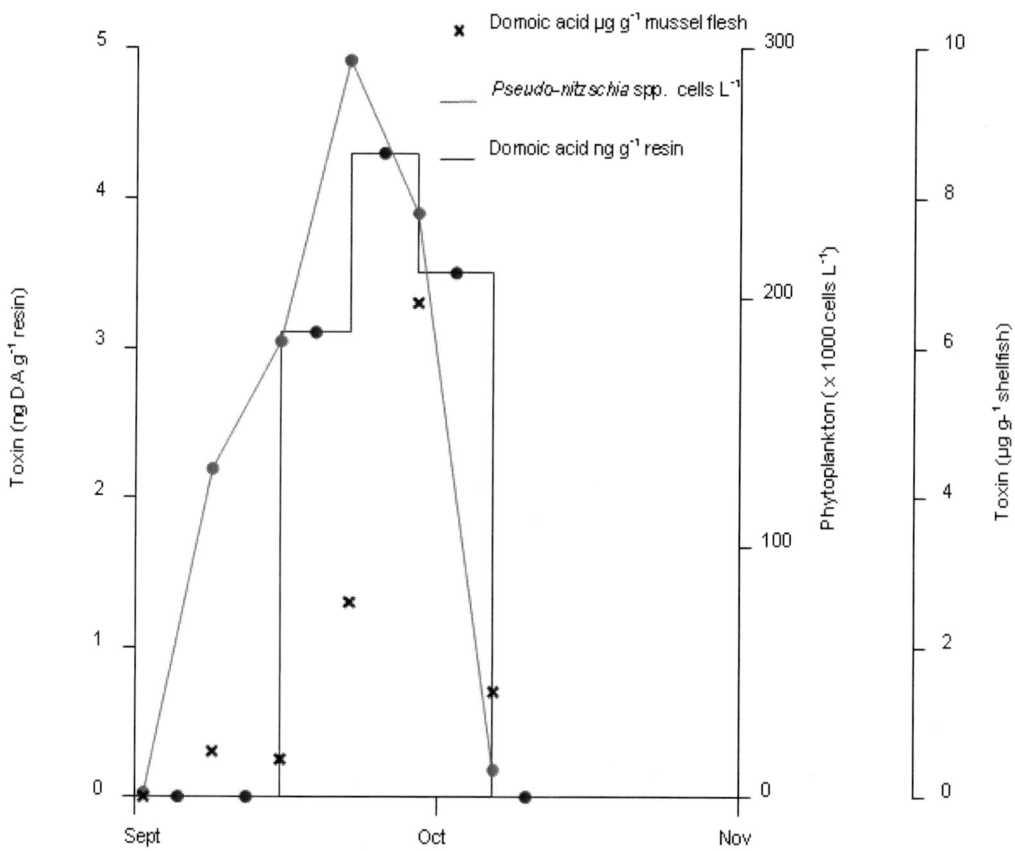

Figure 3. Evolution of the concentration of DA in SPATT and mussels in relation to the concentration of *Pseudo-nitzschia spp.* cells detected at Loch Ewe during September-October 2008.

1.4. Cyanobacteria

Large blooms of cyanobacteria producing potent and environmentally persistent biotoxins constitute an emerging global human and animal health problem (Figure 4) in freshwater and estuarine habitats (Codd et al. 2005).

Since the first report of toxin production by the cyanobacterium *Microcystis aeruginosa* in 1946 (Ashworth 1946), additional toxic species have been described, characterised by structural, functional and phylogenetic diversity and variable toxicological features (Carmichael 1992). Microcystins are the most frequently occurring and widespread of the cyanotoxins. Exposure to microcystins in food and drinking water can cause serious and sometimes fatal hepatotoxicity and possible tumour induction in humans and animals (Carmichael et al. 2001). Based on toxicity data, the World Health Organisation (WHO) recommended a tolerable daily intake value for microcystin-LR not exceeding 0.04 $\mu g.kg^{-1}$ body weight and a drinking water safety guideline of 1.0 $\mu g.L^{-1}$. Some studies have shown that tap water may contain microcystins even after passing through water treatment plants (Rapala et al. 2002).

As a consequence, monitoring these particular cyanotoxins in the aquatic environment becomes essential. Although there are no obligatory guidelines for other cyanobacteria, the presence of these compounds in water is highly undesirable and tools to monitor their presence are necessary. The majority of conventional sampling and concentration/clean-up techniques utilised to monitor the presence of cyanotoxins in water bodies use grab sampling coupled to solid phase extraction but there are some associated limitations. Grab samples do not give a complete picture of the cyanotoxins concentration and their capacity is limited (Kohoutek et al. 2010). Sample volume sizes are generally 1 to 2 L which consequently limits the detection limit. On the other hand, passive solid phase adsorption has the capacity to handle large volumes of water during periods ranging from several days to several weeks. Associated with appropriate analytical detection methods (Mountford et al. 2005), SPATT has the potential to monitor the presence of cyanotoxins in both fresh and marine waters as well as in water treatment plants.

Figure 4. Warning sign in a cyanobacteria contaminated freshwater area (New Zealand).

To date, only a couple of studies have described the testing of several candidate adsorbents for using in SPATT to track a variety of different toxins produced by cyanobacteria.

The first study looking at cyanotoxins and passive samplers (Kohoutek et al. 2008) aspired to develop and evaluate the use of passive samplers for microcystins and to compare the ability of different configurations of membrane/adsorbent to effectively accumulate these cyanotoxins. They created in-house passive samplers using four different types of membranes (polycarbonate, polyethersulfonate, polyester and nylon) and two different sorbents (Oasis HLB and Bondesil-LMS) in a design similar to the POCIS. Laboratory trials revealed the highest sampling rates (expressed as the volume of water completely cleared of cyanotoxins per day) for microcystins RR and LR were obtained with the sampler made with the polycarbonate membrane and containing Oasis HLB (2.75 mg sorbent/cm^2). This passive sampler was able to adsorb 40% of the total microcystins in the water during a 14 days experiment. The microcystins were extracted from the passive sampler following a two steps extraction procedure. In the first stage, the sorbent was extracted twice with 20 mL of methanol:water (90:10, v/v) acidified with 0.1% trifluoroacetic acid in an ultrasonic bath for 15 minutes. The supernatants were pooled and diluted tenfold with deionised water. In the second stage, the microcystins were concentrated by solid phase adsorption using a C18 cartridge. The microcystins were eluted from the cartridge using the same extraction solvent as in the first stage. The eluate was finally evaporated to dryness and reconstituted in 50% methanol:water before analysis by high performance liquid chromatography (HPLC) and diode array detection. Although the authors of the study do not state it, it is assumed that all the microcystins are entirely recovered from the Oasis HLB sorbent. This extraction procedure could possibly be shortened if it was combined with LC-MS. After extraction of the microcystins from the sorbent, the extract could be evaporated to dryness and reconstituted in a small volume of 50% methanol:water before direct analysis by LC-MS. This analytical instrumentation is highly sensitive and selective and the improved specificity of the MS delivers precision that is less matrix dependent than with UV detection (Karlsson et al. 2005).

The results of the study carried out in New-Zealand (Wood et al. 2007) showed that the SPATT technique could be used to collect and concentrate anatoxin-a (ATX) and homoanatoxin-a (HTX), two major toxins released by benthic cyanobacteria and widely distributed in rivers (Wood et al. 2007, Cadel-Six et al. 2007, Heath et al. 2010). Fifteen adsorbents were assessed in laboratory trials for their ability to bind ATX in a solid phase extraction (SPE) format type. After 24 hours exposure, between 20% and 30% of the ATX in spiked water was adsorbed onto SPATT bags filled with powdered activated carbon (PAC) G-60, Strata-X (polymeric resin), AG 50W-X4 (strong cation exchange) and Amberlite IRP-64 (weak cation exchange). Those were the adsorbents with the most interesting results. Although only a very low percentage of ATX could be recovered from those four adsorbents on average (4-7%), the authors were satisfied by the reproducibility of the adsorption/desorption process and by the very good sensitivity of the ATX detection method using LC-MS. Field trials were conducted using SPATT bags held in holding tubes and filled with Strata-X and PAC as showed in Figure 2E. The selected adsorbents were both found to be effective, and after only four hours deployment several anatoxin analogues were detected at low levels in all SPATT bags. Of the four water samples analysed after grab sampling and carried out for comparison purposes, only one contained detectable amounts of toxins which highlights the benefits of the continuous and integrated sampling SPATT technique. The PAC material seemed to be able to adsorb anatoxin and analogues for longer compared to the Strata-X material, as saturation or equilibration occurred with the latter after four hours deployment which did not seem to happen with the former material.

Another study (Miller et al. 2010) provided the first documentation of microcystin intoxication in southern sea otters from Monterey Bay in central California (USA) as a direct consequence of the land-sea flow of microcystins with trophic transfer through marine invertebrates. To confirm the presence of microcystins, SPATT bags containing 3g of HP-20 resin were deployed at the freshwater outflows of selected local rivers and along the central California coast. Prior to field deployments, the SPATT bags were evaluated in laboratory experiments using LC-MS for adsorption of microcystins. It was found that hundreds of ppb of microcystins (as often observed in contaminated freshwater) contained in a filtered freshwater sample were adsorbed by the SPATT in less than 24 hours following an exponential adsorption curve. Total recovery of the microcystins was achieved with a sequential 50% methanol extraction. The deployment of SPATT to monitor the presence of microcystins during summer and fall 2009 at different locations in Monterey Bay allowed this team of scientists to positively suggest the main source of these toxins which caused the death of the southern sea otters was of freshwater origin. Furthermore, compared to periodic grab sampling, SPATT showed much higher sensitivity for field adsorption of microcystins in both fresh and salt water which confirmed previous findings (Kohoutek et al. 2010).

1.5. Lipophilic Marine Biotoxins and SPATT Field Trials

1.5.1. New Zealand

The year 2004 (MacKenzie et al. 2004) showed the first study reporting the extensive use of SPATT in laboratory experiments and in field trials during diarrhetic shellfish poisoning (DSP) and yessotoxins (YTX) bloom events caused by the presence of two marine dinoflagellates, *Dinophysis acuminata* and *Protoceratium reticulatum* respectively. Three different adsorbents were tested (Table 1) and the DIAION® HP-20 resin was found to be the most effective for the adsorption of the lipophilic toxins (OA, PTX2, DTX1 and YTX) released during the trials toxic episodes. The SPATT bags extraction procedure started with a soaking step in MilliQ water to remove marine salts. This was followed by 2 hours soaking in methanol (80 mL) to desorb the toxins from the resin. The extract was washed and filtered through glass wool and anhydrous $NaSO_4$ with an additional portion of methanol (80 mL). The sample was then evaporated to dryness, re-suspended in methanol (15 mL), reduced again to dryness, re-suspended in 80% methanol (5 mL) and filtered (0.2 μm) before an aliquot was taken for analysis by LC-MS/MS and ELISA. Application of this extraction method to the removal of OA, DTX1 and PTX2 from the HP-20 resin led to good efficiency recoveries (80-115%).

The main findings from the field trials were that significant amounts of OA, PTX2, DTX1 and PTX2sa were recovered from the SPATT bags deployed for less than 2 days even though the concentration of *Dinophysis* cells (ca. 200 cells/L) in the water column was low. Furthermore, quantifiable amounts of YTX were recovered from the resin while the concentration in the water samples of *P. reticulatum*, the dinoflagellate known as producing YTX in these waters was less than 100 cells/L, which is below the limit of detection of the conventional monitoring method used in New Zealand. This highlights the high sensitivity of the SPATT technique due to the toxin accumulation process and shows its potential as an early warning system for upcoming HAB events.

During a toxic event where the SPATT bags were replaced daily, it was noticed that the amount of toxins (OA+DTX1, PTX2 and YTX) recovered from the resin followed a similar pattern. The concentration of toxins recovered from the resin was at its maximum a week after the cell number of toxin dinoflagellate producers in the water column had reached its peak.

During another experiment, adsorption of the toxins described above by the SPATT bags and two types of mussels (Blue mussels and Greenshell™ mussels) was compared. The results showed that the relationships varied according to the toxin type and shellfish species. During the first 8 days of the exposure period, a good linear relationship ($R^2 > 0.86$) between SPATT bags and both mussel species was observed for the uptake of OA and DTX1. For PTX2 in Greenshell™ mussels, the relationship was still linear ($R^2 \sim 0.84$) while in Blue mussels, the concentration of PTX2 stayed relatively constant despite continuous linear adsorption by the SPATT bags during this period. For YTX in both species, the relationship was an exponential increase in levels relative to the SPATT bags.

1.5.2. Scotland

Laboratory experiments were carried out in Scotland (SPIES-DETOX 2007) to assess and compare the capacity of adsorption of OA by two different resins, DIAION® HP20 and SEPABEADS® SP700. It was found that SP700 appeared to be superior to HP20 for rapid adsorption of OA from seawater and deionised water. After 1 minute exposure, 32.6% of OA in seawater was adsorbed onto the SP700 resin while only 13.5% of OA was adsorbed by HP20. After 72 hours, nearly all available OA was adsorbed regardless of the type of resin and after a week, residual OA was not detected in either seawater or deionised water and appeared to have reached a steady state. Due to the rapid adsorption of OA and the ease in which the SP700 resin beads could be manipulated in the laboratory without the need of solvent activation (as opposed to HP20), the SP700 resin beads were used in further laboratory experiments and field trials. Recovery of OA, PTX2, AZA1 and YTX from SP700 was then investigated. Using methanol as extracting solvent, results showed that OA was the most easily recovered toxin (62%) while YTX (47%), AZA1 (41%) and PTX2 (22%) appeared to bind more strongly to the resin. As a result, it was considered that 100 mL methanol was the volume needed for the optimum recovery of the lipophilic toxins (LSTs) from the SP700 resin. Other solvents such as isopropanol and chloroform were tested to improve the recovery of the LSTs but were not successful. Seawater temperature was a parameter thought to have a potential influence on the uptake kinetics of the LSTs by the SP700 resin. The adsorption of OA and PTX2 at different temperatures (6°C, 10°C and 18°C) covering a range of coastal waters temperatures across Europe was assessed. Results suggested that SP700 resin behaved in the same way and adsorption of the toxins was not temperature dependent. Additional laboratory experiments involved the study of the adsorption and recovery of spirolides (SPXs) to SP700. *Alexandrium ostenfeldii* known to produce SPXs is widely distributed in north temperate waters and SPXs have been reported in plankton and shellfish from European waters (Villar Gonzalez et al. 2006, Aasen et al. 2005). The only SPX standard available at the time, 13-desmethyl SPX C was spiked in seawater containing SP700 resin beads and very rapid adsorption (100%) observed after just one hour exposure. Using the elution procedure (100 mL methanol) previously used with the LSTs, a maximum recovery of 54% was achieved for 13-desmethyl SPX C.

Following the previous laboratory investigations, field studies were undertaken at two aquaculture sites on the west coast of Scotland (Loch Ewe and Loch Shieldaig). Optimised SPATT bags containing ca. 12 g SP700 resin were suspended at 7 m to a weighted line. On a weekly basis, the bags were retrieved and replaced by new ones. Integrated water samples for phytoplankton counts and identification were collected at the monitored location as well as mussels (*Mytilus edulis*). SPATT bags were returned to the laboratory and stored frozen at -20ºC prior to methanolic (100 mL) extraction and analysis by LC-MS. Since April 2005 at Loch Ewe, OA and PTX-2 have been detected weekly in the SPATT extracts demonstrating that these toxins were ubiquitous at this specific location even in the absence of known causative phytoplankton in the water column (Figure 5). Other LSTs frequently detected were DTX1 and DTX2 while YTX and AZA1 were less frequently recovered from the SPATT bags.

Blooms of *D. acuminata* (ca. 2500 cells.L-1) and *D. norvegica* (ca. 1500 cells.L-1) which reached a peak in June 2005 and July 2006 respectively were followed few days after by DSP peaks of toxicity in SPATT bags. Furthermore, after the *D. norvegica* bloom reached its peak at the end of July 2006, results showed that the peak of DSP toxicity in tested mussels was reached after the peak of DSP toxicity in SPATT bags which was a promising result in relation to a possible use of early warning system for SPATT. Unfortunately from a scientific point of view, the 2006 bloom of *D. norvegica* was the last "major" *Dinophysis* toxic event which took place at Loch Ewe. However, monitoring at this location is still on-going at the time of writing. It is hoped that the data collected from SPATT, mussels and phytoplankton together will help giving a better picture of the state of the environment in relation to marine biotoxins at this specific location.

1.5.3. Australia

SPATT bags identical to the ones used in New Zealand were deployed for several months from August 2004 at two sites in the waters of Southeast Queensland around North Stradbroke Island to assess the presence of dissolved toxins in the water column (Takahashi et al. 2007). A number of LSTs (OA, PTX2) and gymnodimine (GD) were recovered from the SPATT bags. PTX2 was found to be the most abundant toxin recovered from the resin during the period November 2004-March 2005 while GD was the most abundant toxin during the period August-September 2004. The SPATT-water partition coefficient (K_{SW}) for OA was determined experimentally after taking the assumption that SPATT resin reached equilibrium within the 7 days of deployment and was used to estimate the K_{SW} values for PTX2 and GD. The calculated estimated mean concentration of toxins present in water obtained from the respective K_{SW} was 0.3, 0.06 and 0.04μg.L^{-1} for PTX2, GD and OA respectively while the maximum estimated toxin concentration for the same toxins was 1.1, 0.3 and 0.2 μg.L^{-1}. This is the first field trial to report the adsorption of gymnodimine by HP20 resin filled SPATT.

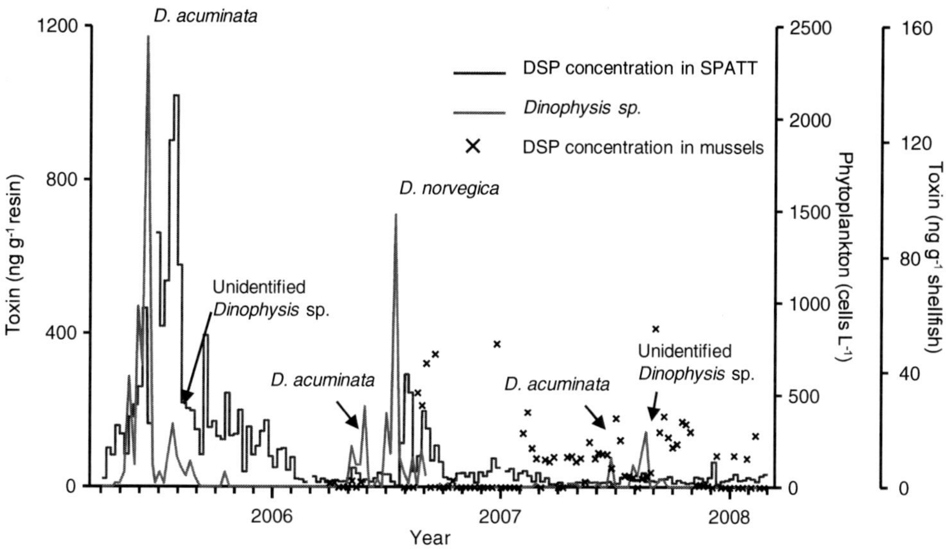

Figure 5. DSP toxins (OA+DTX1+DTX2) in SPATT relative to *Dinophysis* cell numbers and DSP toxins detected in mussels at Loch Ewe.

1.5.4. Ireland

The uptake and desorption behaviours of OA and DTX1 from *Prorocentrum lima* cultures by five styrene-divinylbenzene based polymeric resins (Sepabeads® SP850 and SP825L, Amberlite® XAD4, Dowex® Optipore® L-493 and Diaion® HP20) in SPATT bags (Table 1) were evaluated in a mesocosm study (Fux et al. 2008). All resins were able to accumulate OA and DTX1 following a 12 hours exposure to the *P. Lima* culture. SP825L and SP850 were found to be the resins that adsorbed OA and DTX1 the most rapidly up to 48 hours. HP20 was found to have a linear uptake for OA and DTX1 up to 72 hours with correlation coefficients above 0.98 for both toxins, while the other resins seemed to have reached saturation more rapidly. However after 72 hours, HP20 was the adsorbent which accumulated the largest amount of OA and DTX1 compared to all other resins. This could be explained by the larger pore size of the HP20 resin that may govern the capacity and equilibrium of toxin adsorption. Elution profiles of OA and DTX1 determined for HP20, SP825L and XAD4 resins using methanol elution (1 $mL.min^{-1}$) had a similar Gaussian type profile, whereas the desorption of OA and DTX1 from SP850 and L-493 did follow a slightly different trend. Excellent recoveries were obtained for the five resins using 23 mL methanol, ranging from 96 to 99% for both OA and DTX1. Field trials using SPATT discs filled with the five polymeric resins were carried out in August 2006 for a period of 2 weeks. The SPATT discs were replaced after exposure for one week. It was found that HP20 discs always accumulated more OA than the other resins. Additionally, AZA1 and PTX2 were recovered from all the resins but DTX2 was only recovered from 3 resins (HP20, SP825L and SP850). This study confirmed that HP20 was the resin of choice as previously reported (MacKenzie et al. 2004) to rapidly obtain lipophilic toxin profiles.

A different study (Fux et al. 2009) described the use of the SPATT approach to evaluate the presence of lipophilic toxins in three shellfish production sites on the West Coast of Ireland (Bantry Bay, Killary Harbour and McSwynes Bay) during summer 2005. The toxin profiles and concentrations of the regulated toxins belonging to the OA, AZA, PTX, YTX groups and the unregulated spirolides (SPX) group adsorbed on the deployed SPATT discs, were compared to the toxins bio-accumulated in both locally grown and transplanted mussels (*M. edulis*) which were originally free of toxins. Toxin quantitation was performed by either LC-MS/MS or ultra-performance LC-MS/MS (UPLC-MS/MS) for all matrices. This study confirmed that the SPATT discs were very sensitive detectors of lipophilic marine toxins. Similarly to the observation described in previous studies (MacKenzie et al. 2004, SPIES-DETOX 2007), OA was detected in all SPATT bags retrieved weekly from the three monitored locations even when no *Dinophysis* species were detected in the water. Transplanted mussels deployed at Bantry Bay did not accumulate OA group toxins during the entire sampling period, while *Dinophysis* spp. were not detected in water samples taken at the same location. However, relatively high concentrations of OA and DTX2 (2 $\mu g.g^{-1}$ resin) were detected in the SPATT. This suggests that contamination of mussels by DSP toxins comes solely from feeding on toxin-producing phytoplankton and that dissolved DSP toxins are not accumulated by mussels. Azaspiracid contamination occurred in Bantry and McSwynes Bay during the monitored period. AZA1 and AZA2 were mainly found in SPATT while AZA3 was only detected in low amounts in SPATT discs deployed at McSwynes Bay. On the other hand, AZA3 was the predominant AZA in the transplanted mussels. These observations led to suggest that AZA1 and AZA2 are toxins produced by the causative organism whereas AZA3 is more likely to be a mussel metabolite. This finding emphasises the potential of the SPATT technique to produce additional information leading to a better understanding of the origin of marine biotoxins in relation to shellfish and toxin-producing phytoplankton. Some of the SPATT extracts analysed using UPLC-MS contained small quantities of YTX (in the apparent absence of known YTX producing organisms) and SPX-13-desMeC which was the first time these toxins were detected in Irish waters. Unlike in the study reported from New Zealand (MacKenzie et al. 2004), data obtained from Killary harbour showed that the emergence of *D. acuta* in the water column was followed a couple of days later by the simultaneous detection of OA in both SPATT and mussels. The ability for the SPATT to forecast shellfish contamination was consequently not observed at this location. However, it is possible to suggest that by increasing the SPATT sampling frequency, an increase in the concentration of toxins in the SPATT would have been recorded before an increase of the toxin levels in the mussels.

1.5.5. Spain

During November 2005, a dense bloom of *D. acuta* was studied in the Galician Rias Baixas in North-West Spain (Pizarro et al. 2008). Duplicate SPATT bags made of 77 μm plankton mesh and containing HP20 (3 g) as described previously (MacKenzie et al. 2004) were deployed for a week at three different depths (3, 7 and 12 m) on a rope hung by a mussel raft in Ria de Pontevedra. The toxins were extracted with methanol (70 mL) after the resin was washed with deionised water (MacKenzie et al. 2004). The toxin extracts were evaporated then reconstituted in methanol (500 μL) before being filtered (0.45 μm) and analysed by LC-MS. Large quantities of OA, DTX2, PTX2 and PTX2sa (6.0 μg, 3.4 μg, 2.5 μg and 1.5 μg per resin holder respectively) were recovered from SPATT bags deployed at 3 m depth. The recovered quantities of toxins became smaller the deeper the SPATT bags were deployed. A small amount of what was presumably PTX1 (0.1 μg per resin holder) was also recovered from the SPATT at the depth of 3 m. The highest toxin values contained in the SPATT bags at 3 m compared to the lower depths was effectively confirmed by a higher presence of *D. acuta* in the first 5 m of the water column.

1.5.6. Norway

The suitability of the SPATT approach for monitoring algal toxins in Norwegian waters was investigated in 2005 at Flødevigen, on the south coast of Norway (Rundberget et al. 2009). SPATT discs containing HP20 resin were deployed weekly at 1 m depth and results of the resin extracts obtained after LC-MS analysis were compared to those from shellfish analyses and phytoplankton monitoring. During the thirteen weeks monitoring period which took place during the July-October period, three OA/DTXs events associated with increases of *D. acuta* and *D. acuminata* could be observed. In the first event around week 30, there was an increase in *Dinophysis* cell densities in the water column which was followed a few days after by an increase in OA/DTX levels in the SPATT discs. However, toxin levels in mussels did not rise owing probably to the fact that the number of *Dinophysis* cells in the water column (ca. 100-200 cells.L^{-1}) was extremely low compared to the number of other micro-algae present during this period (ca. 3 x 10^6 cells.L^{-1}). In the second event around week 33-34, levels of OA/DTXs in the SPATT discs and in the mussels increased and reached nearly simultaneously peak levels, ca. 650 ng.g^{-1} of resin and ca. 200 ng.g^{-1} mussels respectively. These events were preceded the week before by a slight increase in *Dinophysis* cell counts (ca. 100 cells.L^{-1}). An interesting point to note is that around week 32-33, the toxicity in the SPATT disc was increasing whereas the toxicity in mussels was slightly decreasing. This discrepancy could again be explained by the presence of high numbers of other flagellates in the water. The third event occurred towards the end of the field trial. During week 39, levels of OA/DTXs reached a maximum in the mussels and was preceded a week before by a peak in Dinophysis cells (ca. 140 cells.L^{-1}). The peak level of OA/DTXs in the SPATT was reached around week 37. This third event was difficult to explain. The concentration of other algae had decreased (down to ca. 10^6 cells.L^{-1}) which might explain the higher toxicity in the mussels. Overall, based on this field trial, the authors found it difficult to recommend the SPATT technique as an early warning tool. The concentration of *Dinophysis* cells was low during the monitored period and with the exception of weeks 29 and 30, levels of *Dinophysis* were below 200 cells.L^{-1} which might have accounted for the lack of correlation between Dinophysis cell counts and the levels of OA/DTXs in the SPATT discs and mussels.

During the same trial, low levels of AZA1, AZA2, AZA3 and AZA6 were detected in the mussels while in SPATT discs, only AZA1 and AZA2 were observed. This is similar to the findings in Ireland (Fux et al. 2008) and seems to validate the hypothesis that AZA3 and AZA6 are produced by shellfish metabolism after ingestion of AZA1 and AZA2. Additionally, a spirolide analogue, most probably 20-methylSPX-G, was detected at low levels (ca. 5-40 ng.g^{-1} estimated) in the SPATT discs throughout the summer. Finally, PTX2 and PTX2sa were also detected in the SPATT discs, endorsing previous observations (MacKenzie et al. 2004) that PT2sa could also be produced in *Dinophysis* after enzymatic action.

1.6. SPATT Monitoring and Food Safety

The SPATT results obtained after a wide range of laboratory experiments and field trials in Europe, the USA and in the Southern Hemisphere emphasised the great potential of this passive sampling technique allowing the provision of detailed time-averaged information on the profile of a wide range of marine and freshwater toxins. The SPATT samplers are cheap to produce and extremely convenient to use. When combined with LC-MS or ELISA analysis, this provides a highly sensitive monitoring technique which has the potential to give additional information to the traditional phycotoxins monitoring programs.

The first application of passive sampling to marine biotoxins was described in 2004 (MacKenzie et al. 2004) and the initial field trials demonstrated the potential for this technique to be a useful warning tool. A gradual increase in the concentration of toxins in the water column was detected in deployed SPATT bags before shellfish toxicity reached its peak. However, field trials results from Norway and Ireland using the SPATT technique did not confirm the findings made in New Zealand regarding the ability to provide an early warning system. In Norway (Rundberget et al. 2009), interpretation of the results of the SPATT trial was hindered by the very high concentration of other flagellates and algae compared to the *Dinophysis* spp. cell numbers in the water column. In Ireland at Killary Harbour (Fux et al. 2009), occurrence of *D. acuta* and DTX2 appeared at the same time in the SPATT samplers and the mussels.

A very good understanding of the geography, meteorology and hydrography dynamics at the location and surrounding vicinity of the shellfish producing areas is needed to help monitor the emergence and evolution of dissolved toxins and associated toxin phytoplankton producers. As an example for the SPATT survey carried out in Ireland, it would have been interesting to put some SPATT discs upstream of the monitored Killary Middle location, at the mouth of the Killary fjord (Killary Outer) 5 kms away. Marine biotoxin contaminations of shellfish on the west coast of Ireland have been linked to the mass transport of toxic cells into these areas through coastal currents (Raine et al. 1998). SPATT samplers located at the mouth of the Killary fjord would have accumulated toxins before the SPATT discs installed downstream of the monitored area. Such a SPATT arrangement would perhaps have had a greater chance to warn of the arrival of the *Dinophysis* toxic event.

In the New Zealand SPATT trial (MacKenzie et al. 2004), the dinoflagellate bloom was originated and developed within the surrounding area of the sampling location. In this case, the SPATT bags installed at the same location as the monitored mussels were able to provide an early warning because they managed to pick up the regular increase in the toxins present in the water column before the increased toxicity in the shellfish. Continuous monitoring using the SPATT technique is of paramount importance since it is the increase in toxin concentrations above their respective background levels that signify the onset of blooms. To improve the capacity to provide an early warning of shellfish contamination, SPATT sampling frequency would need to be increased, from a weekly basis down to every 2-3 days.

As a consequence, toxins would have less time to accumulate in the SPATT adsorbents which could become an issue if the techniques used to quantify the toxins are not sensitive enough. However, most of the laboratories involved with the analysis of SPATT extracts have been using LC-MS. This is a technique of choice for toxin quantification in shellfish because it provides high selectivity and sensitivity towards one or several classes of toxins. ELISA is also a specific and sensitive technique which possesses the advantage in giving quick quantitative results. Combined with SPATT, this could potentially monitor effectively the evolution of toxicity at shellfish production sites, giving the bivalve growers a more reactive approach to their business.

Conclusion

This review has gathered strong evidence that solid phase passive adsorption has good potential to be used as a tool to monitor the presence of marine and freshwater toxins in the water column (MacKenzie 2010). SPATT has the potential to be a welcome addition to the traditional marine biotoxins official monitoring programs currently using shellfish testing and phytoplankton identification and could help reduce labour and costs associated with traditional surveillance programs. In addition to its simplicity and low cost, the SPATT methodology acts as an artificial time and spatially integrated sampling device directly targeting phycotoxins which can be deployed in environments unfavourable to sentinel shellfish. Furthermore, the toxins trapped by the adsorbents contained in the SPATT bags do not undergo any biotransformation unlike those accumulated in shellfish. This provides a very useful account of the toxin dynamics in relation to phytoplankton blooms and has the potential to shed new light on the variations in the specific toxicity of producers. The SPATT technique is also very sensitive as revealed by the continual detection of OA and PTX2 in SPATT bags deployed on a weekly basis all year long at a monitoring site on the west coast of Scotland, even in the absence of the associated toxic phytoplankton species, thus providing the opportunity for lengthy early warning periods. Furthermore, the ability of SPATT to adsorb a wide range of marine biotoxins potentially in large quantities could be used as a cheap tool to help with the production of standards. It is recognised that the lack of toxin standards holds back the development of new methods for the detection and quantification of marine toxins. Dialysis bags filled with specific resins (Rodriguez et al. 2011) and disposed into large toxin-producing phytoplankton cultures could be a good way to accumulate a wide range of toxins which could then be easily extracted and further purified.

Shellfish flesh testing is still a necessity in marine biotoxins monitoring programs but the addition of SPATT could have the potential to reduce the need to carry out routine shellfish monitoring tests as a large percentage of samples tested are negative. SPATT extracts could be analysed using rapid screening methods (e.g. ELISA, protein phosphatase 2 assay, surface plasmon resonance biosensor assay) which would give food safety authorities the assurance that the shellfish growing areas are not contaminated. In the event of a SPATT result exceeding a defined action level, shellfish testing could be carried out to quantify the degree of contamination. However, for the SPATT technique to be recognised by the competent regulatory authorities as a food safety monitoring tool, requirements will be needed to set up a standardised passive sampler design and to agree on the nature and amount of the adsorption material in relation to the toxin groups targeted. Additionally, validated sampling units applicable to all algal toxins groups will have to be set up and this will likely involve the setting up of inter laboratory exercises as well as field trials at location where harmful algal blooms take place. This could be a long lasting process due to the irregularity of the toxic episodes. However, good collaboration between the validation participants would involve keeping a close eye on any incoming toxic events which could increase the likelihood of a successful and swift validation process.

Although the majority of studies to date have focused on the development of SPATT for the detection of the widespread marine lipophilic toxins in coastal waters, additional research is still needed to find suitable adsorption substrates for the more hydrophilic toxins such as domoic acid and PSP toxins. Likewise, an increase in the geographical occurrence of previously low-profile marine toxins (e.g. brevetoxins, palytoxins and ciguatera toxins) should drive the search for suitable adsorption materials with the ability to efficiently bind those emerging toxins. Further work could look into scaling down the SPATT extraction step by using a small portion (ca. 100 mg) of the adsorbent material extracted with a reduced volume of solvent thus providing a more economical and environmentally friendly technique.

Furthermore, the downsized SPATT extraction combined with a rapid testing method (e.g. ELISA, lateral flow immunoassay, surface Plasmon resonance biosensor assay) could prove a useful and relatively simple combination for shellfish producers who would have the potential to carry out "on the spot" SPATT testing with little resources. This interesting combination could improve harvesting time management by providing shellfish producers with a useful method helping them to get organised before the arrival of a toxic event notified by an increase in toxins captured by SPATT. An increase in the frequency of the SPATT deployments and subsequent rapid testing throughout the harmful algal bloom would provide crucial information with regards to the evolution of the bloom and would notify its termination.

Acknowledgments

The author would like to thank Jennifer Graham at Marine Scotland Science for proof reading.

References

Aasen, J; MacKinnon, SL; LeBlanc, P; Walter, JA; Hovgaard, P; Aune, T; Quilliam, MA. Detection and Identification of Spirolides in Norwegian Shellfish and Plankton. *Chem. Res. Toxicol.*, 2005 18, 509-515.

Ashworth, CT; Mason, MF. Observations on the Pathological Changes Produced by a Toxic Substance Present in Blue-Green Algae (Microcystis aeruginosa). *Am. J. Pathol.*, 1946 22, 369-380.

Bates, SS; Garrison, DL; Horner, RA. Bloom dynamics and physiology of domoic acid-producing *Pseudonitzschia* species. In: Anderson DM, Cembella AD, Hallegraeff GM editors. *Physiological Ecology of Harmful Algal Blooms.* Heidelberg: Springer-Verlag; 1998; 267-292.

Cadel-Six, S; Peyraud-Thomas, C; Brient, L; Marsac, NT; Rippka, R; Mejean, A. Different genotypes of anatoxin-producing cyanobacteria coexist in the Tarn River. *Appl. Environ. Microbiol.,* 2007 73, 7605-7614.

Carmichael, WW. Cyanobacteria secondary metabolites: the cyanotoxins. *J. Appl. Bacteriol.,* 1992 72, 445-459.

Carmichael, WW; Azevedo, S; An, JS; Molica, RJ; Jochimsen, EM; Lau, S; Rinehart, KL; Shaw, GR; Eaglesham, GK. Human fatalities from cyanobacteria: chemical and biological evidence for cyanotoxins. *Environ. Health Perspect.,* 2001 109, 663-668.

Codd, GA; Morrison, LF; Metcalf, JS. Cyanobacterial toxins: risk management for health protection. *Toxicol. Appl. Pharmacol.,* 2005 203, 264-272.

Comerton, AM; Andrews, RC; Bagley, DM. Practical overview of analytical methods for endocrine-disrupting compounds, pharmaceuticals and personal care products in water and wastewater. *Philos. Trans. R. Soc. A,* 2009 367, 3923-39.

Fux, E; Marcaillou, C; Mondeguer, F; Bire, R; Hess, P. Field and mesocosm trials on passive sampling for the study of adsorption and desorption behaviour of lipophilic toxins with a focus on OA and DTX1. *Harmful Algae,* 2008 7, 574-583.

Fux, E; Bire, R; Hess, P. Comparative accumulation and composition of lipophilic marine biotoxins in passive samplers and in mussels (*M. edulis*) on the West Coast of Ireland. *Harmful Algae,* 2009 8, 523-537.

Gorecki, T; Namiesnik, J. Passive sampling. *Trends Anal. Chem.,* 2002 21, 276-291.

Hallegraeff, GM. Harmful Algal Blooms : a global overview. In: Hallegraeff GM, Anderson DM, Cembella AD editors. *Manual on harmful marine microalgae*, vol 11, 2nd ed. Paris: IOC-UNESCO; 2003; 25-49.

Harvell, CD; Kim, K; Burkholder, JM; Colwell, RR; Epstein, PR; Grimes, DJ; Hofmann, EE; Lipp, EK; Osterhaus, ADME; Overstreet, RM; Porter, JW; Smith, GW; Vasta, GR. Emerging marine diseases-climate links and anthropogenic factors. *Science,* 1999 285, 1505.

Heath, M; Wood, SA; Ryan, KG.. Polyphasic assessment of fresh-water benthic mat-forming cyanobacteria isolated from New Zealand. *FEMS Microbiol. Ecol.,* 2010 73, 95-109.

Huckins, JN; Tubergen, MW; Manuweera, GK. Semipermeable membrane devices containing model lipid: A new approach to monitoring the availability of lipophilic contaminants and estimating their bioconcentration potential. *Chemosphere,* 1990 20, 533-552.

Huckins, JN; Manuweera, GK; Petty, JD; Mackay, D; Lebo, JA. Lipid-containing semipermeable membrane devices for monitoring organic contaminants in water. *Environ. Sci. Technol.,* 1993 27, 2489-2496.

Karlsson, KM; Kankaanpaa, H; Huttunen, M; Meriluoto, J. First observation of microcystin-LR in pelagic cyanobacterial blooms in the northern Baltic Sea. *Harmful Algae,* 2005 4, 163-166.

Kohoutek, J; Babica, P; Blaha, L; Marsalek, B. A novel approach for monitoring of cyanobacterial toxins - development and evaluation of the passive sampler for microcystins. *Anal. Bioanal. Chem.,* 2008 390, 1167-1172.

Kohoutek, J; Marsalek, B; Blaha, L. Evaluation of the novel passive sampler for cyanobacterial toxins microcystins under various conditions including field sampling. *Anal. Bioanal. Chem.,* 2010 397, 823-828.

Lane, JQ; Meiling Roddam, C; Langlois, GW; Kudela, RM. Application of Solid Phase Adsorption Toxin Tracking (SPATT) for field detection of the hydrophilic phycotoxins domoic acid and saxitoxin in coastal California. *Limnol. Oceanogr.: Methods,* 2010 8, 645-660.

Lefebvre, KA; Bill, BD; Erickson, A; Baugh, KA; O'Rourke, L; Costa, PR; Nance, S; Trainer, VL. Characterization of intracellular and extracellular saxitoxin levels in both field and cultured Alexandrium spp. samples from Sequim Bay, Washington. *Mar. Drugs,* 2008 6, 103-116.

Li, H; Helm, PA; Metcalfe, CD. Sampling in the Great Lakes for pharmaceuticals, personal care products, and endocrine-disrupting substances using the passive polar organic chemical integrative sampler. *Environ. Toxicol. Chem.,* 2010 29, 751-762.

MacKenzie, L; Beuzenberg, V; Holland, P; McNabb, P; Selwood, A. Solid phase adsorption toxin tracking: a new monitoring tool that simulates the biotoxin contamination of filter feeding bivalves. *Toxicon,* 2004 44, 901-918.

MacKenzie, L. *In situ* passive solid-phase adsorption of micro-algal biotoxins as a monitoring tool. *Curr. Opin. Biotechnol.,* 2010 21, 326-331.

Miller, MA; Kudela, RM; Mekebri, A; Crane, D; Oates, SC; Tinker MT ; Staedler M ; Miller WA ; Toy-Choutka S; Dominik C; Hardin D; Langlois G; Murray M; Ward K; Jessup DA. Evidence for a novel marine harmful algal bloom: Cyanotoxin (microcystin) transfer from land to sea otters. *PLoS ONE,* 2010 5, 1-11.

Mountford, DO; Holland, P; Sprosen, J. Method for detecting classes of microcystins by combination of protein phosphatase inhibition assay and ELISA: comparison with LC-MS. *Toxicon,* 2005 45, 199-206.

Namiesnik, J; Zabiegala, B; Kot-Wasik, A; Partyka, M; Wasik, A. Passive sampling and/or extraction techniques in environmental analysis: a review. *Anal. Bioanal. Chem.,* 2005 381, 279-301.

Paerl, HW, Huisman, J. Climate: blooms like it hot. *Science*, 2008 320, 57-58.

Paz, B; Riobo, P; Fernandez, AL; Fraga, S; Franco, JM. Production and release of yessotoxins by the dinoflagellates *Protoceratium reticulatum* and *Lingulodinium polyedrum* in culture. *Toxicon,* 2004 44, 251-258.

Pizarro, G; Escalera, L; Gonzalez-Gil, S; Franco, JM; Reguera, B. Growth, behaviour and toxin content of *Dinophysis acuta* Ehrenberg during a daily cycle. *Mar. Ecol.: Prog. Ser.,* 2008 353, 89-105.

Quilliam, MA; Xie, M; Hardstaff, WR. Rapid extraction and cleanup for liquid chromatographic determination of domoic acid in unsalted seafood. *AOAC Int.,* 1995 78, 543–554.

Raine, R; McMahon, T. Physical dynamics on the continental shelf off southwestern Ireland and their influence on coastal phytoplankton blooms. *Cont. Shelf Res.,* 1998 18, 883-914.

Rapala, J; Lahti, K; Rasanen, LA; Esala, AL; Niemela, SI; Sivonen, K. Endotoxins associated with cyanobacteria and their removal during drinking water treatment. *Water Res.,* 2002 36, 2627-2635.

Robertson, A; Reeves, KL; Capling, J; Garnett, C; Quilliam, MA. In *12th International Conference on Harmful Algae Programme and Abstracts*; ISSHA, Copenhagen, 2006, p 51.

Rodriguez, P; Alfonso, A; Turrell, E; Lacaze, J-P, Botana, LM. Study of solid phase adsorption of paralytic shellfish poisoning toxins (PSP) onto different resins. *Harmful Algae,* 2011, *in press.*

Rundberget, T; Gustad, E; Samdal, IA; Sandvik, M; Miles, CO. A convenient and cost-effective method for monitoring marine algal toxins with passive samplers. *Toxicon,* 2009 53, 543-550.

Shumway, SE; Davis, C; Downey, R; Karney, R; Kraeuter, J; Parsons, J; Rheault, R; Wikfors, G. *World Aquaculture,* 2003 34,15-17.

SPIES-DETOX; Active biological monitoring and removal of toxins in aquaculture ecosystems and shellfish - including the development of a Solid-Phase *In-Situ* Ecosystem Sampler (SPIES) and detoxification of shellfish (DETOX), FP6 Collective research Project, 12-month Periodic activity report, 2007, pp A3-A11.

Takahashi, E; Yu, Q; Eaglesham, G.; Connell, DW; McBroom, J; Costanzo, S; Shaw, G.R. Occurrence and seasonal variations of algal toxins in water, phytoplankton and shellfish from North Stradbroke Island, Queensland, Australia. *Mar. Environ. Res.,* 2007 64, 429-442.

Turrell, E. In *Report of the ICES-IOC working group on harmful algal Bloom dynamics*; ICES CN 2008/OCC:03 section 7.1.9, 2008, pp 23-24.

Villar Gonzalez, A; Rodriguez-Velasco, ML; Ben-Gigirey, B; Botana, LM. First evidence of spirolides in Spanish shellfish. *Toxicon,* 2006 48, 1068-1074.

Wood, SA; Selwood, AI; Rueckert, A; Holland, PT; Milne, JR; Smith, KF; Smits, B; Watts, LF; Cary, SC. First report of homoanatoxin-a and associated dog neurotoxicosis in New Zealand. *Toxicon,* 2007 50, 292-301.

Zabiegala, B; Kot-Wasik, A; Urbanowicz, M; Namiesnik, J. Passive sampling as a tool for obtaining reliable analytical information in environmental quality monitoring. *Anal. Bioanal. Chem.,* 2010 396, 273-296.

In: New Trends in Marine and Freshwater Toxins
Editors: A. G. Cabado and J. M. Vieites
ISBN: 978-1-62257-787-3

Chapter 5

PHYCOTOXINS BIOTRANSFORMATIONS, SHELLFISH DETOXIFICATION AND INDUSTRIAL APPLICATION

Alberto Otero, María José Chapela, Miroslava Atanassova and Ana G. Cabado
Microbiology and Biotoxins Area,
Spanish National Association of Sea Food Producers,
Technological Centre (ANFACO-CECOPESCA), Spain

ABSTRACT

In this chapter knowledge regarding phycotoxins mitigation and shellfish detoxification mechanisms will be considered from two points of view. On the one hand information will be presented regarding phycotoxins natural detoxification mechanisms: enzymatic biotransformations as detoxification, tissue distribution and changes in the shellfish toxic profile. Knowledge of these transformations among analogues may have important implications for removal of toxins from a contaminated product as well as for the detection and characterization of their toxicity. On the other hand, studies related to human detoxification procedures will be gathered, such as molluscs feeding with non-toxic dinoflagellates or with detoxifying bacteria as well as other treatments as biological or chemical control. We mainly refer to industrial detoxification as potential human detoxification mechanisms by means of hepatopancreas ablation, molluscs processing (cooking, steaming, sterilization), supercritical fluids and acetyl-cysteine treatment, among others.

1. INTRODUCTION

The threat of shellfish toxins is not only a major cause of concern for human health but it is also detrimental to the economy and tourism. Naturally occurring freshwater or marine biotoxins can be structurally modified by several biological factors. In some cases, these biotransformations result in new toxins that cannot be biosynthesized by cyanobacteria or microalgaes alone. Moreover, less toxic toxins may be converted into analogues with greater toxicity or vice-versa. Knowledge about mollusc toxin degradation physiology is very scarce. Detoxification or elimination of the toxins from shellfish is dependent on the toxin profile since some analogues eliminate faster than others and also on the bivalve. For instance, DTX-2 shows slower elimination in relation to OA and PTX-2 eliminates quicker in mussels than in cockles. Both non-esterified OA and DTX-2 seem to be eliminated slower than their respective esterified forms from mussel and *Donax* clams [Vale, 2004]. Lindegarth et al. conclude that differential rates of assimilation and/or biotransformation of the OA-group and PTX explain some of the observed differences in retention and toxin profiles between the bivalves, rather than differences in elimination rates. From an industrial point of view, these authors report that oysters may be regarded as a low-risk species for DSP contamination, which should be taken into consideration by regulatory authorities, at least in sampling and monitoring programs. In mussels and oysters the free form of OA was eliminated at a faster rate than free DTX-1 and DTX-2. PTX compounds were rapidly eliminated in both species. Mussels rapidly accumulate OA-group toxins, whereas oysters never reached this limit and the way to eliminate these toxins together with PTX is different dependent on the bivalve [Lindegarth et al., 2009]. Enzymatic activity of shellfish and some marine bacteria, both play a role in interconversion or transformation of toxins after accumulation in shellfish. Shellfish bacteria can mediate biotransformations due to their diverse metabolic capacities. A deep understanding of these biotransformations might allow for a mechanism of detoxification to be established and utilized in the water supply, shellfish farming industry or depuration plants.

1.1. Enzymatic Biotransformations, Tissue Distribution and Changes in the Shellfish Toxic Profile

1.1.1. Lipophilic Toxins

1.1.1.1. Okadaic Acid-Group Toxins, Pectenotoxins and Yessotoxins

The contamination of shellfish by diarrheic shellfish poisoning (DSP) toxins is the most important biotoxin contamination event responsible for closure of European Union coastal shellfish production areas [Artigas et al., 2007]. DSP toxins are produced by dinoflagellates of the genera *Dinophysis* and *Prorocentrum* [Murata et al., 1987; Yasumoto et al., 1978] and bivalves accumulate the toxins by algal ingestion from the water column [Morono et al., 2003]. The lipophilic polyether compounds associated with the DSP toxin complex are okadaic acid (OA), dinophysis toxins (DTXs), pectenotoxins (PTXs) and yessotoxins (YTXs) [Dominguez et al., 2010].

The possible lipophilic toxin transformations taking place in bivalve tissues upon accumulation and resulting analogues are reviewed in detail in the next few paragraphs.

Okadaic Acid-Group Toxins

It is well established that OA and DTXs, as well as other lipophilic toxins, tend to accumulate in visceral tissues of bivalve molluscs. Due to enzymatic transformations in the digestive gland (hepatopancreas), a significant portion of the total toxin content found in them consists of acylated forms of the OA and derivatives – 7-O-acyl DTX compounds (often designated as DTX-3). The degree to which toxins are transformed vary between species, and is also depending on the parent toxin involved. Esterification is believed to occur in bivalve tissue as part of a general enzymatic system of fatty acid synthesis and breakdown. The biotransformation of DTX to 7-O-acyl-DTX has been so far demonstrated to occur in the scallop *Patinopecten yessoensis* [Suzuki et al., 1999], in mussels (*Mytilus edulis* and *Mytilus galloprovincialis*), in some species of clams (*Donax trunculus*), in cockles and oysters [Torgersen et al., 2008]. This process is to a high extent species specific, for instance oysters accumulate most of the toxin as DTXs while in mussels and clams it might be about a half of the total. Recently, in a study of 7-O-acyl esters of OA in *M. edulis*, DTX-1 and DTX-2 profiles were reported to be similar in samples collected from different locations at different times and to correspond with the general lipid profiles from mussels [Torgersen et al., 2008]. The acylated derivatives show an increased liposolubility compared to their parent, unesterified compounds and possess toxic activity upon hydrolysis in the human gastrointestinal tract.

Pectenotoxin Group

In the case of pectenotoxins, the products of the metabolization by shellfish are the pectenotoxin seco acids (PTX-SA) where the characteristic lactone group from the PTX structure has been opened giving the free acid form of the toxins. The first examples isolated have been PTX-2-SA together with 7-epi-PTX-2-SA, from the mussel *Perna canaliculus* in Irish waters [Daiguji et al., 1998a; James et al., 1999]. Afterwards, PTX-SAs have been detected in mussels from Croatia [Pavela-Vrancic et al., 2001], in New Zealand [Suzuki et al., 2001a; Suzuki et al., 2001b] in mussels and in a common cockle from Portugal [Vale and Sampayo, 2002], in Norway [Miles et al., 2004] and in Japanese scallops [Goto et al., 2001]; [Suzuki et al., 2001a; Suzuki et al., 2001b]. Wilkins and colleagues have identified and reported in 2006 fatty acid esters of PTX-2-SA among which the most abundant form has been the 37-O-acyl ester of PTX-2-SA, followed by the corresponding 11-O-acyl and 33-O-acyl esters [Wilkins et al., 2006]. The above fatty acid derivatives resemble these reported for OA/DTX, brevetoxins and spirolides and are product of bivalve metabolism since no such substances have ever been observed in phytoplankton [Aasen et al., 2006].

Yessotoxin Group

A large number of yessotoxin analogues have also been found in shellfish, the first being the 45-hydroxy-YTX, detected in Japanese scallop *P. yessoensis* [Yasumoto et al., 1989]. Afterwards, 45,46,47-trinor-YTX was identified from the same species [Satake et al., 2006] followed by other analogues found in mussels (*Mytillus galloprovincialis*) from Adriatic sea like the homoYTX, 45-hydroxyhomo-YTX [Satake et al., 1996], adriatoxin [Ciminiello et al., 1998], carboxy-YTX [Ciminiello et al., 2000a], carboxyhomo-YTX [Ciminiello et al., 2000b], noroxohomo-YTX [Ciminiello et al., 2001] and noroxo-YTX [Ciminiello et al., 2002]. In *M. edulis* from Norway 1-desulfo-YTX [Daiguji et al., 1998b], 41a-homoYTX, 44, 55-dihydroxy-YTX [Mackenzie et al., 2002] and 45-hydroxycarboxy-YTX [Aasen et al., 2005] have been discovered. The most recent analogues identified in moluscs (Adriatic Sea mussels) are 1-desulfocarboxy homo-YTX and 4-desulfocarboxyhomo-YTX [Ciminiello et al., 2007]. In spite of the high number of analogues found, the most abundant one is typically 45-hydroxy-YTX, followed by carboxy-YTX [Mackenzie et al., 2002]; [Aasen et al., 2005].

Recent studies indicate that moluscs rapidly oxidize YTX to 45-OH-YTX and more slowly to carboxy-YTX. Afterwards, 45-OH-YTX is possibly metabolized to 45-OH carboxy-YTX [Aasen et al., 2005].

The lipophilic nature of OA, DTXs, PTXs and YTXs causes their common tissue distribution and similar elimination mechanisms, therefore, these toxin characteristics will be discussed together.

Several mechanisms are supposed to be involved in the compartmentalization of lipophilic toxins in bivalves [Blanco et al., 2007]. Toxins are absorbed from the water column/toxic phytoplankton most likely in the digestive gland of the bivalve and depending on their affinity for the constituents of this organ and on the existence of transport mechanisms they might be transported to other organs. If toxins reach non visceral organs they might be subjected to different chemical and biochemical environments, with different enzymatic and excretory capabilities, favoring certain types of transformations more than others. The accumulation of different forms of the original toxins in different tissues has been previously demonstrated for other groups of lipophilic toxins as azaspiracids [James et al., 2002; Magdalena et al., 2003]. A clear tissue compartmentalization, preferential in terms of concentration for the digestive gland, has been established so far in various bivalve species like the bay scallop *Argopecten irradians* [Bauder et al., 2001], the king scallop *Pecten maximus* [Hess et al., 2003] and the mussel *Mytilus edulis* [Pillet et al., 1995]. Thus, most of the total toxin body burden in both juvenile and adult *A. irradians* scallops has been confined to viscera (71 % in adults and 85 % in juveniles), with 4 % in adult gonads and only very low toxin amounts detected in gills, mantle and adductor muscle. *P. maximus* may have concentrations of OA in the gonads which are roughly 2 % of those found in the digestive gland [Hess et al., 2003]. In the mussel *M. galloprovincialis* OA-group toxins are also preferentially accumulated in the digestive gland and the lack of transport to non visceral tissues has been reported by Blanco et al. providing as possible explanation the very strong bond between the toxins and some cellular components of digestive gland cells [Blanco et al., 2007]. In *Pecten fumatus* removal of the digestive gland has been shown to reduce the toxin concentration of the soft tissues by only 22 %, indicating that this species is capable of accumulating a large proportion of OA in non visceral tissues [Madigan et al., 2006]. The pectenotoxin and yessotoxin groups have also been shown to accumulate in the digestive gland as for example in the scallop *Patinopecten yessoensis* [Suzuki et al., 2005].

Although excretion and depuration of lipophilic toxins has been extensively studied during the last two decades, their species-specific kinetics, metabolism and mechanism of depuration are not well known [Duinker et al., 2007]. Different hypothesis exist regarding the influence of various environmental and physiological factors on depuration rates. A consistent difference in depuration rates has been seen between seasons [Svensson, 2003], with faster depuration rates found in summer compared to autumn. The influence of environmental conditions such as temperature, water salinity, light transmission and fluorescence, which affect the metabolic processes rates and food availability, as well as the individual body weight have been proven to affect toxin depuration in bivalves [Blanco et al., 1999].

Depending on affinities of the different bivalve organs towards lipophilic toxins once the toxin reaches the tissue it is released either very slowly or very quickly. Thus, in *Argopecten irradians* scallops a lot higher detoxification (toxin release) rates from gills, mantle, adductor muscle and gonads than from viscera have been observed. Toxin loss from viscera followed a biphasic pattern in this species, with a rapid release during the first 3 days (77 % of the original toxin load) and then a much more gradual detoxification during the ensuing weeks. Besides the tissue effect on detoxification of different toxin forms (derivatives of the parent toxin) bring to different release/decontamination rates. *In vitro* decontamination experiments in common cockle (*Cerastoderma edule*) have demonstrated that total okadaic acid toxins (OA) and total dinophysistoxin-2 (DTX-2) are eliminated equally, while in the blue mussel (*Mytillus galloprovincialis*) total OA is eliminated faster than total DTX-2 [Vale, 2004]. Torgessen et al. studied profiles and levels of fatty acid esters of OA, DTX-1, DTX-2 and PTX2 SA in blue mussels (*Mytilus edulis*) and European flat oysters (*Ostrea edulis*), collected during a bloom of *Dinophysis* spp. and after 3 and 6 weeks of depuration. Calculations of depuration rates for all individual esters of each parent compound showed that the esters of DTX-1 depurated significantly slower from both mussels and oysters compared to esters of OA, DTX-2 and PTX-2 SA, but overall the depuration rates of esters of both toxin groups were highly similar for both species. This is an indication that differences in depuration rates are not causing the large species-specific differences in levels and profiles of these toxins. Instead, the results for the OA-group toxins suggest that a higher rate of esterification in oysters is the main factor causing the observed differences in the proportion of esters to free toxin. For PTX-2 SA, the important differences in ester profiles and higher proportions of esters of PTX-2 SA in mussels compared to oysters suggest differential assimilation and metabolic rate processes for the PTXs compared to OA-group toxins between species. Hence, although produced by the same *Dinophysis* species, conclusions about the dynamics of one toxin group based on results from the other group should be avoided [Torgersen et al., 2008].

Lindegarth et al. [Lindegarth et al., 2009] presented results from an experiment related to the comparison of the uptake and elimination of diarrhetic shellfish toxins (DST) of the okadaic acid (OA) and pectenotoxin (PTX) groups between blue mussels (*Mytilus edulis*) and European oysters (*Ostrea edulis*). Caged mussels and oysters were suspended in the water column and exposed to a dense bloom of *Dinophysis acuta* (500-2000 cells /L) for 4 weeks, which was followed by detoxification in the laboratory during 7 weeks. Weekly sampling and analysis of OA-group toxins including fatty acid esters (DTX-3) as well as PTX in individual shellfish and plankton samples were performed. The results showed that mussels rapidly accumulated OA-group toxins to levels about 10 times above the regulation limit (160 μg OA/kg) whereas oysters never reached these concentrations during the field exposure. Overall, levels were 10-50 times greater in mussels. The OA-group toxins were mainly in the form of esters (> 90 %) in oysters, whereas in mussels, the esters constituted only a minor proportion of total OA toxin levels. Reduction rates were estimated for OA toxin to evaluate if faster elimination could explain the lower toxin retention in oysters. However, no consistent species specific difference in reduction rates was observed, but esters of OA appeared to be reduced at a faster rate in oysters (elimination half life ($t_{1/2}$) of 23 days) compared with mussels ($t_{1/2}$ = 35 days). In both species, the free form of OA was eliminated at a faster rate ($t_{1/2}$ = 15-17 days) compared with free DTX1 ($t_{1/2}$ = 23-31 days) and DTX2 ($t_{1/2}$ = 28-33 days). Slightly slower elimination rates were estimated for the ester forms ($t_{1/2}$ = 23-42 days). Regarding PTX, PTX-2 seco acid (PTX-2 SA) was the major PTX detected in both species, but small amounts of PTX-2, PTX-12 and PTX-12 SA were also found. As for the OA-group toxins, oysters generally contained lower total amounts of PTX compared with mussels, but the difference was much less apparent. Estimation of reduction rates of the different PTX compounds showed that these toxins were rapidly eliminated in both oysters and mussels

($t_{1/2}$= 6-13 days). The cited study [Lindegarth et al., 2009] suggested that differential rates of gut assimilation and/or biotransformation of the OA-group and PTX could explain some of the observed differences in retention and toxin profiles between the bivalves, rather than differences in elimination rates. From the industrial perspective, these results suggest that *O. edulis* may be regarded as a low-risk species for DSP contamination, which should be taken into consideration by regulatory authorities in charge of sampling frequencies and monitoring programs for shellfish toxins.

From a food safety point of view, the PTX-compounds are of minor importance compared to the OA-group considered all together, taking into account the uncertainty regarding the oral toxicity of the PTXs, the preferential ingestion of OA-group toxins compared to PTXs and the faster depuration of PTXs.

Recent studies on the lipophilic toxin group have focused on the elucidation of the mechanisms of interaction at cellular level between bivalve tissues and toxins. In this sense, Sugiyama was the first in isolating and characterizing two okadaic acid binding proteins from the marine sponge *Halichondria japonica* (OABP1 and OABP2, the last one being a multimer of two subunits) involved in the detoxification of OA [Sugiyama et al., 2007]. Another recent study identified a target protein in mussels serving for decontamination from azaspiracids [Nzoughet et al., 2008]. Based on these advances a current study of Rossignoli and Blanco provided evidence that in *M. galloprovincialis* the 91 % of the OA could be found in the soluble fraction of digestive gland cells, free or bound to biomolecules. 32 % were quantified as belonging to the free OA citosol fraction. All findings suggested that about 70 % of the total OA contained in the digestive gland had been bound to a soluble receptor larger than 30 kDA but smaller than 300 kDA. As a final conclusion of this study, it was suggested that the intracellular immobilization of OA takes place by storage in a high density lipoprotein [Rossignoli and Blanco, 2010]. Further research efforts are necessary for the complete elucidation of the mode of action at cellular level and the structure of the protein receptor of OA in shellfish.

1.1.1.2. Azaspiracids

Azaspiracids is the shellfish-associated toxin group most recently discovered. Its existence was revealed after a poisoning episode that occurred in 1995 in the Netherlands, caused by toxic mussels from Killary Harbour, Ireland. The causative toxin was isolated and structurally elucidated, being designated azaspiracid in reference to its chemical structure [Satake et al., 1997; Satake et al., 1998]. Subsequent episodes allowed the identification of other new toxins from the group: AZA-2 and AZA-3 [Ofuji et al., 1999], AZA-4 and 5 [Ofuji et al., 2001], AZA-6 to AZA-12 [Díaz Sierra et al., 2003; James et al., 2003b]. Up to 27 different naturally occurring analogues of AZA1 have been identified, as well as 4 methyl ester analogues that are artefacts of storage in methanolic solution [Rehmann et al., 2008]. Only AZA-1, 2 and 3 in much lower concentrations have been detected in water column or phytoplankton, suggesting that the other analogues isolated from shellfish are product of biotransformations once the former three analogues have been ingested by bivalves.

The identification of the azaspiracids producing organism has been a difficult task. James and colleagues detected the presence of AZA-1, 2 and 3 in extracts from the heterotrophic dinoflagellate *Protoperidium cassipes* [James et al., 2003a]. However, the heterotrophic nature of the species, together with the fact that the attempts to confirm its azaspiracids production in laboratory cultures manipulating environmental variables were not successful, suggested that the real potential role of this organism is to act as a vector when feeding on the actual producer species rather than being a progenitor of these toxins. Finally, a small photosynthetic dinoflagellate designed as *Azadinium spinosum* was found to produce AZA-1, 2 and an AZA-2 isomer when cultured in pure culture, and it was also present in field samples rich in azaspiracids [Krock et al., 2009; Tillmann et al., 2009].

The investigation on proteins related to mussel (*M. edulis*) hepatopancreas contaminated with azaspiracids toxins revealed that these were weakly bound to a protein with a molecular weight of 45 kDa. This protein was also present in control non-contaminated mussels, albeit at much lower concentration [Nzoughet et al., 2008]. The presence of this protein could explain the prolonged retention time of the azaspiracids in shellfish. Migration of toxin into other tissues different from hepatopancreas, reduced metabolic activity during periods with reduced temperatures and the sporadic or continued presence of azaspirazid-producing species have been suggested as causes of this long detoxification period [Twiner et al., 2008].

When studying the anatomical distribution of azaspiracids in scallops (*P. maximus*), substantial differences in azaspiracids tissue content was found. The azaspiracids composition was determined in several tissues, finding that about 85 % of the toxin was concentrated in the hepatopancreas [Braña Magdalena et al., 2003]. Hess and colleagues discovered that, when comparing the azaspiracids content in the digestive gland to that in whole flesh of *M. edulis*, azaspiracids accumulated in the digestive gland, similar to other lipophilic toxins. The ratio of toxin in the digestive gland compared to the whole mussel was on average circa 5, both for a bulk sample collected in Norway and for 28 samples collected over a 3-year period in Ireland [Hess et al., 2005]. The same authors also observed a 2-fold increase of apparent content of azaspiracids upon steaming fresh mussels that affected both the whole flesh and digestive gland content. This effect was apparently caused by an indirect concentration through the loss of water from the matrix during the steaming step, phenomenon that has been associated to other lipophilic toxins.

Recently, McCarron and colleagues studied the actual cause of the increase in the azaspiracids content in samples after heat treatment. They observed an increase in the content of AZA-3 in samples naturally contaminated with azaspirazids that had been homogenized and heated in sealed containers to prevent the loss of water, whereas the concentrations of AZA-1 y 2 were unaffected. This experiment demonstrated that the increase in AZA content was not related to water loss. They ruled out that this increment could be related to a release from the matrix or caused by matrix effect during the LC-MS analysis. Finally, they demonstrated that toxin conversion via decarboxylation of AZA-17 was the responsible for the increase of AZA-3. In the same way other carboxy-AZA like AZA-19, 21 and 23 were converted after heating into their decarboxylated analogs (AZA-6, 4 and 9 respectively). Taking into account that only AZA-1 and 2 seem to be the former analogues produced by the primary causative organism, McCarron and colleagues proposed that the oxidative metabolism of AZA-1 and 2 in shellfish would result in 22-carboxylated metabolites (AZA-17 and 19) which undergo decarboxylation when heated to form AZA-3 and 6, respectively. The metabolic grid consisting in oxidation of AZA-1 and 2 at C3, C23 and 22-methyl together with nonenzymatic decarboxylation at C22 is capable of accounting for all the currently identified AZAs in shellfish [McCarron et al, 2009].

1.1.1.3. Brevetoxins

Brevetoxins is a group of more than ten natural neurotoxins produced by the marine dinoflagellate *Karenia brevis.* Other species of the genus *Karenia* and species of raphidophytes (e.g., *Chantonella spp.*) have also been documented as brevetoxin producers [Bourdelais et al., 2002]. At least nine brevetoxins are known to be produced by *K. brevis*, grouped according to their backbone structure into type A and B brevetoxins. Principal A backbone types are PbTX-1 and PbTX-7 and B-type are PbTX-2, PbTX-3 and PbTX-9. PbTX-2 is the most abundant toxin of the group, while PbTX-1 is the most toxic [Landsberg, 2002].

Studies in vitro exposing oyster to pure algal brevetoxins revealed that PbTX-2 was extensively metabolized to the reduction product PbTX-3 and two cysteine conjugates named cysteine- PbTX and cysteine- PbTX-3 sulfoxide [Plakas et al., 2002]. Corresponding cysteine conjugates with A-type backbone structure, probable derivatives of PbTX-1, were confirmed to be present in naturally polluted oysters, as well as other peptide conjugates, both with A- and B-type backbone, like glycine-cysteine-PbTX, gamma-glutamyl-cysteine-PbTX, and glutathione-PbTX, and a series of fatty acids – aminoacids conjugates [Wang et al., 2004]. Additionally, oxidized forms of both type-A and type-B brevetoxin parent molecules, PbTX-1 and PbTX-2 respectively, have been reported, in which the R-group on the molecule is converted to a carboxylic acid from an aldehyde [Ishida et al., 2004; Plakas et al., 2004; Wang et al., 2004]. Other brevetoxin derivatives modified on the A-ring lactone of the molecule are more polar and have been confirmed as hydrolyzed products through opening the A-ring of the parent toxins. These derivatives have been isolated by in vitro metabolic studies [Radwan et al., 2005], from *K. brevis* cultures, natural bloom samples and eastern oysters exposed to *K. brevis* [Abraham et al., 2006], yet their toxic potency relative to the parent compounds remains unknown [Roth et al., 2007].

As all these metabolites vary widely regarding polarity and hydrophobicity, intra- and inter-specific differences in their tissue distribution and rates of elimination would be expected. Dickey et al. found that Eastern oysters remain toxic up to 75 days after dissipation of *K. brevis* bloom [Dickey et al., 1999]. Temporal patterns of toxicity assessed by mouse bioassay and cytotoxicity assays suggested variable rates of elimination of individual metabolites. While results monitoring the elimination of brevetoxins and its metabolites in cockle and greenshell mussels exposed to *K. brevis* cultures showed rates of elimination generally consistent with relative polarity of these substances [Ishida et al., 2004], the characterization of the rate of elimination in eastern oyster made by Plakas et al. did not offer consistent results with polarity and hydrofobicity [Plakas et al., 2004]. Additionally, the lack of information on the toxicity of the different metabolites due to its difficult extraction and synthesis increase the difficulty on assessing the total shellfish toxicity (reviewed in [Plakas and Dickey, 2010]).

A particular fact concerning brevetoxins is that *K. brevis* is an easily lysable unarmored cell that releases toxins to the water column after cell lysis, causing composition changes. These toxins may remain dissolved or in association with small particles or micelles due to their hydrophobic nature. The ratio of intra/extra-cellular amount of brevetoxins is a good indicator of the stage of the algal bloom, since at the beginning of a bloom most of the toxic content remains intracellular while the presence of abundant extra-cellular toxin indicates changes in the bloom dynamics. The ratio PbTX-2/PbTX-3 is also indicative of the bloom stage, with the growth phase represented by a high ratio PbTX-2/PbTX-3 while the reduction of this ratio indicates a mature bloom or its later phase with many lysed cells (reviewed in [Pierce et al., 2008]).

Biotransformations known to date of the main lipophilic toxin groups in molluscs are summarized in Table 1.

Table 1. Schematic presentation of known biotransformations of the main lipophilic toxin groups in molluscs

Initial toxin form	Derivative after biotransformation	References
Diol esters of the OA OA DTX1 DTX2	Enzymatic Esterification: DTX3: 7-O-Acyl-OA 7-O-Acyl-DTX1 7-O-Acyl-DTX2	Suzuki et al., 1999 Torgersen et al., 2008
PTX2	Enzymatic acylation/esterification: PTX-2-SA 37-O-acyl PTX-2-SA 33-O-acyl PTX-2-SA 11-O-acyl PTX-2-SA	Daiguji et al., 1998a James et al, 1999 Pavela-Vrancic et al., 2001 Vale&Sampayo, 2002b Miles et al., 2004c Willkins et al., 2006
PTX2	Enzymatic-mediated oxidation: PTX1 PTX3 PTX6	Vale&Sampayo, 2002b Fernandez-Puente et al., 2004
PTX2	Non-enzymatic hydrolysis: 7-epi-PTX 2SA	Daiguji et al., 1998a James et al, 1999
YTX homoYTX 45-OH-YTX	Enzymatic oxidation: 45-OH-YTX 45-OHhomo-YTX carboxy-YTX 45-OHcarboxy-YTX	Yasumoto&Takizawa, 1997 Dominguez et al., 2010 Aasen et al., 2005 Samdal, 2005
AZA1-6	Enzymatic hydroxylation: AZA7-11	Díaz Sierra et al, 2002; James et al., 2003 Ofuji et al., 2001
AZA1, 2	Enzymatic hydroxylation: AZA 7, 8, 11, 12, 15, 16, 17, 19, 21, and 23	McCarron et al, 2009
AZA 17, 19, 21, and 23	Nonenzymatic decarboxylation AZA 3, 6, 4 and 9	McCarron et al, 2009

Table 1. (Continued)

PbTX-1, PbTX-2 PbTX-1, PbTX-2	Aminoacid conjugation BTX-B1,-B2, other conjugates Fatty acid esterification BTX-B3	Ishida et al., 1995; Morohashi et al., 1995; Murata et al., 1998; Wang et al., 2004 Morohasi et al., 1995
PbTX-2 BTX-B2 PbTX-2 PbTX-2 BTX-B2	N-myristoyl and N-palmitoyl conjugation BTX-B4 Oxidation BTX-B5 Reduction PbTX-3, -9 S-desoxy-BTX-B2	Morohasi et al., 1999 Ishida et al., 2004 Plakas et al.,2002

1.1.2. Hydrophilic Toxins

1.1.2.1. Saxitoxin and Analogues

Paralytic shellfish poisoning is one of the most frequent types of human poisoning caused by seafood and has therefore been extensively studied over the last forty years. The increasing exploitation of wild fisheries for marine bivalve moluscs and the intensification of aquaculture, as well as potential spreading of the causative organisms to new habitats due to the global climate change have all contributed to the global increase in paralytic shellfish poisoning events. Paralytic shellfish toxins (PSP) are a group of more than 57 analogues of the saxitoxin (STX), the first toxic molecule from the group isolated from the siphon of the Alaska butter clam *Saxidomus giganteus* [Schantz and Widholm, 2001]. These analogues differ in side group moieties, and this characteristic has been used to group them. They may be non-sulfated (STX and Neo STX), mono-sulfated (GTXs 1-6) or di-sulfated (C1-C4 toxins). In addition these analogues could be decarbamoylated, increasing the variants of the group, and can be classified into N-sulfocarbamoyl, decarbamoyl and carbamoyl each with increasing toxicity in bioassays. Recently, the increase in screening efforts and the use of improved methods for detection and structure elucidation have resulted in an increase of the number of PSP toxins analogues reported in literature, becoming a challenge to PSP identification and monitoring activities [Wiese et al., 2010].

In marine enviroments, PSPs are generally produced by phytoplanktonic dinoflagelate species from the genera *Alexandrium*, *Pyrodinium* and *Gymnodinium* while in freshwater medium these toxins are produced by several genera of cyanobacteria. The toxin profiles of the causative dinoflagellate species and strains are quite variable [Oshima et al., 1990], reflecting upon the toxicity of the contaminated shellfish that ingest them. However, due to the enzymatic activities and chemical transformations in shellfish tissues species-specific modifications in these toxin profiles are observed. On the other hand, it has been demonstrated that PSPs persist in the water column for more than 3 months at 25° C and certain chemical transformations occur after the beginning of the toxic event. For example, in the case of the cyanobacterial species *Anabaena circinalis*, the predominantly produced carbamate toxins (C-toxins) are converted to the highly toxic decarbamoyl dc-GTXs, increasing five to six fold original toxicity [Jones and Negri, 1997].

In previous studies of PSPs chemical transformations of marine systems five types of reactions have been recognized, most of them taking place in marine shellfish tissues once PSPs have been accumulated. They are the following:

Reaction 1) hydrolysis of carbamoyl or N-sulfocarbamoyl group to form decarbamoyl analogues ([Sullivan et al., 1983]; [Oshima, 1995]; [Cembella et al., 1993]; [Negri and Jones, 1995]; [Negri et al., 1997]).

Reaction 2) hydrolysis of N-sulfocarbamoyl group to form carbamate analogues ([Cembella et al., 1993]; [Bricelj et al., 1991]).

Reaction 3) reductive cleavage of C_{11}-sulfate groups to form desulfated analogues ([Shimizu and Yoshioka, 1981]; [Kotaki et al., 1985]; [Oshima, 1995]; [Cembella et al., 1993]; [Negri and Jones, 1995]).

Reaction 4) $\beta - \alpha$ epimerization ([Bricelj et al., 1991]; [Oshima et al., 1990];[Lassus et al., 1992]; [Oshima, 1995]; [Negri et al., 1997]).

Reaction 5) reduction of the N_1 hydroxyl group ([Shimizu and Yoshioka, 1981]; [Kotaki et al., 1985]; [Oshima et al., 1990]; [Oshima, 1995]).

The reaction 1 is a simple chemical step that takes place at different environmental conditions with different speed (rapidly at neutral or mild acidic pH and 90° C and much slower at 25° C). When boiled with strong acid at low pH conditions (2-4), the N-sulfo-carbamoyled toxins partially transform into corresponding more toxic carbamate analogues. Most of these chemical changes can occur spontaneously at certain environmental conditions and go until reaching equilibrium. Due to the availability of natural reducers like glutathione and cysteine, commonly found in shellfish, elimination of the N_1 hydroxyl group (reaction type 5) and also elimination of the *O*-sulfate group at position 11 from the toxin molecule (reactions type 1 and 2) can occur in shellfish tissues [Artigas et al., 2007]. Transformations might be due to shellfish enzymatic activity, since carbamoylase enzymes catalyze the hydrolysis of N-sulfocarbamoyl or carbamate moieties of PSP toxins, thus rendering them into the corresponding decarbamoyl analogues. In the work of Artigas and colleagues the toxin profiles of 8 different shellfish species (blue mussel *M. galloprovincialis*, common cockles *Cerastoderma edule*, clams *Ruditapes decussata*, *Venerupis pullastra*, *Scrobicularia plana*, razor clams *Solen marginatus*, oysters *Crassostrea japonica* and offshore clams *Spisula solida*) were monitored during a 2005 *G. catenatum* bloom at the NW Portuguese coast. All bivalves monitored, with the exception of clams *S. plana* and *S. solida,* contained similar toxin profiles namely C1+2, C3+4, GTX 1+2+3+4, STXs, B1, B2, neoSTX (NEO), dcGTXs, dcSTXs, typical for the algal producer. However, in *S. plana* traces of N-sulfocarbamoyl toxins were detectable such as C1/2 and B1, while in *S. solida* the profile of whole flesh tissues contained only decarbamoyl gonyautoxins (dcGTX1–dcGTX4) and decarbamoyl saxitoxins (dcSTX and/or dcNEO). This was attributed to the very rapid transformation of the initial carbamate toxins into the dercarbamate analogues, showing the high enzymatic (carbamoylase) activities in *S. solida*. In *S. plana* these activities were also present but were much lower due to minor enzymatic specificity. No relevant carbamoylase activities were found in *M. edulis*, *C. edule*, *C. gigas* and clam *Donax trunculus*.

The molecular characteristics of the carbamoylases involved in the metabolism of PSTs in shellfish are still not well studied largely because of the difficulties in isolation caused by instability and low natural amounts of the molecules. Carbamoylase activities of similar character have so far been found in few species like the little neck clam *Protothaca staminea* [Sullivan et al., 1983], two Japanese bivalves – *Mactra chinensis* and *Peronidia venulosa* [Oshima, 1995], North American Atlantic surf clams *Spissula solidissima* [Bricelj and Cembella, 1995] and the above cited Portuguese species surf clam *S. solida* and peppery furrow shell *S. plana* [Artigas et al., 2007]. Recently, a sulfocarbamoylase I has been discovered in Japanese bivalve *P. venulosa*, a novel enzyme hydrolyzing the carboxyl bond in N-sulfocarbamoyl moieties of PSPs [Cho et al., 2008]. This enzyme has been fully characterized as a multimer of two subunits with total molecular weight of 300 kDa.

Transformations due to boiling in acidic conditions should be considered, since the AOAC mouse bio-assay official method allows an extraction pH between 2 and 4, and this variation of 2 units may lead to discrepancies during PSP controls [Vale et al., 2008]. Vale et al studied the changes in potency *in vitro* and in the toxic profile of two natural contaminated samples with C toxins and C plus GTX6 respectively, boiled at different acidic conditions. They observed that extraction at lower pH increased the toxicity of the samples. The sample with C toxins profile increased its in vitro potency by 4, 10 and 50 times after boiling for 10 minutes at pH 4, 2.5 and 1.1 respectively, while the sample with C plus GTX6 toxic profile increased by 2 and 4 times after heating at pH 2.5 and 1.1, with very small effect at pH 4. After heating, an increase in the presence of GTX3, 4, 5, STX and NeoSTX was observed in the first sample, which explains the increase of in vitro potency according to toxicity equivalency factors. In the case of the second sample, the rise in potency *in vitro* is explained by a low increase in the presence of NeoSTX and GTX1 [Vale et al., 2008].

Differences in tissue distribution have been observed among shellfish carbamoylases. In general the highest rates of toxin bioconversion are associated with metabolically active tissues such as the digestive gland and gill, which first come into contact with toxins liberated from the dinoflagellate cells. In the case of the bivalve *P. venulosa*, enzymatic activity has been observed in all tissues [Cho et al., 2008]. Its digestive gland has shown the highest activity (28 %), followed by the mantle (21 %), others (20 %), adductor muscle (11 %), muscle (8 %), crystalline style (buried in the digestive gland) (6 %), and gills (5 %). Similarly, all tissue homogenates (digestive gland, gill, mantle, siphon and adductor muscle) of *P. staminea* have shown decarbamoylation activity with relatively high conversion rates, more specific for C2 than for C1 (90 % conversion in digestive gland within 4 hours), thus demonstrating the different affinities of the carbamoylases depending on the orientation of the 11-OH-sulfate group in the toxins [Fast et al., 2006]. In the case of *P. staminea,* NEO and STX biotransformations to dcSTX were also detected but only in digestive gland homogenates. In contrast, most of the enzymatic activity (73 %) in *M. chinensis* has been observed in the digestive gland [Lin et al., 2004]. The enzymatic activity in other *Mactra* tissues has been reported as very low (crystalline style (7 %), gills (6 %), mantle and others (4 %), foot (3 %), adductor muscle (2 %) and siphon (1 %)).

It is generally known that mussels and oysters reflect more the PSP toxin profile composition of the causative plankton than clams such as *Meretrix lamarckii*, *Pseudocardium sachalinensis* and *Protothaca staminea* that often show different compositions [Samsur et al., 2007]. In *P. venulosa* a disappearance of the initial C1/C2 toxin concentrations and appearance of dcGTX2/3 was observed as well as the reductive elimination of 11-O-sulfate from GTX2/3 and N1-reduction of NEO [Cho et al., 2008].

In a similar way, in the short-necked clam *Tapes japonica* fed with toxic algae from the species *Gimnodinium catenatum* it was observed that the toxin composition clearly differed from that of the algal producer which consisted only of N-sulfocarbamoyl toxins, C1, C2, GTX5 and GTX 6, with predominance of C2 [Samsur et al., 2007]. After 30 minutes of algal supply to the clams, the ratio of C2 to C1 in clam tissues was much lower and further decreased up to 72 hours. In addition, the proportion of C1+2 also decreased gradually during the same period. On the other hand, decarbamoyl toxins (dcSTX and dcGTX2, 3) in *T. japonica* first appeared at 3 hours and their proportion increased to a maximum (50 %) at 72 hours. Small amounts of GTX2,3 were also detected during the 12-144 hours after the beginning of the supply with *G. catenatum.* A fluctuation in the proportion of GTX5+6 had been seen throughout the whole rearing period. The decline of the ratio of C2 to C1 observed was attributed to β - α epimerization and the appearance of dc and carbamate toxins – to the enzymatic hydrolysis of N- sulfocarbamoyl toxins (C1, 2 and GTX5). The result concerning the gradual decline of the C1+2 and the gradual increase of the dc toxins in this study suggested that the decarbamoyl toxins were retained longer in bivalve tissues, whereas C1, 2 were released more rapidly into water, possibly through brachial respiration or by elution, since carbamate toxins are the most hydrophilic analogues from the PSP group. At the end of the experimental period (168 hours) only 1 % of the initially supplied total toxin concentrations were detected.

The reasons for the respective differences among tissues and species in relative carbamoylase activity cannot be readily explained. Nevertheless, it might be attributed to the absence of enzymes or the presence of multiple carbamoylases or to differences in specificity and affinity of a single broad-spectrum enzyme with respect to various substrates. That is a hypothesis already proposed by Fast and colleagues in 2006. It has also been pointed out that the functional role of carbamoylase enzymes in some bivalves is intriguingly elusive, and is not obviously related to a detoxification process.

The species specific differences in the toxin distribution in bivalves, together with the variability of carbamoylase activities and the bioconversion of the initial PSP toxins to more toxic analogues within bivalve tissues bring to differences in shellfish species total toxicity and detoxification kinetics. In general the detoxification kinetics in bivalves appears to be biphasic and must be described by means of two- or multi-compartmental models [Blanco et al., 2003]. In a study of 24 shellfish species from the Guandong coast [Lin et al., 2004] it was found that two species were the most toxic – scallops *Chlamys nobilis* and green-lipped mussels *Perna viridis.* In the study of Choi and colleagues under laboratory conditions depuration, in these two species was characterized by a rapid loss within the first day, followed by a secondary slower loss of toxins [Choi et al., 2003]. The total toxicity decreased on day 1 and increased again on day 2, the depuration rate constants being for the fast phase of depuration (from day 0 to day 1) 1.16 and 0.87 per day for scallops and mussels respectively and for the slow depuration phase (from day 2 to day 13) 0.040 and 0.063 respectively. The initial rapid depuration phase represented the gut evacuation of the unassimilated toxin or toxigenic algal cells, whereas the slower depuration phase represented the loss of toxins that had been assimilated and incorporated into tissues. During the 13-day

Table 2. Toxicity Equivalency Factors of 14 STX analogues based on acute *i.p.* toxicity in mice proposed by EFSA (EFSA 2009)

Toxin	TEF	Toxin	TEFs
STX	1	C2	0.1
NeoSTX	1	C4	0.1
GTX1	1	dc-STX	1
GTX2	0.4	dc-NeoSTX	0.4
GTX3	0.6	dc GTX2	0.2
GTX4	0.7	GTX3	0.4
GTX5	0.1	11-hydroxy-STX	0.3
GTX6	0.1		

depuration experiment in the non-visceral tissues of the mussel (including adductor muscle, gill, foot and other soft tissues) and the adductor muscle of scallops about 3 % of the total toxin burden was detected while the resting 97 % were localized in the visceral tissues. The percentage of toxin in the digestive gland of mussels increased with the decrease of toxin burden in other tissues, a fact that was explained by the observed tendency in mussels to redistribute accumulated toxins from other tissues back to digestive gland for detoxification. Thus, green-lipped mussels are considered quick detoxifiers and are known to remove most of the accumulated toxins within few weeks while the scallops are slow detoxifiers and tend to retain toxins in their body for several months to years [Choi et al., 2003]. The results obtained in this study for the depuration rates in *Chlamys nobilis* were much higher than those in *Placopecten magellanicus*, but comparable to those measured in *Patinopecten yessoensis* [Bricelj and Shumway, 1998].

Toxicity Equivalency Factors (TEFs) should be taken into account when considering total toxicity of a sample based on its content of individual toxins. Table 2 illustrates the TEFs as proposed by EFSA (EFSA, 2008).

1.1.2.2. Domoic Acid

The amnesic shellfish poisoning (ASP) is caused by a single toxin, domoic acid and its isomers, which belong to a group of neurotoxic aminoacids called kainoids. Domoic acid is the main form present both in phytoplankton and shellfish but isomers and 5'-epi-domoic acid have also been found in small amounts [Walter et al., 1994; Wright et al., 1990]. Limited references on the presence of domoic acid isomers in shellfish can be found in literature. Isodomoic acid C was found in New Zealand contaminated shellfish and was shown to be produced by a local strain of *Pseudo-nitzschia australis*. However, its affinity for specific receptors was small, indicating low neurotoxic potential [Holland et al., 2005]. A strain of the diatom *Nitzschia navis-varingica* from Philippines Islands was found to produce isodomoic acids A and B when isolated and mass cultured [Kotaki et al., 2005]. Experiments demonstrated that exposure of domoic acid to simulated sunlight modifies its chemical structure, causing both reversible isomerization reactions and irreversible decarboxylation reactions, producing a suite of isomers and decarboxylated derivatives [Bouillon et al., 2008]. The acute intraperitoneal toxicity of domoic acid isomers A, B and C was found to be lower than that of domoic acid. Furthermore, the severities of the behavioral changes induced by these isomers were all much lower than that of domoic acid itself, suggesting that these substances would pose relatively little risk to human or animal health [Munday et al., 2008].

During toxic episodes, the level of domoic acid accumulated in bivalve tissues depends both on the amount and toxicity of the cells present in the water column and on the balance of the different metabolic mechanisms involved in uptake and excretion of domoic acid. These mechanisms differ greatly between different bivalve species. Mussels (*M. edulis*) fed with toxigenic diatom *Nitzschia pungens* accumulated domoic acid at higher rates than scallops (*P. magellaicus*) in the same conditions, and the depuration rate after the cessation of feeding on *Nitzschia* was also higher [Wohlgeschaffen et al., 1992]. Oysters seemed to accumulate lower amounts of domoic acid than other bivalves in the same conditions, and rarely attained the regulatory level (reviewed in [Mafra et al., 2010]).

Tissular retention of domoic acid presents important inter-species differences. While some species like the mussels *M. edulis, M californianus, M. edulis* or the softshell clam *Mya arenaria* rapidly eliminate domoic acid, other bivalves such as the horse mussel *Volsella modiolus*, the sea scallop *P. magellanicus* and especially the king scallop *P. maximus* and the razor clam *Siliqua patula* accumulate it (Reviewed in [Mauriz and Blanco, 2010]), even over regulatory level, for very long periods, that vary from weeks to years after the toxic episode ends [Blanco et al., 2002a; Blanco et al., 2006; Drum et al., 1993].

Domoic acid elimination kinetics has not been demonstrated for most important commercial species. While a single-compartment kinetics model, where domoic acid elimination occurs at a constant decay rate over the entire depuration period, is applicable for 90 % of *M. edulis* and for *Pecten sp.* [Blanco et al., 2002a; Novaczek et al., 1992; Wohlgeschaffen et al., 1992], a two-compartment kinetics model, where an initial phase of rapid domoic acid elimination is followed by a period of slower toxin loss, has been demonstrated for *M. galloprovincialis* [Blanco et al., 2002b] and is applicable for other species like 10 % of *M. edulis* [Novaczek et al., 1992], *S. patula* [Drum et al., 1993] and the mussel *Volsella modiolus* [Gilgan et al., 1990].

Intracellular distribution of domoic acid and its possible binding to specific sites inside the cell that could make difficult the elimination of the toxin should be considered when discussing toxin depuration rates. Intracellular distribution of domoic acid in the digestive gland of naturally contaminated scallops *P. maximus* was studied by Mauriz and Blanco, concluding that nearly all of the toxin was found in a free form in the cytosol of the cells and not bound to any compound [Mauriz and Blanco, 2010]. These results are similar to those obtained for *M. edulis* [Madhyastha et al., 1991; Novaczek et al., 1991], species with a faster domoic acid depuration rate. Therefore the difference in the depurative capability of both species should probably be originated from differences in the membrane transporters, as domoic acid cannot freely pass the cell membrane [Mauriz and Blanco, 2010].

Anatomical distribution is an important fact taking into account that in some species only specific tissues are intended for human consumption. The different rates of accumulation of domoic acid in the tissues of bivalves is affected by several factors: the presence of specific binding sites for domoic acid in tissues, the capacity of its transport across the gastrointestinal membrane after ingestion and the capacity of transfering the toxin from visceral to other tissues via circulatory system [Mafra et al., 2010]. It has been observed that domoic acid mainly accumulates in visceral tissues, but the ratio of toxin present in visceral tissues/whole

Table 3. Schematic presentation of known biotransformations of the main hydrophilic toxin groups in molluscs

Initial toxin form	Derivative after biotransformation	References
Carbamoyl and Carbamoyl-N sulfated toxins (C1/2) GTX2,3 GTX1,4	Enzymatic decarbamoylation: dcGTX2, 3 dcGTX1, 4	Artigas et al., 2007 Asakawa et al., 1995 Bricelj et al., 1990, 1991, Bricelj et al., 1998 Cho et al., 2008 Jones &Negri, 1997 Samsur et al., 2007
GTX STX NEO	dcSTX neoSTX dcNEO	Baron et al., 2006 Fast et al., 2006 Weise et al., 2010
GC1-3 (mono-hydroxy-benzoate analogues)	dcSTX	Vale, 2008 Weise et al., 2010
Carbamoyl and Carbamoyl-N sulfated toxins (C1/2) GTX5 C1/2	Desulfonation: STX	Jones &Negri, 1997 Cembella et al., 1993 Bricelj et al., 1991
GTX neoSTX NEO	Reductive cleavage: GTX5 NEO 11-OH-STX STX	Artigas et al., 2007 Blanco et al., 2003 Sato et al., 2000 Shimizu&Yoshioka, 1981
C2 GTX3 dcGTX3	$\beta - \alpha$ epimerisation: C1 GTX2 dcGTX2	Bricelj et al., 1991 Cho et al., 2008 Fast et al., 2006
DA	Isomerisation Iso- DA-D, E and F	Bouillon et al., 2008 Kotaki et al., 2005 Munday et al., 2008

body vary between species: 94-99 % in *P. maximus* [Blanco et al., 2002a; Blanco et al., 2006; Bogan et al., 2007a; Bogan et al., 2007b; Campbell et al., 2003], 93 % in *M. edulis* [Grimmelt et al., 1990] or 70 % in *C. virginica* [Roelke et al., 1993].

Biotransformations known to date of the main hydrophilic toxin groups in molluscs are summarized in Table 3.

1.2. Bacterial Degradation of Toxins

Certain bacterial species associated with toxic dinoflagellates not only affect the modulation of their toxicity and growth through the production of vitamins, iron chelators and cytokinins but they also can be involved in toxin production increasing or decreasing toxins originated by dinoflagellates. Moreover, many reports related to the possible role of some bacteria in the production of toxins have been published lately. Tetrodotoxin (TTX) seems to have its origin in species of dinoflagellates such as *Alexandrium tamarense* [Kodama et al., 1996; Sato et al., 2000], but also some freshwater cyanobacteria have been shown to produce it [Carmichael and Mahmood, 1984]. A recent paper shows that TTX is originated primarily by marine bacteria identified as *Vibrio alginolyticus*, other *Vibrio* strains, *Shewanella alga, S putrefaciens* and *Alteromonas tetraodonis*, among others and pufferfish bioaccumulate it via the food chain that begins with these bacteria [Noguchi and Arakawa, 2008]. The first confirmation of *Vibrio harveyi* as a TTX-producer in an *in vitro* environment was published by Campbell et al and this seems to be the main source of TTX in the pufferfish [Campbell et al., 2009].

Other reports show that Palitoxin or Palitoxin-like compounds are produced by some bacteria as *Bacillus cereus* as well as others belonging to the genera *Brevibacterium* or *Acinetobacter* [Seemann et al., 2009].

It seems that some cyanobacteria of the genus *Hydrocoleum* are involved in the biogenesis of Paralytic Shellfish Toxin (PST), neurotoxins and Ciguatoxin (CTX)-like compounds [Laurent et al., 2008]. In particular, Perez-Guzman et al indicate that the Cytophaga- Flavobacter-Bacteroides belonging to the superphyllum Bacteroidetes could be a candidate involved in ciguatoxin production [Perez-Guzman et al., 2008].

On the contrary, it has been reported that some marine bacteria or bacteria isolated from shellfish are able to metabolize biotoxins. Bio-transformations of toxins caused by bacteria was first mentioned by Kotaki et al who proposed that some marine bacteria as *Vibrio* and *Pseudomonas* are able to metabolize Paralytic Shellfish Poisoning (PSP) toxins [Kotaki et al., 1985]. In other study the natural elimination of domoic acid (ASP) from certain bivalve species was attributed to the presence of marine bacteria [Stewart et al., 1998].

The bacterial degradation of PSP toxins was further described by Smith et al showing that the microflora of shellfish possesses adequate enzymatic capacity to carry out extensive side chain modification of these toxins [Smith et al., 2001].

The ability of marine bacteria to use toxins can be developed as a metabolic adaptation, as Donovan et al suggest [Donovan et al., 2008]. Shellfish containing PSP toxins create an enriched environment for bacteria that are able to utilize the toxins. Very elegant experiments were designed in order to prove this hypothesis. These studies were done selecting marine bacteria capable of reducing toxicity of PST within a reasonably short period of time. In LC-FLD and LC-MS analyses a decrease in toxin peaks was most often observed without the appearance of new peaks. This research shows the ability of bacterial isolates from PSP-contaminated mussels to degrade or sometimes obliterate relatively high concentrations of PSP toxins within a short time frame (2-3 days). This study was done in aerobic conditions although reductive eliminations for some bacteria have been shown to proceed more readily under anaerobic conditions [Donovan et al., 2008]. They further evaluate the phenotypic and taxonomic characterization of these isolates suggesting that seven isolates, all novel strains of *Pseudoalteromonas haloplanktis*, reduced the toxicity of PST [Donovan et al., 2009].

A recent report shows that probiotic bacteria strains are effective in elimination of different cyanobacterial toxins from solutions. A combination of the mixture of three strains enhanced the removal capacity up to 80 % as compared to the properties of the individual strains [Nybom et al., 2008].

In the frame of a project supported and funded by the 6th Framework Programme of the European Community (SPIES DETOX, Collective Research Project 0302790-2, years 2007-2009) studies were done to evaluate the possibility that marine bacteria could detoxify shellfish. Microencapsulation of these algal toxin-degrading bacteria was proposed as a method of delivering a concentrated pulse of bacteria to the digestive system of the shellfish to assist in the break-down of accumulated toxins. It was previously shown in the SPIES-DETOX project that it is possible to successfully encapsulate marine bacteria using alginate as a matrix and that the microcapsules are readily filtered and ingested by shellfish. In this project, capsules in a range of sizes (10 to 250 μm) were fed to mussels and scallops to determine the optimum size of bacteria microcapsules which could be actively ingested by shellfish.

Then, these preliminary investigations suggest that microencapsulation of toxin-degrading bacteria could be used to develop a practical method to detoxify shellfish of PSP toxins. In initial trials some degradation of PSP toxins was observed in shellfish fed with microencapsulated bacteria compared to control shellfish. In a new proposal BEADS "Bio-engineered micro Encapsulation of Active agents Delivered to Shellfish" (Grant Agreement Number 262649) it is suggested that bacteria demonstrating degradation of the algal toxins can be considered for using as part of a feeding ('probiotic') regime to clear contaminated shellfish of the toxins. An aim of this new project is to incorporate toxin degrading isolates into a 'probiotic' which may be administered in vivo to bivalve molluscs in commercial depuration systems. Work is now required to determine the feasibility of feeding microencapsulated bacteria diets to shellfish to degrade algal toxins at semi-industrial/industrial scales.

1.3. Human Detoxification

1.3.1. Non-Toxic Algal Feeding Detoxification

Studies have been done to determine the impact of non-toxic algal food in detoxification rate. Field detoxification by moving mussels from toxic to non-toxic environments (relaying) was performed by several authors. In fact the availability of non-toxic food has been proposed by several authors to be the main factor affecting DSP detoxification in mussels [Medhioub et al., 2010]

Concerning ASP, Wohlgeschaffen et al. [Wohlgeschaffen et al., 1992] found no difference in the detoxification rate of the hydrophilic neurotoxin domoic acid among *Mitylus edulis* fed with *Nitzschia* or starved mussels.

Regarding the effect of environmental factors on PSP detoxification rates in *Mytilus galloprovincialis,* phytoplankton concentration seemed to have no particular effects [Blanco et al., 1997]. Accordingly, Chen and Chou studied the accumulation of PSP toxins produced by a toxic strain of the dinoflagellate *Alexandrium minutum* Halim in purple clams (*Hiatula rostrata*) for subsequent study of toxin distribution during depuration (detoxification by a nontoxic microalgal diet or starvation). Results obtained confirm the high toxicity of the digestive gland, and that the depuration efficiency between feed with nontoxic microalgae and starvation is similar [Chen and Chou, 2001].

Other authors studied the use of diets based on non-toxic flagellates or diatoms to detoxify experimental PSP contaminated adult Pacific oysters (*Crassostrea gigas*) by the toxic dinoflagellate *Alexandrium minutum* Halim with levels above the safety threshold. Results showed that despite large individual variations in toxin levels at the end of the contamination period, detoxification times were of the same order of magnitude (3 to 4 days), reaching a toxin level equal to or less than the safety threshold. These variations were most likely related to marked individual variability in valve and/or clearance activities. No significant differences in detoxification rates were found when oysters were fed with *Isochrysis galbana*, *Tetraselmis suecica*, *Thalassiosira weissflogii*, or *Skeletonema costatum* [Lassus et al., 2000].

Taking these results into account, it seems that in shellfish contaminated with PSP toxins a non-toxic algae feeding do not cause a significant increase in detoxification rate. However, some papers show that detoxification of bivalves fed on unharmful microalgae is possible after several days of treatment. In fact, Gueguen [Gueguen et al., 2008] found that a *Skeletonema costatum* diet doubled detoxification rates in Pacific oysters *Crassostrea gigas*. Also, it was recently reported that natural detoxification processes in *Mytilus chilensis* and *Aulacomya atra* appeared to be mainly related to elapsed time after the bloom and maximum PSP concentration acquired by the shellfish; minor factors, but still significant, were environmental characteristics as the zone, temperature and salinity [Molinet et al., 2010].

Regarding other groups of toxins, detoxification of tetrodotoxin, that it is accumulated in the pufferfish, was studied by Noguchi and Arakawa. Results showed that fish becomes non-toxic when is fed with TTX-free diets in an environment in which the invasion of TTX-bearing organisms has been eliminated [Noguchi and Arakawa, 2008].

Medhioub and colleagues studied the detoxification kinetics of fast-acting toxins (FAT) in marine molluscs fed with specific diets of non-toxic algae (*Isochrysis galbana*). To find an optimal detoxification method, they performed two experiments in which clams were first fed with the toxic dinoflagellate *Karenia selliformis* to artificially contaminate them with gymnodimine thus simulating the effect of natural toxic episodes. As a second step, the same clams were fed with a non-toxic algae, *I galbana*, to speed up the detoxification process. The first results revealed faster detoxification rates in the digestive gland when clams were fed on *I galbana* compared with a starved control (no food) and a typical detoxification pattern, i.e. a rapid drop in toxin content within the first days followed by a secondary slower decrease [Medhioub et al., 2010].

1.3.1.1. The Role of Organic Matter in the Detoxification Process

Detoxification of Pacific oyster *Crassostrea gigas* fed on two different diets, one of *Skeletonema costatum* and a second one that used the same *S. costatum* diet but supplemented with silt particles was studied by some authors. Results showed that the *S. costatum* diet significantly reduced the time needed for oysters to reach the sanitary threshold, but no effect of the silt supplement could be demonstrated conclusively [Gueguen et al., 2008]. As, Guéguen point out in this work, it would probably be necessary to test different silt concentrations, and to compare detoxification kinetics between oysters kept in particle-free seawater and in seawater with silt but no algae. Also, it would be worth finding out if the effect of *S. costatum* on detoxification rates could be improved by using other microalgae species as complementary food.

1.4. Industrial Detoxification

Apart from the mitigation of some of the harmful algae effects achieved by a toxic episode, algal blooms could be prevented by early detection and tracking. This will be the aim of chapter 4 that will further describe the, so far, reported mechanisms and tools that contribute to early toxic episodes detection.

In this part of the chapter, several reports will be compiled concerning different mechanisms of potential industrial detoxification, since currently there is not an industrial detoxification protocol available.

1.4.1. Decrease of Lipophilic Toxins

Only natural detoxification is authorized for these toxins, although since they are lipophilic compounds their removal from the shellfish and, more specifically, from the digestive gland looks much more difficult that the removal of hydrophilic toxins. Nevertheless some efforts were done in order to get rid of these toxins that induce one of the major economical problems in some European countries. In a recent study, mussels naturally contaminated with okadaic acid and dinophisistoxins were treated under several conditions; freezing, ozonization, thermal processing or soy lecithin. No effect was observed with any treatment [Reboreda et al., 2010].

However, it has been reported that it is possible to reduce levels using supercritical CO_2 although the final product is not commercially acceptable. With this method most of the toxin is eliminated (up to 90 %), and the biological activity against its target enzyme is also severely affected (up to 70 % reduction) [Gonzalez et al., 2002]. It has also been tried, due to the lipophilic nature of the toxin, the use of amphoterous compounds as a feed to ''wash'' the toxin. The results showed a decrease in the time needed, but it is still too long for industrial implementation [Kroken et al., 2006a].

A third way has also been explored which is the manipulation of mussel metabolism: addition of N-acetylcisteine (NAC), a glutation precursor, would increase the velocity of natural detoxification mechanisms. Depuration of bivalve molluscs with NAC enhanced antioxidant defenses and the detoxification of organic contaminants. Specifically, NAC treatment increased glutathione (GSH) synthesis, raised the glutathione redox potential (GSH/GSSG ratio), and induced the glutathione S-transferase (GST) and glutathione reductase (GR) enzyme activities in the digestive gland of bivalves, allowing them to eliminate contaminants faster than with conventional depuration [Peña-Llopis et al., 2006].

In addition there are some papers that point out the potential of other mechanisms that could play a role in detoxifying molluscs. It was proved that including an emulsifier in the diet is useful for the detoxification of shellfish of lipophilic toxins, reducing the toxin concentration from 10000 μg/Kg to 160 μg/Kg in 24 days [Kroken et al., 2006b]. Moreover, a recent study shows the potential of Philippine clay minerals to physiologically remove toxic phytoplankton cells under laboratory conditions [Padilla et al., 2010].

1.4.2. Decrease of PSP Toxins

Taking advantage of the hydrosolubility of saxitoxin and derivatives, a series of cooking and washes followed by sterilization have been tried to reduce the toxicity of shellfish. In fact, a Commission Decision establishes special conditions for the harvesting and processing of certain molluscs coming from areas where the PSP level exceeds the limit laid down in the European Legislation [E.C., 1996]. Spain can authorize the harvesting of the bivalve *Acanthocardia tuberculatum* in production areas where the PSP level in the edible parts of these molluscs is higher than 800 μg/Kg but lower than 3000 μg/Kg under special conditions. These bivalves have to be processed according to a protocol defined in this Decision [E.C., 1996].

At an experimental level, the use of alkaline solutions followed by cooking and washing have also demonstrated the reduction of PSP level [Lagos et al., 2001; Vieites, 1999].

The effect of steam cooking in PSP profiles of scallop *Patinopecten yessoensis* was studied by some authors. Toxins analysis showed that steam cooking induced significant loss of PSP-toxins from viscera (16 %), adductor muscle (24 %), gill and mantle (11 %) while 32 % of the toxins were retained inside viscera and adductor muscle. Overall, 51 % of PSP-toxins leaked out from scallop tissues during steam cooking but there was no significant loss of PSP-toxins from gonad. In conclusion, samples analysed in this study did not differ significantly between raw and steamed samples. Results of this work also showed that steam cooking process did not initiate prominent toxins transformation [Wong et al., 2009].

Three different types of natural contaminated shellfish, mussels, clams and cockles were analyzed by other authors. Thermal processing of mussels with a level of 405 μg eq STX/Kg whole body induced a reduction of the toxin concentration below the detection limit in mouse bioassay (MBA). PSP levels were not affected by freezing and thawing in clams containing a level of PSP close to the legal limit (800 μg eq STX/Kg whole meat), although using the thermal processing after that reduced the PSP content below the limit of detection. In cockles with a level of 673 μg eq STX/Kg, freezing and thawing decreased PSP concentration that was further reduced by thermal processing [Reboreda et al., 2010].

1.4.3. Decrease of Domoic Acid Content

Detoxification of scallops contaminated with domoic acid can be reached by hepatopancreas ablation. In this context, a Commission Decision that establishes special conditions for the harvesting and processing of certain bivalve with a level of ASP exceeding the limit in the EC that consists mainly in hepatopancreas removal was published [E.C., 2002]. A study related to decrease of phycotoxin in bivalves by industrial processes was recently reported. In this work, a possible removal of hydrophilic and lipophilic toxins in different species of bivalves was studied [Reboreda et al., 2010]. For domoic acid detoxification, apart from hepatopancreas ablation, other treatments were considered; mollusc freezing, thawing and thermal processing. The best result was achieved by removing the digestive gland from scallops reaching levels below the legal limit in Europe (20 mg/Kg of whole body) when bivalves contained levels of domoic acid around 100 mg/Kg. In addition, a combination of thermal processing with freezing/thawing is able to reduce limits below 20 mg/Kg. However, it is worth to mention that is not economically feasible to remove the digestive gland in some bivalves that has a considerable hepatopancreas due to the huge amount of meat that is lost by applying this procedure.

Table 4. Schematic presentation of known detoxification mechanisms of the main toxin groups in molluscs

Toxin group	Detoxification mechanisms	References
PSP	Bacteria (Vibrio, Pseudomonas, Pseudoalteromonas haloplanktis…) Alkaline Solutions-Cooking-Washing Steam cooking Termal processing	Kotaki et al., 1985 Smith et al., 2001 Donovan et al., 2008 Donovan et al., 2009 Nybom et al., 2008 Vieites, 1999 Lagos, 2001 Wong et al., 2009 Reboreda et al., 2010
ASP	Marine bacteria Hepatopancreas ablation Freezing Thawing Termal processing	Stewart et al., 1998 Reboreda et al., 2010
Lipophilic toxins	Non-toxic feed Supercritical CO_2 Amphoterous compounds N-acetylcisteine Emulsifiers	Walid et al, 2010 González et al, 2002 Kroken et al., 2006a Peña-Llopis et al., 2006 Kroken et al., 2006b
TTX	Non-toxic feed	Noguchi & Arakawa, 2008
Fast Actin Toxins	Non-toxic feed	Medhioub et al, 2010

Detoxification mechanisms known to date of the main toxin groups in molluscs are summarized in Table 4.

CONCLUSION

The development of shellfish aquaculture has been hampered by repeated contamination of products with naturally occurring potent algal toxins derived from phytoplankton that are harmful to human health. Shellfish that filter feed on toxic phytoplankton are capable of retaining the toxin within different tissues in a species specific pattern and depending on the nature of each toxin. The toxins accumulated are subjected to various types of biotransformations, chemical or enzymatic, that leads to the appearance of novel forms (analogues) of the initial toxin molecules and imply changes in the total toxicity burden. After toxic episode remission, toxin concentrations in molluscs tend to diminish, however, the rates of natural detoxification depend on the species and the chemical character of the toxins. Detoxification periods might be prolonged in time causing negative economic impact on the aquaculture industry. Due to this fact, methods that could improve these natural depuration rates are required and various different detoxification methods have been tested. Only for some toxins, such as PSP and domoic acid, existing detoxification methods reach reduction in toxin content that allows mollusc commercialization. However, in most cases no efficient detoxification method exists. The development of methods for rapid detoxification of commercial bivalves is necessary for the shellfish harvesting sector in order to diminish economical losses due to harvesting areas closure but protecting consumer's health. As chemical pollution and harmful algal blooms represent a threat of increasing importance to human health, the achievement of such methodology would open new avenues for marine organisms detoxification.

ACKNOWLEDGMENTS

This work was supported by the Ministry of Science and Innovation of Spain AGL2009-13581-C02-02; ATLANTOX "Advanced Test about New Toxins appeared in the Atlantic Area (2008-1/003), PHARMATLANTIC "Knowledge Transfer Network for Prevention of Mental Diseases and Cancer in the Atlantic Area (2009-1/117) and BEADS "Bio-engineered micro Encapsulation of Active agents Delivered to Shellfish " (262649).

REFERENCES

Aasen, J., Samdal, I.A., Miles, C.O., Dahl, E., Briggs, L.R., and Aune, T., (2005). Yessotoxins in Norwegian blue mussels (*Mytilus edulis*): uptake from *Protoceratium reticulatum*, metabolism and depuration. *Toxicon* 45:265-272.

Aasen, J.A.B., Hardstaff, W., Aune, T., and Quilliam, M.A., (2006). Discovery of fatty acid ester metabolites of spirolide toxins in mussels from Norway using liquid chromatography/tandem mass spectrometry. *Rapid Communications in Mass Spectrometry* 20:1531-1537.

Abraham, A., Plakas, S.M., Wang, Z.H., Jester, E.L.E., El Said, K.R., Granade, H.R., Henry, M.S., Blum, P.C., Pierce, R.H., and Dickey, R.W., (2006). Characterization of polar brevetoxin derivatives isolated from *Karenia brevis* cultures and natural blooms. *Toxicon* 48:104-115.

Artigas, M.L., Vale, P.J.V., Gomes, S.S., Botelho, M.J., Rodrigues, S.M., and Arnorim, A., (2007). Profiles of paralytic shellfish poisoning toxins in shellfish from Portugal explained by carbamoylase activity. *Journal of Chromatography A* 1160:99-105.

Bauder, A.G., Cembella, A.D., Bricelj, V.M., and Quilliam, M.A., (2001). Uptake and fate of diarrhetic shellfish poisoning toxins from the dinoflagellate *Prorocentrum lima* in the bay scallop *Argopecten irradians. Marine Ecology Progress Series* 213:39-52.

BEADS. Bio-engineered micro Encapsulation of Active agents Delivered to Shellfish. Grant Agreement Number 262649. Years 2010-2012.

Blanco, J., Acosta, C.P., Bermúdez de la Puente, M., and Salgado, C., (2002a). Depuration and anatomical distribution of the amnesic shellfish poisoning (ASP) toxin domoic acid in the king scallop *Pecten maximus. Aquatic Toxicology* 60:111 - 121.

Blanco, J., Acosta, C.P., Marino, C., Muniz, S., Martin, H., Morono, A., Correa, J., Arevalo, F., and Salgado, C., (2006). Depuration of domoic acid from different body compartments of the king scallop *Pecten maximus* grown in raft culture and natural bed. *Aquatic Living Resources* 19:257-265.

Blanco, J., Bermúdez de la Puente, M., Arevalo, F., Salgado, C., and Morono, A., (2002b). Depuration of mussels (*Mytilus galloprovincialis*) contaminated with domoic acid. *Aquatic Living Resources* 15:53-60.

Blanco, J., Fernandez, M.L., Miguez, A., and Morono, A., (1999). Okadaic acid depuration in the mussel *Mytilus galloprovincialis*: one- and two-compartment models and the effect of environmental conditions. *Marine Ecology Progress Series* 176:153-163.

Blanco, J., Mariño, C., Martín, H., and Acosta, C.P., (2007). Anatomical distribution of diarrhetic shellfish poisoning (DSP) toxins in the mussel *Mytilus galloprovincialis. Toxicon* 50:1011 - 1018.

Blanco, J., Morono, A., Franco, J., and Reyero, M.I., (1997). PSP detoxification kinetics in the mussel *Mytilus galloprovincialis*. One- and two-compartment models and the effect of some environmental variables. *Marine Ecology Progress Series* 158:165-175.

Blanco, J., Reyero, I., and Franco, J., (2003). Kinetics of accumulation and transformation of paralytic shellfish toxins in the blue mussel *Mytilus galloprovincialis. Toxicon* 42:777-784.

Bogan, Y.M., Harkin, A.L., Gillespie, J., Kennedy, D.J., Hess, P., and Slater, J.W., (2007a). The influence of size on domoic acid concentration in king scallop, *Pecten maximus* (L.). *Harmful Algae* 6:15-28.

Bogan, Y.M., Kennedy, D.J., Harkin, A.L., Gillespie, J., Vause, B.J., Beukers-Stewart, B.D., Hess, P., and Slater, J.W., (2007b). Variation in domoic acid concentration in king scallop (*Pecten maximus*) from fishing grounds around the Isle of Man. *Harmful Algae* 6:81-92.

Bouillon, R.C., Kieber, R.J., Skrabal, S.A., and Wright, J.L.C., (2008). Photochemistry and identification of photodegradation products of the marine toxin domoic acid. *Marine Chemistry* 110:18-27.

Bourdelais, A.J., Tomas, C.R., Naar, J., Kubanek, J., and Baden, D.G., (2002). New fish-killing alga in coastal Delaware produces neurotoxins. *Environmental Health Perspectives* 110:465-470.

Braña Magdalena, A., Lehane, M., Moroney, C., Furey, A., and James, K.J., (2003). Food safety implications of the distribution of azaspiracids in the tissue compartments of scallops (*Pecten maximus*). *Food Additives and Contaminants* 20:154-160.

Bricelj, V.M. and Cembella, A.D., (1995). Fate of gonyautoxins in surf clams, *Spisula solidissima*, grazing upon toxigenic *Alexandrium*, p. 413-418. In: Lassus, P., Arzul, G., Erard, E., Gentien, P., and Marcaillou, C. (Eds.), *Harmful Marine Algal Blooms. Technique et Documentation*, Lavoisier/Intercept, Paris.

Bricelj, V.M., Lee, J.H., and Cembella, A.D., (1991). Influence of dinoflagellate cell toxicity on uptake and loss of paralytic shellfish toxins in the northern quahog *Mercenaria mercenaria*. *Marine Ecology Progress Series* 74:33-46.

Bricelj, V.M. and Shumway, S.E., (1998). Paralytic shellfish toxins in bivalve molluscs: occurrence, transfer kinetics and biotransformation. *Reviews in Fisheries Science* 6:315-383.

Campbell, D.A., Kelly, M.S., Busman, M., Wiggins, E., and Fernandes, T.F., (2003). Impact of preparation method on gonad domoic acid levels in the scallop, *Pecten maximus* (L.). *Harmful Algae* 2:215-222.

Campbell, S., Harada, R.M., DeFelice, S.V., Bienfang, P.K., and Li, Q.X., (2009). Bacterial production of tetrodotoxin in the pufferfish *Arothron hispidus*. *Natural Product Research* 23:1630-1640.

Carmichael, W.W. and Mahmood, N.A., (1984). Toxins from Fresh-Water Cyanobacteria. *American Chemical Society Symposium Series* 262:377-389.

Cembella, A.D., Shumway, S.E., and Lewis, N.I., (1993). Anatomical distribution and spatiotemporal variation in paralytic shellfish toxin composition in two bivalve species from the Gulf of Maine. *Journal of Shellfish Research* 12:389-403.

Chen, C.Y. and Chou, H.N., (2001). Accumulation and depuration of paralytic shellfish poisoning toxins by purple clam *Hiatula rostrata* Lighttoot. *Toxicon* 39:1029-1034.

Cho, Y., Ogawa, N., Takahashi, M., Lin, H.P., and Oshima, Y., (2008). Purification and characterization of paralytic shellfish toxin-transforming enzyme, sulfocarbamoylase I, from the Japanese bivalve *Peronidia venulosa*. *Biochimica Et Biophysica Acta-Proteins and Proteomics* 1784:1277-1285.

Choi, M.C., Hsieh, D.P.H., Lam, P.K.S., and Wang, W.X., (2003). Field depuration and biotransformation of paralytic shellfish toxins in scallop *Chlamys nobilis* and green-lipped mussel *Perna viridis*. *Marine Biology* 143:927-934.

Ciminiello, P., Dell'Aversano, C., Fattorusso, E., Forino, M., Grauso, L., Magno, S.G., Poletti, R., and Tartaglione, L., (2007). Desulfoyessotoxins from adriatic mussels: A new problem for seafood safety control. *Chemical Research in Toxicology* 20:95-98.

Ciminiello, P., Dell'Aversano, C., Fattorusso, E., Forino, M., Magno, S., and Poletti, R., (2002). The detection and identification of 42,43,44,45,46,47,55-heptanor-41-oxoyessotoxin, a new marine toxin from Adriatic shellfish, by liquid chromatography-mass spectrometry. *Chemical Research in Toxicology* 15:979-984.

Ciminiello, P., Fattorusso, E., Forino, M., Magno, S., Poletti, R., and Viviani, R., (1998). Isolation of adriatoxin, a new analogue of yessotoxin from mussels of the Adriatic sea. *Tetrahedron Letters* 39:8897-8900.

Ciminiello, P., Fattorusso, E., Forino, M., and Poletti, R., (2001). 42,43,44,45,46,47,55-Heptanor-41-oxohomoyessotoxin, a new biotoxin from mussels of the northern Adriatic sea. *Chemical Research in Toxicology* 14:596-599.

Ciminiello, P., Fattorusso, E., Forino, M., Poletti, R., and Viviani, R., (2000a). A new analogue of yessotoxin, carboxyyessotoxin, isolated from Adriatic Sea mussels. *European Journal of Organic Chemistry* 2:291-295.

Ciminiello, P., Fattorusso, E., Forino, M., Poletti, R., and Viviani, R., (2000b). Structure determination of carboxyhomoyessotoxin, a new yessotoxin analogue isolated from adriatic mussels. *Chemical Research in Toxicology* 13:770-774.

Daiguji, M., Satake, M., James, K.J., Bishop, A., MacKenzie, L., Naoki, H., and Yasumoto, T., (1998a). Structures of new pectenotoxin analogs, pectenotoxin-2 seco acid and 7-epi-pectenotoxin-2 seco acid, isolated from a dinoflagellate and greenshell mussels. *Chemistry Letters*:653-654.

Daiguji, M., Satake, M., Ramstad, H., Aune, T., Naoki, H., and Yasumoto, T., (1998b). Structure and fluorometric HPLC determination of 1-Desulfoyessotoxin, a new yessotoxin analog isolated from mussels from Norway. *Natural Toxins* 6:235-239.

Díaz Sierra, M., Furey, A., Hamilton, B., Lehane, M., and James, K.J., (2003). Elucidation of the fragmentation pathways of azaspiracids, using electrospray ionisation, hydrogen/deuterium exchange, and multiple-stage mass spectrometry. *Journal of Mass Spectrometry* 38:1178-1186.

Dickey, R., Jester, E., Granade, R., Mowdy, D., Moncreiff, C., Rebarchik, D., Robl, M., Musser, S., and Poli, M., (1999). Monitoring brevetoxins during a Gymnodinium breve red tide: Comparison of sodium channel specific cytotoxicity assay and mouse bioassay for determination of neurotoxic shellfish toxins in shellfish extracts. *Natural Toxins* 7:157-165.

Dominguez, H.J., Paz, B., Daranas, A.H., Norte, M., Franco, J.M., and Fernandez, J.J., (2010). Dinoflagellate polyether within the yessotoxin, pectenotoxin and okadaic acid toxin groups: Characterization, analysis and human health implications. *Toxicon* 56:191-217.

Donovan, C., Ku, J., Quilliam, M., and Gill, T., (2008). Bacterial degradation of paralytic shellfish toxins. *Toxicon* 52:91 - 100.

Donovan, C.J., Garduno, R.A., Kalmokoff, M., Ku, J.C., Quilliam, M.A., and Gill, T.A., (2009). Pseudoalteromonas bacteria are capable of degrading paralytic shellfish toxins. *Applied and Environmental Microbiology* 75:6919-6923.

Drum, A.S., Siebens, T.L., Crecelius, E.A., and Elston, R.A., (1993). Domoic acid in the Pacific razor clam *Siliqua patula* (Dixon, 1789). *Journal of Shellfish Research* 12:443-450.

Duinker, A., Bergslien, M., Strand, O., Olseng, C.D., and Svardal, A., (2007). The effect of size and age on depuration rates of diarrheic shellfish toxins (DST) in mussels (*Mytilus edulis* L.). *Harmful Algae* 6:288-300.

E.C., (1996). Commission Decision 96/77/EC of 18 January 1996 establishing the conditions for the harvesting and processing of certain bivalve molluscs coming from areas where the paralytic shellfish poison level exceeds the limit laid down by Council Directive 91/492/EEC. *Official Journal of European Communities* 15:46 - 47.

E.C., (2002). Commission Decision 2002/226/EC of 15 March 2002 establishing special health checks for the harvesting and processing of certain bivalve molluscs with a level of amnesic shellfish poison (ASP) exceeding the limit laid down by Council Directive 91/492/EEC *Official Journal of European Communities* 75:65 - 66.

Fast, M.D., Cembella, A.D., and Ross, N.W., (2006). In vitro transformation of paralytic shellfish toxins in the clams *Mya arenaria* and *Protothaca staminea. Harmful Algae* 5:79-90.

Gilgan, M., Burns, B.G., and Landry, G., (1990). Distribution and magnitude of domoic acid contamination of shellfish in Atlantic Canada during 1988., p. 171–188. In: Graneli, E., Anderson, D., Edler, L., and Sunstrom, B. (Eds.), *Toxic Marine Phytoplankton*. Elsevier, New York.

Gonzalez, J.C., Fontal, O.I., Vieytes, M.R., Vieites, J.M., and Botana, L.M., (2002). Basis for a new procedure to eliminate diarrheic shellfish toxins from a contaminated matrix. *Journal of Agricultural and Food Chemistry* 50:400-405.

Goto, H., Igarashi, T., Yamamoto, M., Yasuda, M., Sekiguchi, R., Watai, M., Tanno, K., and Yasumoto, T., (2001). Quantitative determination of marine toxins associated with diarrhetic shellfish poisoning by liquid chromatography coupled with mass spectrometry. *Journal of Chromatography A* 907:181-189.

Grimmelt, B., Nijjar, M.S., Brown, J., Macnair, N., Wagner, S., Johnson, G.R., and Amend, J.F., (1990). Relationship between domoic acid levels in the blue mussel (*Mytilus edulis*) and toxicity in mice. *Toxicon* 28:501-508.

Gueguen, M., Bardouil, M., Baron, R., Lassus, P., Truquet, P., Massardier, J., and Amzil, Z., (2008). Detoxification of Pacific oyster *Crassostrea gigas* fed on diets of *Skeletonema costatum* with and without silt, following PSP contamination by *Alexandrium minutum. Aquatic Living Resources* 21:13-20.

Hess, P., McMahon, T., Slattery, D., Swords, D., Dowling, G., McMarron, M., Clarke, D., Gibbons, W., Silke, J., and O'Cinneide, M., (2003). Use of LC-MS testing to identify lipophilic toxins to establish local trends and interspecies differences and to test the comparability of LC-MS testing with the mouse bioassay: an example from the Irish biotoxin monitoring program 2001, p. 57-66. In: Villalba, A., Reguera, B., Romalde, J. L., and Beiras, R. (Eds.), *Molluscan Shellfish Safety*. Xunta de Galicia and UNESCO, Spain.

Hess, P., Nguyen, L., Aasen, J., Keogh, M., Kilcoyne, J., McCarron, P., and Aune, T., (2005). Tissue distribution, effects of cooking and parameters affecting the extraction of azaspiracids from mussels, *Mytilus edulis*, prior to analysis by liquid chromatography coupled to mass spectrometry. *Toxicon* 46:62-71.

Holland, P.T., Selwood, A.I., Mountfort, D.O., Wilkins, A.L., McNabb, P., Rhodes, L.L., Doucette, G.J., Mikulski, C.M., and King, K.L., (2005). Isodomoic acid C, an unusual amnesic shellfish poisoning toxin from *Pseudo-nitzschia australis. Chemical Research in Toxicology* 18:814-816.

Ishida, H., Nozawa, A., Nukaya, H., Rhodes, L., McNabb, P., Holland, P.T., and Tsuji, K., (2004). Confirmation of brevetoxin metabolism in cockle, *Austrovenus stutchburyi*, and greenshell mussel, *Perna canaliculus*, associated with New Zealand neurotoxic shellfish poisoning, by controlled exposure to *Karenia brevis* culture. *Toxicon* 43:701-712.

James, K.J., Bishop, A.G., Draisci, R., Palleschi, L., Marchiafava, C., Ferretti, E., Satake, M., and Yasumoto, T., (1999). Liquid chromatographic methods for the isolation and identification of new pectenotoxin-2 analogues from marine phytoplankton and shellfish. *Journal of Chromatography A* 844:53-65.

James, K.J., Lehane, M., Moroney, C., Fernandez-Puente, P., Satake, M., Yasumoto, T., and Furey, A., (2002). Azaspiracid shellfish poisoning: unusual toxin dynamics in shellfish and the increased risk of acute human intoxications. *Food Additives and Contaminants* 19:555-561.

James, K.J., Moroney, C., Roden, C., Satake, M., Yasumoto, T., Lehane, M., and Furey, A., (2003a). Ubiquitous 'benign' alga emerges as the cause of shellfish contamination responsible for the human toxic syndrome, azaspiracid poisoning. *Toxicon* 41:145-151.

James, K.J., Sierra, M.D., Lehane, M., Magdalena, A.B., and Furey, A., (2003b). Detection of five new hydroxyl analogues of azaspiracids in shellfish using multiple tandem mass spectrometry. *Toxicon* 41:277-283.

Jones, G.J. and Negri, A.P., (1997). Persistence and degradation of cyanobacterial paralytic shellfish poisons (PSPs) in freshwaters. *Water Research* 31:525-533.

Kodama, M., Sato, S., Sakamoto, S., and Ogata, T., (1996). Occurrence of tetrodotoxin in *Alexandrium tamarense*, a causative dinoflagellate of paralytic shellfish poisoning. *Toxicon* 34:1101-1105.

Kotaki, Y., Furio, E.F., Satake, M., Lundholm, N., Katayama, T., Koike, K., Fulgueras, V.P., Bajarias, F.A., Takata, Y., Kobayashi, K., Sato, S., Fukuyo, Y., and Kodama, M., (2005). Production of isodomoic acids A and B as major toxin components of a pennate diatom *Nitzschia navis-varingica*. *Toxicon* 46:946-953.

Kotaki, Y., Oshima, Y., and Yasumoto, T., (1985). Bacterial transformation of paralytic shellfish toxins in coral-reef crabs and a marine snail. *Bulletin of the Japanese Society of Scientific Fisheries* 51:1009-1013.

Krock, B., Tillmann, U., John, U., and Cembella, A.D., (2009). Characterization of azaspiracids in plankton size-fractions and isolation of an azaspiracid-producing dinoflagellate from the North Sea. *Harmful Algae* 8:254-263.

Kroken, G., Loken, J., and Myklebust, B., (2006a), Detoxification of shellfish, WIPO, Norway, PCT/NO2006/000101, WO/2006/098640.

Kroken, G., Loken, J.P., and Myklebust, B., (2006b), Feed composition containing standard feed components, useful for detoxifying bivalves and other shellfish, comprises an emulsifier, As, F. T., WO2006098640-A1; NO200705348-A; EP1865793-A1; US2009005345-A1.

Lagos, N., Rutman, M., Blamey, J., Ocaranza, M.P., Chiong, M., Hinrichsen, J.P., and Lopez, C., (2001), Procedure for detoxification of shellfish, contaminated with paralytic shellfish toxins, United States, 6171626,

Landsberg, J.H., (2002). The effects of harmful algal blooms on aquatic organisms. *Reviews in Fisheries Science* 10:113-390.

Lassus, P., Bardouil, M., Ledoux, M., Murail, I., Bohec, M., Truquet, P., Frémy, J.-M., and Rohmer, V., (1992). Paralytic phycotoxin uptake by scallops (*Pecten maximus*). *Aquatic Living Resources* 5:319-324

Lassus, P., Bardouil, M., Massselin, P., Naviner, M., and Truquet, P., (2000). Comparative efficiencies of different non-toxic microalgal diets in detoxification of PSP-contaminated oysters (*Crassostrea gigas* Thunberg). *Journal of Natural Toxins* 9:1-12.

Laurent, D., Kerbrat, A.S., Darius, H.T., Girard, E., Golubic, S., Benoit, E., Sauviat, M.P., Chinain, M., Molgo, J., and Pauillac, S., (2008). Are cyanobacteria involved in Ciguatera Fish Poisoning-like outbreaks in New Caledonia? *Harmful Algae* 7:827-838.

Lin, H.P., Cho, Y., Yashiro, H., Yamada, T., and Oshima, Y., (2004). Purification and characterization of paralytic shellfish toxin transforming enzyme from *Mactra chinensis*. *Toxicon* 44:657-668.

Lindegarth, S., Torgersen, T., Lundve, B., and Sandvik, M., (2009). Differential retention of okadaic acid (OA) group toxins and pectenotoxins (PTX) in the blue mussel, *Mytilus edulis* (L.), and european flat oyster, *Ostrea edulis* (L.). *Journal of Shellfish Research* 28:313-323.

McCarron, P., Kilcoyne, J., Miles, C.O. and Hess, P.(2009). Fromation of Azaspiracids-3, -4, -6 and 9 via decarboxylation of carboxyazaspirazids metabolites fron shellfish.*Journal of Agricultural and Fodd Chemistry* 57:160-169.

McKenzie, L., Holland, P., McNabb, P., Beuzenberg, V., Selwood, A., and Suzuki, T., (2002). Complex toxin profiles in phytoplankton and Greenshell mussels (*Perna canaliculus*), revealed by LC-MS/MS analysis. *Toxicon* 40:1321-1330.

Madhyastha, M.S., Novaczek, I., Ablett, R.F., Johnson, G., Nijjar, M.S., and Sims, D.E., (1991). In vitro study of domoic acid uptake by digestive gland tissue of blue mussel (*Mytilus edulis*, L). *Aquatic Toxicology* 20:73-81.

Madigan, T.L., Lee, K.G., Padula, D.J., McNabb, P., and Pointon, A.M., (2006). Diarrhetic shellfish poisoning (DSP) toxins in South Australian shellfish. *Harmful Algae* 5:119-123.

Mafra, L.L., Bricelj, V.M., and Fennel, K., (2010). Domoic acid uptake and elimination kinetics in oysters and mussels in relation to body size and anatomical distribution of toxin. *Aquatic Toxicology* 100:17-29.

Magdalena, A.B., Lehane, M., Moroney, C., Furey, A., and James, K.J., (2003). Food safety implications of the distribution of azaspiracids in the tissue compartments of scallops (*Pecten maximus*). *Food Additives and Contaminants* 20:154-160.

Mauriz, A. and Blanco, J., (2010). Distribution and linkage of domoic acid (amnesic shellfish poisoning toxins) in subcellular fractions of the digestive gland of the scallop *Pecten maximus*. *Toxicon* 55:606-611.

Medhioub W., Gueguen M., Lassus P., Bardouil M., Truquet P., Sibat M., Medhioub N., Soudant, P., Kraiem M., and Azmil Z. (2010). Detoxification enhancement in the gymnodimine-contaminated grooved carpet shell, *Ruditapes decussatus* (Linné). *Harmful Algae* 9:200-207.

Miles, C.O., Wilkins, A.L., Munday, R., Dines, M.H., Hawkes, A.D., Briggs, L.R., Sandvik, M., Jensen, D.J., Cooney, J.M., Holland, P.T., Quilliam, M.A., MacKenzie, A.L., Beuzenberg, V., and Towers, N.R., (2004). Isolation of pectenotoxin-2 from *Dinophysis acuta* and its conversion to pectenotoxin-2 seco acid, and preliminary assessment of their acute toxicities. *Toxicon* 43:1-9.

Molinet, C., Niklitschek, E., Seguel, M., and Diaz, P., (2010). Trends of natural accumulation and detoxification of paralytic shellfish poison in two bivalves from the Northwest Patagonian inland sea. *Revista De Biologia Marina Y Oceanografia* 45:195-204.

Morono, A., Arevalo, F., Fernandez, M.L., Maneiro, J., Pazos, Y., Salgado, C., and Blanco, J., (2003). Accumulation and transformation of DSP toxins in mussels *Mytilus galloprovincialis* during a toxic episode caused by *Dinophysis acuminata*. *Aquatic Toxicology* 62:269-280.

Munday, R., Holland, P.T., McNabb, P., Selwood, A.I., and Rhodes, L.L., (2008). Comparative toxicity to mice of domoic acid and isodomoic acids A, B and C. *Toxicon* 52:954-956.

Murata, M., Kumagai, M., Lee, J.S., and Yasumoto, T., (1987). Isolation and structure of yessotoxin, a novel polyether compound implicated in diarrhetic shellfish poisoning. *Tetrahedron Letters* 28:5869-5872.

Negri, A.P. and Jones, G.J., (1995). Bioaccumulation of paralytic shellfish poisoning (PSP) toxins from the cyanobacterium *Anabaena circinalis* by the fresh-water mussel *Alathyria condola*. *Toxicon* 33:667-678.

Negri, A.P., Jones, G.J., Blackburn, S.I., Oshima, Y., and Onodera, H., (1997). Effect of culture and bloom development and of sample storage on paralytic shellfish poisons in the cyanobacterium *Anabaena circinalis*. *Journal of Phycology* 33:26-35.

Noguchi, T. and Arakawa, O., (2008). Tetrodotoxin - Distribution and accumulation in aquatic organisms, and cases of human intoxication. *Marine Drugs* 6:220-242.

Novaczek, I., Madhyastha, M.S., Ablett, R.F., Donald, A., Johnson, G., Nijjar, M.S., and Sims, D.E., (1992). Depuration of domoic acid from live blue mussels (*Mytilus edulis*). *Canadian Journal of Fisheries and Aquatic Sciences* 49:312-318.

Novaczek, I., Madhyastha, M.S., Ablett, R.F., Johnson, G., Nijjar, M.S., and Sims, D.E., (1991). Uptake, disposition and depuration of domoic acid by blue mussels (*Mytilus edulis*). *Aquatic Toxicology* 21:103-118.

Nybom, S.M.K., Salminen, S.J., and Meriluoto, J.A.O., (2008). Specific strains of probiotic bacteria are efficient in removal of several different cyanobacterial toxins from solution. *Toxicon* 52:214-220.

Nzoughet, K.J., Hamilton, J.T.G., Floyd, S.D., Douglas, A., Nelson, J., Devine, L., and Elliott, C.T., (2008). Azaspiracid: First evidence of protein binding in shellfish. *Toxicon* 51:1255-1263.

Ofuji, K., Satake, M., McMahon, T., James, K.J., Naoki, H., Oshima, Y., and Yasumoto, T., (2001). Structures of azaspiracid analogs, azaspiracid-4 and azaspiracid-5, causative toxins of azaspiracid poisoning in Europe. *Bioscience Biotechnology and Biochemistry* 65:740-742.

Ofuji, K., Satake, M., McMahon, T., Silke, J., James, K.J., Naoki, H., Oshima, Y., and Yasumoto, T., (1999). Two analogs of azaspiracid isolated from mussels, *Mytilus edulis*, involved in human intoxication in Ireland. *Natural Toxins* 7:99-102.

Oshima, Y., (1995). *Chemical and enzymatic transformation of paralytic shellfish toxins in marine organisms*. Lavoisier Intercept, Paris.

Oshima, Y., Sugino, K., Itakura, H., Hirota, M., and Yasumoto, T., (1990). Comparative-Studies on Paralytic Shellfish Toxin Profile of Dinoflagellates and Bivalves, p. 391-396. In: Graneli, E., Sundstrom, B., Edler, L., and Anderson, D. M. (Eds.), *Toxic Marine Phytoplankton*. Elsevier, New York.

Padilla, L.V., San Diego-McGlone, M.L., and Azanza, R.V., (2010). Exploring the potential of clay in mitigating *Pyrodinium bahamense* var. *compressum* and other harmful algal species in the Philippines. *Journal of Applied Phycology* 22:761-768.

Pavela-Vrancic, M., Mestrovic, V., Marasovic, I., Gillman, M., Furey, A., and James, K.K., (2001). The occurrence of 7-epi-pectenotoxin-2 seco acid in the coastal waters of the central Adriatic (Kastela Bay). *Toxicon* 39:771-779.

Peña-Llopis, S., Peña, J.B., Serrano, R., and Pitarch, E., (2006), N-Acetylcysteine boots detoxification and depuration of organic contaminants in shellfish, University Institute of Pesticides and Water, Jaume I University, Valencia (Spain).

Perez-Guzman, L., Perez-Matos, A.E., Rosado, W., Tosteson, T.R., and Govind, N.S., (2008). Bacteria associated with toxic clonal cultures of the dinoflagellate *Ostreopsis lenticularis*. *Marine Biotechnology* 10:492-496.

Pierce, R., Henry, M., and Blum, P., (2008). Brevetoxin abundance and composition during ECOHAB-Florida field monitoring cruises in the Gulf of Mexico. *Continental Shelf Research* Volume 28, Ecology and Oceanography of Harmful Algal Blooms in Florida:45-58.

Pillet, S., Pereira, A., Brackman, J.-C., and Houvenaghel, G., (1995). Patterns in long term accumulation of okadaic acid and DTX-1 in blue mussels, *Mytilus edulis*, experimentally fed with the DSP-containing alga *Prorocentrum lima.*, p. 487-492. In: Lassus, P., Arzul, G., Erard, E., Gentien, P., and Marcaillou-Le Baut, C. (Eds.), *Harmful Marine Algal Blooms. Tecnique et Documentation.* Lavoisier, Intercept Ltd., Paris.

Plakas, S.M. and Dickey, R.W., (2010). Advances in monitoring and toxicity assessment of brevetoxins in molluscan shellfish. *Toxicon* 56:137-149.

Plakas, S.M., El Said, K.R., Jester, E.L.E., Granade, H.R., Musser, S.M., and Dickey, R.W., (2002). Confirmation of brevetoxin metabolism in the Eastern oyster (*Crassostrea virginica*) by controlled exposures to pure toxins and to *Karenia brevis* cultures. *Toxicon* 40:721-729.

Plakas, S.M., Wang, Z.H., El Said, K.R., Jester, E.L.E., Granade, H.R., Flewelling, L., Scott, P., and Dickey, R.W., (2004). Brevetoxin metabolism and elimination in the Eastern oyster (*Crassostrea virginica*) after controlled exposures to *Karenia brevis. Toxicon* 44:677-685.

Radwan, F.F.Y., Wang, Z.H., and Ramsdell, J.S., (2005). Identification of a rapid detoxification mechanism for brevetoxin in rats. *Toxicological Sciences* 85:839-846.

Reboreda, A., Lago, J., Chapela, M.J., Vieites, J.M., Botana, L.M., Alfonso, A., and Cabado, A.G., (2010). Decrease of marine toxin content in bivalves by industrial processes. *Toxicon* 55:235-243.

Rehmann, N., Hess, P., and Quilliam, M.A., (2008). Discovery of new analogs of the marine biotoxin azaspiracid in blue mussels (*Mytilus edulis*) by ultra-performance liquid chromatography/tandem mass spectrometry. *Rapid Communications in Mass Spectrometry* 22:549-558.

Roelke, D.L., Fryxell, G.A., and Cifuentes, L.A., (1993). Effects on the oyster *Crassostrea virginica* caused by exposure to the toxic diatom *Nitzschia pungens f. multiseries. Journal of Shellfish Research* 12:143.

Rossignoli, A.E. and Blanco, J., (2010). Subcellular distribution of okadaic acid in the digestive gland of *Mytilus galloprovincialis*: First evidences of lipoprotein binding to okadaic acid. *Toxicon* 55:221-226.

Roth, P.B., Twiner, M.J., Wanga, Z., Dechraoui, M.Y.B., and Doucette, G.J., (2007). Fate and distribution of brevetoxin (PbTX) following lysis of *Karenia brevis* by algicidal bacteria, including analysis of open A-ring derivatives. *Toxicon* 50:1175-1191.

Samsur, M., Takatani, T., Yamaguchi, Y., Sagara, T., Noguchi, T., and Arakawa, O., (2007). Accumulation and elimination profiles of paralytic shellfish poison in the short-necked clam *Tapes japonica* fed with the toxic dinoflagellate *Gymnodinium catenatum. Journal of the Food Hygienic Society of Japan* 48:13-18.

Satake, M., Eiki, K., Ichimura, T., Ota, S., Sekiguchi, K., and Oshima, Y., (2006). Structure of 45,46,47-trinorhomoyessotoxin, a new yessotoxin analog, from *Protoceratium reticulatum* which represents the first detection of a homoyessotoxin analog in Japan. *Harmful Algae* 5:731-735.

Satake, M., Ofuji, K., James, K., Furey, A., and Yasumoto, T., (1997). New toxic events caused by irish mussels, p. 468-469. In: Reguera, B., Blanco, J., Fernández, M. L., and Wyatt, T. (eds.), VIII International Conference on Harmful Algae. UNESCO and Xunta de Galicia, Vigo, Spain.

Satake, M., Ofuji, K., Naoki, H., James, K.J., Furey, A., McMahon, T., Silke, J., and Yasumoto, T., (1998). Azaspiracid, a new marine toxin having unique spiro ring assemblies, isolated from irish mussels, *Mytilus edulis*. *Journal of the American Chemical Society* 120:9967-9968.

Satake, M., Terasawa, K., Kadowaki, Y., and Yasumoto, T., (1996). Relative configuration of yessotoxin and isolation of two new analogs from toxic scallops. *Tetrahedron Letters* 37:5955-5958.

Sato, S., Ogata, T., Borja, V., Gonzales, C., Fukuyo, Y., and Kodama, M., (2000). Frequent occurrence of paralytic shellfish poisoning toxins as dominant toxins in marine puffer from tropical water. *Toxicon* 38:1101-1109.

Schantz, S.L. and Widholm, J.J., (2001). Cognitive effects of endocrin-disrupting chemicals in animals. *Environmental Health Perspectives* 109:1197-1206.

Seemann, P., Gernert, C., Schmitt, S., Mebs, D., and Hentschel, U., (2009). Detection of hemolytic bacteria from *Palythoa caribaeorum* (Cnidaria, Zoantharia) using a novel palytoxin-screening assay. *Antonie Van Leeuwenhoek International Journal of General and Molecular Microbiology* 96:405-411.

Shimizu, Y. and Yoshioka, M., (1981). Transformation of Paralytic Shellfish Toxins as Demonstrated in Scallop Homogenates. *Science* 212:547-549.

Smith, E.A., Grant, F., Ferguson, C.M.J., and Gallacher, S., (2001). Biotransformations of paralytic shellfish toxins by bacteria isolated from bivalve molluscs. *Applied and Environmental Microbiology* 67:2345-2353.

SPIES-DETOX. Active biological monitoring and removal of toxins in aquaculture ecosystems and shellfish - including the development of a Solid-Phase In-situ Ecosystem Sampler (SPIES) and detoxification of shellfish (DETOX). Collective Research Project 0302790-2, Years 2007-2009.

Stewart, J.E., Marks, L.J., Gilgan, M.W., Pfeiffer, E., and Zwicker, B.M., (1998). Microbial utilization of the neurotoxin domoic acid: blue mussels (*Mytilus edulis*) and soft shell clams (*Mya arenaria*) as sources of the microorganisms. *Canadian Journal of Microbiology* 44:456-464.

Sugiyama, N., Konoki, K., and Tachibana, K., (2007). Isolation and characterization of okadaic acid binding proteins from the marine sponge *Halichondria okadai*. *Biochemistry* 46:11410-11420.

Sullivan, J.J., Iwaoka, W.T., and Liston, J., (1983). Enzymatic transformation of PSP toxins in the littleneck clam (*Protothaca staminea*). *Biochemical and Biophysical Research Communications* 114:465-472.

Suzuki, T., Igarashi, T., Ichimi, K., Watai, M., Suzuki, M., Ogiso, E., and Yasumoto, T., (2005). Kinetics of diarrhetic shellfish poisoning toxins, okadaic acid, dinophysistoxin-1, pectenotoxin-6 and yessotoxin in scallops *Patinopecten yessoensis*. *Fisheries Science* 71:948-955.

Suzuki, T., Mackenzie, L., Stirling, D., and Adamson, J., (2001a). Conversion of pectenotoxin-2 to pectenotoxin-2 seco acid in the New Zealand scallop, *Pecten novaezelandiae*. *Fisheries Science* 67:506-510.

Suzuki, T., Mackenzie, L., Stirling, D., and Adamson, J., (2001b). Pectenotoxin-2 seco acid: a toxin converted from pectenotoxin-2 by the New Zealand Greenshell mussel, *Perna canaliculus*. *Toxicon* 39:507-514.

Suzuki, T., Ota, H., and Yamasaki, M., (1999). Direct evidence of transformation of dinophysistoxin-1 to 7-O-acyl-dinophysistoxin-1 (dinophysistoxin-3) in the scallop *Patinopecten yessoensis*. *Toxicon* 37:187-198.

Svensson, S., (2003). Depuration of Okadaic acid (Diarrhetic Shellfish Toxin) in mussels, *Mytilus edulis* (Linnaeus), feeding on different quantities of nontoxic algae. *Aquaculture* 218:277-291.

Tillmann, U., Elbrachter, M., Krock, B., John, U., and Cembella, A., (2009). *Azadinium spinosum* gen. et sp nov (*Dinophyceae*) identified as a primary producer of azaspiracid toxins. *European Journal of Phycology* 44:63-79.

Torgersen, T., Sandvik, M., Lundve, B., and Lindegarth, S., (2008). Profiles and levels of fatty acid esters of okadaic acid group toxins and pectenotoxins during toxin depuration. Part II: Blue mussels (*Mytilus edulis*) and flat oyster (*Ostrea edulis*). *Toxicon* 52:418-427.

Twiner, M.J., Rehmann, N., Hess, P., and Doucette, G.J., (2008). Azaspiracid shellfish poisoning: A review on the chemistry, ecology, and toxicology with an emphasis on human health impacts. *Marine Drugs* 6:39-72.

Vale, C., Alfonso, A., Vieytes, M.R., Romaris, X.M., Arevalo, F., Botana, A.M., and Botana, L.M., (2008). In vitro and in vivo evaluation of paralytic shellfish poisoning toxin potency and the influence of the pH of extraction. *Analytical Chemistry* 80:1770-1776.

Vale, P., (2004). Differential dynamics of dinophysistoxins and pectenotoxins between blue mussel and common cockle: a phenomenon originating from the complex toxin profile of *Dinophysis acuta*. *Toxicon* 44:123-134.

Vale, P. and Sampayo, M.A.D., (2002). Evaluation of marine biotoxin's accumulation by *Acanthocardia tuberculatum* from Algarve, Portugal. *Toxicon* 40:511-517.

Vieites, J.M., Botana, L.M., Vieytes, M.R., Leira, F.J., (1999). Canning process that diminishes paralytic shellfish poison in naturally contaminated mussels (*Mytilus galloprovincialis*). *Journal of Food Protection* 62:515-519.

Walter, J.A., Falk, M., and Wright, J.L.C., (1994). Chemistry of the shellfish toxin domoic acid: characterization of related compounds. *Canadian Journal of Chemistry* 72:430.

Wang, Z.H., Plakas, S.M., El Said, K.R., Jester, E.L.E., Granade, H.R., and Dickey, R.W., (2004). LC/MS analysis of brevetoxin metabolites in the Eastern oyster (*Crassostrea virginica*). *Toxicon* 43:455-465.

Wiese, M., D'Agostino, P.M., Mihali, T.K., Moffitt, M.C., and Neilan, B.A., (2010). Neurotoxic Alkaloids: Saxitoxin and Its Analogs. *Marine Drugs* 8:2185-2211.

Wilkins, A.L., Rehmann, N., Torgersen, T., Rundberget, T., Keogh, M., Petersen, D., Hess, P., Rise, F., and Miles, C.O., (2006). Identification of fatty acid esters of pectenotoxin-2 seco acid in blue mussels (*Mytilus edulis*) from Ireland. *Journal of Agricultural and Food Chemistry* 54:5672-5678.

Wohlgeschaffen, G.D., Mann, K.H., Rao, D.V.S., and Pocklington, R., (1992). Dynamics of the phycotoxin domoic acid - accumulation and excretion in 2 commercially important bivalves. *Journal of Applied Phycology* 4:297-310.

Wong, C.K., Hung, P., Lee, K.L.H., Mok, T., and Kam, K.M., (2009). Effect of steam cooking on distribution of paralytic shellfish toxins in different tissue compartments of scallops *Patinopecten yessoensis*. *Food Chemistry* 114:72-80.

Wright, J.L., Bird, C.J., de Freitas, A.S., Hampson, D., McDonald, J., and Quilliam, M.A., (1990). Chemistry, biology, and toxicology of domoic acid and its isomers. *Canada Diseases Weekly Report* 16 Suppl 1E:21-26.

Yasumoto, T., Murata, M., Lee, J.S., and Torigoe, K., (1989). Polyether toxins produced by dinoflagellates, p. 375-382. In: Natori, S., Hashimoto, K., and Ueno, Y. (Eds.), *Mycotoxins and Phycotoxins' 88*. Elsevier, Amsterdam.

Yasumoto, T., Oshima, Y., and Yamaguchi, M., (1978). Occurrence of a new type of shellfish poisoning in the Tohoku district. *Bulletin of the Japanese Society for the Science of Fish* 44:1249-1255.

In: New Trends in Marine and Freshwater Toxins
Editors: A. G. Cabado and J. M. Vieites
ISBN: 978-1-62257-787-3

Chapter 6

ADVANCES IN KNOWLEDGE OF PHYCOTOXINS, NEW INFORMATION ABOUT THEIR TOXICOLOGY AND CONSEQUENCES ON EUROPEAN LEGISLATION

***Maria del Carmen Louzao*[1], *Jorge Lago*[2], *Martiña Ferreira and Amparo Alfonso*[1]**

[1]Department of Pharmacology,
Veterinary Faculty University of Santiago de Compostela, Lugo, Spain
[2]Spanish National Association of Sea Food Producers,
Technological Centre, Vigo, Spain

ABSTRACT

Marine biotoxins may pose an extreme risk to seafood consumers, making necessary the establishment of regulations on safe maximum legal limits to the toxin content in shellfish and other seafood, in order to ensure the protection of public health. The European Union has one of the most complete and strict legislations in food safety, based on risk assessment which is in turns supported on independent scientific advice, mostly undertaken, supervised or collected by the European Food Safety Authority (EFSA). Thus, rules on food safety related to marine biotoxins are mainly based on toxicological data from in vivo and in vitro experiments, as well as clinical observations. Periodic revisions of the available scientific information are required to update regulations, modify legal limits or introduce new rules e. g. on emerging toxins or those commonly found in imported seafood from third countries. This chapter offers a review of the latest information available on the toxicology of the main groups of marine biotoxins (domoic acid group, saxitoxin group, okadaic acid group, azaspiracids group, pectenotoxin group, yessotoxin group, palytoxin group, cyclic imines, brevetoxin group and ciguatoxin group), providing data on the mechanisms of action at cellular and molecular levels, effects in experimental animals or observations on wild fauna, and toxicological reports from human cases. This information is subsequently related to the legislation on marine biotoxins in force in the European Union, paying attention to the establishment of acute reference doses and the subsequent maximum legal limits set from lowest observed adverse effect levels and no observed adverse effect levels. Current or foreseeable legislative changes derived from new toxicological data or the development of new detection methods for toxins are also discussed.

1. Introduction

The European Union legislation in food safety is considered one of the most strict and protective to consumers worldwide. The current European Union regulation on food control (European Parliament, 2002), in force since February 2002, was adopted with the aim of substituting the different laws existing in each Member State and harmonising concepts, principles and procedures and establish common food and feed safety requirements within the European Union.

European legislation foresees that *"measures adopted by the Member States and the Community governing food and feed should generally be based on risk analysis except where this is not appropriate to the circumstances or the nature of the measure"* (European Parliament, 2002). Risk assessment is therefore the main basis on which risk management decisions on food safety are made in the European Union, even though societal, economic, traditional, ethical or environmental factors relevant to the matter under consideration may also be taken into account. In this sense, Regulation (EC) No. 178/2002 established the creation of the European Food Safety Authority (EFSA) as an independent source of scientific advice and communication on risks associated with the food chain. EFSA undertakes a large part of its work in response to specific requests for scientific assessments from the European Commission, the European Parliament and Member States.

As the risk assessor, EFSA produces scientific opinions and advice to provide foundation for European policies and legislation and for taking effective and timely risk management decisions. These may involve adoption or revision of European legislation on food or feed safety, opinion on regulated substances such as pesticides or food additives, or developing new regulatory frameworks and policies. EFSA's work is carried out by the Scientific Committee and the Scientific Panels, composed by experts in risk assessment of proven scientific excellence, that tackle assessment in their respective field. In the case of marine biotoxins, the Contaminants Panel is responsible to provide scientific support in order to review current regulations or establish new ones derived from the existing toxicological and experimental data, the most recent of which are reviewed in the first part of this chapter. In the second part, this information is discussed in the context of European legislation on marine biotoxins.

1.1. New Trends Related to Toxicology

1.1.1. Hydrophilic Toxins

1.1.1.1. Domoic Acid Group

Domoic acid (DA) and its isomers cause amnesic shellfish poisoning (ASP) in humans. These toxins are produced by the marine macroalgae *Chondria armata* and several species of the diatom genus *Pseudo-nitzschia*. DA is a naturally occurring tricarboxylic amino acid, with a molecular weight (MW) of 311 Da, analogue of the excitatory amino acid (EAA) neurotransmitter L-glutamate. DA is structurally similar to the toxin kainic acid, a ligand for the ionotropic class of glutamate receptors. Several isomers of DA, A to H, can be found, though they are present at very low concentrations in shellfish and display a much lower toxicity (Doucette et al., 2008). DA is unstable at high temperatures and extreme pHs and can be degraded or converted to epi-DA during storage (Doucette et al., 2008). Shellfish such as crabs, mussels, razor clams and scallops can accumulate DA either by direct filtration of the plankton or by feeding directly on contaminated organisms. This toxin can also be accumulated in certain filter feeding fish, such as anchovies, sardines, mackerel, and albacore, though levels are usually lower than in shellfish (Costa et al., 2010). Cooking does not significantly reduce toxin levels. DA was for the first time identified as the toxic substance responsible for a 1987 outbreak of gastrointestinal and neurological illness and death in people from Prince Edward Island, Canada, who had consumed toxic cultured mussels (Wright et al., 1990). Consumption of DA contaminated shellfish or fish leads to gastrointestinal symptoms and at higher doses, neurological toxicity in higher species, including humans.

DA is a potent neurotoxin recognized as an agonist of non-N-methyl-D-aspartate (non-NMDA) glutamate receptors including both α-amino-3-hydroxy-5-methyl-4-isoxazolepropionate (AMPA) and kainite receptors (Doucette et al., 2008). Glutamate is a major excitatory neurotransmitter in the brain, and the action of DA on non-NMDA receptors perturbs neurotransmission. Non-NMDA and kainate subfamilies of glutamate receptors comprise tetrameric proteins formed by either identical or, less frequently, different subunits (Lerma, 2006). Nine different subunits have been identified, four of which are components of AMPA receptors (subunits GluR1-4), whereas the remaining five subunits represent the monomers of kainate receptors (subunits GluR5-7 and KA1 and KA2). DA has been shown to be a ligand for the AMPA/kainate subtypes receptors and a selective agonist to a subtype of glutamate receptor subunit GluR6 (Johansen et al., 1993).

DA toxicity has been extensively studied either *in vitro* and *in vivo*. The toxin produces acute cytotoxicity in primary cultures of cerebellar granule cells. This effect is associated with a cytosolic acidification and an increase in intracellular free Ca^{2+} concentration depending on the extracellular Ca^{2+} and mediated by the activation of the AMPA/kainate subtype of glutamate receptors (Vale-Gonzalez et al., 2006). Administration of DA to rodents produces a reproducible pattern of behavioral toxicity that culminates in seizures and is similar, but not identical, to that seen following administration of kainic acid. The toxicity of DA has been studied in several species, at different doses and by different routes: oral, intraperitoneal or intravenous administrations. After oral administration death occurred at concentrations over 50 μg/kg b.w. (body weight) in mice and rats. The intraperitoneal injection of DA to mice and rats has been associated with specific signs, such as a unique scratching of the shoulders by the hind leg, followed by convulsions and often death. LD_{50} in female mice is ranged between 5.8 and 2.9 mg DA/kg b.w. depending on the purity of extract tested (EFSA, 2009g). In rats clinical signs appeared after administration of 1 mg DA/kg b.w. At doses over 4 mg/kg b.w. histopathological lesions in several areas of central nervous system had been referred. The intravenous administration of 0.5-1.0 mg/kg b.w. of DA displayed seizure discharges in the hippocampus, tonic-clonic convulsions, and death within a few days (EFSA, 2009g).

In addition DA is a potent toxicant to newborn animals, which display both immediate and permanent toxicity when exposed to doses below those considered toxic in adult animals. A timed dependent neuroexcitotoxicity involving hyperactivity, stereotypic scratching, convulsions, and death has been reported in newborn rats given DA intraperitoneally. The LD_{50} values for postnatal day two and day ten rats were 0.25 and 0.7 mg/kg b.w., respectively (Xi et al., 1997). For comparison with adult animals, studies in mature rats reported signs of moderate toxicity at intraperitoneal dose of approximately 2.0 mg/kg b.w., whereas in adult mice double dose is necessary (Tryphonas et al., 1990). The chronic administration of sub-toxic dose did not cause clinical abnormalities nor morphological features on diverse organs. However electron microscopy analyses revealed some changes in central nervous system tissues (Pulido, 2008). A daily dose of 0.1 mg/kg b.w. did not produce any effect, this dose has been identified as the NOAEL (no-observed-adverse-effect-level) in rats (Truelove et al., 1996). There are no published reports on the chronic toxicity or carcinogenicity of DA. Reports on the genotoxicity of DA are very limited and the available data are contradictory.

There is little knowledge about the *in vivo* toxicity of DA isomers. It seems that DA isomers are less toxic than DA and the information on the relative potency of epi-DA *in vivo* is lacking. However, because the occurrence data did not make a distinction between the concentration of parent DA and epi-DA and results reported are a sum of both, it has been assumed that DA and epi-DA are equally toxic. Since DA isomers occur at much lower concentrations and less toxic than DA, EFSA Contaminants Panel has concluded that TEFs are not required for DA isomers (EFSA, 2009g).

The acute reference dose based on data on acute toxicity in humans from the 1987 intoxication has been calculated. First a LOAEL (lowest-observed-adverse-effect-level) of 0.9 mg/kg b.w. for clinical symptoms has been established. And then from this value the acute reference dose of 30 μg/kg b.w. was calculated.

1.1.1.2. Saxitoxin Group

Paralytic shellfish poisoning (PSP) toxins are a group of potent neurotoxic compounds produced by dinoflagellates belonging to the genus *Alexandrium*, *Gymnodinium* and *Pyrodinium*. Saxitoxin (STX) was the first toxin identified and characterized in this group, even more than 57 different compounds with similar structure have been described to date, among others neosaxitoxin (NeoSTX) decarbamoylsaxitoxin (dc-STX), decarbamoylneosaxitoxin (dc-NeoSTX), gonyautoxins 1 to 6 (GTX) and their decarbamoyl analogs and C toxins (Wiese et al., 2010). These products are water soluble, heat stable and stable under acidic conditions but quickly degraded at alkaline pH, even at room temperature (Alfonso et al., 1994; Louzao et al., 1994a; Louzao et al., 1994b). STX is an alkaloid of 299 Da molecular weight, composed by a 3,4-propinoperhydropurine tricyclic system. It belongs to the family of guanidinium-containing marine natural products with two guanidine groups responsible for its high polarity (Wiese et al., 2010). The accumulation of PSP toxins in shellfish represents an important public health problem and affects fisheries industry. When PSP toxins are accumulated in bivalves they can be gradually transformed from the GTX group to STX (Kodama et al., 2008). PSP poisoning exhibits symptoms including tinkling sensation or numbness around the lips, numbness of extremities, gastrointestinal problems and difficulty in breathing, and in fatal cases death.

These toxins bind to the receptor site 1 of the alfa-subunit of the voltage-gated Na^+ channel (Hille, 1975; Cestèle et al., 2000). This interaction plugs the channel and blocks its ion conductance (Hille, 1968; Louzao et al., 2001). In excitable cells the blockade of Na^+ conductance prevents membrane depolarization and the transmission of action potential and as a consequence there is a progressive loss of neuromuscular functions and subsequently different neurotoxic symptoms that can produce asphyxia and death. Additionally, Na^+ channels and STX binding are also related with other physiological functions since voltage-gated Na^+ channels play a key role in the pathogenesis of neuropathic pain; therefore, drugs that block these channels are potentially therapeutic (Amir et al., 2006). In addition, it has been described high affinity of STX for Ca^{2+} and K^{+} channels, neuronal nitric oxide synthase, STX metabolizing enzymes and two circulatory fluid proteins found in the blood of puffer fish (Llewellyn, 2006). Therefore, even the molecular target for STX group are the voltage-gated Na^{+} channels all those actions should be taken into account when explaining the toxicological effects of this group of toxins (Suárez-Isla, 2008).

The toxicity of PSP compounds is usually referred to STX toxicity. The mouse intraperitoneal lethal dose 50 % (LD_{50}) of STX is 10 μg/kg b.w., the mouse intravenous (i.v.) LD_{50} is 3 μg/kg b.w. and after oral administration LD_{50} is 260 μg/kg b.w. These values can oscillate depending on the assay technique, the purity of the toxin and also the strain of mice or the animal used. Prior exposure to non-lethal doses of PSP toxins seems to reduce the susceptibility of rats to lethal doses of these toxins (EFSA, 2009b).

The ratio of the toxicity potency of each compound compared to the potency of the reference compound is referred to as the Toxic Equivalency Factor (TEF, see Chapter 5 and (Botana et al., 2010). Although overall values of STX analogues TEFs seem to be similar, several studies show some discrepancies about the equivalence between mousse units and STX concentration or with the toxicity of some analogues as dc-STX or GTX1 (EFSA, 2009b). Most of these studies were conducted with non-certified standards when the use of qualified pure toxins is of primary importance. The Working Group on marine biotoxins, as part of EFSA Contaminants Panel (EFSA, 2009b), has defined, based on current knowledge, the TEF that should be used for conversion of analytical results into toxic concentrations in PSP group. The Panel, based on pure reference material indicated a TEF of 1.0 for STX, NeoSTX, dc STX and GTX1, 0.7 for GTX4 and 0.6 for GTX 3. A TEF of 0.4 for GTX2, dc-NeoSTX and dc-GTX3, 0.3 for 11-hydroxy-STX, 0.2 for dc-GTX3 and 0.1 for GTX5, GTX6, C2 and C4 (Vale et al., 2008a; EFSA, 2009b). The Panel proposed TEFs are based on acute effects after intraperitoneal administration, therefore these effects are not related to oral absorption, due to the large variations of results available in the bibliography. The impact of applying the new TEFs proposed by the Panel to data obtained by a chemical detection method appears to depend on the toxin profile of sample, mainly in relation to the proportion of the dc-STX analogue. The overall impact of the new TEFs is expected to be limited, except for situations where dc-STX is present in significant amounts (EFSA, 2009b).

1.1.2. Lipophilic Toxins

1.1.2.1. Okadaic Acid and Analogues

Okadaic acid (OA) is a cytotoxic polyether derivative of a C38 fatty acid. OA and its analogues, such as the dinophysis toxins (for instance DTX1, DTX2, and DTX3), and OA esters together form the group of OA-toxins, which are lipophilic and heat stable. These toxins are produced as secondary metabolites by dinoflagellates belonging to the genera *Dinophysis* and *Prorocentrum* (Yasumoto et al., 1984). To date, 26 OA analogues have been identified from the dinoflagellates of the genus *Prorocentrum*.

OA-group toxins cause Diarrhoeic Shellfish Poisoning (DSP), which is a specific type of food poisoning characterized by symptoms such as diarrhea, nausea, vomiting and abdominal pain (Dominguez et al., 2010). These symptoms may occur in humans shortly after ingestion of filter feeding bivalve mollusks like mussels, scallops, oysters or clams that accumulated the toxin mainly in their digestive gland (EFSA, 2008a). The wide geographic occurrence of this poisoning poses serious threats to both human health and shellfish industries.

OA and its analogues act by binding and inhibiting serine/threonine protein phosphatases 1 (PP1) and 2A (PP2A), which results in the hyperphosphorylation of many cell proteins (Takai et al., 1987; Honkanen et al., 1994; Dawson et al., 1999). Therefore OA has the potential to disturb the equilibrium of the phosphorylation and dephosphorylation status important for nearly all regulatory processes in the cell (Ehlers et al., 2011). This also includes changes in metabolism, induction of cell motility, loss of adhesion and cytotoxicity in mammalian cells (Estevez et al., 1994; Lerga et al., 1999; Louzao et al., 2003; Cabado et al., 2004; Louzao et al., 2005; Santaclara et al., 2005). In addition OA has an impact on cytoskeletal dynamics (Berven et al., 2001; Vilarino et al., 2008; Espina et al., 2010b). Moreover, this toxin is known to act as a potent neurotoxin for cultured neurons (Fernandez et al., 1991) and to induce apoptotic events in various cell lines like intestinal, neuronal or leukemic cells (Rossini et al., 2001; Leira et al., 2002a; Lago et al., 2005; Jayaraj et al., 2009; Ravindran et al., 2010).

The toxicological database for OA-group toxins is limited and comprises mostly studies on their acute toxicity. Acute toxicity data on OA-group toxins are presented in Table 1. OA is primarily considered an enterotoxin (Edebo et al., 1988; Ito et al., 2002b). The most important acute effects of OA-group toxins in mice and rats are lethality, intestinal injury including diarrhea and liver injury. Essentially the same toxicological endpoints have been reported following intraperitoneal and oral exposure to the toxins (EFSA, 2008a). The mechanism by which OA induces diarrhea in animals and humans was originally proposed to involve hyperphosphorylation of proteins that control Na^+ secretion by intestinal cells (Cohen et al., 1990). Subsequent studies indicates that OA does not act as a secretagogue in the intestine but disrupts the barrier function of intestinal cells and increases paracellular permeability indicating that this alteration is the most likely cause of diarrhea in animals and humans that have ingested OA (Tripuraneni et al., 1997). Injection of OA and other diarrhoeic shellfish toxins into mice causes vessel congestion and extravasation into the lamina propria (Terao et al., 1986), and instillation of DTX1 in ligated loops of the rat intestine leads to rapid fluid accumulation (Edebo et al., 1988). In line with these observations *in vivo* sprinkling of rat colonic mucosa with OA causes a significant decrease in transepithelial electrical resistance without any measurable effect on ion currents of the tissue and in the absence of cell lysis. Also OA was found to cause mucosal edema and submucosal fluid accumulation (Hosokawa et al., 1998). Based on the available data, the mechanism by which OA induces diarrhea in animals and humans includes submucosal fluid accumulation in the intestine wall, the fluid then crosses the epithelial barrier by paracellular pathway and is eventually secreted into the intestinal lumen (EFSA, 2008a).

The extent of the OA-induced injuries and the toxin organotrophicity are dose-related and may be determined by the administration route. The lethal oral dose of OA-group toxins in mice may be 2-10 times higher than the intraperitoneal lethal dose. *In vivo* studies in mice have reported the distribution and excretion of OA following oral administration, as well as morphofunctional changes of organs targeted by the toxin. Following oral administration the critical effect of OA in mice was diarrhea. Also in mice, oral doses of OA induced hypersecretion in the small intestine. The amount of fluid became prominent after 15 minutes, and reached a maximum after 60 minutes at the lower dose. After about 1 hour, severe mucosal injury in the small intestine was seen; extravasion of serum into lamina propria of villi, degeneration of absorptive epithelium of iliac villi, and desquamation of the degenerated epithelium from the lamina propria (EFSA, 2008a). Mice administered DTX3 orally showed light diarrhea within 3 hours, and slightly reduced body weight at 24 hours. After 24 hours, the stomach had light erosions, the small intestine displayed signs of the last phase of recovery, and after 48 hours recovered. The intestinal changes associated with these levels of DTX3, both injury and recovery, were almost the same as those described for OA, but less prominent. The injuries attributed to DTX3 were restricted to the gastrointestinal organs. Rats are more tolerant to OA than mice, but the small intestine is their most sensitive organ (EFSA, 2008a). Liver injury was observed at 1,000-2,000 μg OA/kg b.w. in mice. For DTX1 liver injury did not occur in mice and rats up to an oral dose of 750 μg/kg b.w. whereas for DTX3 the same dose caused liver injury in mice and rats (EFSA, 2008a). After intravenous administration, OA acts as a hepatotoxin with undetectable effects on the intestine.

The toxicological basis to establish TEFs in the evaluation of the combined acute toxicity of toxins of the OA group is that these toxins share a common biochemical mechanism of action (EFSA, 2008a). This is supported by the fact that the relative potency of LD_{50} in mice following intraperitoneal administration of OA and DTX2 is similar to their relative inhibitory effect on PP2A (Aune et al., 2007). It is presumed that the combined exposure to two or more toxins will be additive with respect to dose (dose-addition), and the relative acute oral toxicities are assumed to mirror the relative acute toxicity following intraperitoneal administration. This assumption is not valid for the acylated toxins (DTX3). Human data on DTX3 toxicity following consumption of crabs indicate a slightly lower toxicity for DTX3 than for OA. Since the DTX3 is a very weak inhibitor of PP2A (Takai et al., 1992), hydrolysis to free the corresponding unesterified parent toxins (OA, DTX1 and DTX2) will most likely be a rate limiting step for exerting the toxic effects. Based on those considerations, the following TEFs were established: OA = 1, DTX1 = 1, DTX2 = 0.6. For DTX3 the TEF values are equal to those of the corresponding unesterified toxins (OA, DTX1, and DTX2) (EFSA, 2008a).

No long-term toxicity/carcinogenicity experiments have been reported for OA-group toxins, but OA is identified as a tumor promoter in various organs and embryotoxic (Fujiki et al., 2009; Ehlers et al., 2010). Following the European Regulation (EC) the maximum limit of OA is 160 μg/kg (EFSA, 2008a), this means that small quantities of OA may be present in mollusks that have passed legal controls before its marketing, and therefore chronic exposure to this toxin may exist in regular consumers (Valdiglesias et al., 2010). In view of the tumor promoting effects of OA it was studied whether there may be a link between cancer risk and exposure to OA-group toxins in humans (Ehlers et al., 2011). It was suggested a significant association between consumption of mollusks and digestive cancer along the coasts of France and Spain (Cordier et al., 2000).

Non-standard *in vitro* assays have shown some evidence for unspecific DNA-adduct formation in mammalian cell lines (Fessard et al., 1996) indicating genotoxicity and a clastogenic well as an neurogenic impact on different cell lines (Tohda et al., 1993; Traore et al., 2001; Le Hegarat et al., 2006). For DTX2 and DTX3 no genotoxicity data are available. Recent studies concluded that the genotoxicity of OA is chiefly cell type dependent and concentration dependent and may be related to cytotoxicity of OA, giving sense to controversial genotoxicity data found in the literature (Souid-Mensi et al., 2008; Valdiglesias et al., 2010).

Table 1. Acute toxicity of OA-group toxins; i.p.: intraperitoneal administration

Compound	Route	Toxicity µG/KG B.W.	Reference
OA	i.p.	LD_{50} 192-225	Terao et al (1986); Tubaro et al (2003); Aune et al (2007)
OA	oral	LD_{50} 400-2000	Ito et al (2002); Tubaro et al (2003); Le Hégarat et al (2006)
DTX1	i.p.	LD_{50} 160	Ito et al (2002)
DTX1	oral	LD_{50} 300	Ogino et al (1997)
DTX2	i.p.	LD_{50} 350	Aune et al (2007)
DTX3	i.p.	LD_{50} 200-500	Ito et al (2002)

In summary, it appears that OA is not mutagenic *per se*, but induces changes at the chromosome level and is aneugenic at least *in vitro;* these effects may be related to toxicity.

1.1.2.2. Azaspiracids Group

Azaspiracids (AZAs) are a group of polyether marine neurotoxins described in shellfish associated to human intoxications. To date, 32 different AZA analogues have been identified, AZA1 to 32 (Twiner et al., 2008). AZA1 structure was originally proposed by Satake, Yasumoto and co-workers (Satake et al., 1998) and reassigned by Nicolaou and co-workers after its total synthesis (Nicolaou et al., 2003a; Nicolaou et al., 2003b). AZAs structures are characterized by a trioxadispiroacetal system fused onto a tetrahydrofuran ring (ABCD domain), an azaspiro-ring system fused onto a 2,9-dioxabicyclo [3.3.1] nonane system (FGHI domain), and a terminal carboxylic acid moiety (James et al., 2008). The predominant natural analogues are AZA1, known as azaspiracid, AZA2, 8-methylazaspiracid, and AZA3, 22-demethylazaspiracid. The symptoms associated to AZAs consumption are very similar to those of DSP, i.e., vomiting, stomach cramps and severe diarrhea. Neurological symptoms with progressive paralysis, fatigue, breathing difficulties and death have been reported after intraperitoneal injection of AZA-contaminated shellfish extracts into mice or rats. In addition to pathological effects, some histological alterations in the pancreas, spleen, liver, thymus and small intestine have been described (Ito et al., 2000). Besides, the chronic exposure to the toxin was associated with the development of lung tumors in mice (Ito et al., 2002a).

Since AZA1 is the most abundant in nature, most research was centered on its intracellular target. However, neither its mechanism of action nor its pharmacokinetic behavior have been elucidated, principally because of the lack of standards and reference tissues, even recent advances in purification methodology and toxin stability have been achieved (Alfonso et al., 2008b; Alfonso et al., 2008c). Several approaches have been undertaken to know the biological target of these toxins. Mostly AZA1 increases second messenger levels, cAMP and cytosolic Ca^{2+}, and does not affect intracellular pH. The increase in cytosolic Ca^{2+} is dependent on both the release of Ca^{2+} from intracellular pools and the influx from extracellular media (Roman et al., 2002; Vale et al., 2007a). AZA2 has the same effect than AZA1, while AZA3 does not empty intracellular stores but increases cytosolic Ca^{2+} levels. AZA2 did not modify intracellular pH, but AZA3 slightly alkalinizes the cytosol. These toxins increase cAMP levels and the modulation of this pathway inhibits AZA2- and AZA3-evoked Ca^{2+} increase and AZA3-induced pH rise (Román et al., 2004).

Surprisingly, AZA4 did not modify cytosolic Ca^{2+} in resting cells; however, toxin dose-dependently inhibited the increase in cytosolic Ca^{2+} levels induced by other drugs as thapsigargin or maitotoxin. In addition, this analogue increases cAMP levels and inhibits intracellular alkalinization independently of the Ca^{2+} presence (Alfonso et al., 2005a). AZA5 does not modulate cytosolic Ca^{2+} or intracellular pH (Alfonso et al., 2006). Clearly, each AZA analogue shows different effect on the same intracellular targets, depending on their structure (Botana et al., 2007). The importance of Ca^{2+}, anionic fluxes and intracellular pH changes in AZAs cytotoxicity is still unknown. However, some of these primary targets can be involved in the mechanism of action of this group of natural toxins. Other long term *in vitro* experiments reveal that AZAs toxicity is related to the alteration of the actin cytoskeleton arrangement accompanied by changes in cell shape and loss of cell adherence to the substrate. AZA effects on the actin cytoskeleton are irreversible, take place after long incubation periods and are not related to early activation pathways (Vilariño et al., 2007; Vilariño, 2008; Vilariño et al., 2008). In addition, AZAs treatment increases the phosphorylation state of Jun-N-terminal kinase (JNK) and causes nuclear accumulation of phosphorylated JNK. This pathway is associated to the cytotoxic effect of AZAs (Vale et al., 2007a; Vale et al., 2008b). These results could provide the basis to identify the mechanism of action of this group of toxins.

The intraperitoneal minimal lethal dose of AZA1 in mice is 200 μg/kg b.w., 110 μg/kg b.w. of AZA2 and 140 μg/kg b.w. of AZA3 (Satake et al., 1998; Ofuji et al., 1999). No data about AZA4 and AZA5 are available although less toxicity is predicted (Furey et al., 2010). This group of toxins shows high oral toxicity. Damage in different tissues (epithelial cells, lamina propia, T and B cells) and organs (small intestine or liver) has been shown after oral administration of dose between 200 and 900 μg/kg b.w. of AZA1 within 4 to 24 hours of toxin administration (Ito et al., 2000; Ito et al., 2002a). The lethal oral dose in mice was estimated to be higher than 700 μg/kg b.w. Different studies about oral toxicity of AZA-1 indicate that single oral doses causing lethality vary from 250 to 600 μg/kg b.w. (EFSA, 2008c). Chronic oral administration of 20-50 μg/kg b.w. of AZA1 produces toxic effects on small intestine and liver. In addition this toxin has been shown to be a tumour initiator (Ito, 2008).

The TEFs relative to AZA1 calculated from values obtained by LC-MS/MS and mousse bioassay proposed by EFSA Contaminants Panel are AZA1=1, AZA2=1.8, AZA3=1.4, AZA4=0.4, AZA5=0.2 (EFSA, 2008c). However, due to the lack of information about the toxicity of each analogue, robust TEF values can not be established and the values published should be revised when more acute toxicity assays for all five AZA analogues will be available (EFSA, 2008c).

1.1.2.3. Pectenotoxins

The pectenotoxins (PTXs) are a family of polyether macrolactones originally isolated from scallops *Patinopecten yessoensis* but actually produced by the *Dinophysis* dinoflagellates. They have been detected in microalgae and bivalve mollusks in Australia, Japan, New Zealand and in a number of European countries (Krock et al., 2008; Pizarro et al., 2008; Torgersen et al., 2008). Their presence in shellfish was discovered due to their acute toxicity in the mouse bioassay after intraperitoneal injections of lipophilic extracts of Shellfish (EFSA, 2009f).

PTXs mechanism of action involves the interaction with the actin cytoskeleton (Zhou et al., 1994; Spector et al., 1999; Leira et al., 2002b; Ares et al., 2005; Espina et al., 2008; Espina et al., 2010a). To date more than 20 analogues have been isolated and characterized in which the cyclic structure is important for both toxicity and activity on the actin cytoskeleton. For instance, pectenotoxin 1 (PTX1), pectenotoxin 2 (PTX2), pectenotoxin 6 (PTX6) and pectenotoxin 11 (PTX11) modify actin cytoskeleton and morphology of neuroblastoma cells (Ares et al., 2007). However, pectenotoxin-2 secoacid (PTX2 SA), a non-cyclic PTX2 derivative found in many shellfish worldwide, has low toxicity and no cytoskeletal effect (Miles et al., 2004; Ares et al., 2007). Therefore from a structure–activity viewpoint, the lactone ring is essential for the action of PTXs on actin cytoskeletal dynamics (Ares et al., 2005; Allingham et al., 2007; Ares et al., 2007).

PTXs are cytotoxic. PTX2, considered as the major compound of this group, displays selective and potent cytotoxicity against human cells of lung, colon and breast cancer and human leukemia cells (Kim et al., 2008). It has been established that PTX2 triggers cell death by apoptosis, on the basis of morphological changes, a reduction of the mitochondrial membrane potential, an increase in the level of cytoplasmic cytochrome *c* and Smac/DIABLO, and the activation of caspase-3 and caspase-9 (Fladmark et al., 1998; Chae et al., 2005; Kim et al., 2008).

Cytotoxic agents have been shown to play important roles in anticancer therapy due to their ability to interfere with actin cytoskeleton dynamics. PTXs may be potent anti-cancer drug candidates since they clearly alter actin cytoskeleton, furthermore morphological assessments indicated a higher sensitivity of cancer cell line towards PTX1, PTX2 and PTX11 (Espina et al., 2008). Another study indicate that the anticancer effect of PTX2 is due to disruption of the actin cytoskeleton through the inhibition of actin polymerization, forming a complex with G-actin (Saito et al., 1996). The mechanism whereby PTX2 induces cell death and cell cycle-dysfunction in leukemia cells is related to G_2/M phase arrest, endoreduplication, and apoptosis through the ERK and JNK signal pathway via actin depolymerization (Moon et al., 2008). In this case F-actin depolymerization induced by PTX-group toxins is the causative event of subsequent cell death.

The adverse health effects of PTXs in humans are unknown. Therefore the toxicological database for PTXs comprises mostly results on their acute toxicity in mice. Interestingly, the differences in potency between analogues are much more severe at cellular level than in the whole organism. Cytotoxicity of PTXs toward hepatocytes decreased as the level of oxidation at C-43 increased (PTX2>PTX1>PTX6>PTX9). Therefore, successive oxidation events in the Japanese scallop can be considered as a detoxification cascade (Espina et al., 2010a). At the cellular level concentrations about 1μM are required to induce significant depolymerizing effect on the actin cytoskeleton of primary hepatocytes and neuroblastoma whereas other cellular-damage indicator parameters such as percent of attachment, cytosolic Ca^{2+} levels, or cell proliferation were unaffected (Leira et al., 2002b; Ares et al., 2007; Espina et al., 2010a). PTX9 was even less potent than PTX6 in primary hepatocytes. PTX9 is a 7-*S* isomer of PTX6 and Yasumoto already pointed out that the 7-*S*-epimers of the PTX molecules seem to be less toxic toward mice than their corresponding 7-*R*-epimers (Yasumoto et al., 1984).

On the basis of the available information it could be concluded that the toxin-actin interaction and the consequent perturbation of the actin cytoskeleton could be the molecular basis of the cell damage in biological systems exposed to PTX-group toxins. Histopathological studies have shown that intraperitoneal injection of PTX1 damage the liver in a dose dependent way (Terao et al., 1986; Burgess et al., 2001), but no pathological changes were detected in the intestine or other visceral organs. *In vivo* the acute toxicities of PTX1, PTX2, PTX3 and PTX11 by intraperitoneal administration to mice were similar (Miles et al., 2004; Suzuki et al., 2006; Ares et al., 2007) and the symptoms of intoxication were indistinguishable (animals became hunched and lethargic, showed labored respiration, ataxia and cyanosis) (Munday, 2008a). The toxicity of PTX4 and PTX6 is lower, the lethal dose to mice was 500 µg/kg b.w. PTX6 by intraperitoneal injection, with liver bleeding and injuries at the gastric organs and the kidney (Ito et al., 2008). For the other analogues tested no lethality was observed (lethal doses>5000 µg/kg b.w.). Acute toxicity data on PTX-group toxins are presented in Table 2.

Oral toxicity of PTX-group toxins studied generally appears to be much lower than toxicity following intraperitoneal administration. Following oral administration PTX-group toxins show low systemic absorption and reported toxicity is mainly restricted to the intestinal tract. It was reported that in mice, single oral administration of PTX2 at doses of 400 µg/kg b.w. and above resulted in tissue injury with vacuole formation in epithelial cells and fluid accumulation in the small intestine (EFSA, 2009f). No effects were observed at 300 µg/kg b.w. In a recent study mice and rats were given single oral doses of PTX6 by gavage (Ito, 2008), 2 mg/kg did not have diarrheic activity; however, the middle-lower small intestine (jejunum-ileum) was eroded at villi by edema. In mice, doses up to 5000 µg/kg b.w. only caused a slight ultrastructural and transient injury at the small intestinal villus tops. Neither was other mouse organs affected. In the same study PTX2 caused intestinal fluid secretion in mice and slight fluid secretion in rats at single doses of 500 and 1500 µg/kg b.w., respectively. There are no data on possible effects of PTX-group toxins following repeated oral administration.

The available data on lethality in mice only comprise information following intraperitoneal injection and are not sufficient to establish robust TEFs. In order to be prudent it was proposed a provisional TEF value of 1 to be used for PTX1, PTX2, PTX3, PTX4, PTX6 and PTX11, until more robust data become available. PTX7, PTX 8, PTX 9, PTX2 SA and 7-epi-PTX2 SA are much less toxic and were not assigned TEFs (EFSA, 2009f).

Table 2. Acute toxicity of PTX-group toxins

Compound	Route	Toxicity µg/kg b.w.	Reference
PTX1	i.p.	MLD 250	Yasumoto et al (1985)
PTX2	i.p.	LD_{50} 219-411	Miles et al (2004)
PTX3	i.p.	MLD 350	Murata et al (1986)
PTX4	i.p.	MLD 770	Yasumoto et al (1989)
PTX6	i.p.	MLD 500	Yasumoto et al (1989)
PTX7	i.p.	MLD > 5000	Sasaki et al (1998)
PTX8	i.p.	MLD > 5000	Sasaki et al (1998)
PTX9	i.p.	MLD > 5000	Sasaki et al (1998)
PTX11	i.p.	LD_{50} 244	Suzuki et al (2006)
PTX2 SA	i.p.	MLD > 5000	Miles et al (2004)
7-epi-PTX2 SA	i.p.	MLD > 5000	Miles et al (2006)

Notes: i.p.: intraperitoneal administration; MLD: minimum lethal dose.

1.1.2.4. Yessotoxins Group

Yessotoxins (YTXs) are polyether compounds produced by phytoplanktonic dinoflagellates *Proteceratium reticulatum* and *Lingulodinium polyedrum*, and originally isolated from *Patinopecten yessoensis* (Satake et al., 1997). YTX planar structure was determined in 1987 by NMR techniques (Murata et al., 1987) and later on the absolute configuration was reported (Satake et al., 1996; Takahashi et al., 1996). There are several natural compounds belonging to the YTX class, all of them having 11 contiguous ether rings, an unsaturated side chain and one or more sulfate groups.The existence of this toxin group was discovered due to their high acute toxicity in mice after intraperitoneal injection of lipophilic extracts of mollusks. However, much lower toxicity has been reported after oral administration (Terao et al., 1990; Ogino et al., 1997; Aune et al., 2002) and no reports about human intoxications caused by YTXs have been published. More than 90 YTX analogues have been described, even structures for most of them have not been determined and only 30 of them have been isolated (Miles et al., 2005; Paz et al., 2008).

YTX modulates cytosolic Ca^{2+} levels through Ca^{2+} influx channels (De La Rosa et al., 2001a; De La Rosa et al., 2001b). The chemical structure of YTX resembles those of brevetoxins and ciguatoxins, whose mechanism of action is mediated through voltage-gated Na^+ channels. However, YTX did not induce any direct effect on Na^+ channels (Inoue et al., 2003). On the other hand, YTX induces a dose-dependent decrease of cAMP and cGMP levels after 10 minutes of incubation of in vitro experimental setups. These effects are Ca^{2+}-dependent and can be modified by specific phosphodiesterase (Alfonso et al., 2008a). Cells regulate the levels of these second messengers by a balance between adenylyl cyclases (synthesis) and phosphodiesterases (PDEs) (hydrolysis). PDEs are a group of isozymes that includes several families with different substrate specificity, affinity, sensitivity to inhibitors and tissue localization. YTX induces a dose-dependent increase in PDEs activity. In parallel, the toxin decreases cAMP levels and increases the rate of hydrolysis of this second messenger. All these effects can be mimicked by PDEs activators and are modulated by enzyme inhibitors (Alfonso et al., 2003). These results point out PDEs as a cellular target for YTX and Ca^{2+} as key factor for toxin effect. The interaction between these enzymes and YTX was demonstrated by immobilizing PDEs in a biosensor surface. The value of the kinetic equilibrium dissociation constant (K_D) for YTX-PDEs association is 3.74×10^{-6} M (Pazos et al., 2004). The K_D value increases when YTX molecule is modified, indicating a structure-activity relationship. Under the same conditions, the K_D for hydroxy-YTX-PDEs interaction is 7.36×10^{-6} M, and that for carboxy-YTX-PDEs interaction is 23×10^{-6} M (Pazos et al., 2005). These results point out a structure-selectivity of YTX-PDEs association, and agree with the decrease in toxic effect observed with some YTX analogs (Tubaro et al., 2003). The PDEs-YTX interaction was later on confirmed by measuring changes in fluorescence polarization of an enzyme-dye conjugate in the presence of YTX (Alfonso et al., 2005b). By using different enzyme families in a sensor surface and by measuring changes in fluorescence polarization, it was concluded that YTX binds to cyclic nucleotide PDE 1, PDE 3 and PDE4, and shows high affinity for exonuclease PDE I (Alfonso et al., 2005b; Pazos et al., 2005). On the other hand, YTX increases interleukin-2 production in human lymphocytes after 24 hours incubation (Alfonso et al., 2003), and inhibits protein phosphatases, but the effect is four orders of magnitude lower than the one induced by diarrheic shellfish poisoning (DSP) (Ogino et al., 1997). DSP toxins are specific and potent inhibitors of Ser/Thr protein phosphatases PP1 and PP2A. These enzymes play a critical role in phosphorylation/dephosphorylation processes within eukaryotic cells. Therefore even YTXs often coexist with DSP toxins, their effects and mechanism of action are different.

Several cytotoxic effects of YTXs have been reported after *in vitro* experiments in different cellular models. YTX induces cells detachment from culture dishes (Ogino et al., 1997). However no effects on F-actin levels were detected (Leira et al., 2003; Ares et al., 2005). On the other hand, an accumulation of a 100 kDa fragment of E-cadherin without a parallel loss of the intact protein due to YTX effect was described (Pierotti et al., 2003). The collapse of E-cadherin system happened after 2-5 days YTX treatment (sub-nanomolar range) (Ronzitti et al., 2004). However YTX did not induce any effect on E-cadherin system in *in vivo* experiments (Callegari et al., 2006). YTX induced apoptotic events in different cellular lines (Alfonso et al., 2008a).

No reports of human poisoning induced by YTXs have been reported. Several studies have been published about *in vivo* toxicity. YTX has been extensively reported as lethal for mice after intraperitoneal injection. However, a range from 80 to 700 µg/kg b.w. of toxin has been published as LD_{50} after intraperitoneal administrations. This wide range can be due to different experimental conditions, animals or toxins purity. The LD_{50} of some analogues has also been studied and it is in the range described for YTX, 444 µg/kg b.w. for homoYTX or 301 µg/kg b.w. for di-desulfo-YTX (Terao et al., 1990; Tubaro et al., 2003). Cardiac cellular damages have been found after YTX intraperitoneal administration (Terao et al., 1990; Aune et al., 2002), however no effects in liver, pancreas, lungs, adrenals, kidney, spleen or thymus have been reported. With the discrepancies in LD_{50} of YTX is not easy to establish TEF of analogues, however according with toxicological available data the following TEFs have been proposed by EFSA Contaminants Panel: YTX=1, 1a-homoYTX=1, 45-hydroxyYTX=1 and 45-hydroxy-1a homoYTX=0.5 (EFSA, 2008b).

No lethality or changes in mice behaviour were observed after oral administration of doses up to 54 mg/kg b.w. (Tubaro et al., 2010), only some effects in cardiac cells were observed (Aune et al., 2002). In addition no changes at any level were observed after daily oral administration, 7 days, only some ultrastructural effect in cardiac cells were reported (Tubaro et al., 2010). After oral administration, most of the toxin is recovered from lower intestine and faeces and only trace amount are found in blood urine and tissues (Munday et al., 2008).

1.1.2.5. Palytoxins

Palytoxins (PlTXs) are a group of extremely potent, marine natural products originally isolated in Hawaii and Japan from zoanthids belonging to the genus *Palythoa*, but subsequently identified worldwide in red algae, sea anemones, and several dinoflagellates such as *Ostreopsis* (Moore et al., 1971; Beress et al., 1983; Ishida et al., 1983; Cagide et al., 2009). From these primary sources, palytoxin and analogues move up the food chain, and have been found in fish, crabs and mollusks often at high concentrations, being responsible for human illness and death (Aligizaki et al., 2011). Assays on the exposure of some marine invertebrates to these toxins also give indications of the high impact that these compounds may have on natural food webs (Louzao et al., 2010b).

Palytoxin (PlTX) is a large (MW 2680 Da), very complex molecule with a long polyhydroxylated and partially unsaturated aliphatic backbone that may have more than 10^{21} stereoisomers (Katikou, 2007). Only the chemical structures of some of the PlTX analogs have been elucidated, e.g., ostreocin (Ukena et al., 2001), mascarenotoxins and ovatoxin

(Ciminiello et al., 2008). PlTXs exert their potent biological activity by altering mechanisms of ion homeostasis in excitable and non-excitable tissues. Na^+,K^+-ATPase is the molecular target and receptor of the toxin in sensitive systems (Habermann, 1989). PlTX binds to the pump and transforms it into a non-selective cation channel, allowing passive flow of ions following their concentration gradients and triggering membrane depolarization (Artigas et al., 2004; Louzao et al., 2006; Rossini et al., 2010). As a result palytoxin causes a wide spectrum of secondary pharmacological actions such as cytoskeletal filamentous actin disassembly and cytomorphologic alterations (Valverde et al., 2008; Vilarino et al., 2008; Louzao et al., 2010a). Cytotoxicity of PlTX, ostreocin-D and ovatoxin has been observed in a variety of experimental systems *in vitro*. The chemical composition of palytoxin and its analogue ostreocin-D differs in that two methyls, two hydroxyls and one proton present in the first molecule are substituted by four protons and one hydroxyl in the second molecule (Ukena et al., 2002). However, these changes do not seem to affect essentially their biological potencies, for instance both compounds displayed similar effects upon the cytoskeleton, causing depolymerisation of filamentous actin and redistribution of globular actin (Ares et al., 2005; Louzao et al., 2007; Ares et al., 2009).

As a consequence of its molecular action skeletal, heart and smooth muscle cells are the major targets of PlTX among excitable cells, responding with contraction induced both directly, following the increased intracellular Ca^{2+} concentrations (Ito et al., 1977), and indirectly, through the Ca^{2+}-induced release of neurotransmitters (Nagase et al., 1987; Karaki et al., 1988; Vale et al., 2006). An increased metabolism of arachidonic acid and the production of eicosanoids has long been recognized as a cellular response to PlTX in different cells, such as rat liver cells, mouse clavarie, mouse 3T3 fibroblasts (Miura et al., 2006). The increased release of prostaglandin from the endothelium and smooth muscle cells has been shown to determine norepinephrine release and contraction of the rabbit aortas (Nagase et al., 1987). A search for different biochemical mediators of palytoxin led also to the identification of mitogen activated protein (MAP) kinases (Vale et al., 2007b; Wattenberg, 2007). The central role of MAP kinases in regulating a variety of critical cellular functions, ranging from enzyme activity to gene expression and ultimately to proliferation may help explain how palytoxin can stimulate carcinogenic effects being identified as a tumor promoter (Wattenberg, 2011).

The occurrence of palytoxin-producing species along the Mediterranean coasts (France, Greece, Italy, and Spain) causes serious concerns to human health since shellfish species intended for consumption may result contaminated (Deeds et al., 2010). Among the PlTX responses found in animal models, vasoconstriction is the cause of heart failure and the consequent death of animals administered with the toxin (Wiles et al., 1974). Evaluation of PlTX toxicity using various animal models determined that their acute toxicity is strongly dependent on the route of administration (Ito et al., 2009). Acute toxicity data on PlTX-group toxins are presented in Table 3.

PlTX was extremely potent through intravenous, intraperitoneal, and intratracheal (i.t.) exposure but was less potent by direct intragastric exposure as well as by sub-lingual administration (Wiles et al., 1974; Habermann, 1989). Following intravenous administration, the rabbit, dog and monkey are most susceptible and the mouse the least susceptible species. Dogs and monkeys given PlTX have early intoxication signs that were ataxia, drowsiness, and limb weakness, but appeared to die from rapid cardiac failure due to profound coronary vasoconstriction (EFSA, 2009e). For all species, histological damage was observed in various organs including liver, lungs, kidneys, brain, and gastrointestinal tract (Deeds et al., 2010). By intramuscular (i.m.) exposure the LD_{50}s in dogs and rats were approximately 2.5-fold higher compared to i.v. exposure. Local irritation and swelling occurred at the site of injection and the onset of symptoms was delayed compared to i.v. After the subcutaneous injection the LD_{50}s in mice and rats were around 4.5-fold higher compared to i.v. exposure. Onset of symptoms was delayed compared both to i.v. and i.m. exposure (Deeds et al., 2010).

The estimated LD_{50} after intraperitoneal injection of palytoxin, is 25 ng/kg b.w. for rabbits (the most sensitive species) and values ranging between 0.31-1 μg/kg b.w. for mice (Ito et al., 1996; Riobo et al., 2008). Toxic signs are characteristic with stretching of hind limbs, lower back and concave curvature of the spinal column, muscle spasms, respiratory distress dyspnoea and progressive muscular paralysis (Ito et al., 1996; Riobo et al., 2008). Speed of symptom onset was similar to those for intramuscular and subcutaneous administration. Lymphoid tissues, including spleen and thymus showed necrosis, and observations of peritonitis and congestion, bleeding and severe toxicity to the mucosa in the small intestine were made (Terao et al., 1992; Ito et al., 1996). Ostreocin-D seems to be less toxic than PlTX by intraperitoneal administration (a relative potency ranging from 0.4 to 1.0) (Ito et al., 2009). By intratracheal administration of PlTX toxic symptoms were similar to those for intraperitoneal administration with paralytic signs and lung pathology including extensive bleeding in the alveoli, edema around blood vessels, gastrointestinal erosions and atrophy of kidney glomeroli except with more pronounced respiratory difficulty and wheeling.

PlTX is much less toxic to animals by the oral route (LD_{50} in mice was 549-1039 μg/kg b.w.), even though consumption of seafood contaminated with palytoxin has been implicated in several incidents of human intoxication (Tubaro et al., 2011. In press). Toxicity signs in mice were from scratching to spasms, paralysis mainly in hind limbs, respiratory distress and jumping. Histopathological alterations were observed in the forestomach, liver and pancreas. Dose related ultrastructural changes were seen in heart and skeletal muscle cells with rounded mitochondria and fiber degeneration (Sosa et al., 2009). Mice receiving ostreocin-D by gavage had slight and transient injuries in the stomach and the lung (Ito et al., 2009). Ostreocin-D appeared to be only slightly less toxic than PlTX.

Since PlTX and ostreocin-D only show minor structural differences, cause similar signs of toxicity and act on the same target, they could be considered to be equipotent by the oral route. The severe effects, including mortalities, could be explained by the PlTXs absorption during ingestion prior to acid decomposition in the stomach. Among other responses, vasoconstriction is the cause of heart failure and the consequent death of animals administered with the toxin (Wiles et al., 1974; Habermann, 1989; Ito et al., 2009). PlTX also caused significant, non-lethal effects through dermal and ocular exposure (Nordt et al., 2009). Recent evidence also suggests that humans are negatively impacted through palytoxin exposure by inhalation (Durando et al., 2007) where people showed symptoms of a serious respiratory distress (Ciminiello et al., 2010).

1.1.2.6. Cyclic Imines

The cyclic imines are a heterogeneous group of marine natural products sharing common macrocyclic features and the presence of an active imine moiety (Otero et al., 2011). This group includes spirolides (SPXs), gymnodimines (GYMs), pinnatoxins, prorocentrolides, pteriatoxins and spiro-prorocentrimine (Cembella et al., 2008).

Table 3. Acute toxicity of PlTX-group toxins

Compound	Route	Toxicity µg/kg b. w.	Reference
PlTX	i.v. rabbit	LD_{50} 0.025	Wiles et al (1974)
	i.v. dog	LD_{50} 0.033-0.06	Wiles et al (1974); Ito et al (1982)
	i.v. monkey	LD_{50} 0.078	Wiles et al (1974)
	i.v. rat	LD_{50} 0.089	Wiles et al (1974)
	i.v. guinea pig	LD_{50} 0.11	Wiles et al (1974)
	i.v. mouse	LD_{50} 0.15-0.45	Wiles et al (1974); Moore and Scheuer (1971)
	i.m. rat	LD_{50} 0.24	Wiles et al (1974)
	i.m. dog	LD_{50} 0.080	Wiles et al (1974)
	s.c. rat	LD_{50} 0.40	Wiles et al (1974)
	s.c. mouse	LD_{50} 1.39	Wiles et al (1974)
	i.p. rat	LD_{50} 0.63	Wiles et al (1974)
	i.p. mouse	LD_{50} 0.31-1	Ito et al (1996); Moore and Scheuer (1971); Riobo et al (2008)
	i.t. rat	LD_{50} 0.36	Wiles et al (1974)
	oral mouse	LD_{50} 510-1039	Sosa et al (2009)
	oral rat	$LD_{50} > 40$	Wiles et al (1974)
Ostreocin-D	i.p. mouse	LD_{50} 0.75	Ito el al (2009)

Notes: i.v.: intravenous; i.p.: intraperitoneal; i.m.: intramuscular; s.c.: subcutaneous.

SPXs are produced by the dinoflagellate *Alexandrium ostenfeldii*, whereas GYMs are produced by the dinoflagellate *Karenia selliformis* (Hu et al., 2001). Prorocentrolides and spiro-prorocentrimine are produced by members of dinoflagellate genus *Prorocentrum* (Cembella et al., 2008). The organism producing pinnatoxins and pteriatoxins has not been identified but is suspected to be a marine dinoflagellate (Cembella, 1998). All macrocyclic imines have a six or seven membered imino ring associated by a spiro linkage to a cyclohexenyl ring, a macrocycle comprising 16-27 carbon atoms with a five- or six-membered cycloether. The cyclohexenyl ring has a different side chain depending on the toxin group (Cembella et al., 2008). SPX are the most extensively studied subgroup of imine toxins and as consequence the reference group in this big family. These toxins are divided in three groups with a polyether ring system and a specific imine ring. In the first group the imine ring is heptacyclic and includes spirolides A, B, C, D and some analogues. In the second group the imine ring is not cyclic and includes E and F analogues and the third group is composed by spirolide G and one analogue with the imine ring intact and an unusual 5:5:6-trispiroketal ring system, never observed before in other marine toxins group (Hu et al., 2001; Pelc et al., 2005). A new subclasse of SPX marine toxins, represented by SPX H and I, has recently been isolated from the dinoflagellate *A. ostenfeldii* (Roach et al., 2009). These analogues are structurally distinct from other SPX in a 5:6 dispiroketal ring system rather than the trispiroketal ring system characteristic of previously isolated spirolides. Pinnatoxins group includes four different analogues A to C while pteriatoxins and GYMs groups include three analogues each (A to C). Prorocentrolides are the largest cyclic imine toxins, 1 kDa molecular weight, and only one spiro-prorocentrimine has been so far described.

No data about toxic effects in humans due to ingestion of SPX have been reported. However, general symptoms such as tachycardia and gastric distress were reported following consumption of contaminated shellfish from Nova Scotia during the spring and summer of 1991, when SPX were detected in mollusc from this region (Richard et al., 2001; Sleno et al., 2005). Although the mechanism of action of cyclic imines in cells is not yet totally understood, it has been suggested that the muscarinic acetylcholine receptors might be implicated in its mode of action (Gill et al., 2003). This group of toxins binds in the subnanomolar range with subtype selectivity to nicotinic acetylcholine receptors. SPXs and GYMs are potent antagonists of those receptors (Kharrat et al., 2008; Vilariño et al., 2009; Bourne et al., 2010; Fonfría et al., 2010). The integrity of the molecules seems to be important to keep this mode of action. Particularly, the imine ring is required for the biological activity of pinnatoxin A (PnTX A), since an amino ketone analog does not display show affinity for nicotinic acetylchlonile receptors (Araoz et al., 2011).

13-desmethyl SPX C and 13,19-didesmethyl SPX C can be detected in blood, urine and faeces of mice following oral administration. These toxins can be detected in blood after fifteen minutes of a single oral dose after, while after 1 hour administration, no quantifiable levels were found in blood. At this time high levels of both toxins were found in urine and faeces (manuscript in preparation). In addition, the rapid recovery seen in animals following a sub-lethal dose of GYM or SPX suggests a rapid detoxification or excretion of the toxins in animal species (Richard et al., 2001; Munday et al., 2004). As for *in vitro* toxicological data, micromolar concentrations of GYM or pinnatoxins produce some decrease in cell viability, while no remarkable *in vitro* cytotoxic effects of other groups have been published (Munday et al., 2004).

The LD_{50} of GYM after intraperitoneal injection is 96 μg/kg b.w. and after oral administration 755 μg/kg b.w. In both cases no histological changes have been observed in any tissue tested (Munday, 2008b). The LD_{50} of SPX after intraperitoneal administration is around 40μg/ml. The toxicity can be higher, 6.9 and 8 μg/kg b.w., depending on the purity and analogue tested. By the gavage route the LD_{50} is between 53 and 176 μg/kg b.w. of the most toxic analogues and in feeding experiments the range is between 500 and 1005 μg/kg b.w. (EFSA, 2010d) No macroscopic or histological changes have been observed in mice after SPX administration. The symptoms associated to SPX or GYM lethal doses either after intraperitoneal or oral routes are uncoordinated movements, lacrimation, exophtalmia, mouth breathing and death. At sub-lethal dose levels after prostration and respiratory distress the animals recover fully and their subsequent appearance and behaviour are normal (Munday, 2008b). In the case of pinnatoxins, LD_{50} values ranged between 16 and 50 μg/kg b.w. have been published depending on the analogue tested (EFSA, 2010d). After a lethal dose of pinnatoxins, mice were hyperactive followed by an abrupt decrease in activity, abdominal breathing, and extension of the hind legs and, in some cases, slight exophthalmia although respiration rate was still normal and 25 minutes approx. after dosing, mice died with a brief period of running movements just before death associated with severe exophthalmia. At sub-lethal doses, after abdominal breathing and inaction, the animal is full recovered within 2 hours (Selwood et al., 2010). After intraperitoneal administration, the acute toxicity of pteriatoxins oscillates between 100 and 8 μg/kg b.w. depending on the analogue of pteriatoxin (Munday, 2008b). No details about acute toxicity of prorocentrolides are available.

There are no long-term studies on the groups of cyclic imines in experimental animals that would allow establishing a TEF. Since these toxins are characterised by binding to and blocking of nicotinic receptors in the central and peripheral nervous system including neuromuscular junctions, an acute reference dose should be established for the different groups of cyclic imines. However, due to the lack of adequate quantitative data on acute oral toxicity this is not possible.

1.1.2.7. Brevetoxins

Brevetoxin (BTX) is primarily produced by a dinoflagellate *Karenia brevis* and accumulates in shellfish and fish. BTX-group toxins are lipid-soluble cyclic polyether compounds, which are grouped in two types of chemical structures (A and B), based on their backbones. BTX-1 (type A) is reported to be the most potent while BTX-2 (type B) is the most abundant in *K. brevis*. BTX-1 and BTX-2 are considered the parent toxins from which other BTX-group toxins derive (Baden et al., 2005). BTXs are extensively metabolized in shellfish and fish. Consumers of contaminated seafood are thus primarily exposed to metabolites rather than parent algal BTX.

BTX-group toxins bind to site 5 and activate the voltage-gated Na^+ channels in cell membrane thereby causing depolarization of neuronal and muscle cell membranes (Louzao et al., 2004; Louzao et al., 2006; Watkins et al., 2008; Meunier et al., 2009). The activation of receptor site 5 of the voltage-gated Na^+ channels by BTX-group toxins leads to acute neural injury and cell death through an increase in intracellular Ca^{2+} with a 2-fold larger response for BTX-1 compared to BTX-2 or BTX-3 (Lepage et al., 2003). The potency of BTX-group toxins depends on two factors: the affinity of the toxin for its target and the efficacy of that binding to elicit a response in target cells (Ramsdell, 2008). Neurologic and respiratory problems associated with BTX-group toxins are also due to opening of Na^+ channels and persistent membrane depolarization.

Brevetoxins can cause significant mortalities of fish and other aquatic animals, including marine mammals, through direct exposure during harmful algal blooms, or indirectly in the food web. Many filter-feeding molluscan are known to accumulate brevetoxins with no obvious adverse effects, however do pose a significant human health risk (Plakas et al., 2010). Human exposure to BTX may occur via ingestion of brevetoxin-contaminated shellfish causing neurologic (neurotoxic) shellfish poisoning (NSP). Little epidemiological data exist for this poisoning due to the cases of NSP are likely underreported and misdiagnosed (Watkins et al., 2008). Symptoms and signs are mainly gastrointestinal (e.g., abdominal pain, nausea, vomiting, diarrhea) and neurological (e.g., ataxia, myalgia, paresthesia, and reversal of temperature sensation, cramps, paralysis, seizures and coma) (Plakas et al., 2010). They typically occur within 30 minutes to 3 hours of consuming contaminated shellfish and last for a few days. Persistent symptoms and fatalities have not been reported. Although ingestion is associated with more severe effects due to exposure to larger doses, inhalation of aerosolized BTX during red tide events in the coastal areas around the world is significantly more widespread (Milian et al., 2007). Inhalation of red tide aerosols is associated with both upper and lower respiratory symptoms including bronchoconstriction (Backer et al., 2005; Fleming et al., 2007; Hilderbrand et al., 2010) and persons with asthma are particularly susceptible to those effects (Fleming et al., 2009). Also a dermal exposure can result in irritant effects.

The toxicological database for BTX-group toxins is limited and comprises only studies on their acute toxicity following intravenous, intraperitoneal and oral administration. Acute toxicity data on BTX-group toxins and metabolites are limited and are presented in Table 4.

Table 4. Acute toxicity of BTX-group toxins after oral, intraperitoneal (i.p.) and intravenous (i.v.) administration in mice

Compound	Route	Toxicity µG/KG B.W.	Reference
BTX-2	i.v.	LD_{50} 200	Baden & Mende (1982)
BTX-2	i.p.	LD_{50} 200	Baden & Mende (1982)
BTX-2	oral	LD_{50} 6600	Baden & Mende (1982)
BTX-3	i.v.	LD_{50} 94	Baden & Mende (1982)
BTX-3	i.p.	LD_{50} 170-250	Baden & Mende (1982) Selwood et al (2008)
BTX-3	oral	LD50 520	Watkins (2008)
BTX-B2	i.p.	LD_{50} 400	Selwood et al (2008)
S-deoxy-BTX-B2	i.p.	LD_{50} 211	Selwood et al (2008)
BTX-B1	i.p.	MLD 50	Selwood et al (2008)
BTX-B2	i.p.	MLD 306	Murata et al (1998)
BTX-B3	i.p.	MLD>300	Morohashi et al (1995)
BTX-4	i.p.	MLD 100	Morohashi et al (1995)
BTX-5	i.p.	MLD 300-500	Ishida et al (2004)

Notes: MLD: minimum lethal dose.

At lethal intraperitoneal doses, signs of BTX intoxication included: immobility after 15 minutes, respiration paralysis, exophthalmia and rapid flicking movements of the hind legs immediately before death. No macroscopic lesions were observed at necropsy. At sublethal doses, fast abdominal breathing was noticed directly after injection followed by a precipitate drop of the respiration rate, limbs appeared to be completely paralyzed, but movement was regained after 3-5 hours. No gross pathological abnormalities were recorded at the end of the 7-day observation period (Baden, 1989; Selwood et al., 2008). After intravenous administration toxic signs and death were observed almost instantly. After oral administration toxic signs occurred approximately after 5 hours. The oral potency of BTX-2 is more than one order of magnitude lower than that of BTX-3. The difference in oral toxicity between BTX-2 and BTX-3 might be due to differences in their rate of absorption rather than first pass metabolism of BTX-2 in the liver. Based on intraperitoneal toxicity BTX-2, BTX-3, BTX-B2 and S-deoxy-BTX-B2 appear to have similar toxic potencies. However, based on the limited oral toxicity data it appears that the toxicity of BTX-3 is about 10 fold higher than that of BTX-2 (EFSA, 2010c).

There are several indications of clastogenic activity (chromosomal aberrations and DNA damage) of BTX-group toxins *in vitro*. BTX-2 also induced DNA damage *in vivo*. Neither BTX-2 nor BTX-6 was mutagenic in a reverse mutation assays, but there is evidence that BTX-2 forms DNA adducts in isolated rat lung cells treated *in vitro* and in lung tissue following intratracheal administration to rats. These observations raise concern about potential carcinogenicity of BTX-2 and its consequential long term effects (Radwan et al., 2006; EFSA, 2010c). There are no long term studies on BTX-group toxins in experimental animals.

NSP appears to be limited to the Gulf of Mexico, the east coast of the United States of America (U.S.A.), and the New Zealand Hauraki Gulf region. To date BTX have not been reported in shellfish or fish from Europe. However, the discovery of new BTX-group toxin producing algae and the apparent trend towards expansion of algal bloom distribution, suggest that BTX could also emerge in Europe. Due to the lack of occurrence data on shellfish or fish in Europe, the limited data on acute toxicity and the lack of data on chronic toxicity, it is difficult to evaluate the risk associated with the BTX-group toxins in shellfish and fish that could reach the European market. Currently there are no regulatory limits for BTX-group toxins in shellfish or fish in Europe.

1.1.2.8. Ciguatoxins

Ciguatoxin (CTX)-group toxins are lipid-soluble polyether marine biotoxins (Isobe et al., 2010). They are classified based on the place in which they were found as Pacific (P), Caribbean (C) and Indian Ocean (I) CTX-group toxins (Pearn, 2001; Ting et al., 2001). It have been identified the chemical structures of more than 20 analogues of P-CTX-group toxins, two C-CTX-group toxins (C-CTX-1 and C-CTX-2) with several analogues and four closely related I-CTX-group toxins. All ciguatoxin congeners differ slightly in their structures and toxicities, in which the variability of symptoms following exposure in the different regions may underlie (Lewis, 2001). The predominant congener found in fish flesh of the Pacific, Pacific ciguatoxin 1 (P-CTX-1), can cause human illness at 0.1 ppb and is roughly 10 times more potent than the most common Caribbean congener, Caribbean ciguatoxin 1 (C-CTX-1) (Lehane et al., 2000).

CTX are secondary metabolites produced by the benthic dinoflagellate of the genus *Gambierdiscus* growing predominantly in association with macroalgae in coral reefs in tropical and subtropical climates (Friedman et al., 2008). The toxin is transferred through the food web as the algae is consumed by herbivorous fish, which are eaten by carnivorous fish, which are in turn ingested by humans (Lehane et al., 2000; Pearn, 2001). Elimination of ciguatoxins from fish is reported to be slow, which may serve as a reservoir for toxin accumulation (Lehane et al., 2000). In certain cases, enzymatic modification of the polyether backbone by the fish can lead to additional congeners of the natural products (Nicolaou et al., 2008).

New observations suggest expansion of the biogeographical range of *Gambierdiscus* spp. and ciguatoxic fish (Dickey et al., 2010). Recently CTX-group toxins were identified for the first time in fish caught in Europe e.g. in Canary Islands and Madeira (Perez-Arellano et al., 2005; Boada et al., 2010; Otero et al., 2010). Additionally, identification of species of dinoflagellates of the genus *Gambierdiscus*, which are potential CTX producers, has been recorded in the Mediterranean Sea since 2003 and for the first time in the waters of the Canary Islands in 2004 (Caillaud et al., 2010). It was suggested that harmful algae bloom (HAB) intensification and expansion are linked to anthropogenic (Ruff, 1989) and naturally occurring environmental changes, including global warming and increased nutrient loading (Dickey et al., 2010).

The primary mechanism of ciguatoxin action is the high affinity binding and activation of site 5 on voltage-gated Na^+ channels in cell membranes (Bidard et al., 1984; Lombet et al., 1987; Gawley et al., 1992; Dechraoui et al., 1999) with a consequent increase of ion permeability. CTX induced a Na^+ dependent, and tetrodotoxin sensitive, excitable cell depolarization in a variety of isolated nerve and muscle tissue and cell preparations (Ohshika, 1971; Terao, 2000; Louzao et al., 2004; Louzao et al., 2006). Ciguatoxins cause spontaneous, and enhance evoked, action potentials by lowering activation thresholds and delayed repolarization of voltage-gated Na^+ channels (Bidard et al., 1984; Molgo et al., 1990). Further CTX cellular effects include elevation of intracellular Ca^{2+} levels (Molgo et al., 1993) stimulation of spontaneous and evoked neurotransmitter release from synaptosomes and motor nerve terminals (Molgo et al., 1990); axonal and Schwann cell edema (Dickey et al., 2010); induction of tetrodotoxin-sensitive leakage current in dorsal root ganglion neurons (Strachan et al., 1999); and blockade of voltage-gated K^+ channels (Birinyi-Strachan et al., 2005).

CTX induced depolarization of nerve cells is believed to cause the array of neurological signs associated with ciguatera fish poisoning (CFP) (Cameron et al., 1991). CFP is a human intoxication primarily caused by the consumption of large predator fishes that have accumulated the CTX-group toxins by feeding on smaller contaminated coral reef fish. CFP is a complex syndrome characterized by a wide variety of symptoms and signs such as gastrointestinal (e.g. vomiting, diarrhea, nausea), neurological (e.g. tingling, itching) and cardiovascular (e.g. hypotension, bradycardia) disturbances and general symptoms of fatigue, weakness and peripheral pain that, in extreme cases, can persist for years (Matta et al., 2002; Shoemaker et al., 2010). In severe cases the symptoms may begin as soon as 30 minutes after ingestion of contaminated fish, while in milder cases they may be delayed for 24 to 48 hours. Fatalities may occur due to cardiorespiratory failure. The frequency of CFP varies by region throughout the world (Lewis, 2001). Currently there are no regulatory limits for CTX-group toxins in the European Union (EU), but the EU regulation states that controls are to take place to ensure that fishery products containing biotoxins such as ciguatoxin are not placed on the market. The collection of CFP epidemiological data is inefficient, and the public health impact of this disease is likely significantly underestimated (EFSA, 2010a).

The altered ion conductance and increased neurotransmitter secretion represent the molecular basis for the CTX induced loss of cell excitability in nervous and muscular systems which may lead to paralysis in animals exposed to CTX-group toxins (EFSA, 2010a). The toxicological database for CTX-group toxins is limited and comprises mostly studies on their acute toxicity following intraperitoneal and oral administration. Acute toxicity data on CTX-group toxins are presented in Table 5.

The CTXs are acutely toxic upon intraperitoneal administration. The main signs of toxicity for P-CTX-1 are hypothermia, piloerection, diarrhea, lacrymation, hypersalivation, dyspnoea, wobbly upright gait, gasping and terminal convulsions with tail arching and death from respiratory failure. For P-CTX-2 and P-CTX-3 progressive hind limb paralysis is seen in addition. At doses near the LD_{50} the minimum time to death varies from a half to one hour for the three P-CTX toxins (EFSA, 2010a). Upon intraperitoneal injection of crude extracts of the initially characterized I-CTX, signs similar to those of PCTX- and C-CTX-group toxins were observed (Hamilton et al., 2002). Effects of single and repeated intraperitoneal exposures to CTX at a single dose (Terao et al., 1991)or multiple dose for 15 days (Terao et al., 1992)were examined in male mice. Mice received a single dose of 0.7 μg/kg b.w. and showed laborious movements and lumbar muscle contraction followed by severe watery diarrhea that lasted for only 90 minutes, and followed by an apparent recovery after a few hours (Terao et al., 1991). Then, suddenly severe dyspnoea and cyanosis appeared and 70 % of the mice died within 24 hours. The remaining 30 % showed paralyzed paws, and in half of them, penis erection and dilated and filled urinary bladder. Dilated heart and lung edema and marked congestion of the organs were observed. Upon histopathology, necrotic cells were seen in the heart and by electron microscopy, characteristic ultrastructural changes in the heart were rounded mitochondria and marked edema between myofibrils and other organelles. Degeneration of adrenal medulla and erect penises with cavernous thrombi were observed. The livers were congested with the presence of thrombi and the toxins caused swollen synapses in the smooth muscle layer of the vas deferens. Similar changes were observed in the smooth muscular layer of the small intestine, but despite severe diarrhea no changes were seen in the mucosal layer (EFSA, 2010a). A single dose of 0.1 µg/kg b.w. of P-CTX-1 did not cause any morphological effect in the heart either seen at macroscopic, microscopic or ultrastructural examination (Terao et al., 1992). 15 repeated doses of 0.1 µg/kg b.w. caused effects of similar severity as those seen after a single dose of 0.7 µg/kg b.w. Within a month after the last dose, the myocytes and capillaries appeared normal whereas bundles of dense collagen in the interstitial spaces persisted at 14 months. No differences in clinical signs or histopathology between P-CTX-1 and P-CTX-4C were observed (EFSA, 2010a).

Experiments where mice were exposed by oral savage to different doses of CTX result that the lethal dose and clinical signs were almost the same as those seen following intraperitoneal administration, except that diarrhea only occurred after parenteral administration (Ito et al., 1996). The mice receiving a single oral dose of 0.7 µg/kg b.w. showed similar symptoms and histopathological changes as those seen after intraperitoneal administration (Terao et al., 1991). Atropine pretreatment prevented the diarrhea indicating that it was caused by actions on the autonomic nerve system in the intestinal wall. However, atropine did not prevent cardiac injuries. No differences in clinical signs or histopathology between P-CTX-1 and P-CTX-4C were observed. There are no long term studies in experimental animals (EFSA, 2010a).

The different CTX-group toxins share the same molecular receptor and dose addition is presumed to be the result of simultaneous exposure to multiple analogues. Therefore it was adopted the following TEFs for CTX-group toxins based on their acute *intraperitoneal* LD_{50} in mice: P-CTX-1 = 1, P-CTX-2 = 0.3, P-CTX-3 = 0.3, P-CTX-3C = 0.2, 2,3-dihydroxy PCTX-3C = 0.1, 51-hydroxy P-CTX-3C = 1, P-CTX-4A = 0.1, P-CTX-4B = 0.05, C-CTX-1 = 0.1 and C-CTX-2 = 0.3. These TEFs should be applied to express individual analogues identified with quantitative detection methods as P-CTX-1 equivalents (EFSA, 2010a).

Table 5. Acute toxicity of CTX-group toxins after intraperitoneal (i.p.) administration in mice

Compound	Route	toxicity µG/KG B.W.	Reference
P-CTX-1	i.p.	LD_{50} 0.25	Lewis et al. (1991); Dickey (2008)
P-CTX-2	i.p.	LD_{50} 2.3-0.9	Lewis et al. (1991); Lewis (2001)
P-CTX-3	i.p.	LD_{50} 0.9	Lewis et al. (1991)
P-CTX-3C	i.p.	LD_{50} 1.3-2	Satake et al (1993); Lewis (2001)
2,3-dihydroxy P-CTX-3C	i.p.	LD_{50} 1.8	Satake et al (1998)
51-hydroxy P-CTX-3C	i.p.	LD_{50} 0.27	Satake et al (1998)
P-CTX-4A	i.p.	LD_{50} 2	Satake et al (1997)
P-CTX-4B	i.p.	LD_{50} 4	Murata et al (1990); Satake et al (1997)
C-CTX-1	i.p.	LD_{50} 3.6-3.7	Vernoux and Lewis (1997); Lewis et al (1998); Dickey (2008)
C-CTX-2	i.p.	LD_{50} 1	Vernoux and Lewis (1997)

1.2. European Union Regulation of Marine Biotoxin Contents in Shellfish. Risk Assesment and Establishment of Acute Reference Doses and Legal Limits

The control of the safety of live bivalve shellfish in the European Union is established in a series of regulations that set maximum legal limits of biotoxin content, designate responsible authorities for the control of the presence of biotoxins in bivalve molluscs and establishes official detection methods. Regulation (EC) No. 853/2004 (European Parliament, 2004a), laying down specific hygiene rules for foodstuffs of animal origin, establishes the maximum levels of ASP, PSP and DSP, measured in whole body or any part edible separately, that live bivalve molluscs placed on the market for human consumption must not exceed, as follows:

Toxin or toxin group	Maximum legal limits in the European Union
Domoic acid	20 mg of domoic acid per kg
Saxitoxin group	800 μg per kg
Okadaic acid, dinophysistoxins and pectenotoxins together	160 μg of okadaic acid equivalents per kg
Yessotoxins	1 mg of yessotoxin equivalent per kg
Azaspiracids	160 μg of azaspiracid equivalents per kg

Each EU Member State designates competent authorities, which are responsible for monitoring the presence of these regulated marine biotoxins in live bivalves, as well as the presence of toxic plankton species in production areas (European Parliament, 2004b) The recognised detection methods for regulated marine biotoxins were set in the Regulation (European Commission, 2005) According to these rules, HPLC was established as the official detection method for ASP, whereas PSP and DSP must be detected by mouse bioassay (MBA). Nevertheless, this Regulation foresees the use of other recognised methods, especially in order to replace bioassays, provided that their implementation would ensure a level of public health protection equivalent to that of official detection methods. Hence, Regulation (EC) No. 2074/2005 was further amended to accept the Lawrence method (AOAC, 2005) for the detection of PSP (European Commission, 2006) and of the ELISA method (AOAC, 2006) for screening purposes in the detection of ASP (European Parliament, 2007).When results are challenged, mouse bioassay and HPLC will prevail as the reference methods for PSP and ASP detection respectively. Recently, Regulation (EC) No. 2074/2005 was modified with the aim of replacing the mouse bioassay with a LC-MS/MS method validated in an inter-laboratory study carried out by Member States. LC-MS/MS will be established after June 2011 as the reference method, although mouse bioassay may still be used during the period of adaptation to the new method, that will extend no longer than 31 December 2014 (European Commission, 2011).

EU legislation/concern on marine biotoxins is currently undergoing a revision in depth, motivated by three main reasons (Paredes, 2011):

1. Opinions on the need for reassessment of the current maximum legal limits of marine biotoxins in bivalves, according to the existing data on their acute and chronic toxicity to humans and experimental animals.
2. The increasing concerns about the ethical implications of bioassays, and the growing pressure to replace them by alternative detection methods.
3. The appearance of biotoxins so far unreported in the EU areas of bivalve production, which may require additional regulations to guarantee the human health safety in relation to consumption of bivalve shellfish.

European Regulations No. 853/2004 and No. 854/2004 establish that legislation on food safety must have a sound scientific basis and that for that purpose the European Food Safety Agency should be consulted whenever necessary. In this sense, since 2007 EFSA's Panel on Contaminants in the Food Chain (CONTAM) has issued a series of scientific opinions upon request from the European Commission, in order to revise the EU current limits of regulated marine biotoxins and to provide guidance on risk assessment associated to emerging ones. According to the provisions made in food safety regulations (European Parliament, 2002; European Parliament, 2004b), EFSA opinions may determine modifications of current legislation on marine biotoxins.

In order to calculate the concentration of marine biotoxins to which consumers may be potentially exposed, it is necessary to know the habits of shellfish consumption. This is usually subjected to large seasonal variations, therefore one-year basis or longer studies are preferable to determine the frequency of consumption, the size of the shellfish portion consumed in one meal and the number of consumers. Nevertheless, this type of information is very scarce and available data are limited to a few surveys, mostly aimed at the evaluation of nutrient intake or contaminant exposure of sub-populations or ethnical groups within a country (Tsuchiya, 2008; Whyte, 2009; Fialkowski, 2010).

The EU approach to diet studies is completely different. Dietary surveys in the EU have been mainly addressed to obtain information about short-term food consumption, using dietary records rather than diet history methodologies, in the opinion that the former can be better implemented and they are less of a burden for the subjects, therefore encouraging participation. The main drawback of diet surveys carried out so far in the EU is undoubtedly the great methodological heterogeneity among different studies; nevertheless, these are the main source of information on which EU risk assessment on marine biotoxins is done. Recently, EFSA has released reports and scientific opinions highlighting the necessity of harmonized methods for the collection of data on food habits within the EU, but recommended protocols for dietary surveys still rely mainly on short-term consumption data. According to EFSA's guidelines, information on short-term diet habits can also be appropriate to estimate long-term food consumption, provided a sufficient number of replicates are available to accurately estimate within-person variability, and provided this type of surveys are complemented with questionnaires to obtain information about rarely consumed foods (EFSA, 2009c; EFSA, 2010b).

Since the effects of chronic exposure to marine biotoxins in humans are still virtually unknown due to the insufficient available data, the scientific assessment of risks associated to these compounds has been based mainly on data referred to acute toxicity. For this reason, currently it is not possible to establish tolerable daily intake (TDI) values, and risk assessment focuses on the establishment of acute reference doses (ARfDs), which are defined as the amount of substance in food –in the present case, a marine biotoxin-, normally expressed on a body weight basis (mg/kg or μg/kg of body weight), that can be ingested in a period of 24 hours without appreciable health risk to the consumer on the basis of all known facts at the time of evaluation (FAO/IOC/WHO, 2004).

In order to calculate the concentration of one or a group of marine biotoxins to which a consumer may be potentially exposed, EFSA's Contaminants Panel has established the size of the reference shellfish portion that can be consumed in one meal. From the data provided by five Member States (France, Germany, The Netherlands, United Kingdom and Italy) at the time of writing of the first scientific opinion on marine biotoxins -okadaic acid and analogues- (EFSA, 2008a), the Contaminants Panel identified 400 g as the size of a large portion of shellfish meat consumed in Europe and used this figure for the risk assessment of shellfish consumption with regard to the exposure to all the regulated marine biotoxins in the EU (okadaic acid and analogues, domoic acid, saxitoxin-group toxins, azaspiracid-group toxins, yessotoxin-group toxins and pectenotoxin-group toxins). The 400 g figure was calculated from the 95th percentile values of shellfish consumption provided by Member States and assumed to be a realistic estimate of the portion size for high consumers. Controversy has arisen in some Member States, particularly the shellfish producer ones, which consider this figure as too high and not representative of the actual shellfish portion that is usually consumed in one meal. Nevertheless, EFSA recently confirmed their earlier decision after receiving new data on shellfish consumption from Belgium, France, Portugal and Spain, and 400 g remains as the appropriate figure to calculate ARfDs and to protect high consumers against acute effects of marine biotoxins (EFSA, 2010e). In addition, this figure is in good agreement with the report of the Joint FAO/IOC/WHO ad hoc expert consultation on marine biotoxins (FAO/IOC/WHO, 2004), where 380 g was reported as the 97.5th percentile largest portion size for consumers only.

1.3. Revision of Toxicity Data and of European Union Maximum Legal Limits for Marine Biotoxins

Based on the existing toxicity data, the Contaminants Panel established ARfDs for regulated marine biotoxins -okadaic acid (OA) and analogues, azaspiracid (AZA)-group toxins, yessotoxin (YTX)-group toxins, saxitoxin (STX)-group toxins, pectenotoxin (PTX)-group toxins and domoic acid (DA). Taking ARfDs into account, the use of the 400 g portion to calculate the potential exposure to marine biotoxins present in bivalve shellfish involves that, under the European Union current maximum legal limits, high consumers may be exposed to concentrations higher than ARfDs for okadaic acid and analogues, domoic acid, and saxitoxin-group toxins and azaspiracid-group toxin. For yessotoxins and pectenotoxins, current European Union limits provide enough protection. Due to the existing probability of exceeding ARfDs when a portion of 400 g shellfish meat is consumed, EFSA has established recommendations of lowering maximum legal limits for some biotoxins. These conclusions have been presented on a series of scientific opinions released between 2008 and 2010 (EFSA, 2008a; EFSA, 2008b; EFSA, 2008c; EFSA, 2009f; EFSA, 2009g; EFSA, 2009b; EFSA, 2009a).

1.3.1. Hydrophilic Toxins

1.3.1.1. Domoic Acid Group

Domoic acid (DA) and its isomer epi-domoic acid (epi-DA) are the causative agents of Amnesic Shellfish Poisoning (ASP). The current limit in the European Union and other countries with regulation on ASP (Rodríguez-Velasco, 2008) and chapters 10 and 11 in this book of the sum of DA and epi-DA in shellfish was established as 20 mg/kg of shellfish meat, and is in agreement with regulatory limits in Canada and the Pacific Northwest region of US (Lefebvre, 2010) set between 1988 and 1992 after a series of ASP outbreaks in these areas. Recently, EFSA Contaminants Panel revised existent data on the toxicity of DA and established an ARfD of 30 µg of sum of DA and epi-DA/kg b.w., which would involve that the current legal limit of 20 mg/kg could result on a DA exposure exceeding 4-fold the ARfD (EFSA, 2009g).

Data on the toxicity of DA in humans are very limited, and the majority of the existent data on its acute effects were collected after a toxic episode occurred in 1987 when over 107 people became ill and 3 died after consuming DA-contaminated mussels harvested from the eastern coast of Prince Edward Island, Canada (Bates et al., 1989; Wright et al., 1989; Perl, 1990). Since this event, many countries worldwide have established monitoring and assessment programs through their food safety regulatory agencies in order to control the presence of DA in seafood. The increased awareness of risks and occurrence of DA, as well as the implementation of control measures, have been clearly decisive, and since Prince Edward Island episode no human ASP outbreaks have been documented in countries with active preventive programs against DA poisoning (Todd, 1993). Nevertheless, DA toxic events may occur in underdeveloped countries or coastal areas where monitoring coverage is sporadic or non-existent (Lefebvre, 2010). Poisoning caused by the ingestion of prey contaminated with DA has also been identified in sea birds and sea lions along the west coast of United States and Mexico (Sierra-Beltrán, 1997; Goldstein, 2008). Observations in sea lions have revealed the existence of two different syndromes, derived from acute and chronic toxicosis (Goldstein, 2008).

The calculation of the exposure to DA of affected people in the Canada toxic event was based on DA concentrations in samples of mussel left over from meals, the number of mussels each individual recalled eating or average portion size, and average body weights for females and males (Todd, 1993). From the available data, it was estimated that exposure of 0.2-0.3 mg/kg b.w. did not cause symptoms, whereas ingestion of 0.9-2.0 mg/kg b.w. resulted mainly in gastrointestinal symptoms and patients exposed to the highest reported doses, 4.1 and 4.2 mg DA/kg b.w., required intensive care hospitalisation and suffered permanent neurological symptoms (Perl, 1990; Todd, 1993).

EFSA Contaminants Panel considered appropriate the establishment of an acute reference dose based on the LOAEL of 0.9 mg/kg b.w. derived from the observations after the outbreak in Canada. From this value, a NOAEL was calculated as 0.3 mg/kg b.w., by applying a factor of 3 for extrapolation from the LOAEL. Then, the Contaminants Panel decided to apply an additional factor of 10 to allow for human variability and also for the fact that sensitive methods for the detection of neurotoxic effects had not been used in the investigation of the affected individuals. As a result, an ARfD of 30 µg DA/kg b.w. was established by the Contaminants Panel (EFSA, 2009g). Taking into account the current European Union limit for the sum of DA and epi-DA (20 mg/kg of shellfish meat), the Contaminants Panel concluded that the consumption of a 400 g shellfish portion could result on a dietary exposure of approximately 8 mg toxin. For an adult of 60 kg, this is equivalent to 130 μg/kg b.w., approximately 4-fold higher than the ARfD. In order not to exceed ARfD, for an adult of 60 kg a 400 g portion should not contain more than 1.8 mg of DA, which correspond to 4.5 mg/kg of shellfish meat.

Since the current legal limit for DA and epi-DA could cause dietary exposures higher than the ARfD, a survey was conducted by EFSA to determine the DA and epi-DA concentration of commercialized shellfish samples. 43,000 samples from 10 Member States were analyzed between 1999 and 2008. This study estimated the 95th percentile of the sum of DA and epi-DA in shellfish samples reaching the market as 2.5 mg/kg of shellfish meat. Despite this data show that DA concentrations in the great majority of marketed shellfish are far below the legal limit, 3.5 % of the analyzed samples were above the value calculated in order not to exceed ARfD, i. e. 4.5 mg/kg of shellfish meat. These occurrence data, together with data of shellfish consumption, were used in a Monte Carlo simulation to obtain a probabilistic estimate of the dietary exposure to DA and epi-DA. This estimation concluded that the chance for a 60 kg adult of exceeding the ARfD of 30 μg/kg b.w. (corresponding to 1.8 mg DA) when consuming a 400 g portion of shellfish available on the European market is about 1 %.

The availability of sensitive methods of detection allows an accurate quantification of the concentration of DA and the exhaustive control of the shellfish marketed in the European Union, guaranteeing that it complies with maximum legal limits. Estimates of occurrence and dietary exposure suggest that, even under EFSA recommendations on the reduction of the legal limit for DA from 20 mg/kg to 4.5 mg/kg of shellfish meat, the percentage of marketed shellfish exceeding the recommended value is only 3.5 %, and the probability of exceeding the ARfD after consuming a large portion of shellfish is 1 %. Furthermore, in its scientific opinion on domoic acid (EFSA, 2009g), EFSA concludes that the assessment of acute risk is likely to be conservative, i. e., to overestimate the risk, what would involve a high level of protection to shellfish consumers.

1.3.1.2. Saxitoxin Group

The regulatory limit for the Paralytic Shellfish Poisoning toxin group was initially established in California in 1930s as 800 μg/kg (originally, 80 μg/100 g) of shellfish meat, based in bioassays of toxic activity in mice (Wekell et al., 2004). This limit was introduced when little was known about the chemical structure and the toxicology of the saxitoxin (STX) group, but it was adopted by many countries with regulation on marine biotoxins and is currently in force in the European Union (European Parliament, 2004a), United States, Canada, Australia, New Zealand, Chile, Mexico, Peru, Uruguay, Venezuela, Morocco and Singapore, although some of these countries have set exceptions for certain bivalve species to be canned or gastropods after siphon removal (Rodríguez-Velasco, 2008). The original legal limit has been modified in some cases with the introduction of the expression "800 μg of STX equivalents", since at least 24 saxitoxin analogues with different levels of toxicity have been identified (Etheridge, 2010). European Union limit is not expressed in STX equivalents (European Parliament, 2004a), but EFSA scientific opinions refer to μg of STX-eq, i.e. STX equivalents (EFSA, 2009b).

The level of protection set in USA in the 1930s and adopted by other countries has proven effective over more than 60 years, since during this period no cases of PSP caused by shellfish from properly monitored areas have been reported (Wekell et al., 2004). On the other hand, PSP outbreaks derived from personal shellfish harvesting or from unregulated areas have been occurring worldwide. From these episodes some quantitative estimations on exposure to PSP have been obtained and used to determine acute toxicity on humans, although uncertainties exist, particularly relating to the amounts of consumed shellfish, as well as information about its origin or cooking. Just like for other groups of toxins, it is not possible to establish a tolerable daily intake (TDI) for STX-group toxins due to the lack of data on repeated exposure (EFSA, 2009b).

The sensitivity to STX seems to be highly variable among individuals, since different works revised by EFSA to elaborate their hazard characterization of PSP reported in occasions wide ranges of toxin intake for each of the levels of poisoning determined, from "slight" to "extremely severe" (Langeland, 1984; Kuiper-Goodman, 1991) and other references cited in (EFSA, 2009b). Overall, the revision of several hundred cases of human PSP episodes indicates a LOAEL about 1.5 μg STX-eq/kg b.w. From the available toxicological data, EFSA reported that doses associated with severe illness ranged from 5.6 to 2058 μg STX-eq/kg b. w., whereas toxin intakes that did not cause symptoms were estimated to range between 0.3 and 90 μg STX-eq/kg b.w. (EFSA, 2009b). Since these data show that many individuals did not suffer adverse reactions at much higher STX intakes than the LOAEL, and given the high number of reported cases –around 500- it is expected that this value is very close to the threshold for effects in the most sensitive individuals. Therefore, EFSA Contaminants Panel concluded that no calculation factors were necessary to consider variations in sensitivity among humans, and only a factor of 3 was applied in order to obtain a NOAEL of 0.5 μg STX-eq/kg b.w.; hence, in the case of STX-group toxins NOAEL value is equivalent to the ARfD. The recommended ARfD is in the range of the value established in the FAO/IOC/WHO ad hoc Expert Consultation held in 2004, where 0.7 μg STX-eq/kg b.w. was established as the ARfD, based on an estimated LOAEL of 2 μg/kg b.w. (FAO/IOC/WHO, 2004).

Under the current European Union regulation, an adult of 60 kg consuming the calculated large shellfish portion of 400 g (EFSA, 2009a; EFSA, 2010e) could be exposed to a maximum of 5.3 μg STX-eq/kg b.w., more than 10-fold the ARfD. Consequently, a 400 g shellfish portion should not contain more than 30 μg STX-eq., or 75 μg STX-eq/kg of shellfish meat, in order not to exceed the ARfD for a 60-kg adult.

The occurrence of STX-group toxins in shellfish was evaluated by EFSA in samples from France, Germany, Italy, Portugal, Spain, UK and Norway analyzed between 2000 and 2008 by mouse bioassay or HPLC-FLD. Difficulties arose from the lack of quantitative data for a great number of these samples. This required an approach to attribute numerical values to samples reported below limit of detection (LOD) or limit of quantification (LOQ); in this situation, the assignment of minimum, maximum or half-way possible values to non-quantified samples leads to different scenarios that must be considered in the evaluation of toxin occurrence. Consequently, the dietary exposure assessment is fully dependent on the chosen approach (lower, middle or upper values). On the basis of the currently available data, EFSA concluded that a reliable estimation of the dietary exposure to STX-group toxins is not possible.

The current scenario does not allow an exhaustive risk assessment for STX-group toxins in the European Union. Even though toxicological data seem to be accurate enough to establish an ARfD of 0.5 μg STX-eq/kg b.w., there is a lack of information on the occurrence of these toxins in shellfish from European Union, due to the absence of numerical data for a high percentage of the samples of marketed shellfish. The low sensitivity of the official methods of detection, approximately 370 μg/kg shellfish meat for the mouse bioassay and between 10 and 80 μg/kg for HPLC-based methods (Wekell et al., 2004; EFSA, 2009b) limits the precise quantification of the toxic content of many analyzed samples. Hence a reliable exposure assessment that includes the calculation of probability of exceeding legal limits or estimated ARfD is not feasible. Likewise, the reduction of the regulatory limit derived from EFSA recommendations on the ARfD of 0.5 μg/kg b.w. is currently not possible due to the limitations of the available detection methods. Whereas legal limits below the range of 370 μg/kg shellfish meat would result in the MBA not being applicable anymore, HPLC-based methods still need development, including the supply of standards of all STX analogues; in this sense, other methos have proven to achieve excellent detection and quantification limits, although they lack to be internationaly validated(Garet E, 2010).

It can be expected that regulation of STX-group toxins in shellfish will not undergo significant changes in the following years, and modifications will be a consequence of the advances on the development of more sensitive HPLC detection techniques that allow the quantification of low toxin concentrations. In this situation, the replacement of mouse bioassay by instrumental analysis is the most foreseeable outcome, whereas the reduction of maximum limits will likely require further evaluation of toxin occurrence and toxicological data, since the regulatory limit of 800 μg STX-eq/kg b.w. seems to have been protective enough so far.

1.3.2. Lipophilic Toxins

1.3.2.1. Okadaic Acid and Analogues

Okadaic acid and its analogues, namely dinophysis toxins (DTX1, DTX 2and DTX3) (OA-group) cause Diarrhoeic Shellfish Poisoning (DSP), the first syndrome associated to lipophilic toxins intoxication, which gave name to the different lipophilic toxins until relatively few years. DTX3 are in fact a wide range of derivatives which result from esterification of OA, DTX1 and DTX2 with fatty acids. Toxins causing DSP were first reported in Japan in 1978 and since then, they have been reported worldwide, especially in Europe and Japan.

Toxicological data in humans are referred to the acute intoxication an there is no data referred to chronic exposure. TEFs established by the Contaminants Panel were based on LD_{50} experiments following intraperitoneal injection, mainly in mice. These TEFs are OA = 1, DTX1 = 1, DTX2 = 0.6. For DTX3 the TEF values are equal to those of the corresponding unesterified toxins (OA, DTX1, and DTX2). The lethal oral dose is from 2 to 10 times higher than the intraperitoneal lethal dose.

Toxic effects in humans are referred to acute intoxication, producing gastrointestinal symptoms, sometimes accompanied by fever, chill and headache which usually disappear after 2-3 days. Data from reports many times is not accompanied by analytical data or if they are, control was performed by MBA and no information regarding toxin concentration nor toxin profile is available. Taking into account the valid data regarding intoxication and food analysis, a LOAEL of 50 μg OA-eq/person (0.8 μg OA-eq/kg b.w.) was estimated.

Current legislation assumes that mammalian bioassays (MBA and RBA) are able to detect the presence of OA-group at the regulatory limit; based on the equivalence of one mouse unit (MU, the minimum amount of toxin administered intraperitoneally wich kills a 20 g mouse in 24 h) is 4 μg OA as stated by Yasumoto (Yasumoto et al., 1978). Nevertheless the EFSA opinion on okadaic acid and analogues (EFSA, 2008a), based on data from Aune et al performed at single laboratory and single mouse strain, which achieved 4 μg OA as the LD_{50} (amount of OA which kills half of the injected mice) (Aune et al., 2007) indicated that the probability of achieving a positive result performing the MBA with hepatopacreas is around 40-50 % at the regulatory level of 160 μg OA/kg of whole flesh and 90 % at contamination level of 200 μg OA/kg of whole flesh.

The study of OA group toxins prevalence data requested by EFSA included data recorded from 2001 to 2006 in several countries and showed that this group of toxins exhibits a huge variation along the year, showing samples higher than the legal limit. This is not necessarily a matter or worry, since the prevalence study included pre market sampling, and the study revealed that pre-market control prevents batches with high levels of toxins to reach the market.

The key factor of exposure assessment and risk management is the high portion size, established as 400 g. Taking this portion size into account, the deterministic estimate of dietary exposure to OA-group toxins leads to conclude that a meal with molluscs at the current regulatory limit of 160 μg/kg would result in an exposure of 64 μg OA equivalents per person. A probabilistic estimation of the exposure, performed by a Monte Carlo simulation led the expert panel of EFSA to conclude that the chance to exceed an intake of 64 μg OA equivalents is about 4 %. After established a LOAEL of 0.8 μg OA-eq/kg b.w., and taking into account that symptoms are relatively mild and reversible, an uncertainty factor of 3 was applied to obtain a NOAEL of 0.3 μg OA-eq/kg b.w.

The Monte Carlo simulation calculated that there is a 20 % chance to exceeding the intake of ARfD of 0.3 μg OA-eq/kg b.w.

A 400 g portion of molluscs at the regulated limit of 160 μg OA equivalents would result in an intake of 64 μg OA-eq, which means, for a 60 kg b.w. average adult, an intake of approximately 1 μg OA-eq/kg b.w., in the region of LOAEL. Hence, according to the Panel, to ensure population health, shellfish should no exceed a limit of 45 μg OA-eq/kg, which corresponds to a 3.5 times reduction in current limits.

1.3.2.2. Azaspiracids Group

Azaspiracids (AZAs) are a group of lipophilic shellfish toxins causing AZA poisoning, characterized by nausea, vomiting diarrhoea and stomach cramps. Around 20 analogues have been indentified, being AZA1, AZA2 and AZA3 the most important ones based on occurrence and toxicity (EFSA, 2008c) AZAs were first detected in mussels from Ireland, and from then, these toxins have been identified in other filter feeding mollusks in United Kingdom, and Norway in significant levels, as well as in France, Spain, Portugal and Morocco. Although crustaceans are not affected by current European Union marine biotoxins legislation, several samples of crabs (*Cancer pagurus*) showed relatively important levels, even above the regulated level for mollusk (EFSA, 2008c).

The EFSA study of AZAs occurrence in European waters employed data (12270 samples) collected from 2003 to 2007 by Germany, Ireland, Norway, Spain and UK; the reported found concentrations ranged from not detected to a maximal concentration of 1630 μg/kg, and 84.1 % of analysed samples were below limit of detection.

Taking these data into account, the deterministic estimate of dietary exposure to AZAs in a 400 g meal of shellfish would be, for the 95th percentile, i.e. with a concentration of 40 μg/kg, of 16 μg AZA1 equivalents per person (equivalent to 0.25 μg/kg b.w. for a 60 kg adult). For concentrations at the current regulatory limit of 160 μg AZA1 equivalents/kg shellfish meat consumption of a 400 g portion would result in an exposure of 64 μg AZA1 equivalents per person (equivalent to 1 μg/kg b.w. for a 60 kg adult).

The probabilistic estimate of dietary exposure to AZAs using a Monte Carlo simulation led to the calculation that the probability of exceeding the deterministic dietary exposure of 64 μg AZA1 equivalents per person is 0.13 %; the probability of exceeding the intake of the established ARfD of 12 μg AZA1 equivalents for a 60 kg adult is about 4 % (EFSA, 2008c).

There have been very few clinical cases in consumers due to AZAs consumption. The first cases, due to the toxin was still unknown, it was not possible to correlate to a level of toxin intake. First risk assessment was performed by the Food Safety Authority of Ireland (FSAI), who estimated the intake of AZAs to be among 6.7 and 24.9 μg per person, with a median value of 14.5 μg per person. This first attempt was performed assuming equivalent toxicities of AZA1, AZA2 and AZA3 and similar proportions in the sample, as well as that a inactivation of the toxicity took place during cooking. A review of the assessment taking into account later experimental data including that a concentration of AZAs took place due to cooking because of loss of water in the mussels changed the estimation of toxin intake to be among 50.1 and 253.3 μg per person, with a median value of 113.4 μg per person (1.9 μg AZA1-eq/kg b.w. for a 60 kg adult). Since the 160 μg/kg limit for AZAs was established, the only one reported outbreak was due a batch exceeding legal limit (EFSA, 2008c).

Since there is not available data related to continuous intake, a tolerable daily intake (TDI) cannot be established and a ARfD of 0.2 μg AZA1-eq/kg b.w. (which leads to a ARfD of 12 μg AZA1-eq/person for a 60 kg adult) has been proposed.

A 400 g meal of mollusks at the legal limit of 160 μg AZA1-eq/kg would mean an intake of 64 μg AZAs (approximately 1 μg AZA1-eq/kg b.w. for a 60 kg adult). This intake exceeds about 5 times the ARfD of 0.2 μg/AZA1. As we told earlier, the probability of exceeding the intake of the established ARfD of 12 μg AZA1 equivalents for a 60 kg adult with a 400 g meal of mollusk flesh is about 4 % (EFSA, 2008c).

1.3.2.3. Pectenotoxins

Pectenotoxin (PTX)-group toxins are a group of polyether-lactone toxins which have been detected in microalgae and bivalve molluscs in Australia, Japan and New Zealand and in a number of European countries. Their presence in shellfish was discovered due to their acute toxicity in the mouse bioassay after intraperitoneal injections of lipophilic extracts of shellfish. This group of toxins appears always associated to the OA-group toxins, and this is probably the reason because they are grouped in European legislation (European Parliament, 2004a). Nevertheless, since they do not share the same mechanism of action, the CONTAM Panel concluded that pectenotoxins should not be included in the regulatory limit for OA-group toxins (EFSA, 2009f).

Once again, toxicological data are limited and mainly referred to MBA. Oral toxicity seems to be lower than intraperitoneal toxicity.

Indeed, there are no reports of PTX toxins induced human illness, in part due to that these toxins co-occur with OA. Nevertheless, taking into account that some reported human cases where PTX were present, OA levels were similar to its LOAEL, it is possible that PTX did not contribute to the symptoms. An ARfD of 0.8 µg PTX2-eq/kg b.w. was established.

The EFSA survey of PTX–group toxins occurrence studied data from the period 2005-2008 from Germany, Italy, Spain, Norway and United Kingdom and rendered little data regarding PTX other than PTX2 (the only one which had certified calibrant solution available at that moment). For this reason, occurrence and exposure calculations were performed using PTX2 data from Germany and Norway, considering that those data submitted by, Italy, Spain and UK (whose monitoring program relied at the moment on MBA) were too scarce. Analysis of reported data revealed a low amount of PTX2 in shellfish from European waters. Although maximum detected values reached 418 µg/kg, an important number of samples were under limit of detection or limit of quantification, and highest detected values were from targeted samples. The limited number of samples made comparisons among species, origin of samples or analytical methods (MBA vs LC-MS/MS) a difficult issue.

The deterministic estimation of exposure to PTX2 led to a calculation of (for an average 60 kg adult and a 400 g of shellfish meal) 0.5 µg PTX2/kg b.w. This represents less than half of the exposure (1.1 µg PTX2/kg b.w.) of a person eating a 400 g portion at the level of 160 µg of sum of OA, DTX and PTX2/kg whole shellfish meat (current European Union limit), and is also less than the proposed ARfD of 0.8 µg PTX2 equivalents/kg b.w.

The probabilistic estimation of dietary exposure to PTX2 has been performed by a Monte Carlo simulation led to a median value of approximately 0.05 µg PTX2/kg b.w.), and a 95th percentile of approximately 0.2 µg PTX2/kg b.w.) and a 99th percentile of 0.45 µg PTX2/kg b.w., which means that the chance to exceed the ARfD for PTX-group toxins of 0.8 µg PTX2-eq/kg b.w. is about 0.2 %.

1.3.2.4. Yessotoxins Group

Yessotoxins (YTX) have not been associated with illness in humans. TEF have been obtained by lethality in mice obtained by intraperitoneal injection. Indeed, no lethality and no clinical signs of toxicity were observed by oral administration in mice. Indeed, no data regarding chronic exposure, carcinogenicity or genotoxicity are available. Taking these data into account, the Panel decided to stablish a NOAEL of 5 mg/kg b.w., which leads to a ARfD of 25 µg YTX equivalents/kg b.w. Taking into account the 400 g portion that EFSA decided to establish as an average meal, the intake of 400g of shellfish flesh at the current legislated level of 1 mg/kg means an intake of 6.7 µg/kg b.w. in a 60 kg adult, which is below the ARfD of 25 µg YTX equivalents/kg b.w. (EFSA, 2008b).

A survey of yessotoxin incidence in the European waters and market, collecting existing data along 2000-2008, showed a different range of concentrations depending on the geographical harvesting area, with maximum at the Adriatic Sea, which represents the worst scenario for YTX along the European coast, followed by Norway. Furthermore, distribution among species also showed important differences, where mussels showed highest YTX concentrations. From the worst situation, mussels from the Adriatic coast, 4.5 % (1 out of 22) of the reported samples showed a concentration above the limit value of 1 mg/kg. Taking into account the data obtained in the survey, in the worst scenario, a Monte Carlo simulation led to a exposure of 158 μg YTX equivalents per portion at the 95th percentile and a deterministic probability of 0.5 % of a 400 μg YTX intake (corresponding to a 400 g portion containing YTXs at the current level of 1 mg/kg) (European Parliament, 2004a).

1.3.2.5. Palytoxins

Palytoxin (PlTX) and PlTX-like compounds are a group of toxins not yet legislated in European regulations. Nevertheless, some episodes have alredy happened in European coasts. PlTX-group toxins have mainly been detected in soft corals of the genus *Palythoa* and in algae of the genus *Ostreopsis* blooms of *Ostreopsis* spp. can lead to contamination of shellfish and have recently been reported in some European countries: France, Greece, Italy and Spain. The occurrence data reported by European countries were limited and comprised only PlTX and ovatoxin-A in mussels and sea urchins (EFSA, 2009e).

Palytoxin-group are insoluble in nonpolar solvents, and the European Union harmonised protocol for lipophilic toxins by MBA does not efficiently extract PlTX-group toxins; nevertheless, there are reported LD_{50} of PlTX in mice following MBA, which range from 150 to 720 ng/kg b.w.

Toxic effects on humans are not well defined. In some cases there has been skeletal muscle injury, with release of myocyte content into plasma, leading to rhabdomyolysys; others include myalgia and weakness, possibly accompanied by fever, nausea and vomiting. Fatalities appear to be rare although there are reports of severe cases, in which patients died after about 15 hours. In other cases, simptomatology is respiratory, in seaside regions of Italy, without clear relationship with seafood comsumption. In other cases the poisoning took place through injured skin or associated to ciguatoxins.

Reported cases of PlTX-group toxins do not allow to infer NOAEL or LOAEL in humans, since there are no available data on the concentration of the toxin in food. Hence ARfD has been inferred from toxicological data in animal models, taking into account that LOAEL was likely to be close to a NOAEL, the differences in toxicity depending on the administration route, and the animal specie. These considerations derived in an oral ARfD of 0.2 μg/kg b.w. which applies to the sum of PlTX and ostreocin-D. For a 60 kg adult to avoid exceeding the ARfD of 0.2 μg/kg b.w., a 400 g portion of shellfish meat should not contain more than 12 μg of the sum of PlTX and ostreocin-D, which would correspond to 30 μg/kg shellfish meat (EFSA, 2009e).

The EFSA scientific opinion on PlTX-group toxins was performed with very few data related to incidence in European waters, due to this toxin group is relatively new in European coast and it is not under monitoring surveillance. Indeed, part of the managed data came from intentionally contaminated shellfish in research projects. Due to the low number of data, statistical and exposure assessment must be considered very carefully. The worst case scenario considered, with a 400 g mussels intake in one meal and contamination levels of 462 μg/kg shellfish meat, leaded to a deterministic exposure estimation of 185 μg of PlTX-group toxins, corresponding to about 3 μg/kg b.w. for a 60 kg person. Due to lack of data regarding occurrence of PlTX-group toxins in shellfish intended for human consumption, a probabilistic estimation was not feasible.

1.3.2.6. Cyclic Imines

Cyclic imines (CIs), encomprise several subgroups: spirolides (SPXs), gymnodimines (GYMs), pinnatoxins (PnTXs), pteriatoxins (PtTXs) and other toxins in shellfish. A survey of

CIs occurrence in Europe led to a little amount of data provided mainly by France, Italy, and the Netherlands with results obtained regarding SPXs, with punctual data regarding PnTX G and A. Concentration of SPXs ranged from "not detected" to 105 μg sum of SPXs/kg. The percentage of analytical results <LOD or <LOQ was relatively high. With the low number of analytical data it is quite difficult to draw a risk assessment for these toxins, but a deterministic estimate of dietary exposure to SPXs concluded that consumption of a 400 g portion of shellfish meat containing SPXs at the 95th percentile of the obtained concentrations (8.9 μg sum of SPXs/kg shellfish meat) would result in a single dietary exposure of 0.06 μg sum of SPXs/kg b.w. for 60 kg person (i.e. 3.6 μg sum of SPXs/person). The probablilistic stimate of dietary exposure concluded that the chance to exceed the deterministic dietary exposure estimate of 3.6 μg sum of SPXs per person corresponding to a consumption of a portion of 400 g of shellfish containing the 95th percentile SPX concentration is about 4 % (EFSA, 2010d). Since there are no data regarding clinical cases in humans, an ARfD could not be calculated, and based on animal data, the Contaminants Panel concluded that current estimated exposure to SPXs does not raise concern for the health of the consumer (EFSA, 2010d).

1.3.2.7. Brevetoxins

Brevetoxin-(BTX) group are marine biotoxins which can accumulate in shellfish and fish and cause neurologic shellfish poisoning (NSP). So far, there has not been reports on clinical cases or toxin occurrence in Europe, but some trends to expansion of some producing algae, draw a far to moderate risk of this toxin group to appear in European coasts, and hence the interest in pay attention to data obtained in other parts of the world. Where detected, concentration in shellfish have been reported to range from 880 to 49 000 μg BTX-2 equivalents/kg, using the conversion of (1 MU = 4 μg) BTX-2 equivalents, although values depend on the analytical method employed (EFSA, 2010c). The very few available data do not allow to perform any risk assessment for European population.

1.3.2.8. Ciguatoxin

As stated earlier, ciguatoxins (CTX)-group toxins are a group of toxins which derived from biotransformation of gambiertoxins in fish. They are mainly found in Pacific, Caribbean and Indian Ocean. Nevertheless, CTX-group toxins have been recently identified in fish caught in Europe. Indeed, *Gambierdiscus* sp., which is the microalgae responsible of gambiertoxins production, has been identified in European waters (Mediterranean Sea and Canary Islands) (EFSA, 2010a).

Occurrence data of (CTX)-group toxins in Europe are very scarce, limited to two outbreak in Madeira and the Canary Islands and analytical data are limited to three fishes that had been consumed by the patients. With these data it was impossible to perform an estimation of the occurrence of CTX-group.

Any risk assessment of (CTX)-group toxins needs an estimation of fish consumption in Europe. There are available data regarding fish consumption from 16 European countries. However, the methodology applied in the consumption surveys is not uniform and fish consumption habits differ among countries. A 350 g fish per meal was considered to reflect high consumers (EFSA, 2010a).

Table 6. Maximum concentration of marine biotoxins based on 400 g portion and risk assessment

Toxin	Current European Union limits	Exposure by eating 400 g portion at the European Union limit	ARfD	Max. concentration based on 400 g portion
OA	160 µg OA-eq/kg SM	64 µg OA-eq/person (1 µg OA-eq/kg b.w.)	0.3 µg OA-eq/kg b.w.	45 µg OA-eq/kg SM
AZA	160 µg AZA-eq/kg SM	64 µg AZA1-eq/person (1 µg AZA1-eq/kg b.w.)	0.2 µg AZA1-eq/kg b.w.	30 µg AZA1-eq/kg SM
PTX	160 µg OA-eq/kg SM	64 µg PTX2/person (1 µg PTX2-eq/kg b.w.)	0.8 µg PTX2-eq/kg b.w.	120 µg PTX2-eq/kg SM
YTX	1 mg YTX-eq/kg SM	400 µg YTX-eq/person (6.7 µg YTX-eq/kg b.w.)	25 µg YTX-eq/kg b.w.	3.75 mg YTX-eq/kg SM
STX	800 µg PSP/kg SM	320 µg STX-eq/person (5.3 µg STX-eq/kg b.w.)	0.5 µg STX-eq/kg b.w.	75 µg STX -eq/kg SM
DA	20 mg DA/kg SM	8 mg DA/person (130 µg DA/kg b.w.)	30 µg DA/kg b.w.	4.5 mg DA/kg SM

Notes: Adapted from (Efsa, 2009c; Paredes, 2011). SM: shellfish meat.

From the currently existing data, it is impossible to stablish neither a TDI nor a ARfD. In order to establish a safe limit of (CTX)-group toxins in fish flesh, and taking into account that a number of publications state that cases of CFP in the Pacific mostly occur following the consumption of fish containing the equivalent of 0.1-5 µg P-CTX-1/kg of fish flesh and, in line with the approach of AO (FAO, 2004), the Contaminants Panel applied a safety factor of 10 to the lowest concentration 0.1 µg equivalents of P-CTX-1/kg in fish associated with mild symptoms to indicate a concentration of 0.01 µg equivalents of P-CTX-1/kg of fish, which is expected not to exert effects in sensitive individuals.

1.3.3. Processing of Bivalve Mollusks and Its Effect on the Level of Lipophilic Marine Biotoxins

The most important effect of food processing on lipophilic marine biotoxins regarding the raw product is a concentration of OA, and AZAs due to the water loss of the flesh during the processing (boiling or steaming). There are no specific data regarding yessotoxins, but it can be assumed that a similar concentration takes place (EFSA, 2009d).

Besides this concentration there is a redistribution of OA from digestive gland to the whole flesh and a minor degradation of OA and DTX2, which can not counteract the concentration due to water loss and hence, the net effect is the increase of toxin concentration. Regarding AZAs, AZA1 and AZA2 are not affected by high temperature, although it seems that AZA3 can be increased due to a conversion of the analogue AZA17 (EFSA, 2009d) The effect of processing on these toxins and other is revised in deep in chapter 5.

Conclusion

Legal limits for marine biotoxins in targeted seafood have been implemented along the years, as new knowledge or new toxins appeared (for instance, the domoic acid oubreak in 1987).

Since last harmonization efforts on European Legislation on food control (European Parliament, 2002), changes have been focussed on analytical metodologies (European Commission, 2006; European Parliament, 2007; European Commission, 2011) or to regulate safe methodologies to get safe products for consumers from certain bivalves that do not comply the general rule (European Commission, 2002). Legislated levels have remained unchanged and have protected consumers' health.

Taking into account the exposed EFSA scientific opinions, which tend to be very consevative when establishing ARfDs, and the statement of the 400 g meal, it exists the possibility of a change in legislated limits. The improvement of new official analytical methodology for PSP toxins and lipophilic toxins, (European Commission, 2006; European Commission, 2011) may open the gate to reduce current levels. Table 6 summarizes current legal status and the new scenario if EFSA's recommendations grew to a legislative change.

Acknowledgments

This work was funded with the following FEDER cofunded-grants: From Ministerio de Ciencia e Innovación of Spain AGL2007-60946/ALI, SAF2009-12581 (subprograma NEF), AGL2009-13581-CO2-01, AGL2009-13581-C02-02, TRA2009-0189, AGL2010-17875.; From Xunta de Galicia, Spain: GRC 2010/10, and PGIDT07CSA012261PR, PGDIT 07MMA006261PR, PGIDIT (INCITE) 09MMA003261PR, 2009/XA044, 2009/053 (Consell. Educación), 2008/CP389 (EPITOX, Consell. Innovación e Industria, programa IN.CI.TE.), 10PXIB261254 PR. From EU VIIth Frame Program: 211326 - CP (CONffIDENCE), 265896 BAMMBO, 265409 μAQUA, and 262649 BEADS From the Atlantic Area Programme (Interreg IVB Trans-national) 2008-1/003 ATLANTOX ",(2009-1/117) PHARMATLANTIC

References

Alfonso, A.; Louzao, M.C.; Vieytes, M.R. & Botana, L.M., (1994). Comparative study of the stability of saxitoxin and neosaxitoxin in acidic solutions and lyophilized samples. *Toxicon,* 32, 1593-1598.

Alfonso, A.; de la Rosa, L.A.; Vieytes, M.R.; Yasumoto, T. & Botana, L.M., (2003). Yessotoxin a novel phycotoxin, activates phosphodiesterase activity. Effect of yessotoxin on cAMP levels in human lymphocytes. *Biochem Pharmacol,* 65, 193-208.

Alfonso, A.; Román, Y.; Vieytes, M.R.; Ofuji, K.; Satake, M.; Yasumoto, T. & Botana, L.M., (2005a). Azaspiracid-4 inhibits Ca^{2+} entry by stored operated channels in human T lymphocytes. *Biochem Pharmacol.,* 69, 1627-1636.

Alfonso, A.; Vieytes, M.R.; Ofuji, K.; Satake, M.; Nicolaou, K.C.; Frederick, M.O. & Botana, L.M., (2006). Azaspiracids modulate intracellular pH levels in human lymphocytes. *Biochem Biophys Res Commun,* 346, 1091-1099.

Alfonso, A. & Alfonso, C. Yessotoxin: Pharmacology and Mechanism of action. Biological Detection. In: Botana, L. M. (Ed.), *Seafood and Freshwater Toxins. Second Edition.* London: CRC Press. Taylor and Francis Group; 2008a; p. 315-327

Alfonso, C.; Alfonso, A.; Vieytes, M.R.; Yasumoto, T. & Botana, L.M., (2005b). Quantification of yessotoxin using the fluorescence polarization technique, and study of the adequate extraccion procedure. *Anal. Biochem,* 344, 266-274.

Alfonso, C.; Alfonso, A.; Otero, P.; Rodríguez, P.; Vieytes, M.R.; Elliot, C.; Higgins, C. & Botana, L.M., (2008b). Purification of five azaspiracids from mussel samples contaminated with DSP toxins and azaspiracids. *J Chromatogr B Analyt Technol Biomed Life Sci.,* 865, 133-140.

Alfonso, C.; Rehmann, N.; Hess, P.; Alfonso, A.; Wandscheer, C.B.; Abuín, M.; Vale, C.; Otero, P.; Vieytes, M.R. & Botana, L.M., (2008c). Evaluation of various pH and temperature conditions on the stability of Azaspiracids and their importance in preparative isolation and toxicological studies. *Anal Chem,* 80, 9672-9680.

Aligizaki, K.; Katikou, P.; Milandri, A. & Diogene, J., (2011). Occurrence of palytoxin-group toxins in seafood and future strategies to complement the present state of the art. *Toxicon,* 57, 390-399.

Allingham, J.S.; Miles, C.O. & Rayment, I., (2007). A structural basis for regulation of actin polymerization by pectenotoxins. *J Mol Biol,* 371, 959-970.

Amir, R.; Argoff, C.E.; Bennett, G.J.; Cummins, T.R.; Durieux, M.E.; Gerner, P.; Gold, M.S.; Porreca, F. & Strichartz, G.R., (2006). The role of sodium channels in chronic inflammatory and neuropathic pain. *J Pain.,* 7.

AOAC, (2005). Official Method 2005.06. Paralytic Shellfish Poisoning Toxins in Shellfish. Prechromatographic Oxidation and Liquid Chromatography with fluorescence detection. First Action 2005. *Journal AOAC International,* 88, 1714.

AOAC, (2006). Official Method 2006.02. Domoic acid toxin in shellfish. . *Official Methods of Analysis of AOAC INTERNATIONAL, 19th Ed.*

Araoz, R.; Servent, D.; Molgó, J.; Iorga, B. I.; Fruchart-Galliard, C.; Benoit, E.; Gu, Z.; Stivala, C. & Zakarian, A. (2011). Total synthesis of pinnatoxins A and G and a revision of the mode of action of pinnatoxin A. J Am Chem Soc, DOI: 10.1021/ja201254c.

Ares, I.R.; Louzao, M.C.; Vieytes, M.R.; Yasumoto, T. & Botana, L.M., (2005). Actin cytoskeleton of rabbit intestinal cells is a target for potent marine phycotoxins. *J Exp Biol,* 208, 4345-4354.

Ares, I.R.; Louzao, M.C.; Espina, B.; Vieytes, M.R.; Miles, C.O.; Yasumoto, T. & Botana, L.M., (2007). Lactone ring of pectenotoxins: a key factor for their activity on cytoskeletal dynamics. *Cell Physiol Biochem,* 19, 283-292.

Ares, I.R.; Cagide, E.; Louzao, M.C.; Espina, B.; Vieytes, M.R.; Yasumoto, T. & Botana, L.M., (2009). Ostreocin-D impact on globular actin of intact cells. *Chem Res Toxicol,* 22, 374-381.

Artigas, P. & Gadsby, D.C., (2004). Large diameter of palytoxin-induced Na/K pump channels and modulation of palytoxin interaction by Na/K pump ligands. *J Gen Physiol,* 123, 357-376.

Aune, T.; Sorby, R.; Yasumoto, T.; Ramstad, H. & Landsverk, T., (2002). Comparison of oral and intraperitoneal toxicity of yessotoxin towards mice. *Toxicon,* 40, 77-82.

Aune, T.; Larsen, S.; Aasen, J.A.; Rehmann, N.; Satake, M. & Hess, P., (2007). Relative toxicity of dinophysistoxin-2 (DTX-2) compared with okadaic acid, based on acute intraperitoneal toxicity in mice. *Toxicon,* 49, 1-7.

Backer, L.C.; Kirkpatrick, B.; Fleming, L.E.; Cheng, Y.S.; Pierce, R.; Bean, J.A.; Clark, R.; Johnson, D.; Wanner, A.; Tamer, R.; Zhou, Y. & Baden, D.G., (2005). Occupational exposure to aerosolized brevetoxins during Florida red tide events: effects on a healthy worker population. *Environ Health Perspect,* 113, 644-649.

Baden, D.G., (1989). Brevetoxins: unique polyether dinoflagellate toxins. *FASEB J,* 3, 1807-1817.

Baden, D.G.; Bourdelais, A.J.; Jacocks, H.; Michelliza, S. & Naar, J., (2005). Natural and derivative brevetoxins: historical background, multiplicity, and effects. *Environ Health Perspect,* 113, 621-625.

Bates, S.S.; Bird, C.J.; de Freitas, A.S.W.; Foxall, R.; Gilgan, M.; Hanic, L.A.; Johnson, G.R.; McCulloch, A.W.; Odense, P.; Pocklington, R.; Quilliam, M.A.; Sim, P.G.; Smith, J.C.; Subba Rao, D.V.; Todd, E.C.D.; Walter, J.A. & Wright, J.L.C., (1989). Pennate diatom *Nitzschia pungens* as the primary source of domoic acid, a toxin in shellfish from Eastern Prince Edward Island, Canada. *Canadian Journal of Fisheries and Aquatic Sciences,* 46, 1203-1215.

Beress, L.; Zwick, J.; Kolkenbrock, H.J.; Kaul, P.N. & Wassermann, O., (1983). A method for the isolation of the caribbean palytoxin (C-PTX) from the coelenterate (zooanthid) *Palythoa caribaeorum*. *Toxicon,* 21, 285-290.

Berven, G.; Saetre, F.; Halvorsen, K. & Seglen, P.O., (2001). Effects of the diarrhetic shellfish toxin, okadaic acid, on cytoskeletal elements, viability and functionality of rat liver and intestinal cells. *Toxicon,* 39, 349-362.

Bidard, J.N.; Vijverberg, H.P.; Frelin, C.; Chungue, E.; Legrand, A.M.; Bagnis, R. & Lazdunski, M., (1984). Ciguatoxin is a novel type of Na^+ channel toxin. *J Biol Chem,* 259, 8353-8357.

Birinyi-Strachan, L.C.; Gunning, S.J.; Lewis, R.J. & Nicholson, G.M., (2005). Block of voltage-gated potassium channels by Pacific ciguatoxin-1 contributes to increased neuronal excitability in rat sensory neurons. *Toxicol Appl Pharmacol,* 204, 175-186.

Boada, L.D.; Zumbado, M.; Luzardo, O.P.; Almeida-Gonzalez, M.; Plakas, S.M.; Granade, H.R.; Abraham, A.; Jester, E.L. & Dickey, R.W., (2010). Ciguatera fish poisoning on the West Africa Coast: An emerging risk in the Canary Islands (Spain). *Toxicon,* 56, 1516-1519.

Botana, L.M.; Vale, C. & Vilariño, N. Mechanisms of action of palytoxin and azaspiracids. In: Goudey-Perrière, F., Benot, E., Marchot, P., and Popoff, M. R. (Eds.), *Toxins émergentes: noveaux risques*. Paris: Lavoisier; 2007; p. 149-156.

Botana, L.M.; Vilariño, N.; Alfonso, A.; Vale, C.; Louzao, M.C.; Elliot, C.; Kampbell, K. & Botana, A.M., (2010). The problem of toxicity equivalent factors in developing alternative methods to animal bioassay for marine-toxin detection. *Trends in Analytical Chemistry,* 29, 1316-1325.

Bourne, Y.; Radic, Z.; Aráoz, R.; Talley, T.T.; Benoit, E.; Servent, D.; Taylor, P.; Molgó, J. & Marchot, P., (2010). Structural determinants in phycotoxins and AChBP conferring high affinity binding and nicotinic AChR antagonism. *Proc Natl Acad Sci U S A,* 107, 6076-6081.

Burgess, V. & Shaw, G., (2001). Pectenotoxins--an issue for public health: a review of their comparative toxicology and metabolism. *Environ Int,* 27, 275-283.

Cabado, A.G.; Leira, F.; Vieytes, M.R.; Vieites, J.M. & Botana, L.M., (2004). Cytoskeletal disruption is the key factor that triggers apoptosis in okadaic acid-treated neuroblastoma cells. *Arch Toxicol,* 78, 74-85.

Cagide, E.; Louzao, M.C.; Espina, B.; Vieytes, M.R.; Jaen, D.; Maman, L.; Yasumoto, T. & Botana, L.M., (2009). Production of functionally active palytoxin-like compounds by Mediterranean Ostreopsis cf. siamensis. *Cell Physiol Biochem,* 23, 431-440.

Caillaud, A.; de la Iglesia, P.; Darius, H.T.; Pauillac, S.; Aligizaki, K.; Fraga, S.; Chinain, M. & Diogene, J., (2010). Update on methodologies available for ciguatoxin determination: perspectives to confront the onset of ciguatera fish poisoning in Europe. *Mar Drugs,* 8, 1838-1907.

Callegari, F.; Sosa, S.; Ferrari, S.; Soranzo, M.R.; Pierotti, S.; Yasumoto, T.; Tubaro, A. & Rossini, G.P., (2006). Oral administration of yessotoxin stabilizes E-cadherin in mouse colon. *Toxicology,* 227, 145-155.

Cameron, J.; Flowers, A.E. & Capra, M.F., (1991). Electrophysiological studies on ciguatera poisoning in man (Part II). *J Neurol Sci,* 101, 93-97.

Cembella, A. & Krock, B. Cyclic Imine Toxins: Chemistry, Biogeography, Biosynthesis and Pharmacology. In: Botana, L. M. (Ed.), *Seafood and Freshwater Toxins. Second Edition.* London: CRC Press. Taylor and Francis Group; 2008; p. 561-580

Cembella, A.D. Identifying the planktonic origin and distribution of spirolides in coastal Nova Scotian waters. In: Reguera, B., Blanco, J., M.L., F., and Wyatt, T. (Eds.), *Harmful Algae.* Paris: Xunta de Galicia and International Oceanographic Commission of UNESCO; 1998; p. 481-490

Cestèle, S. & Catterall, W.A., (2000). Molecular mechanisms of neurotoxin action on voltage-gated sodium channels. *Biochimie,* 82, 883-892.

Ciminiello, P.; Dell'Aversano, C.; Fattorusso, E.; Forino, M.; Tartaglione, L.; Grillo, C. & Melchiorre, N., (2008). Putative palytoxin and its new analogue, ovatoxin-a, in *Ostreopsis ovata* collected along the Ligurian coasts during the 2006 toxic outbreak. *J Am Soc Mass Spectrom,* 19, 111-120.

Ciminiello, P.; Dell'aversano, C.; Iacovo, E.D.; Fattorusso, E.; Forino, M. & Tartaglione, L., (2010). LC-MS of palytoxin and its analogues: State of the art and future perspectives. *Toxicon*, 57, 376-389.

Cohen, P.; Holmes, C.F. & Tsukitani, Y., (1990). Okadaic acid: a new probe for the study of cellular regulation. *Trends Biochem Sci,* 15, 98-102.

Cordier, S.; Monfort, C.; Miossec, L.; Richardson, S. & Belin, C., (2000). Ecological analysis of digestive cancer mortality related to contamination by diarrhetic shellfish poisoning toxins along the coasts of France. *Environ Res,* 84, 145-150.

Costa, L.G.; Giordano, G. & Faustman, E.M., (2010). Domoic acid as a developmental neurotoxin. *Neurotoxicology,* 31, 409-423.

Chae, H.D.; Choi, T.S.; Kim, B.M.; Jung, J.H.; Bang, Y.J. & Shin, D.Y., (2005). Oocyte-based screening of cytokinesis inhibitors and identification of pectenotoxin-2 that induces Bim/Bax-mediated apoptosis in p53-deficient tumors. *Oncogene,* 24, 4813-4819.

Dawson, J.F. & Holmes, C.F., (1999). Molecular mechanisms underlying inhibition of protein phosphatases by marine toxins. *Front Biosci,* 4, D646-658.

De la Rosa, L.A.; Alfonso, A.; Vilariño, N.; Vieytes, M.R. & Botana, L.M., (2001a). Modulation of cytosolic calcium levels of human lymphocytes by yessotoxin, a novel marine phycotoxin. *Biochem. Pharmacol.,* 61, 827-833.

De la Rosa, L.A.; Alfonso, A.; Vilariño, N.; Vieytes, M.R.; Yasumoto, T. & Botana, L.M., (2001b). Maitotoxin-induced calcium entry in human lymphocytes - Modulation by yessotoxin, Ca^{2+} channel blockers and kinases. *Cell Signal,* 13, 711-716.

Dechraoui, M.Y.; Naar, J.; Pauillac, S. & Legrand, A.M., (1999). Ciguatoxins and brevetoxins, neurotoxic polyether compounds active on sodium channels. *Toxicon,* 37, 125-143.

Deeds, J.R. & Schwartz, M.D., (2010). Human risk associated with palytoxin exposure. *Toxicon,* 56, 150-162.

Dickey, R.W. & Plakas, S.M., (2010). Ciguatera: a public health perspective. *Toxicon,* 56, 123-136.

Dominguez, H.J.; Paz, B.; Daranas, A.H.; Norte, M.; Franco, J.M. & Fernandez, J.J., (2010). Dinoflagellate polyether within the yessotoxin, pectenotoxin and okadaic acid toxin groups: characterization, analysis and human health implications. *Toxicon,* 56, 191-217.

Doucette, T.A. & Tasker, A. Domoic acid: Detection methods, pharmacology and toxicology. In: Botana, L. M. (Ed.), *Seafood and Freshwater Toxins. Second Edition*. London: CRC Press. Taylor and Francis Group; 2008; p. 397-429

Durando, P.; Ansaldi, F.; Oreste, P.; Moscatelli, P.; Marensi, L.; Grillo, C.; Gasparini, R. & Icardi, G., (2007). Ostreopsis ovata and human health: epidemiological and clinical features of respiratory syndrome outbreaks from a two-year syndromic surveillance, 2005-06, in north-west Italy. *Euro Surveill,* 12, E070607 070601.

Edebo, L.; Lange, S.; Li, X.P. & Allenmark, S., (1988). Toxic mussels and okadaic acid induce rapid hypersecretion in the rat small intestine. *APMIS,* 96, 1029-1035.

EFSA, (2008a). Opinion of the Scientific Panel on Contaminants in the Food Chain on a request from the European Commission on marine biotoxins in shellfish – Okadaic acid and analogues. *EFSA Journal,* 859, 1-62.

EFSA, (2008b). Opinion of the Scientific Panel on Contaminants in the Food Chain on a request from the European Commission on marine biotoxins in shellfish – Yessotoxin group. *EFSA Journal,* 907, 1-62.

EFSA, (2008c). Opinion of the Scientific Panel on Contaminants in the Food Chain on a request from the European Commission on marine biotoxins in shellfish – Azaspiracid. *EFSA Journal,* 723, 1-52.

EFSA, (2009a). Scientific Opinion of the Panel on Contaminants in the Food Chain on a request from the European Commission on marine biotoxins in shellfish - Summary on regulated marine biotoxins. *EFSA Journal,* 1306, 1-23.

EFSA, (2009b). Scientific Opinion of the Panel on Contaminants in the Food Chain on a request from the European Commission on marine biotoxins in shellfish – Saxitoxin group. *EFSA Journal,* 1019, 1-76.

EFSA, (2009c). General principles for the collection of national food consumption data in the view of a pan-European dietary survey. *EFSA Journal,* 1435, [51pp].

EFSA, (2009d). Statement of the Panel on Contaminants in the Food Chain on a request from the European Commission on the influence of processing on the levels of lipophilic marine biotoxins in bivalve molluscs. *EFSA Journal,* 1016, 1-10.

EFSA, (2009e). EFSA Panel on Contaminants in the Food Chain (CONTAM); Scientific Opinion on marine biotoxins in shellfish – Palytoxin group. *EFSA Journal,* 7 (12), 1393 [38pp].

EFSA, (2009f). Scientific Opinion of the Panel on Contaminants in the Food Chain on a request from the European Commission on marine biotoxins in shellfish – Pectenotoxin group. *EFSA Journal,* 1109, 1-47.

EFSA, (2009g). Scientific Opinion of the Panel on Contaminants in the Food Chain on a request from the European Commission on marine biotoxins in shellfish – Domoic acid. *EFSA Journal*, 1-61.

EFSA, (2010a). EFSA Panel on Contaminants in the Food Chain; Scientific Opinion on marine biotoxins in shellfish –Emerging toxins: Ciguatoxin group. *EFSA Journal,* 8 (6), 1627 [38 pp].

EFSA, (2010b). EFSA report on data collection: future directions. *EFSA Journal,* 85, 1533 [35 pp.].

EFSA, (2010c). EFSA Panel on Contaminants in the Food Chain (CONTAM); Scientific Opinion on marine biotoxins in shellfish – Emerging toxins: Brevetoxin group. *EFSA Journal,* 8, 1677-1706.

EFSA, (2010d). EFSA Panel on Contaminants in the Food Chain (CONTAM); Scientific Opinion on marine biotoxins in shellfish – Cyclic imines (spirolides, gymnodimines, pinnatoxins and pteriatoxins). *EFSA Journal,* 8, 1-39.

EFSA, (2010e). EFSA Panel on Contaminants in the Food Chain (CONTAM); Statement on further elaboration of the consumption figure of 400 g shellfish meat on the basis of new consumption data. *EFSA Journal,* 8, 1706. [20 pp].

Ehlers, A.; Stempin, S.; Al-Hamwi, R. & Lampen, A., (2010). Embryotoxic effects of the marine biotoxin okadaic acid on murine embryonic stem cells. *Toxicon,* 55, 855-863.

Ehlers, A.; Scholz, J.; These, A.; Hessel, S.; Preiss-Weigert, A. & Lampen, A., (2011). Analysis of the passage of the marine biotoxin okadaic acid through an in vitro human gut barrier. *Toxicology,* 279, 196-202.

Espina, B.; Louzao, M.C.; Ares, I.R.; Cagide, E.; Vieytes, M.R.; Vega, F.V.; Rubiolo, J.A.; Miles, C.O.; Suzuki, T.; Yasumoto, T. & Botana, L.M., (2008). Cytoskeletal toxicity of pectenotoxins in hepatic cells. *Br J Pharmacol,* 155, 934-944.

Espina, B.; Louzao, M.C.; Ares, I.R.; Fonfria, E.S.; Vilarino, N.; Vieytes, M.R.; Yasumoto, T. & Botana, L.M., (2010a). Impact of the pectenotoxin C-43 oxidation degree on its cytotoxic effect on rat hepatocytes. *Chem Res Toxicol,* 23, 504-515.

Espina, B.; Louzao, M.C.; Cagide, E.; Alfonso, A.; Vieytes, M.R.; Yasumoto, T. & Botana, L.M., (2010b). The methyl ester of okadaic acid is more potent than okadaic acid in disrupting the actin cytoskeleton and metabolism of primary cultured hepatocytes. *Br J Pharmacol,* 159, 337-344.

Estevez, M.D.; Vieytes, M.R.; Louzao, M.C. & Botana, L.M., (1994). Effect of okadaic acid on immunologic and non-immunologic histamine release in rat mast cells. *Biochem Pharmacol,* 47, 591-593.

Etheridge, S.M., (2010). Paralytic Shellfish Poisoning: seafood safety and human health perspectives. *Toxicon,* 56, 108-122.

European Commission, (2002). Commission Decision 2002/226/EC of 15 March 2002 establishing special health checks for the harvesting and processing of certain bivalve molluscs with a level of amnesic shellfish poison (ASP) exceeding the limit laid down by Council Directive 91/492/EEC *Official Journal of European Communities,* 75, 65 - 66.

European Commission, (2005). Commission Regulation (EC) No 2074/2005 of 5 December 2005 laying down implementing measures for certain products under Regulation (EC) Nº 853/2004 of the European Parliament and of the Council and for the organisation of official controls under Regulation (EC) Nº 854/2004 of the European Parliament and of the Council and Regulation (EC) Nº 882/2004 of the European Parliament and of the Council, derogating from Regulation (EC) Nº 852/2004 of the European Parliament and of the Council and amending Regulations (EC) Nº 853/2004 and (EC) Nº 854/2004. Official Journal of the European Union, L338, 27-59

European Commission, (2006). Commission Regulation (EC) Nº 1664/2006 of 6 November 2006 amending Regulation (EC) Nº 2074/2005 as regards implementing measures for certain products of animal origin intended for human consumption and repealing certain implementing measures. Official Journal of the European Union, L320, 13-45.

European Commission, (2011). Commission Regulation (EU) Nº 15/2011 of 10 January 2011 amending Regulation (EC) Nº 2074/2005 as regards recognised testing methods for detecting marine biotoxins in live bivalve molluscs. Official Journal of the European Union, L6, 3-6.

European Parliament, (2002). Regulation (EC) Nº 178/2002 of the European Parliament and of the Council of 28 January 2002 laying down the general principles and requirements of food law, establishing the European Food Safety Authority and laying down procedures in matters of food safety. Official Journal of the European Communities, L31: 1-24.

European Parliament, (2004a). Regulation (EC) Nº 853/2004 of the European Parliament and of the Council of 29 April 2004 laying down specific hygiene rules for food of animal origin. Official Journal of the European Union, 139, 55-205.

European Parliament, (2004b). Regulation (EC) Nº 854/2004 of the European Parliament and of the Council of 29 April 2004 laying down specific rules for the organisation of official controls on products of animal origin intended for human consumption. Official Journal of the European Union , L226, 83-127.

European Parliament, (2007). Commission Regulation (EC) 1244/2007 of the European Parliament and of the Council of 24 of October 2007 laying down implementing measures for certain products of animal origin for human consume. Official Journal of the European Union, L281, 12-18.

FAO, (2004). Marine biotoxins. Food andNutrition Paper 80. Available from http://www.fao.org/docrep/007/y5486e/y5486e00.HTM.

FAO/IOC/WHO, (2004). Report of the Joint FAO/IOC/WHO Ad hoc Expert Consultation on Biotoxins in Bivalve Molluscs, Oslo, Norway, September 26-30, 2004. Advance Pre-Publication Copy. Available at URL: ftp://ftp.fao.org/es/esn/food/biotoxin_report_en.pdf., Oslo, Norway.

Fernandez, M.T.; Zitko, V.; Gascon, S. & Novelli, A., (1991). The marine toxin okadaic acid is a potent neurotoxin for cultured cerebellar neurons. *Life Sci,* 49, PL157-162.

Fessard, V.; Grosse, Y.; Pfohl-Leszkowicz, A. & Puiseux-Dao, S., (1996). Okadaic acid treatment induces DNA adduct formation in BHK21 C13 fibroblasts and HESV keratinocytes. *Mutat Res,* 361, 133-141.

Fialkowski, M.K., McCrory, M. A., Roberts, S. M., Tracy, J. K., Grattan, L. M., Boushey, C. J, (2010). Evaluation of dietary assessment tools used to assess the diet of adults participating in the communities advancing the studies of tribal nations across the lifespan cohort. . *J Am Diet Assoc,* 110, 65-73.

Fladmark, K.E.; Serres, M.H.; Larsen, N.L.; Yasumoto, T.; Aune, T. & Doskeland, S.O., (1998). Sensitive detection of apoptogenic toxins in suspension cultures of rat and salmon hepatocytes. *Toxicon,* 36, 1101-1114.

Fleming, L.E.; Kirkpatrick, B.; Backer, L.C.; Bean, J.A.; Wanner, A.; Reich, A.; Zaias, J.; Cheng, Y.S.; Pierce, R.; Naar, J.; Abraham, W.M. & Baden, D.G., (2007). Aerosolized red-tide toxins (brevetoxins) and asthma. *Chest,* 131, 187-194.

Fleming, L.E.; Bean, J.A.; Kirkpatrick, B.; Cheng, Y.S.; Pierce, R.; Naar, J.; Nierenberg, K.; Backer, L.C.; Wanner, A.; Reich, A.; Zhou, Y.; Watkins, S.; Henry, M.; Zaias, J.; Abraham, W.M.; Benson, J.; Cassedy, A.; Hollenbeck, J.; Kirkpatrick, G.; Clarke, T. & Baden, D.G., (2009). Exposure and effect assessment of aerosolized red tide toxins (brevetoxins) and asthma. *Environ Health Perspect,* 117, 1095-1100.

Fonfría, E.S.; Vilariño, N.; Espiña, B.; Louzao, M.C.; Alvarez, M.; Molgó, J.; Aráoz, R. & Botana, L.M., (2010). Feasibility of gymnodimine and 13-desmethyl C spirolide detection by fluorescence polarization using a receptor-based assay in shellfish matrixes. *Anal Chim Acta.,* 657, 75-82.

Friedman, M.A.; Fleming, L.E.; Fernandez, M.; Bienfang, P.; Schrank, K.; Dickey, R.; Bottein, M.Y.; Backer, L.; Ayyar, R.; Weisman, R.; Watkins, S.; Granade, R. & Reich, A., (2008). Ciguatera fish poisoning: treatment, prevention and management. *Mar Drugs,* 6, 456-479.

Fujiki, H. & Suganuma, M., (2009). Carcinogenic aspects of protein phosphatase 1 and 2A inhibitors. *Prog Mol Subcell Biol,* 46, 221-254.

Furey, A.; O'Doherty, S.; O'Callaghan, K.; Lehane, M. & James, K.J., (2010). Azaspiracid poisoning (AZP) toxins in shellfish: toxicological and health considerations. *Toxicon,* 56, 173-190.

Garet E, G.-F.A., Lago J, Vieites JM, (2010). Comparative evaluation of Enzyme-Linked Immunoassay and reference methods for the detection of shellfish hydrophilic toxins in several presentations of seafood. *Toxicon,* 58, 1410-1415.

Gawley, R.E.; Rein, K.S.; Kinoshita, M. & Baden, D.G., (1992). Binding of brevetoxins and ciguatoxin to the voltage-sensitive sodium channel and conformational analysis of brevetoxin B. *Toxicon,* 30, 780-785.

Gill, S.; Murphy, M.; Clausen, J.; Richard, D.; Quilliam, M.; MacKinnon, S.; LaBlanc, P.; Mueller, R. & Pulido, O., (2003). Neural injury biomarkers of novel shellfish toxins, spirolides: a pilot study using immunochemical and transcriptional analysis. *Neurotoxicology,* 24, 593-604.

Goldstein, T., Mazet, J. A. K., Zabka, T. S., Langlois, G., Colegrove, K. M., Silver, M., Bargu, S., Van Dolah, F., Leighfield, T., Conrad, P. A., Barakos, J., Williams, D. C., Dennison, S., Haulena, M. Gulland, F. M. D., (2008). Novel symptomatology and changing epidemiology of domoic acid toxicosis in California sea lions (*Zalophus californianus*): an increasing risk to marine mammal health. *Proceedings of the Royal Society B,* 275, 267-276.

Habermann, E., (1989). Palytoxin acts through Na^+, K^+-ATPase. *Toxicon,* 27, 1171-1187.

Hamilton, B.; Hurbungs, M.; Vernoux, J.P.; Jones, A. & Lewis, R.J., (2002). Isolation and characterisation of Indian Ocean ciguatoxin. *Toxicon,* 40, 685-693.

Hilderbrand, S.C.; Murrell, R.N.; Gibson, J.E. & Brown, J.M., (2010). Marine brevetoxin induces IgE-independent mast cell activation. *Arch Toxicol.*

Hille, B., (1968). Pharmacological modifications of the sodium channels of frog nerve. *J Gen Physiol.,* 51, 199-219.

Hille, B., (1975). The receptor for tetrodotoxin and saxitoxin. A structural hypothesis. *Biophys J.,* 15, 615-619.

Honkanen, R.E.; Codispoti, B.A.; Tse, K.; Boynton, A.L. & Honkanan, R.E., (1994). Characterization of natural toxins with inhibitory activity against serine/threonine protein phosphatases. *Toxicon,* 32, 339-350.

Hosokawa, M.; Tsukada, H.; Saitou, T.; Kodama, M.; Onomura, M.; Nakamura, H.; Fukuda, K. & Seino, Y., (1998). Effects of okadaic acid on rat colon. *Dig Dis Sci,* 43, 2526-2535.

Hu, T.; Burton, I.W.; Cembella, A.D.; Curtis, J.M.; Quilliam, M.A.; Walter, J.A. & Wright, J.L., (2001). Characterization of spirolides a, c, and 13-desmethyl c, new marine toxins isolated from toxic plankton and contaminated shellfish. *J Nat Prod,* 64, 308-312.

Inoue, M.; Hirama, M.; Satake, M.; Sugiyama, K. & Yasumoto, T., (2003). Inhibition of brevetoxins binding to the voltage-gated sodium channel by gambierol and gambieric acid-A. *Toxicon,* 41, 469-474.

Ishida, Y.; Takagi, K.; Takahashi, M.; Satake, N. & Shibata, S., (1983). Palytoxin isolated from marine coelenterates. The inhibitory action on (Na,K)-ATPase. *J Biol Chem,* 258, 7900-7902.

Isobe, M. & Hamajima, A., (2010). Ciguatoxin: developing the methodology for total synthesis. *Nat Prod Rep,* 27, 1204-1226.

Ito, E.; Yasumoto, T. & Terao, K., (1996). Morphological observations of diarrhea in mice caused by experimental ciguatoxicosis. *Toxicon,* 34, 111-122.

Ito, E.; Satake, M.; Ofuji, K.; Kurita, N.; McMahon, T.; James, K. & Yasumoto, T., (2000). Multiple organ damage caused by a new toxin azaspiracid, isolated from mussels produced in Ireland. *Toxicon,* 38, 917-930.

Ito, E.; Satake, M.; Ofuji, K.; Kurita, N.; McMahon, T.; James, K. & Yasumoto, T., (2002a). Chronic effects in mice caused by oral administration of sublethal doses of azaspiracid, a new marine toxin isolated from mussels. *Toxicon,* 40, 193-203.

Ito, E.; Yasumoto, T.; Takai, A.; Imanishi, S. & Harada, K., (2002b). Investigation of the distribution and excretion of okadaic acid in mice using immunostaining method. *Toxicon,* 40, 159-165.

Ito, E. Toxicology of Azaspiracid-1: acute and chronic poisoning, tumorigenicity and chemical structure relationship to toxicity in mouse bioassay In: Botana, L. M. (Ed.), *Seafood and Freshwater Toxins. Second Edition*. London: CRC Press. Taylor and Francis Group; 2008; p. 775-784.

Ito, E.; Suzuki, T.; Oshima, Y. & Yasumoto, T., (2008). Studies of diarrhetic activity on pectenotoxin-6 in the mouse and rat. *Toxicon,* 51, 707-716.

Ito, E. & Yasumoto, T., (2009). Toxicological studies on palytoxin and ostreocin-D administered to mice by three different routes. *Toxicon,* 54, 244-251.

Ito, K.; Karaki, H. & Urakawa, N., (1977). The mode of contractile action of palytoxin on vascular smooth muscle. *Eur J Pharmacol,* 46, 9-14.

James, K.J.; O´Driscoll, D.; García Fernández, J. & Furey, A. Azaspiracids: Chemistry, Bioconversion, and Determination. In: Botana, L. M. (Ed.), *Seafood and Freshwater Toxins. Second Edition*. London: CRC Press. Taylor and Francis Group; 2008; p. 763-773

Jayaraj, R.; Gupta, N. & Rao, P.V., (2009). Multiple signal transduction pathways in okadaic acid induced apoptosis in HeLa cells. *Toxicology,* 256, 118-127.

Johansen, T.H.; Drejer, J.; Wätjen, F. & Nielsen, E.O., (1993). A novel non-NMDA receptor antagonist shows selective displacement of low-affinity [3H]kainate binding. *Eur J Pharmacol.,* 246, 195-204.

Karaki, H.; Nagase, H.; Ohizumi, Y.; Satake, N. & Shibata, S., (1988). Palytoxin-induced contraction and release of endogenous noradrenaline in rat tail artery. *Br J Pharmacol,* 95, 183-188.

Katikou, P. Chemistry of palytoxins and ostreocins. In: Botana, L. M. (Ed.), *Phycotoxins, Chemistry and Biochemistry*. Ames, I.A: Blackwell Publishing; 2007; p. 75-93.

Kharrat, R.; Servent, D.; Girard, E.; Ouanounou, G.; Amar, M.; Marrouchi, R.; Benoit, E. & Molgó, J., (2008). The marine phycotoxin gymnodimine targets muscular and neuronal nicotinic acetylcholine receptor subtypes with high affinity. *J Neurochem,* 107, 952-963.

Kim, M.O.; Moon, D.O.; Kang, S.H.; Heo, M.S.; Choi, Y.H.; Jung, J.H.; Lee, J.D. & Kim, G.Y., (2008). Pectenotoxin-2 represses telomerase activity in human leukemia cells through suppression of hTERT gene expression and Akt-dependent hTERT phosphorylation. *FEBS Lett,* 582, 3263-3269.

Kodama, M. & Sato, S. Metabolism of Paralytic Shellfish Toxins Incorporated into bivalves. In: Botana, L. M. (Ed.), *Seafood and Freshwater Toxins. Second Edition*. London: CRC Press. Taylor and Francis Group; 2008; p. 165-175.

Krock, B.; Tillmann, U.; Selwood, A.I. & Cembella, A.D., (2008). Unambiguous identification of pectenotoxin-1 and distribution of pectenotoxins in plankton from the North Sea. *Toxicon,* 52, 927-935.

Kuiper-Goodman, T., Todd, T., (1991). Health hazard assessment of PSP in Canadian shellfish.

Lago, J.; Santaclara, F.; Vieites, J.M. & Cabado, A.G., (2005). Collapse of mitochondrial membrane potential and caspases activation are early events in okadaic acid-treated Caco-2 cells. *Toxicon,* 46, 579-586.

Langeland, G., Hasselgård, T., Tangen, K., Skulberg, O. M., Hjelle, A., (1984). An outbreak of paralytic shellfish poisoning in western Norway. *Sarsia* 69, 185-193.

Le Hegarat, L.; Jacquin, A.G.; Bazin, E. & Fessard, V., (2006). Genotoxicity of the marine toxin okadaic acid, in human Caco-2 cells and in mice gut cells. *Environ Toxicol,* 21, 55-64.

Lefebvre, K.A. & Robinson, A., (2010). Domoic acid and human exposure risks : A review. *Toxicon,* 56, 218-230.

Lehane, L. & Lewis, R.J., (2000). Ciguatera: recent advances but the risk remains. *Int J Food Microbiol,* 61, 91-125.

Leira, F.; Alvarez, C.; Vieites, J.M.; Vieytes, M.R. & Botana, L.M., (2002a). Characterization of distinct apoptotic changes induced by okadaic acid and yessotoxin in the BE(2)-M17 neuroblastoma cell line. *Toxicol In Vitro,* 16, 23-31.

Leira, F.; Cabado, A.G.; Vieytes, M.R.; Roman, Y.; Alfonso, A.; Botana, L.M.; Yasumoto, T.; Malaguti, C. & Rossini, G.P., (2002b). Characterization of F-actin depolymerization as a major toxic event induced by pectenotoxin-6 in neuroblastoma cells. *Biochem Pharmacol,* 63, 1979-1988.

Leira, F.; Alvarez, C.; Cabado, A.G.; Vieites, J.M.; Vieytes, M.R. & Botana, L.M., (2003). Development of a F actin-based live-cell fluorimetric microplate assay for diarrhetic shellfish toxins. *Anal Biochem,* 317, 129-135.

LePage, K.T.; Baden, D.G. & Murray, T.F., (2003). Brevetoxin derivatives act as partial agonists at neurotoxin site 5 on the voltage-gated Na^+ channel. *Brain Res,* 959, 120-127.

Lerga, A.; Richard, C.; Delgado, M.D.; Canelles, M.; Frade, P.; Cuadrado, M.A. & Leon, J., (1999). Apoptosis and mitotic arrest are two independent effects of the protein phosphatases inhibitor okadaic acid in K562 leukemia cells. *Biochem Biophys Res Commun,* 260, 256-264.

Lerma, J., (2006). Kainate receptor physiology. *Curr Opin Pharmacol.,* 6, 89-97.

Lewis, R.J., (2001). The changing face of ciguatera. *Toxicon,* 39, 97-106.

Lombet, A.; Bidard, J.N. & Lazdunski, M., (1987). Ciguatoxin and brevetoxins share a common receptor site on the neuronal voltage-dependent Na^+ channel. *FEBS Lett,* 219, 355-359.

Louzao, M.C.; Alfonso, A.; Botana, A.M.; Goenaga, X.; Cabado, A.G.; Vieytes, M.R. & Botana, L.M., (1994a). Effect of lyophilization on the stability of gonyautoxins obtained from contaminated mussels. *Toxicon,* 32, 807-817.

Louzao, M.C.; Alfonso, A.; Cabado, A.G.; Botana, A.M.; Goenaga, X.; Vieytes, M.R. & Botana, L.M., (1994b). Study of the stability of gonyautoxins in acidic solution. *Fres J Ana. Chem,* 349, 465-468.

Louzao, M.C.; Vieytes, M.R.; Baptista de Sousa, J.M.V.; Leira, F. & Botana, L.M., (2001). A fluorimetric method based on changes in membrane potential for screening paralytic shellfish toxins in mussels. *Anal Biochem,* 289, 246-250.

Louzao, M.C.; Vieytes, M.R.; Fontal, O. & Botana, L.M., (2003). Glucose uptake in enterocytes: a test for molecular targets of okadaic acid. *J Recept Signal Transduct Res,* 23, 211-224.

Louzao, M.C.; Vieytes, M.R.; Yasumoto, T. & Botana, L.M., (2004). Detection of sodium channel activators by a rapid fluorimetric microplate assay. *Chem Res Toxicol,* 17, 572-578.

Louzao, M.C.; Vieytes, M.R. & Botana, L.M., (2005). Effect of okadaic acid on glucose regulation. *Mini Rev Med Chem,* 5, 207-215.

Louzao, M.C.; Vieytes, M.R.; Yasumoto, T.; Yotsu-Yamashita, M. & Botana, L.M., (2006). Changes in membrane potential: an early signal triggered by neurologically active phycotoxins. *Chem Res Toxicol,* 19, 788-793.

Louzao, M.C.; Ares, I.R.; Vieytes, M.R.; Valverde, I.; Vieites, J.M.; Yasumoto, T. & Botana, L.M., (2007). The cytoskeleton, a structure that is susceptible to the toxic mechanism activated by palytoxins in human excitable cells. *FEBS J,* 274, 1991-2004.

Louzao, M.C.; Ares, I.R.; Cagide, E.; Espina, B.; Vilarino, N.; Alfonso, A.; Vieytes, M.R. & Botana, L.M., (2010a). Palytoxins and cytoskeleton: An overview. *Toxicon,*57, 460-469 .

Louzao, M.C.; Espina, B.; Cagide, E.; Ares, I.R.; Alfonso, A.; Vieytes, M.R. & Botana, L.M., (2010b). Cytotoxic effect of palytoxin on mussel. *Toxicon,* 56, 842-847.

Llewellyn, L.E., (2006). Saxitoxin, a toxic marine natural product that targets a multitude of receptors. *Nat Prod Rep.,* 23, 200-222.

Matta, J.; Navas, J.; Milad, M.; Manger, R.; Hupka, A. & Frazer, T., (2002). A pilot study for the detection of acute ciguatera intoxication in human blood. *J Toxicol Clin Toxicol,* 40, 49-57.

Meunier, F.A.; Mattei, C. & Molgo, J., (2009). Marine toxins potently affecting neurotransmitter release. *Prog Mol Subcell Biol,* 46, 159-186.

Miles, C.O.; Wilkins, A.L.; Munday, R.; Dines, M.H.; Hawkes, A.D.; Briggs, L.R.; Sandvik, M.; Jensen, D.J.; Cooney, J.M.; Holland, P.T.; Quilliam, M.A.; MacKenzie, A.L.; Beuzenberg, V. & Towers, N.R., (2004). Isolation of pectenotoxin-2 from Dinophysis acuta and its conversion to pectenotoxin-2 seco acid, and preliminary assessment of their acute toxicities. *Toxicon,* 43, 1-9.

Miles, C.O.; Wilkins, A.L.; Hawkes, A.D.; Selwood, A.I.; Jensen, D.J.; Munday, R.; Cooney, J.M. & Beuzenberg, V., (2005). Polyhydroxylated amide analogs of yessotoxin from *Protoceratium reticulatum. Toxicon,* 45, 61-71.

Milian, A.; Nierenberg, K.; Fleming, L.E.; Bean, J.A.; Wanner, A.; Reich, A.; Backer, L.C.; Jayroe, D. & Kirkpatrick, B., (2007). Reported respiratory symptom intensity in asthmatics during exposure to aerosolized Florida red tide toxins. *J Asthma,* 44, 583-587.

Miura, D.; Kobayashi, M.; Kakiuchi, S.; Kasahara, Y. & Kondo, S., (2006). Enhancement of transformed foci and induction of prostaglandins in Balb/c 3T3 cells by palytoxin: in vitro model reproduces carcinogenic responses in animal models regarding the inhibitory effect of indomethacin and reversal of indomethacin's effect by exogenous prostaglandins. *Toxicol Sci,* 89, 154-163.

Molgo, J.; Comella, J.X. & Legrand, A.M., (1990). Ciguatoxin enhances quantal transmitter release from frog motor nerve terminals. *Br J Pharmacol,* 99, 695-700.

Molgo, J.; Shimahara, T. & Legrand, A.M., (1993). Ciguatoxin, extracted from poisonous morays eels, causes sodium-dependent calcium mobilization in NG108-15 neuroblastoma x glioma hybrid cells. *Neurosci Lett,* 158, 147-150.

Moon, D.O.; Kim, M.O.; Kang, S.H.; Lee, K.J.; Heo, M.S.; Choi, K.S.; Choi, Y.H. & Kim, G.Y., (2008). Induction of G2/M arrest, endoreduplication, and apoptosis by actin depolymerization agent pectenotoxin-2 in human leukemia cells, involving activation of ERK and JNK. *Biochemical Pharmacology,* 76, 312-321.

Moore, R.E. & Scheuer, P.J., (1971). Palytoxin: a new marine toxin from a coelenterate. *Science,* 172, 495-498.

Munday, J.S. Toxicology of the pectenotoxins. In: Botana, L. M. (Ed.), *Seafood and Freshwater toxins: Pharmacology, Physiology and Detection*. Boca Raton, Florida: CRC Press (Taylor and Francis Group); 2008a; p. 371-380.

Munday, R.; Towers, N.R.; Mackenzie, L.; Beuzenberg, V.; Holland, P.T. & Miles, C.O., (2004). Acute toxicity of gymnodimine to mice. *Toxicon,* 44, 173-178.

Munday, R. Toxicology of Cyclic Imines: Gymnodimine, Spirolides, Pinnatoxins, Pteriatoxins, Prorocentrolide, Spiro-Prorocentrimine and Symbioimines. In: Botana, L. M. (Ed.), *Seafood and Freshwater Toxins. Second Edition*. London: CRC Press. Taylor and Francis Group; 2008b; p. 581-594

Munday, R.; Aune, T. & Rossini, J.P. Toxicology of the Yessotoxins. In: Botana, L. M. (Ed.), *Seafood and Freshwater Toxins. Second Edition*. London: CRC Press. Taylor and Francis Group; 2008; p. 371-380.

Murata, M.; Kumagai, M.; Lee, J.S. & Yasumoto, T., (1987). Isolation and structure of Yessotoxin, a novel polyether compound implicated in diarrhetic shellfish poisoning. *Tetrahedron Letters,* 28, 5869-5872.

Nagase, H. & Karaki, H., (1987). Palytoxin-induced contraction and release of prostaglandins and norepinephrine in the aorta. *J Pharmacol Exp Ther,* 242, 1120-1125.

Nicolaou, K.C.; Chen, D.Y.; Li, Y.; Qian, W.; Ling, T.; Vyskocil, S.; Koftis, T.V.; Govindasamy, M. & Uesaka, N., (2003a). Total Synthesis of the Proposed Azaspiracid-1 Structure, Part 2: Coupling of the C1-C20, C21-C27, and C28-C40 Fragments and Completion of the Synthesis. *Angew Chem Int Ed Engl,* 42, 3649-3653.

Nicolaou, K.C.; Li, Y.; Uesaka, N.; Koftis, T.V.; Vyskocil, S.; Ling, T.; Govindasamy, M.; Qian, W.; Bernal, F.; Chen, D.Y. & Tachibana, H., (2003b). Total Synthesis of the Proposed Azaspiracid-1 Structure, Part 1: Construction of the Enantiomerically Pure C1-C20, C21-C27, and C28-C40 Fragments. *Angew Chem Int Ed Eng,* 42, 3643-3648.

Nicolaou, K.C.; Frederick, M.O. & Aversa, R.J., (2008). The continuing saga of the marine polyether biotoxins. *Angew Chem Int Ed Engl,* 47, 7182-7225.

Nordt, S.P.; Wu, J.; Zahller, S.; Clark, R.F. & Cantrell, F.L., (2009). Palytoxin poisoning after dermal contact with zoanthid coral. *J Emerg Med.*

Ofuji, K.; Satake, M.; McMahon, T.; Silke, J.; James, K.J.; Naoki, H.; Oshima, Y. & Yasumoto, T., (1999). Two analogs of azaspiracid isolated from mussels, Mytilus edulis, involved in human intoxication in Ireland. *Nat Toxins,* 7, 99-102.

Ogino, H.; Kumagai, M. & Yasumoto, T., (1997). Toxicological evaluation of yessotoxin. *Nat Toxins,* 5, 255-259.

Ohshika, H., (1971). Marine toxins from the Pacific. IX. Some effects of ciguatoxin on isolated mammalian atria. *Toxicon,* 9, 337-343.

Otero, P.; Perez, S.; Alfonso, A.; Vale, C.; Rodriguez, P.; Gouveia, N.N.; Gouveia, N.; Delgado, J.; Vale, P.; Hirama, M.; Ishihara, Y.; Molgo, J. & Botana, L.M., (2010). First toxin profile of ciguateric fish in Madeira Arquipelago (Europe). *Anal Chem,* 82, 6032-6039.

Otero, A.; Chapela, M. J.; Atanassova, M.; Vieites, J. M. & Cabado, A. G. (2011). Cyclic imines: chemistry and mechanisms of action. *Chem Res Toxicol, in press.*

Paredes, I.; Rietjens, I.; Vieites, J.M. & Cabado, A.G., (2011). Update of risk assesment of main marine biotoxins in the European Union *Toxicon, in press.*

Paz, B.; Daranas, A.H.; Norte, M.; Riobó, P.; Franco, J.M. & Fernández, J.J., (2008). Yessotoxins, a group of marine polyether toxins: an overview. *Mar Drugs,* 6, 73-102.

Pazos, M.; Alfonso, A.; Vieytes, M.; Yasumoto, T. & Botana, L., (2004). Resonant mirror biosensor detection method based on yessotoxin-phosphodiesterase interactions. *Anal Biochem,* 335, 112-118.

Pazos, M.; Alfonso, A.; Vieytes, M.; Yasumoto, T. & Botana, L., (2005). Kinetic analysis of the interaction between yessotoxin and analogs and immobilized phosphodiesterases using a resonant mirror optical biosensor. *Chem Res Toxicol,* 18, 1155-1160.

Pearn, J., (2001). Neurology of ciguatera. *J Neurol Neurosurg Psychiatry,* 70, 4-8.

Pelc, M.J. & Zakarian, A., (2005). An approach to the Imine Ring System of Pinnatoxins. *Organic Letters,* 7, 1629-1631.

Perez-Arellano, J.L.; Luzardo, O.P.; Perez Brito, A.; Hernandez Cabrera, M.; Zumbado, M.; Carranza, C.; Angel-Moreno, A.; Dickey, R.W. & Boada, L.D., (2005). Ciguatera fish poisoning, Canary Islands. *Emerg Infect Dis,* 11, 1981-1982.

Perl, T.M.; Bédard, L.; Kosatsky, T; Hockin, J. C.; Todd, E. C. D. & Remis, R. S., (1990). An outbreak of toxic encephalopathy caused by mussels contaminated with domoic acid. *The New England Journal of Medicine,* 322, 1775-1780.

Pierotti, S.; Malaguti, C.; Milandri, A.; Poletti, R. & Paolo Rossini, G., (2003). Functional assay to measure yessotoxins in contaminated mussel samples. *Anal Biochem,* 312, 208-216.

Pizarro, G.; Paz, B.; Franco, J.M.; Suzuki, T. & Reguera, B., (2008). First detection of Pectenotoxin-11 and confirmation of OA-D8 diol-ester in Dinophysis acuta from European waters by LC-MS/MS. *Toxicon,* 52, 889-896.

Plakas, S.M. & Dickey, R.W., (2010). Advances in monitoring and toxicity assessment of brevetoxins in molluscan shellfish. *Toxicon,* 56, 137-149.

Pulido, O.M., (2008). Domoic acid toxicologic pathology: a review. *Mar Drugs,* 6, 180-219.

Radwan, F.F. & Ramsdell, J.S., (2006). Characterization of in vitro oxidative and conjugative metabolic pathways for brevetoxin (PbTx-2). *Toxicol Sci,* 89, 57-65.

Ramsdell, J.S. The molecular and integrative basis to brevetosin toxicity. In: Botana, L. M. (Ed.), *Seafood and Freshwater toxins: Pharmacology, Physiology and Detection*. Boca Raton, FL: CRC Press (Taylor and Francis Group); 2008; p. 519-550.

Ravindran, J.; Gupta, N.; Agrawal, M.; Bala Bhaskar, A.S. & Lakshmana Rao, P.V., (2010). Modulation of ROS/MAPK signaling pathways by okadaic acid leads to cell death via, mitochondrial mediated caspase-dependent mechanism. *Apoptosis,*16,145-161.

Richard, D.; Arsenault, E.; Cembella, A.D. & Quilliam, M. Investigations into the toxicology and pharmacology of spirolides, a novel group of shellfish toxins. In: Hallegraeff, G. M., Blackburn, S. I., Bolch, C. J. and Lewis, R. J. (Eds.), *Harmful Algal Blooms 2000* 2001; p.

Riobo, P.; Paz, B.; Franco, J.M.; Vazquez, J.A.; Murado, M.A. & Cacho, E., (2008). Mouse bioassay for palytoxin. Specific symptoms and dose-response against dose-death time relationships. *Food Chem Toxicol,* 46, 2639-2647.

Roach, J.S.; P., L.; Lewis, N.I.; Munday, R.; Quilliam, M.A. & MacKinnon, S.L., (2009). Characterization of a dispiroketal spirolide subclass from *Alexandrium ostenfeldii. J Nat Prod,* 72, 1237-1240.

Rodríguez-Velasco, M.L. Toxin monitoring programs and regulatory view. In: Botana, L. M. (Ed.), *Seafood and Freshwater Toxins. Pharmacology, physiology and detection.*: CRC Press. Taylor & Francis Group; 2008; p. 919-932.

Roman, Y.; Alfonso, A.; Louzao, M.C.; de la Rosa, L.A.; Leira, F.; Vieites, J.M.; Vieytes, M.R.; Ofuji, K.; Satake, M.; Yasumoto, T. & Botana, L.M., (2002). Azaspiracid-1, a potent, nonapoptotic new phycotoxin with several cell targets. *Cell Signal,* 14, 703-716.

Román, Y.; Alfonso, A.; Vieytes, M.R.; Ofuji, K.; Satake, M.; Yasumoto, T. & Botana, L.M., (2004). Effects of Azaspiracids 2 and 3 on Intracellular cAMP, [Ca^{2+}], and pH. *Chem. Res. Toxicol.,* 17, 1338-1349.

Ronzitti, G.; Callegari, F.; Malaguti, C. & Rossini, G.P., (2004). Selective disruption of the E-cadherin-catenin system by an algal toxin. *Br J Cancer,* 90, 1100-1107.

Rossini, G.P.; Sgarbi, N. & Malaguti, C., (2001). The toxic responses induced by okadaic acid involve processing of multiple caspase isoforms. *Toxicon,* 39, 763-770.

Rossini, G.P. & Bigiani, A., (2010). Palytoxin action on the Na(+),K(+)-ATPase and the disruption of ion equilibria in biological systems. *Toxicon,* 57,429-439.

Ruff, T.A., (1989). Ciguatera in the Pacific: a link with military activities. *Lancet,* 1, 201-205.

Saito, S. & Karaki, H., (1996). A family of novel actin-inhibiting marine toxins. *Clin Exp Pharmacol Physiol,* 23, 743-746.

Santaclara, F.; Lago, J.; Vieites, J.M. & Cabado, A.G., (2005). Effect of okadaic acid on integrins and structural proteins in BE(2)-M17 cells. *Arch Toxicol,* 79, 582-586.

Satake, M.; Terasawa, K.; Kadowaki, Y. & Yasumoto, T., (1996). Relative configuration of yessotoxin and isolation of two new analogs from toxic scallops. *Tetrahedron Lett,* 37, 5955-5958.

Satake, M.; MacKenzie, L. & Yasumoto, T., (1997). Identification of *Protoceratium reticulatum* as the biogenetic origin of yessotoxin. *Nat Toxins,* 5, 164-167.

Satake, M.; Ofuji, K.; Naoki, H.; James, K.J.; Furey, A.; McMahon, T.; Silke, J. & Yasumoto, T., (1998). Azaspiracid, a new marine toxin having unique spiro ring assemblies, isolated from Irish mussels, *Mytilus edulis*. *J. Am. Chem. Soc.,* 120, 9967-9968.

Selwood, A.I.; Ginkel, R.; Wilkins, A.L.; Munday, R.; Ramsdell, J.S.; Jensen, D.J.; Cooney, J.M. & Miles, C.O., (2008). Semisynthesis of S-desoxybrevetoxin-B2 and brevetoxin-B2, and assessment of their acute toxicities. *Chem Res Toxicol,* 21, 944-950.

Selwood, A.I.; Miles, C.O.; Wilkins, A.L.; van Ginkel, R.; Munday, R.; Rise, F. & McNabb, P., (2010). Isolation, structural determination and acute toxicity of pinnatoxins E, F and G. *J Agric Food Chem.,* 58, 6532-6542.

Shoemaker, R.C.; House, D. & Ryan, J.C., (2010). Defining the neurotoxin derived illness chronic ciguatera using markers of chronic systemic inflammatory disturbances: a case/control study. *Neurotoxicol Teratol,* 32, 633-639.

Sierra-Beltrán, A.; Palafox-Uribe, M.; Grajales-Montiel, J. & Cruz-Villacorta, A., (1997). Sea bird mortalily at Cabo San Lucas, Mexico: evidence that toxic diatom blooms are spreading. *Toxicon,* 35, 447-453.

Sleno, L. & Volmer, D.A., (2005). Toxin screening in phytoplankton: detection and quantitation using MALDI triple quadrupole mass spectrometry. *Anal Chem,* 77, 1509-1517.

Sosa, S.; Del Favero, G.; De Bortoli, M.; Vita, F.; Soranzo, M.R.; Beltramo, D.; Ardizzone, M. & Tubaro, A., (2009). Palytoxin toxicity after acute oral administration in mice. *Toxicol Lett,* 191, 253-259.

Souid-Mensi, G.; Moukha, S.; Mobio, T.A.; Maaroufi, K. & Creppy, E.E., (2008). The cytotoxicity and genotoxicity of okadaic acid are cell-line dependent. *Toxicon,* 51, 1338-1344.

Spector, I.; Braet, F.; Shochet, N.R. & Bubb, M.R., (1999). New anti-actin drugs in the study of the organization and function of the actin cytoskeleton. *Microsc Res Tech,* 47, 18-37.

Strachan, L.C.; Lewis, R.J. & Nicholson, G.M., (1999). Differential actions of pacific ciguatoxin-1 on sodium channel subtypes in mammalian sensory neurons. *J Pharmacol Exp Ther,* 288, 379-388.

Suárez-Isla, B.A. Paralytic shellfish toxins. Pharmacology and toxicology. Biological detection methods. In: Botana, L. M. (Ed.), *Seafood and Freshwater Toxins. Pharmacoloy, Physiology and Detection*. London: CRC Press. Taylor and Francis Group; 2008; p. 197-206

Suzuki, T.; Walter, J.A.; LeBlanc, P.; MacKinnon, S.; Miles, C.O.; Wilkins, A.L.; Munday, R.; Beuzenberg, V.; MacKenzie, A.L.; Jensen, D.J.; Cooney, J.M. & Quilliam, M.A., (2006). Identification of pectenotoxin-11 as 34S-hydroxypectenotoxin-2, a new pectenotoxin analogue in the toxic dinoflagellate Dinophysis acuta from New Zealand. *Chem Res Toxicol,* 19, 310-318.

Takahashi, H.; Kusumi, T.; Kan, Y.; Satake, M. & Yasumoto, T., (1996). Determination of the absolute configuration of yessotoxin, a polyether compound implicated in diarrhetic shellfish poisoning, by NMR spectroscopic method using a chiral anisotropic reagent, methoxy-(2-naphthyl)acetic acid. *Tetrahedron Lett.,* 37, 7087-7090.

Takai, A.; Bialojan, C.; Troschka, M. & Ruegg, J.C., (1987). Smooth muscle myosin phosphatase inhibition and force enhancement by black sponge toxin. *FEBS Lett,* 217, 81-84.

Takai, A.; Murata, M.; Torigoe, K.; Isobe, M.; Mieskes, G. & Yasumoto, T., (1992). Inhibitory effect of okadaic acid derivatives on protein phosphatases. A study on structure-affinity relationship. *Biochem J,* 284 (Pt 2), 539-544.

Terao, K.; Ito, E.; Yanagi, T. & Yasumoto, T., (1986). Histopathological studies on experimental marine toxin poisoning. I. Ultrastructural changes in the small intestine and liver of suckling mice induced by dinophysistoxin-1 and pectenotoxin-1. *Toxicon,* 24, 1141-1151.

Terao, K.; Ito, E.; Oarada, M.; Murata, M. & Yasumoto, T., (1990). Histopathological studies on experimental marine toxin poisoning-5. The effects in mice of yessotoxin isolated from *Patinopecten yessoensis* and of a desulfated derivative. *Toxicon,* 28, 1095-1104.

Terao, K.; Ito, E.; Oarada, M.; Ishibashi, Y.; Legrand, A.M. & Yasumoto, T., (1991). Light and electron microscopic studies of pathologic changes induced in mice by ciguatoxin poisoning. *Toxicon,* 29, 633-643.

Terao, K.; Ito, E. & Yasumoto, T., (1992). Light and electron microscopic studies of the murine heart after repeated administrations of ciguatoxin or ciguatoxin-4c. *Nat Toxins,* 1, 19-26.

Terao, K. Ciguatera toxins: toxicology. In: Botana, L. M. (Ed.), *Seafood and Freshwater Toxins*. New York: Marcel Dekker; 2000; p. 449-472

Ting, J.Y. & Brown, A.F., (2001). Ciguatera poisoning: a global issue with common management problems. *Eur J Emerg Med,* 8, 295-300.

Todd, E.C.D., (1993). Domoic acid and amnesic shellfish poisoning: a review. *Journal of Food Protection,* 56, 69-83.

Tohda, H.; Nagao, M.; Sugimura, T. & Oikawa, A., (1993). Okadaic acid, a protein phosphatase inhibitor, induces sister-chromatid exchanges depending on the presence of bromodeoxyuridine. *Mutat Res,* 289, 275-280.

Torgersen, T.; Sandvik, M.; Lundve, B. & Lindegarth, S., (2008). Profiles and levels of fatty acid esters of okadaic acid group toxins and pectenotoxins during toxin depuration. Part II: blue mussels (*Mytilus edulis*) and flat oyster (*Ostrea edulis*). *Toxicon,* 52, 418-427.

Traore, A.; Baudrimont, I.; Ambaliou, S.; Dano, S.D. & Creppy, E.E., (2001). DNA breaks and cell cycle arrest induced by okadaic acid in Caco-2 cells, a human colonic epithelial cell line. *Arch Toxicol,* 75, 110-117.

Tripuraneni, J.; Koutsouris, A.; Pestic, L.; De Lanerolle, P. & Hecht, G., (1997). The toxin of diarrheic shellfish poisoning, okadaic acid, increases intestinal epithelial paracellular permeability. *Gastroenterology,* 112, 100-108.

Truelove, J.; Mueller, R.; Pulido, O. & Iverson, F., (1996). Subchronic toxicity study of domoic acid in the rat. *Food Chem Toxicol.,* 34, 525-529.

Tryphonas, L.; Truelove, J.; Nera, E. & Iverson, F., (1990). Acute neurotoxicity of domoic acid in the rat. *Toxicol Pathol,* 18, 1-9.

Tsuchiya, A., Hardy, J., Burbacher, T. M., Faustman, E. M., Mariën, K., (2008). Fish intake guidelines: incorporating n-3 fatty acid intake and contaminant exposure in the Korean and Japanese communities. *The American Journal of Clinical Nutrition,* 87, 1867-1875.

Tubaro, A.; Sosa, S.; Carbonatto, M.; Altinier, G.; Vita, F.; Melato, M.; Satake, M. & Yasumoto, T., (2003). Oral and intraperitoneal acute toxicity studies of yessotoxin and homoyessotoxins in mice. *Toxicon,* 41, 783-792.

Tubaro, A.; Dell'ovo, V.; Sosa, S. & Florio, C., (2010). Yessotoxins: a toxicological overview. *Toxicon,* 56, 163-172.

Tubaro, A.; Durando, P.; Del Favero, G.; Ansaldi, R.; Icardi, G.; Deeds, J.R. & Sosa, S., (2011). Case definitions for human poisonings postulated to palytoxins exposure. *Toxicon,* 57,478-495.

Twiner, M.J.; Rehmann, N.; Hess, P. & Doucette, G.J., (2008). Azaspiracid shellfish poisoning: a review on the chemistry, ecology, and toxicology with an emphasis on human health impacts. *Mar Drugs,* 6, 39-72.

Ukena, T.; Satake, M.; Usami, M.; Oshima, Y.; Naoki, H.; Fujita, T.; Kan, Y. & Yasumoto, T., (2001). Structure elucidation of ostreocin D, a palytoxin analog isolated from the dinoflagellate *Ostreopsis siamensis*. *Biosci Biotechnol Biochem,* 65, 2585-2588.

Ukena, T.; Satake, M.; Usami, M.; Oshima, Y.; Fujita, T.; Naoki, H. & Yasumoto, T., (2002). Structural confirmation of ostreocin-D by application of negative-ion fast-atom bombardment collision-induced dissociation tandem mass spectrometric methods. *Rapid Commun Mass Spectrom,* 16, 2387-2393.

Valdiglesias, V.; Mendez, J.; Pasaro, E.; Cemeli, E.; Anderson, D. & Laffon, B., (2010). Assessment of okadaic acid effects on cytotoxicity, DNA damage and DNA repair in human cells. *Mutat Res,* 689, 74-79.

Vale-Gonzalez, C.; Alfonso, A.; Suñol, C.; Vieytes, M.R. & Botana, L.M., (2006). Role of the plasma membrane calcium adenosine triphosphatase on domoate-induced intracellular acidification in primary cultures of cerebelar granule cells. *J of Neurosci Res,* 84, 326-337.

Vale, C.; Alfonso, A.; Sunol, C.; Vieytes, M.R. & Botana, L.M., (2006). Modulation of calcium entry and glutamate release in cultured cerebellar granule cells by palytoxin. *J Neurosci Res,* 83, 1393-1406.

Vale, C.; Gómez-Limia, B.; Nicolaou, K.C.; Frederick, M.O.; Vieytes, M.R. & Botana, L., (2007a). The c-Jun-N-terminal kinase is involved in the neurotoxic effect of azaspiracid-1. *Cell Physiol Biochem,* 20, 957-966.

Vale, C.; Gomez-Limia, B.; Vieytes, M.R. & Botana, L.M., (2007b). Mitogen-activated protein kinases regulate palytoxin-induced calcium influx and cytotoxicity in cultured neurons. *Br J Pharmacol,* 152, 256-266.

Vale, C.; Alfonso, A.; Vieytes, M.R.; Romarís, X.M.; Arévalo, F.; Botana, A.M. & Botana, L.M., (2008a). In vitro and in vivo evaluation of paralytic shellfish poisoning toxin potency and the influence of the pH of extraction. *Anal Chem.,* 80, 1770-1776.

Vale, C.; Wandscheer, C.; Nicolaou, K.C.; Frederick, M.O.; Alfonso, C.; Vieytes, M.R. & Botana, L.M., (2008b). Cytotoxic effect of azaspiracid-2 and azaspiracid-2-methyl ester in cultured neurons: involvement of the c-Jun N-terminal kinase. *J Neurosci Res,* 86, 2952-2962.

Valverde, I.; Lago, J.; Reboreda, A.; Vieites, J.M. & Cabado, A.G., (2008). Characteristics of palytoxin-induced cytotoxicity in neuroblastoma cells. *Toxicol In Vitro,* 22, 1432-1439.

Vilarino, N.; Ares, I.R.; Cagide, E.; Louzao, M.C.; Vieytes, M.R.; Yasumoto, T. & Botana, L.M., (2008). Induction of actin cytoskeleton rearrangement by methyl okadaate-comparison with okadaic acid. *FEBS J,* 275, 926-934.

Vilariño, N.; Nicolaou, K.C.; Frederick, M.O.; Vieytes, M.R. & Botana, L.M., (2007). Irreversible cytoskeletal disarrangement is independent of caspase activation during in vitro azaspiracid toxicity in human neuroblastoma cells. *Biochem Pharmacol,* 74, 327-335.

Vilariño, N., (2008). Marine toxins and the cytoskeleton: azaspiracids. *FEBS J,* 275, 6075-6081.

Vilariño, N.; Nicolaou, K.C.; Frederick, M.O.; Cagide, E.; Alfonso, C.; Alonso, E.; Vieytes, M.R. & Botana, L.M., (2008). Azaspiracid substituent at C1 is relevant to in vitro toxicity. *Chem Res Toxicol,* 21, 1823-1831.

Vilariño, N.; Fonfría, E.S.; Molgó, J.; Aráoz, R. & Botana, L.M., (2009). Detection of gymnodimine-A and 13-desmethyl C spirolide phycotoxins by fluorescence polarization. *Anal Chem,* 81, 2708-2714.

Watkins, S.M.; Reich, A.; Fleming, L.E. & Hammond, R., (2008). Neurotoxic shellfish poisoning. *Mar Drugs,* 6, 431-455.

Wattenberg, E.V., (2007). Palytoxin: exploiting a novel skin tumor promoter to explore signal transduction and carcinogenesis. *Am J Physiol Cell Physiol,* 292, C24-32.

Wattenberg, E.V., (2011). Modulation of protein kinase signaling cascades by palytoxin. *Toxicon,* 57, 440-448.

Wekell, J.C.; Hurst, J. & Lefebvre, K.A., (2004). The origin of the regulatory limits for PSP and ASP toxins in shellfish. *Journal of Shellfish Research,* 23, 927-930.

Whyte, A.L.H.; Hook, G. R.; Greening, G. E.; Gibbs-Smith, E. & Gardner, J. P. A., (2009). Human dietary exposure to heavy metals via the consumption of greenshell mussels (*Perna canaliculus* Gmelin 1791) from the Bay of Islands, northern New Zealand. . *Science of the Total Environment,* 407, 4348-4355.

Wiese, M.; D'Agostino, P.M.; Mihali, T.K.; Moffitt, M.C. & Neilan, B.A., (2010). Neurotoxic alkaloids: saxitoxin and its analogs. *Mar Drugs,* 8, 2185-2211.

Wiles, J.S.; Vick, J.A. & Christensen, M.K., (1974). Toxicological evaluation of palytoxin in several animal species. *Toxicon,* 12, 427-433.

Wright, J.L.; Bird, C.J.; de Freitas, A.S.; Hampson, D.; McDonald, J. & Quilliam, M.A., (1990). Chemistry, biology, and toxicology of domoic acid and its isomers. *Can Dis Wkly Rep.,* 16 Suppl 1E, 21-26.

Wright, J.L.C.; Boyd, K.; de Freitas, A.S.W.; M. Falk, M.; Foxall, R.; Jamieson, W.D.; Laycock, A.W.; McCulloch, A.W.; McInnes, A.G.; Odense, P.; Pathak, V.P.; Quilliam, M.A.; Ragan, M.A.; Sim, P.G.; Thibault, J.A.; Walter, J.A.; Gilgan, M.; Richard, D.J.A. & Dewar, D., (1989). Identification of domoic acid, a neuroexcitatory amino acid, in toxic mussels from eastern Prince Edward Island. *Canadian Journal of Chemistry,* 67, 481-490.

Xi, D.; Peng, Y.G. & Ramsdell, J., (1997). Domoic acid is a potent neurotoxin to neonatal rats. *Nat Toxins,* 5, 74-79.

Yasumoto, T.; Oshima, Y. & Yamaguchi, M., (1978). Occurrence of a new type of shellfish poisoning in the Tohoku district. *Bull. Jpn. Soc. Sci. Fish,* 44, 1249-1255.

Yasumoto, T.; Murata, M.; Oshima, Y.; Matsumoto, G.K. & Clardy, J. Diarrhetic Shellfish Poisoning. In: Ragelis, E. P. (Ed.), *Seafood Toxins*. Washington, DC.: American Chemical Society; 1984; p. 207-214.

Zhou, Z.H.; Komiyama, M.; Terao, K. & Shimada, Y., (1994). Effects of pectenotoxin-1 on liver cells in vitro. *Nat Toxins,* 2, 132-135.

In: New Trends in Marine and Freshwater Toxins
Editors: A. G. Cabado and J. M. Vieites
ISBN: 978-1-62257-787-3

Chapter 7

BIOLOGICAL METHODS FOR DETECTION OF PHYCOTOXINS: BIOASSAYS AND *IN VITRO* ASSAYS

Luis M. Botana, Mercedes R. Vieytes[1], Ana M Botana[2], Carmen Vale and Natalia Vilariño

[1]Departamento de Farmacología, Fisiología, Facultad de Ciencias, Lugo, Spain
[2]Química Analítica, Facultad de Veterinaria, Lugo, Spain

ABSTRACT

Marine toxins are an acute threat to consumers of shellfish products, since their toxicity above certain levels may pose a deadly risk. To overcome this problem, all producing countries have implemented a system based, for most of the toxins, on the use of mice to detect the presence of the toxins in seafood products. For several years, ethical and animal welfare concerns have raised the problem of replacing the use of rodents to monitor the presence of marine toxins. In addition to this, an European Directive (86/609/EEC) about animal care have prompted the replacement of animals for marine toxins detection (Communities, E., 1986). The combined situation of a social awareness to prevent animal use, combined with a legal requirement to do so, resulted, upon request by Germany authorities to the European Commission, in the creation of a working group at the European Food Safety Authority (EFSA) on marine biotoxins and shellfish. The outcome of the work of this working group provided several opinions, one for each toxin group (Panel, E.C., 2008c, b, d, a, 2009a, b, d, c, e, 2010b, a, c), that was used as a base to modify the legislation in Europe. As a consequence, a recent Regulation (REGULATION, C., 2011) has modified the current *status quo* in marine toxin control, previously defined by Regulation 2074/2005 (Commission, E., 2006).

1. INTRODUCTION

1.1. Marine Toxins Detection

Regulation 15/2011 has defined liquid chromatography coupled to mass spectrometry (LC-MS) as the reference method for marine toxin detection. This is the first time that marine toxins control is legally required to be done by a non-animal based approach. The replacement will be effective June 2011, and both methods, bioassay and LC-MS will coexist for three years, to them be LC-MS the reference method. To understand the consequences of this change, it is important to discuss the advantages and disadvantages of each approach. From a monitoring point of view, in Europe there are three toxin groups, the domoic acid, also known as ASP (amnesic shellfish poison), that can be monitored by liquid chromatography as reference method (Commission, E., 2005) or ELISA (Commission, E., 2007). Figure 1 shows European official methods for detection of the lipophilic marine toxins and analogs: DTX: dinophysistoxin, OA: okadaic acid, AZA: azaspiracid, YTX: yessotoxin, PTX: pectenotoxin. Finally, the saxitoxin group, also known as PSP (paralytic shellfish poison), that can monitored by mouse bioassay as reference method (Commission, E., 2005), or by HPLC (Commission, E., 2006), although this method is only approved for those toxins and matrices for which it was validated (AOAC, 2005), although an extension of the validationfor dcNEO and dcGTX2,3 has been described (Turner, A.D., Norton, D.M., 2009), and the lipophilic group of toxins (okadaic acid, yessotoxin, pectenotoxin, azaspiracid and their analogs), for which the bioassay is the reference method (Commission, E., 2004, 2005), and that will be replaced by LC-MS as reference method (Regulation, C., 2011).

The mouse bioassay for lipophilic toxins, a method originally developed by Yasumoto et al. (Yasumoto, T., Oshima, Y., 1978), has been used until now as a universal detection method for any liphophilic compound that would show toxicity (Botana, L.M., 2008). Although the method has been useful, its main drawbacks were false positives that would harm the industrial sector. In addition, these false positives were the cause for the inclusion in the legislation of two toxin groups that were proven not to be harmful to humans, namely yessotoxins and pectenotoxins. These two groups of compounds coexist with okadaic acid, but have never been proven to be a cause of human intoxication, plus their toxicities are low. The EFSA working group concluded that and acute reference dose (ARfD), for a 400 g portion of shellfish (far higher than the 100 g upon which Regulation 2074/2005 is based), should be 3.75 mg YTX eq/kg shellfish meat, and 120 μg PTX2 eq./kg. Therefore, there is no clear reason to keep these compounds in the legislation as a harmful threat to consumers. On the other hand, okadaic acid and its analogs the dynophysistoxins(Yasumoto, T., Oshima, Y., 1978), and azaspiracid and its analogs (Ofuji, K., Satake, M., 1999), are harmful compounds that require a premarket control.

European legislation covers therefore domoic acid, the complete group of PSP if using mouse bioassay, and a few of the analogs for okadaic acid, azaspiracids, pectenotoxin and yessotoxin (Regulation, C., 2011), Figure 1. Nevertheless, some toxin groups such as palytoxin and ostreocins, for which the EFSA working group provided an opinion where it suggests an ARfD of 30 μg/Kg shellfish meat, are not covered by the legislacion. Another group to have in mind from a food safety point of view are cyclic imines, for which not enough toxicological studies are available, but having a mechanism of action based on the interaction with nicotinic (Bourne, Y., Radic, Z., 2010) and muscarinic (Wandscheer, C.B., Vilarino, N., 2010) receptors, they are clearly toxins to be cautious about, again, they are not included in the legislation. On the other hand, since European legislation does provide coverage to gasteropods, another drawback to food consumer protection is the fact that tetrodotoxin has been identified as a source of risk, as shown by recent evidence (Rodriguez, P., Alfonso, A., 2008) showing very high levels of tetrodotoxin analogs in gastropods. All of the above suggests that some food safety issues are not covered by current European legislation, and this situation will become less controlled once LC-MS is used as the method of choice for routine analysis, since LC-MS will not track these toxins. LC-MS is a technology very powerful to identify those compounds for which standards are being used, yet this means that all other compounds for which standards are not available, or unknown or unexpected compounds, will not be tracked in the samples. This would mean commercial products being marketed without adequate food safety control.

In addition to the lack of a complete analysis by a LC-MS based monitoring system, there are two more issues to discuss: the toxic limits, and the toxic equivalent factors.

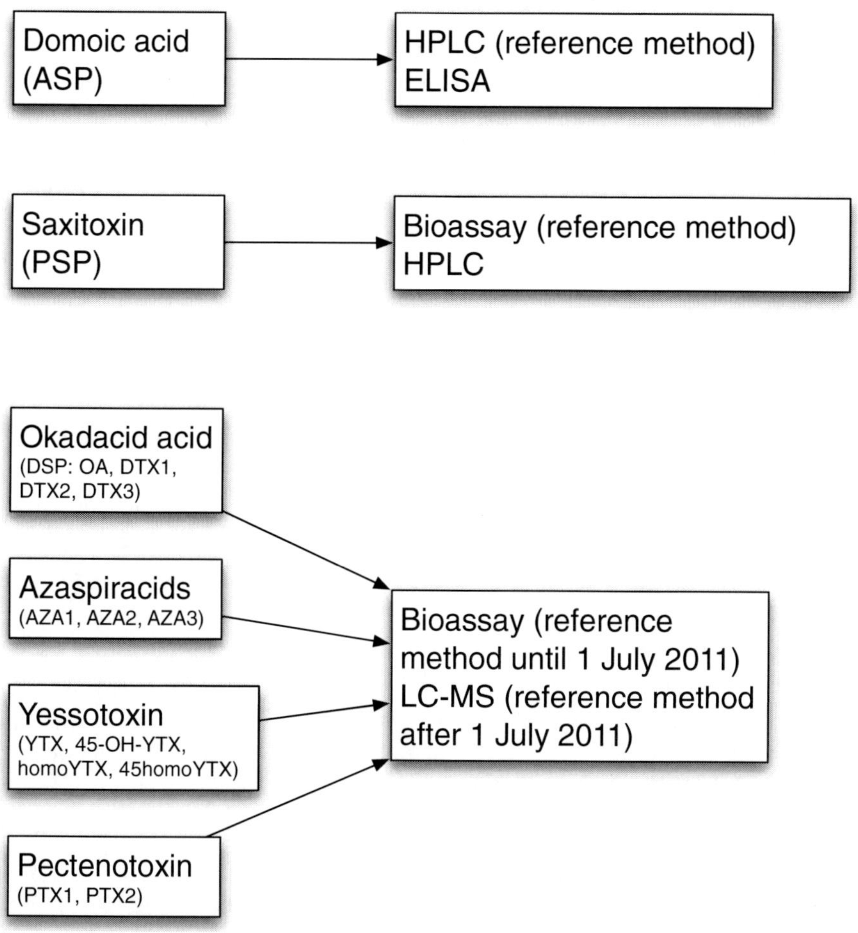

Figure 1. European official methods for detection of marine toxins, and analogs required for control on the lipophilic group. DTX: dinophysistoxin, OA: okadaic acid, AZA: azaspiracid, YTX: yessotoxin, PTX: pectenotoxin.

Reference Toxin TEF=1	Analogue	TEF (EFSA)	TEF 1	TEF 2
Yessotoxin	1a-homo YTX	1		1b
	45-hydroxcyYTX	1	0,38a	0,06b
	45-hydroxy-1a-homoYTX	0,5		
	carboxyYTX		012a	0,02b
Pectenotoxin 2	PTX1	1	0,08c	
	PTX3	1		
	PTX4	1		
	PTX5			
	PTX6	1	0,002c	
	PTX9		0,009c	
	PTX11	1	0,9c	

Figure 2. Example on the use of TEF for toxic conversion. a) (Ferrari, S., Ciminiello, P., 2004). b) (Pazos, M.J., Alfonso, A., 2006) c) (Espina, B., Louzao, M.C., 2010a).

The reason for the European Commission to modify the detection method was the fact that the EFSA working group identified the mouse bioassay as a low sensitivity and low specificity method. Since Regulation 2074/2005 (Commission, E., 2007) is based on a 100 g portion, this does not provide enough protection compared to the larger size figure of 400 g shellfish portion meat identified by the EFSA opinions as the best approach in acute exposures. One of the main reasons for the current toxin regulatory limits for marine toxins is the sensitivity of the mouse bioassay. A change to an analytical method would be coherent with a change in toxin limits to fit a food safety study, yet the only modification has been applied to methodology, not to toxin limits (Regulation, C., 2011). Therefore, 100 g is still considered the intake daily portion. In addition the new legislation (Regulation, C., 2011) requires that total toxicity equivalence shall be calculated using toxicity equivalent factors (TEFs) as recommended by EFSA. This is again an unfortunate situation, as TEFs are by large an uncertainty factor for most of the analogs in the marine toxin spectrum, as recently discussed (Botana, L.M., Vilariño, N., 2010), and therefore all analytical results, that need to be converted to toxic equivalent values of a representative compound in the group, will be affected by the error in the proposed TEF. Figure 2 shows as an example the problem of the TEF to be used. One of the most clear outcomes of the EFSA opinions on marine toxins is the need for further toxicological studies. In the case of the TEF suggested by EFSA opinions, most of the values are equal to 1, given the lack of enough information. Therefore, since TEF are unknown, and those being used refer to intraperitoneal values, the potential source of other mechanisms of action and their TEF opens clear issues of debate (Figure 2). As an example, total toxicity for yessotoxin has been proposed by EFSA to be calculated with a TEF value of 1 for all analogs except 45-OH-homoYTX. These TEF belong to intraperitoneal (ip) calculations (Figure 2), but since YTX has never been reported to be orally toxic, the use of ip TEF does include an obvious error in the calculations, since an analytical result will provide a toxic value of a compound that does not show oral toxicity. Clearly this is a contradictory situation. On the other hand, if E-cadherin is used for the estimation of TEF (Ferrari, S., Ciminiello, P., 2004), or phosphodiesterase activation (Pazos, M.J., Alfonso, A., 2005), then the situation becomes far more complicated, as shown in figure 2. With this regard, okadaic acid does show a matching TEF between whole animal and phosphatase inhibition (Aune, T., Larsen, S., 2007), which is clearly not the case for YTX. A similar situation occurs with pectenotoxins, while EFSA provided a TEF value of 1 for all analogs, the structure activity studies of the interaction of actin and pectenotoxins (Espina, B., Louzao, M.C., 2010a) suggest that this is not a realistic situation. Since actin is well accepted as the target of pectenotoxin (Allingham, J.S., Miles, C.O., 2007), the errors assumed in the use of a general TEF of 1 for all analogs do not seem apropriate, moreover keeping in mind that pectenotoxin, in a similar fashion as yessotoxin, has never been proven to be orally toxic (Miles, C.O., Wilkins, A.L., 2004).

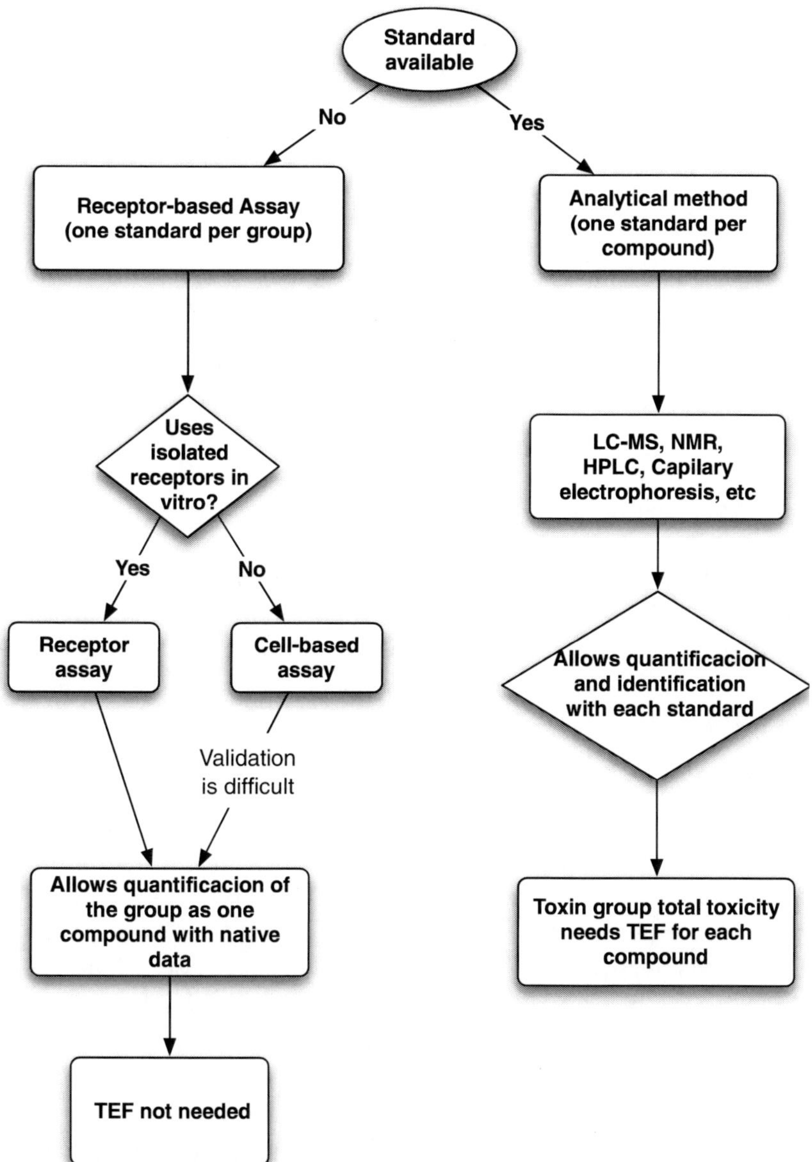

Figure 3. Decission flow to use functional methods or analytical methods.

An important contribution of the EFSA working group opinions, has been to finally classify the toxins according to their mechanism of action, hence the old term diarrheic toxinsto refer to the lypophillic toxin group has been finally forgotten, to now allow the use of the term okadaic acid and analogs, azaspiracid and analogs, yessotoxin and analogs, and pectenotoxin and analogs. Although the mechanism of action of azaspiracids or yessotoxins are not yet well defined (Alfonso, A. and Alfonso, C., 2007, Botana, L.M., Vale, C., 2007, Alfonso, A. and Alfonso, C., 2008, Vale, C., Wandscheer, C., 2008, Vilarino, N., Nicolaou, K.C., 2008, Vale, C., Nicolaou, K.C., 2010), they are clearly different from the mechanism used by other toxin groups, and therefore becomes a group on itself. On the other hand, although diarrhea is a sympton to many toxins (Vale, C. and Botana, L.M., 2008), phosphatase inhibition is only the mechanism of action of the okadaic acid and analogs, formerly identified as part of the diarrheic group, although recent data even question this fact, as the toxic effects of methylokadaic acid are relevant while having a low phosphatase inhibition activity (Espina, B., Louzao, M.C., 2010b).

1.2. *In Vitro* Assays

In vitro methods provide a fast, unexpensive and easy way of quantification of marine toxins. Assuming that LC-MS will be the reference method for now, the obvious approach is to develop in vitro methods that a) detect compounds potentially toxic and present in shellfish, but not regulated, such as palytoxins (Aligizaki, K., Katikou, P., 2008), tetrodotoxin (Rodriguez, P., Alfonso, A., 2008), orcyclicimines (Villar Gonzalez, A., Rodriguez-Velasco, M.L., 2006), or b) methods that detect analogs within regulated groups but not identified by LC-MS due to lack of standards.

The variety of methodologies is very wide, from radioactive methods (Araoz, R., Vilarino, N., 2010) to fluorescent (Leira, F., Alvarez, C., 2003, Louzao, M.C., Vieytes, M.R., 2004), biosensor-based (Pazos, M.J., Alfonso, A., 2004, Campbell, K., Haughey, S.A., 2010), or ELISA (Briggs, L.R., Miles, C.O., 2004, Aasen, J., Samdal, I.A., 2005, Samdal, I.A., Aasen, J.A., 2005). Despite the technology, three basic approaches are being taken: use of receptors, use of antibodies, or use of whole cells. Whole cells are easy to use for the development of sensitive methods for marine toxins, butgiven the complex biochemistry of cells, it is difficult to predict or identify a potential interaction between the targeted toxin group and an unexpected one. Although many methods were developed using whole cells, none has succeeded as a potential alternative solution (Okumura, M., Tsuzuki, H., 2005, Canete, E. and Diogene, J., 2008, Louzao, M.C., Ares, I.R., 2008, Louzao, M.C., Espina, B., 2010). Antibodies are a suitable approach, since they can provide a very sensitive platform for detection of low levels of compounds. Nevertheless, antibodies have a fundamental problem, their crossreactivity, that prevents an adequate quantification to match the toxicicity, unless the crossreactivity is designed to match the relative toxicity of each compound. Although there are some reports of this unique combination of crossreactivity matching relative toxicities (Stewart, L.D., Hess, P., 2009), in general antibodies do not show this feature, and therefore the quantification is prone to very large quantification errors. The use of receptors is clearly the best approach, as the interaction of the analogs of a toxin group does match the relative toxicity, and therefore quantification is a reliable measurement of toxicity. But this does not guarantee a reliable measurement, unless the method has been adequately validated for matrix effect. Although many receptor-based methods were developed (Vieytes, M.R., Fontal, O.I., 1997, Alfonso, C., Alfonso, A., 2005, Vilarino, N., Fonfria, E.S., 2009, Fonfria, E.S., Vilarino, N., 2010a, Fonfria, E.S., Vilarino, N., 2010b), only one went through a full validation study. A colorimetric phosphatase inhibition assay for determination of OA-toxins group in edible molluscs, TOXILINE-DSP.Co (at present Okatest), has been recently validated as an official alternative test for the detection of OA-toxins group, according to the Comission Regulations (EC) No. 2074/2005 and 15/2011. The validation of this test has recently been approved by the VMT. The rationale to decide on the use of one given method is outlined in Figure 3.

Functional assays are often used as a synonym of in vitro assays, but this should not be the case. A functional assay, where the function of the toxins targeting their receptors allows the detection of any compound sharing the same mechanism, and therefore the same toxicity, can only be applied to receptor-based methods. In this case, these methods provide a very useful tool to blindly detect any compound by means of the toxic mechanism. On the other hand, methods based on the use of binders (antibodies, nucleic acids etc), do not guarantee that any given compound with the same mechanism of action would be detected. This is a fundamental difference between functional assays and binder assays. Therefore, receptor-based assays are the best choice. Despide this, their major drawback is their inability to identify or quantify a compound in a mixture, this being the best feature of the analytical LC-MS, which, on the other hand, totally lacks the possibility of detecting unknown mechanism-sharing compounds (Botana, L.M., Alfonso, A., 2009). The best choice is therefore a combined system that uses the broad detection capability of a receptor-based functional assay with the quantification capability of LC-MS. While LC-MS requires one standard per analyte, a functional method requires one standard per toxin group, so the cost of a functional method (not having in mind the high maintenance due to an LC-MS) will always be lower.

CONCLUSION

The change in methodology to monitor the presence of marine toxins does improve animal welfare, but it is not to be considered an improvement over food consumer protection, as only very few compounds will be targeted, allowing non tracked toxins to reach the market, while sustaining a more expensive the complex system.

AKNOWLEDGEMENTS

This work was funded with the following grants:

From Ministerio de Ciencia y Tecnología, Spain: AGL2007-60946/ALI, SAF2009-12581 (subprograma NEF), AGL2009-13581-CO2-01, TRA2009-0189, AGL2010-17875. From Xunta de Galicia, Spain: GRC2010/10, and PGIDT07CSA012261PR, PGDIT 07MMA006261PR, PGIDIT (INCITE) 09MMA003261PR, 2009/XA044, 2009/053 (Consell. Educación), 2008/CP389 (EPITOX, Consell. Innovación e Industria, programa IN.CI.TE.), 10PXIB261254 PR. From EU VIIth Frame Program: 211326 - CP (CONffIDENCE), 265896

BAMMBO, 265409 μAQUA, and 262649 BEADS. From the Atlantic Area Programme (Interreg IVB Trans-national): 2008-1/003 (Atlantox) and 2009-1/117Pharmatlantic.

References

Aasen, J., Samdal, I.A., Miles, C.O., Dahl, E., Briggs, L.R., Aune, T. (2005). Yessotoxins in Norwegian blue mussels (Mytilus edulis): uptake from Protoceratium reticulatum, metabolism and depuration. *Toxicon,* 45, 265-272.

Alfonso, A., Alfonso, C. Pharmacology of Yessotoxin. In: Botana, L.M., editor. *Phytotoxins, chemistry and Biochemistry.* Ames, Iowa: Blackwell Publishing; 2007. p. 203-210.

Alfonso, A., Alfonso, C. Pharmacology and mechanism of action of Yessotoxin: Biological detection. In: Botana, L.M., editor. *Seafood and Freshwater toxins: Pharmacology, Physiology and Detection.* 2nd edition ed. Boca Raton, FL: CRC Press (Taylor and Francys Group); 2008. p. 315-328.

Alfonso, C., Alfonso, A., Vieytes, M.R., Yasumoto, T., Botana, L.M. (2005). Quantification of yessotoxin using the fluorescence polarization technique and study of the adequate extraction procedure. *Anal Biochem,* 344, 266-274.

Aligizaki, K., Katikou, P., Nikolaidis, G., Panou, A. (2008). First episode of shellfish contamination by palytoxin-like compounds from Ostreopsis species (Aegean Sea, Greece). *Toxicon,* 51, 418-427.

Allingham, J.S., Miles, C.O., Rayment, I. (2007). A structural basis for Regulation of actin polymerization by pectenotoxins. *J Mol Biol,* 371, 959-970.

AOAC. (2005). Method 2005.06: Paralytic Shellfish Poisoning Toxins in Shellfish. Prechromatographic Oxidation and Liquid Chromatography with Fluorescence Detection. Official Methods of Analysis of the Association of Official Analytical Chemists., *Method* 2005.06, First Action.

Araoz, R., Vilarino, N., Botana, L.M., Molgo, J. (2010). Ligand-binding assays for cyanobacterial neurotoxins targeting cholinergic receptors. *Anal Bioanal Chem,* 397, 1695-1704.

Aune, T., Larsen, S., Aasen, J.A., Rehmann, N., Satake, M., Hess, P. (2007). Relative toxicity of dinophysistoxin-2 (DTX-2) compared with okadaic acid, based on acute intraperitoneal toxicity in mice. *Toxicon,* 49, 1-7.

Botana, L.M. The mouse bioassay as a universal detector. In: Botana, L.M., editor. *Seafood and Freshwater toxins: Pharmacology, Physiology and Detection.* 2nd edition ed. Boca Raton, FL: CRC Press (Taylor and Francys Group); 2008. p. 149-161.

Botana, L.M., Alfonso, A., Botana, A., Vieytes, M.R., Vale, C., Vilariño, N., Louzao, M.C. (2009). Functional assays for marine toxins as an alternative, high-throughput screening solution to animal tests. *Trends Anal. Chem.*, 28, 603-611.

Botana, L.M., Vale, C., Vilariño, N. Mechanism of action of palytoxin and azaspiracids. In: Goudey-Perriére, F., Benoit, E., Marchot, P., Popoff, M.R., editors. *Toxines émergentes: nouveaux risques.* Paris: Lavoisier. Editions TEC & DOC; 2007. p. 149-156.

Botana, L.M., Vilariño, N., Elliott, C.T., Campbell, K., Alfonso, A., Vale, C., Louzao, M.C., Botana, A.M. (2010). The problem of toxicity equivalent factors in developping alternative methods to animal bioassays for marine toxin detection. *Trends Anal. Chem.*, 29, 1316-1325.

Bourne, Y., Radic, Z., Araoz, R., Talley, T.T., Benoit, E., Servent, D., Taylor, P., Molgo, J., Marchot, P. (2010). Structural determinants in phycotoxins and AChBP conferring high affinity binding and nicotinic AChR antagonism. *Proc Natl Acad Sci U S A*, 107, 6076-6081

Briggs, L.R., Miles, C.O., Fitzgerald, J.M., Ross, K.M., Garthwaite, I., Towers, N.R. (2004). Enzyme-linked immunosorbent assay for the detection of yessotoxin and its analogues. *J Agric Food Chem*, 52, 5836-5842.

Campbell, K., Haughey, S.A., van den Top, H., van Egmond, H., Vilarino, N., Botana, L.M., Elliott, C.T. (2010). Single laboratory validation of a surface plasmon resonance biosensor screening method for paralytic shellfish poisoning toxins. *Anal Chem*, 82, 2977-2988.

Canete, E., Diogene, J. (2008). Comparative study of the use of neuroblastoma cells (Neuro-2a) and neuroblastomaxglioma hybrid cells (NG108-15) for the toxic effect quantification of marine toxins. *Toxicon*, 52, 541-550.

Commission, E. (2004). Regulation (EC) No 853/2004 of the European Parlament and of the Council of 29 April 2004 laying down specific hygiene rules for food of animal origin. *Official Journal of the European Union*, L 226, 22-82.

Commission, E. (2005). Commission Regulation (EC) No 2074/2005 of 5 December 2005 laying down implementing measures for certain products under Regulation (EC) No 853/2004 of the European Parliament. *Official Journal of the European Union*, L 338, 27-59.

Commission, E. (2006). Commission Regulation (EC) No 1664/2006 of 6 November 2006 amending Regulation (EC) No 2074/2005 as regards implementing measures for certain products of animal origin intended for human consumption and repealing certain implementing measures. *Official Journal of the European Union*, L 320, 13-45.

Commission, E. (2007). Commission Regulation (EC) No 2074/2005 of 24 October 2007 amending Regulation (EC) No 2074/2005 as regards implementing measures for certain products of animal origin intended for human consumption and laying down specific rules on official controls for the inspection of meat. *Official Journal of the European Union*, L 281, 12-18.

Communities, E. (1986). COUNCIL DIRECTIVE of 24 November 1986 on the approximation of laws, Regulations and administrative provisions of the Member States regarding the protection of animals used for experimental and other scientific purposes (86/609/EEC). *Official Journal of the European Communities*, L358, 1-29.

Espina, B., Louzao, M.C., Ares, I.R., Fonfria, E.S., Vilarino, N., Vieytes, M.R., Yasumoto, T., Botana, L.M. (2010a). Impact of the pectenotoxin C-43 oxidation degree on its cytotoxic effect on rat hepatocytes. *Chem Res Toxicol*, 23, 504-515.

Espina, B., Louzao, M.C., Cagide, E., Alfonso, A., Vieytes, M.R., Yasumoto, T., Botana, L.M. (2010b). The methyl ester of okadaic acid is more potent than okadaic acid in disrupting the actin cytoskeleton and metabolism of primary cultured hepatocytes. *Br J Pharmacol*, 159, 337-344.

Ferrari, S., Ciminiello, P., Dell'Aversano, C., Forino, M., Malaguti, C., Tubaro, A., Poletti, R., Yasumoto, T., Fattorusso, E., Rossini, G.P. (2004). Structure-activity relationships of yessotoxins in cultured cells. *Chem Res Toxicol*, 17, 1251-1257.

Fonfria, E.S., Vilarino, N., Espina, B., Louzao, M.C., Alvarez, M., Molgo, J., Araoz, R., Botana, L.M. (2010a). Feasibility of gymnodimine and 13-desmethyl C spirolide detection by fluorescence polarization using a receptor-based assay in shellfish matrixes. *Anal Chim Acta,* 657, 75-82.

Fonfria, E.S., Vilarino, N., Molgo, J., Araoz, R., Otero, P., Espina, B., Louzao, M.C., Alvarez, M., Botana, L.M. (2010b). Detection of 13,19-didesmethyl C spirolide by fluorescence polarization using Torpedo electrocyte membranes. *Anal Biochem,* 403, 102-107.

Leira, F., Alvarez, C., Cabado, A.G., Vieites, J.M., Vieytes, M.R., Botana, L.M. (2003). Development of a F actin-based live-cell fluorimetric microplate assay for diarrhetic shellfish toxins. *Anal Biochem,* 317, 129-135.

Louzao, M.C., Ares, I.R., Cagide, E. (2008). Marine toxins and the cytoskeleton: a new view of palytoxin toxicity. *FEBS J,* 275, 6067-6074.

Louzao, M.C., Espina, B., Cagide, E., Ares, I.R., Alfonso, A., Vieytes, M.R., Botana, L.M. (2010). Cytotoxic effect of palytoxin on mussel. *Toxicon,* 56, 842-847.

Louzao, M.C., Vieytes, M.R., Yasumoto, T., Botana, L.M. (2004). Detection of sodium channel activators by a rapid fluorimetric microplate assay. *Chem Res Toxicol,* 17, 572-578.

Miles, C.O., Wilkins, A.L., Munday, R., Dines, M.H., Hawkes, A.D., Briggs, L.R., Sandvik, M., Jensen, D.J., Cooney, J.M., Holland, P.T., Quilliam, M.A., MacKenzie, A.L., Beuzenberg, V., Towers, N.R. (2004). Isolation of pectenotoxin-2 from Dinophysis acuta and its conversion to pectenotoxin-2 seco acid, and preliminary assessment of their acute toxicities. *Toxicon,* 43, 1-9.

Ofuji, K., Satake, M., McMahon, T., Silke, J., James, K.J., Naoki, H., Oshima, Y., Yasumoto, T. (1999). Two analogs of azaspiracid isolated from mussels, Mytilus edulis, involved in human intoxication in Ireland. *Natural Toxins,* 7, 99-102.

Okumura, M., Tsuzuki, H., Tomita, B. (2005). A rapid detection method for paralytic shellfish poisoning toxins by cell bioassay. *Toxicon,* 46, 93-98.

Panel, E.C. (2008a). Influence of processing on the levels of lipophilic marine biotoxins in bivalve molluscs. Satement of the Panel on Contaminants in the Food Chain. *The EFSA Journal* 1016, 1-10.

Panel, E.C. (2008b). Opinion of the Scientific Panel on Contaminants in the Food chain on a request from the European Commission on marine biotoxins in shellfish – azaspiracids. *The EFSA Journal* 723, 1-52.

Panel, E.C. (2008c). Opinion of the Scientific Panel on Contaminants in the Food chain on a request from the European Commission on marine biotoxins in shellfish – okadaic acid and analogues. *The EFSA Journal* 589, 1-62.

Panel, E.C. (2008d). Scientific Opinion on marine biotoxins in shellfish- Yessotoxin group. EFSA Panel on Contaminants in the Food Chain (CONTAM). *The EFSA Journal* 907, 1-62.

Panel, E.C. (2009a). Marine toxins in shellfish. Summary on regulated marine biotoxins. Scientific Opinion of the Panel on Contaminants in the Food chain. *The EFSA Journal* 1306, 1-23.

Panel, E.C. (2009b). Opinion of the Scientific Panel on Contaminants in the Food chain on a request from the European Commission on marine biotoxins in shellfish – Saxitoxin group. *The EFSA Journal* 1019, 1-76.

Panel, E.C. (2009c). Scientific Opinion on marine biotoxins in shellfish- Domoic acid. EFSA Panel on Contaminants in the Food Chain (CONTAM). *The EFSA Journal* 1181, 1-61.

Panel, E.C. (2009d). Scientific Opinion on marine biotoxins in shellfish- Palytoxin group. EFSA Panel on Contaminants in the Food Chain (CONTAM). *The EFSA Journal* 7, 1393 (1338 pp.).

Panel, E.C. (2009e). Scientific Opinion on marine biotoxins in shellfish- Pectenotoxin group. EFSA Panel on Contaminants in the Food Chain (CONTAM). *The EFSA Journal* 1109, 1-47.

Panel, E.C. (2010a). Scientific Opinion on marine biotoxins in shellfish- Cyclic imines (spirolides, gymnodimines, pinnatoxins and pteriatoxins). EFSA Panel on Contaminants in the Food Chain (CONTAM). *The EFSA Journal* 8, 1628 (1639 pp.).

Panel, E.C. (2010b). Scientific Opinion on marine biotoxins in shellfish- Emerging toxins: Ciguatoxin-group toxins. EFSA Panel on Contaminants in the Food Chain. *The EFSA Journal* 8, 1627 (1638 pp).

Panel, E.C. (2010c). Statement on further elaboration of the consumption figure of 400 g shellfish meat on the basis of new consumption data. EFSA Panel on Contaminants in the Food Chain (CONTAM). *The EFSA Journal* 8, 1706 (1720 pp.).

Pazos, M.J., Alfonso, A., Vieytes, M.R., Yasumoto, T., Botana, L.M. (2005). Kinetic analysis of the interaction between yessotoxin and analogues and immobilized phosphodiesterases using a resonant mirror optical biosensor. *Chem Res Toxicol,* 18, 1155-1160.

Pazos, M.J., Alfonso, A., Vieytes, M.R., Yasumoto, T., Botana, L.M. (2006). Study of the interaction between different phosphodiesterases and yessotoxin using a resonant mirror biosensor. *Chem Res Toxicol,* 19, 794-800.

Pazos, M.J., Alfonso, A., Vieytes, M.R., Yasumoto, T., Vieites, J.M., Botana, L.M. (2004). Resonant mirror biosensor detection method based on yessotoxin-phosphodiesterase interactions. *Anal Biochem,* 335, 112-118.

Regulation, C. (2011). Commission Regulation (EU) No 15/2011 of 10 January 2011 amending Regulation (EC) No 2074/2005 as regards recognised testing methods for detecting marine biotoxins in live bivalve molluscs. *Official Journal of the European Communities,* L, 3-4.

Rodriguez, P., Alfonso, A., Vale, C., Alfonso, C., Vale, P., Tellez, A., Botana, L.M. (2008). First toxicity report of tetrodotoxin and 5,6,11-trideoxyTTX in the trumpet shell Charonia lampas lampas in Europe. *Anal Chem,* 80, 5622-5629.

Samdal, I.A., Aasen, J.A., Briggs, L.R., Dahl, E., Miles, C.O. (2005). Comparison of ELISA and LC-MS analyses for yessotoxins in blue mussels (Mytilus edulis). *Toxicon,* 46, 7-15.

Stewart, L.D., Hess, P., Connolly, L., Elliott, C.T. (2009). Development and single-laboratory validation of a pseudofunctional biosensor immunoassay for the detection of the okadaic acid group of toxins. *Anal Chem,* 81, 10208-10214.

Turner, A.D., Norton, D.M., Hatfield, R.G., Morris, S., Reese, A.R., Algoet, M., Lees, D.N. (2009). Refinement and extension of AOAC Method 2005.06 to include additional toxins in mussels: single-laboratory validation. *J AOAC Int,* 92, 190-207.

Vale, C., Botana, L.M. (2008). Marine toxins and the cytoskeleton: okadaic acid and dinophysistoxins. *FEBS J,* 275, 6060-6066.

Vale, C., Nicolaou, K.C., Frederick, M.O., Vieytes, M.R., Botana, L.M. (2010). Cell volume decrease as a link between azaspiracid-induced cytotoxicity and c-Jun-N-terminal kinase activation in cultured neurons. *Toxicol Sci,* 113, 158-168.

Vale, C., Wandscheer, C., Nicolaou, K.C., Frederick, M.O., Alfonso, C., Vieytes, M.R., Botana, L.M. (2008). Cytotoxic effect of azaspiracid-2 and azaspiracid-2-methyl ester in cultured neurons: involvement of the c-Jun N-terminal kinase. *J Neurosci Res,* 86, 2952-2962.

Vieytes, M.R., Fontal, O.I., Leira, F., Baptista de Sousa, J.M., Botana, L.M. (1997). A fluorescent microplate assay for diarrheic shellfish toxins. *Anal Biochem,* 248, 258-264.

Vilarino, N., Fonfria, E.S., Molgo, J., Araoz, R., Botana, L.M. (2009). Detection of gymnodimine-A and 13-desmethyl C spirolide phycotoxins by fluorescence polarization. *Anal Chem,* 81, 2708-2714.

Vilarino, N., Nicolaou, K.C., Frederick, M.O., Cagide, E., Alfonso, C., Alonso, E., Vieytes, M.R., Botana, L.M. (2008). Azaspiracid substituent at C1 is relevant to in vitro toxicity. *Chem Res Toxicol,* 21, 1823-1831.

Villar Gonzalez, A., Rodriguez-Velasco, M.L., Ben-Gigirey, B., Botana, L.M. (2006). First evidence of spirolides in Spanish shellfish. *Toxicon,* 48, 1068-1074.

Wandscheer, C.B., Vilarino, N., Espina, B., Louzao, M.C., Botana, L.M. (2010). Human muscarinic acetylcholine receptors are a target of the marine toxin 13-desmethyl C spirolide. *Chem Res Toxicol,* 23, 1753-1761.

Yasumoto, T., Oshima, Y., Yamaguchi, M. (1978). Ocurrence of a new type of shellfish poisoning in the Tohoku Distric. *Bull. Jpn. Soc. Sci. Fish.,* 44, 1249-1255.

In: New Trends in Marine and Freshwater Toxins
Editors: A. G. Cabado and J. M. Vieites
ISBN: 978-1-62257-787-3

Chapter 8

CHEMICAL METHODS FOR DETECTING PHYCOTOXINS: LC AND LC/MS/MS

Pilar Riobó[1], Elena López[2] and José M. Franco[1]
[1]Instituto de Investigaciones Marinas,
Unidad Asociada de Fitoplancton Tóxico, Vigo, Spain
[2]Laboratorio de Sanidad Exterior,
Ministerio de Política Territorial y Administración Pública, Vigo, Spain

ABSTRACT

Phycotoxins are natural products which are generally synthesized by marine microalgae, especially those belonging to the dinoflagellates group. Approximately 20 species of dinoflagellates and a smaller number of diatoms are currently known to produce phycotoxins and these account for less than 2% of all microalgae species. They are known to produce intoxication syndromes throughout the food chain from tropical to polar latitudes (Hallegraeff, 1993). Marine biotoxins are non-proteinaceous compounds with vastly differing structures, whose molecular weights lie between 250-3500 daltons. Their physico-chemical properties vary according to their polarity, lipophylic nature, thermal stability, sensitivity to pH, oxygen and light, etc. Danger of biotoxins poisoning for humans lies in their acute and chronic effects. Consumption of seafood contaminated with marine biotoxins may cause serious diseases that affect: the nervous system in paralytic shellfish poisoning (PSP), the intestinal system in diarrheic shellfish poisoning (DSP), and loss of memory in amnesic shellfish poisoning (ASP). Other well known toxins that have been identified in seafood from several countries are Brevetoxins (BTXs), Ciguatoxins (CTXs), Palytoxins (PLTXs) and tetrodotoxins (TTXs). Their mode of action is not clearly known, (Hu *et al.*, 2001; Miles *et al.*, 2000) and involves a variety of receptors and metabolic processes, wherein some activate or block sodium channels, while others activate calcium channels. Some of them are glutamate agonists or inhibitors of protein phosphatases.

On the other hand, this group of pharmacologically active marine products includes some very powerful of the so-called "fast acting toxins" such as spirolides (SPXs) and gymnodimines (GYMs). Risk assessment for the presence of phycotoxins has not been an easy matter until now. Pragmatic decisions are normally taken during a bloom. The fact that marine biotoxins pose a risk to human health has led many countries to establish regulations for controlling the presence of these toxins in products of marine origin. Several countries have set specific limits for certain marine toxins in bivalves and fish. Methodology seems to be a common problem when implementing safe and effective control.

The first empirical scientific observations on harmful algal blooms and toxic mussels were reported in 1937. Mouse bioassay was the method developed at the time for detecting toxins produced by such microalgae and has been the standard method applied for years. All analytical methods described below have been therefore evaluated against bioassays.

The mouse bioassay method has an unquestionable value in sanitary control and food safety because it examines the response to toxins in a living organism and permits detection of new toxins. However, there is a growing concern nowadays due to poor specificity of the method and for reasons of animal welfare. Alternative assays and chemical analysis have been developed, which can be valuable tools for determination of biotoxins. The liquid chromatography (LC) method with ultraviolet (UV) and/or fluorescence detection (FLD) was the first analytical technique to be used. During the last decade, the introduction of LC coupled with mass spectrometry (MS) methods has led to significant progress in the field of marine toxins analyses. Optimisation and validation of these methods and development of standards and reference materials is currently the main target in the field of marine toxins.

This chapter focuses on the different chemical methods used for detecting and quantifying biotoxins and furthermore looks into single-laboratory validation and inter-laboratory validation studies that have been carried out or are ongoing.

1. Introduction

1.1. Hydrophilic Toxins

1.1.1. PSP Toxins: STX Group

The earliest documented report of intoxication by PSP occurred in Canada, in 1798, and involved poisoning through mussel consumption in 1793, of several crew members of a ship exploring the British Columbia coastline (Kao, 1993).

PSP is a common seafood toxicity problem with worldwide distribution, and this illness typically arises due to consumption of seafood products contaminated with neurotoxins that are collectively known as saxitoxins (STXs). Shellfish store and concentrate these toxins with no apparent damaging effect on them.

The main producers of STX and derivatives are dinoflagellates and cyanophyceae. Three marine dinoflagellate genera (*Alexandrium*, *Gymnodinium* and *Pyrodinium*) are particularly responsible for PSP in humans. This family of neurotoxins binds to voltage-gated sodium channels, thereby attenuating action potentials by preventing the passage of sodium ions across the membrane.

Symptoms of human PSP intoxication include tingling, numbness, headaches, weakness and difficulty in breathing. Medical treatment consists of providing respiratory support where fluid therapy can be used to facilitate toxin excretion.

Therefore, accumulation of PSP toxins in shellfish leads to a serious public health hazard, which affects the fisheries industry. Thus, many countries have developed monitoring programs and regulatory systems that include routine sampling of bivalves for detecting presence of biotoxins and examination of water samples for presence of toxin producing phytoplankton.

The AOAC mouse bioassay method for PSP toxins is currently the reference method in the European Union (AOAC, 2000). However, controversy still prevails on methodology to be used in monitoring programs for determination of PSP toxins. Regulation nº 853/2004 of the European Union, which defines PSP toxins limits in shellfish for human consumption, provides the possibility of using analytical and biochemical methods instead of the mouse bioassay for evaluation of PSP toxins. The European Union has recently introduced the possibility of using the Lawrence method (Lawrence et al., 1991; EC, 2006), which is based on liquid chromatography with fluorescence detection and pre-column derivatization, for evaluation of PSP toxins. Chromatographic methodologies have the advantages of being highly sensitive and selective for toxin detection. The ones proposed are based on liquid chromatographic separation and fluorescence detection with pre-column or post-column derivatization, as well as methods based on hydrophilic interaction chromatography and mass spectrometry detection.

Chemistry

PSP toxins are a naturally occurring group of neurotoxic alkaloids. Saxitoxin (STX) is the most researched to date, and 57 analogues have been described ever since its discovery in 1957 (Wiese et al., 2010). The PSP toxins can be broadly characterized as hydrophilic or hydrophobic and can be divided into subgroups based on substituent side chains such as carbamate, sulphate, hydroxyl, hydroxybenzoate or acetate. Each moiety then imparts a varying level of toxicity. Only 18 of such analogues clearly present the characteristic biological activity of this syndrome. Recently described analogues include deoxy-decarbamates, types with an alcohol group, etc. and are known to have low toxicity or no toxicity at all.

PSP toxins are soluble in water as well as in mixtures of organic solvents (methanol, ethanol, acetonitrile and water). They are very stable in acidic media and extremely unstable in alkaline conditions or in the presence of oxygen. Their maximum UV absorbance ranges between 200 and 210 nm.

Extraction Procedures

Sample extraction is an important step in any analytical method. The extraction protocol for shellfish samples described in the AOAC (AOAC, 2000) involves extraction with hydrochloric acid followed by boiling. This extraction procedure may lead to partial hydrolysis of certain PSP toxins (C1, C2, C3, C4, GTX5 and GTX6) into more toxic analogues (GTX2, GTX3, GTX1, GTX4, STX and neoSTX). Such conversion process can also occur in the human stomach and therefore this extraction procedure can be considered as a reasonable approach to estimate any potential sample toxicity (CEN, 2003).

The main difficulty faced, when determining these toxins, lies in the extraction procedure used for molluscs due to their complex matrix. Proteins, polysaccharides and pigments contained in shellfish are co-extracted with toxins. These compounds do not interfere in the determination of toxins but rather affect the life of columns, since pigments are fixed in the pre-filters and the column head. Therefore, double peaks or peaks with very exaggerated tails are observed to appear in chromatograms after 50 or 60 sample injections from the LC post-column determination.

A cleanup system based on matrix dispersion and reversed-phase using Oasis HLB or LiChroprep type RP18 phases has recently been tested. Preliminary results show a pigment content reduction by about 50-60% while maintaining recoverability of toxins between 85 and 100% (Riobó *et al.*, 2010) .

Chemical Methods

At present, the mouse bioassay (MBA) is the reference method for analysing PSP toxins. This bioassay unambiguously provides evidence of potential toxicity of the sample and leads us to the conclusion that this method should not be discarded. However, MBA is considered undesirable for ethical reasons, and therefore the importance of developing analytical methods for detecting PSP toxins.

LC-FLD Methods

PSP toxins do not have chromophores and so cannot normally be detected by UV or fluorescence detection methods. In 1975, an alkaline oxidation step leading to purine derivatives was developed for PSP toxins, and this facilitated their detection by fluorescence. Such fluorometric method was thereafter recommended for PSP determination in addition to the mouse bioassay method (Bates et al., 1975). The sample is first acidified and fluorescence intensity of the oxidation products is then measured in solution. The two chromatographic methods currently used, namely; post-column and pre-column oxidation, are based on this principle.

Sullivan´s group (Sullivan et al., 1984) succeeded in separating underivatized PSP toxins by ion-pair chromatography in 1984, through addition of alkyl sulfonic acids followed by post-column oxidation with periodic acid. However, coelution of STX and dc-STX represented an important disadvantage because the toxicity value of STX is double that of dc-STX and gives rise to an incorrect total PSP toxicity.

Therefore, toxins must be individually ascertained when comparing LC results with results from the mouse bioassay method. In 1989, Oshima´s group described an entire separation of the main toxins of the PSP complex (Oshima *et al.*, 1989). However, the method proposed by Oshima requires three separate isocratic runs to analyze most of the PSP toxins and is slow and labour-intensive. These two early works later led to the development of different methodologies, which are basically minor variations of the ion pair and the column(Diener *et al.*, 2006; Franco *et al.*, 1993; Rourke *et al.*, 2008).

The post-column oxidation method described by Franco and Fernández-Villa was standardised as the CEN method 14194 (Van Egmond *et al.*, 1994). In this study the results of three laboratories were selected to demonstrate the validation only for STX and dcSTX. A validation exercise is currently being carried out using the post-column oxidation method proposed by Rourke for the main PSP toxins (Rourke *et al.*, 2008).

Post column oxidation methods are highly specific, sensitive, and MBA extract can be used in this determination. Furthermore, they lead to separation and quantification of the nine toxins which almost represent the total toxicity in mice based on toxic equivalency proposed by Oshima (Oshima *et al.*, 1989): GTX4 (73); GTX1 (99); dcGTX2 (65); dcGTX3 (75); GTX2 (36); GTX3 (63); neoSTX (92); dcSTX (50) and STX (100).

Another chromatographic method in use is the LC with pre-column oxidation. This method is conceptually simple, since oxidation and acidification take place first and theoretically the separation of the reaction products takes place in a LC column (Lawrence *et al.*, 2005) .

The main advantages of the pre-column LC-method are two-fold: it can be automated and it has been successfully validated for 12 STX analogues. But this method presents important disadvantages. Firstly, the MBA extract is not useful for this method and secondly, different peaks are produced by the periodate oxidation coelute like in the case of dc-STX and dc-NEO oxidation peaks. Two additional peaks are generated in the derivatization reaction of GTX1,4 and dc-GTX2,3. These confused chromatograms complicate the proper identification and quantification of the many PSP toxins, which could be a limitation when handling a large number of samples in a monitoring program.

An inter-laboratory exercise was conducted by the Community Reference Laboratory (CRL) to evaluate "fitness for purpose" of the Lawrence method, in order to use it as the official control method for STX-group toxins in EU laboratories (Botana *et al.*, 2007). These results were satisfactory, but emphasised the need for having trained staff and easy availability of standards.

The European Union decided to replace the Oshima relative toxicity values established for each PSP analogue by the toxic equivalency factors (TEFs). These TEFs overestimate toxicity when compared to that obtained by Oshima's method. Assuming a common mode of action, the toxicity of the STX-group toxins is expressed as the sum of STX equivalents determined by LC techniques. Until better information becomes available, the following TEFs are to be adopted, based on acute toxicity following *i.p.* administration in mice: STX = 1, NeoSTX = 1, GTX1 = 1, GTX2 = 0.4, GTX3 = 0.6, GTX4 = 0.7, GTX5 = 0.1, GTX6 = 0.1, C2 = 0.1, C4 = 0.1, dc-STX = 1, dc-NeoSTX = 0.4, dc GTX2 = 0.2, GTX3 = 0.4 and 11-hydroxy-STX = 0.3 (See chapter 5).

LC-MS Methods

Electrospray ionization mass spectrometry (ESI-MS) is effective for detection of polar PSP toxins, which are quite basic and therefore form stable $[M+H]^+$ ions. Thus, direct detection of underivatized PSP toxins is possible using a mass spectrometer as detector (Dell_Aversano *et al.*, 2005; Quilliam, 2003a), by using hydrophilic interaction liquid chromatography (HILIC) and mass spectrometry for the analysis of PSP toxins.

However, GTXs do not separate completely in a TSK-gel Amide-80 column, e.g., GTX-1 coelutes with GTX-2. Thus, the method can only be applied in combination with MS. Furthermore, variable retention times for PSP toxins in different seafood matrixes were observed.

At present, it is not possible achieve complete separation of all relevant PSP toxins for regulatory purposes and furthermore, sensitivity of MS detection is insufficient to control seafood at the regulatory limit established for PSP toxins (800 μg STX equivalents Kg-1 wet weight of soft tissues (Diener *et al.*, 2007).

1.1.2. ASP Toxins: Domoic Acid

A Few Historical Facts

Extracts of the red macroalgae *Chondria armata*, whose Japanese name is hanayanagi or domoi, have been used in traditional Japanese medicine for its anthelmintic properties. Domoic acid was isolated from *Chondria armata* and its chemical structure was described for the first time in 1958 (Takemoto *et al.*, 1958). It was synthetically prepared in 1982 (Ohfune *et al.*, 1982).

Even though domoic acid presents some pharmacological activity, it did not arouse any public health interest until three decades after its isolation. Domoic acid was for the first time associated with toxic syndrome in humans (amnesic shellfish poisoning, ASP) at the end of 1987, in Canada, after a serious food poisoning episode that affected more than a hundred people (Quilliam *et al.*, 1989b). Poisoning in this incident was caused by ingestion of toxic blue mussels (*Mytilus edulis*) and the primary source of domoic acid was found to be the diatom *Pseudonitzchia pungens f. multiseries* (Bates *et al.*, 1989).

Domoic acid producing seaweeds (*Chondria armata, Alsidium corallinum, Pseudonitzschia pungens f. multiseries, Pseudonitzschia Australis, Pseudonitzschia calliantha*) have a worldwide distribution. Therefore, domoic acid has been isolated from many animal species that are present in the human diet such as, anchovies (Fritz, *et al.*), razor clams, dungeness crabs (Wekell *et al.*, 1994), king scallops (Blanco *et al.*, 2002), squid (Bargu *et al.*, 2008) and cuttlefish (Costa *et al.*, 2005) .

Description of Domoic Acid (DA) Structure: Physico-Chemical Characteristics

Domoic acid is a secondary amino acid belonging to the kainic amino acid group. Its chemical structure closely resembles that of kainic acid. Kainic acid was isolated from the seaweed *Diginea simplex* and belongs to the same family as *Chondria armata* (Takemoto, 1978).

Both amino acids present neuroexcitatory and neurotoxic activities. Some of domoic acid's structure is similar to that of glutamic acid. And this is why domoic acid behaves as a glutamate receptor agonist, especially in kainic receptors (Meldrum, 1987) of the central nervous system. Continuous activation of this kind of receptors leads to an excessive accumulation of calcium in the cell. Cellular death is the final output.

Domoic acid is a small molecule (molecular weight 311.14 Da). Its three carboxylic groups make it highly hydrophilic and polar.

Domoic acid shows limited stability when exposed to factors such as high temperatures, extreme pH, and aqueous solutions over time. The main decomposition product after high temperature and time exposure is 5'-epi-domoic acid.

Nine domoic acid isomers have been described to date and just 4 isomers from the entire group have been found in shellfish extracts (5'-epi-domoic acid and isodomoic acid D, E and F) (Walter *et al.*, 1994; Wright *et al.*, 1990).

Chromatographic Detection Methods

Mouse bioassay (MBA) for Domoic Acid (DA) is based on the AOAC mouse bioassay for the saxitoxin (STX)-group of toxins (AOAC Official method 959.08). It involves acidic aqueous extraction through boiling and intraperitoneal injection of 1 mL of the extract into the mouse. Although it was used to detect DA during the first ASP incident in Canada, signs (such as scratching of shoulders using the hind leg, sedation-akinesia, rigidity, convulsions and death) are observed in mice when DA is present at levels starting at approximately 40 mg/kg, even though the regulatory limit is at 20 mg/kg (Ciminiello *et al.*, 2005). Therefore MBA is not capable of checking for compliance with the regulatory limit and thus LC is prescribed as the determination method in the EU legislation.

UV Detection

During the 1987 Canadian food poisoning outbreak investigation (Blanco *et al.*, 2002), a LC-DAD was performed in order to compare both toxic and control shellfish samples. It came to light that within the suspicious samples, there was a compound whose absorption spectrum showed a maximum at 242nm. The compound eluted right after tryptophan which has a maximum absorbance at 220 and 280 nm.

The above meant that monitoring shellfish and phytoplankton samples was necessary in order to protect public health. Quilliam MA *et al.* (Quilliam *et al.*, 1989a) prepared a protocol where extraction was done by boiling shellfish samples in water or by sonication of phytoplankton samples. The LOD was 0.3 ng. A clean-up phase was needed to remove low-polarity substances present in order to avoid problems during the selected LC-DAD chromatographic conditions.

Chromatographic conditions included 10% aqueous acetonitrile containing 0.1% trifluoroacetic acid mobile phase. A mobile phase pH of 2-3 is needed in order to suppress ionization and subsequent tailing due to the characteristics of the three carboxyl groups and one secondary amino group. The stationary phase consisted of a reverse-phased octadecyl silica column such as those previously used in the analysis of other peptides.

The AOAC paralytic shellfish poison extraction procedure based on boiling with 0.1M HCl (Lawrence *et al.*, 1989a) led to worse domoic acid recoveries in both naturally contaminated and spiked samples, than the boiling water extraction approach. The stability of the acid extract however proved not to be as good as the one obtained after boiling water extraction. LOD is 1 mg/kg. A collaborative study was carried out by Lawrence *et al.*, 1991.

Presence of domoic acid could also be confirmed using LC-UV in molluscan samples by derivatization with phenylisothiocyanate (PITC) of the amino group or esterification of the three carboxylic groups (Lawrence *et al.*, 1989b). Prior to derivatization, a clean-up step with SPE cartridges is a must.

Quilliam *et al.*, 1995, showed that a single aqueous 50% methanol extraction provided excellent recovery data (ca. 98%). Moreover, stability of domoic acid in methanolic extracts was better than that obtained in acid extracts. Extraction was followed by a clean-up phase compatible with aqueous methanol extracts and based on strong-anion exchange (SAX). Use of SAX-SPE cartridges was also reported in previous years and gave a high degree of cleanup such that all tryptophan was eliminated from the extracts (Hatfield *et al.*, 1994; Quilliam *et al.*, 1991).

FLD Detection

Even though compliance of set regulatory limits in seafood samples for domoic acid could be easily achieved with the use of LC-UV, there was still a growing interest for decreasing detection and quantification limits of domoic acid applicable to seawater and phytoplankton sample analyses. Such decrease could only be achieved by derivatization of the domoic acid molecule with a fluorescent reagent. None of the LC-FL methods have been subject to an inter-laboratory validation process.

Besides lowering LOD and LOQ, the fluorimetric approach for analysing shellfish samples also removes other compounds with chromatographic properties that are similar to domoic acid. One such is tryptophan, which easily interferes in the LC-UV analysis (James *et al.*, 2000).

9-fluorenylmethylchloroformate (FMOC-Cl) was chosen as the derivative reagent because it quickly reacts with secondary amino acids at the basic pH of seawater (optimal reaction pH is 7.5 ± 0.2) and because derivative stability lasts as long as seven days at room temperature *(Pocklington et al.*, 1990). Reaction time is as little as 45 seconds. The main drawback of this procedure was the extraction of excess FMOC-Cl reagent in order to avoid interference with domoic acid derivative. This was achieved by mixing with ethyl acetate. Fluorescence detector conditions are: λ ex = 264 nm and λ em = 313 nm. The LOD for domoic acid in seawater and phytoplankton aqueous extracts is 15 pg/ml.

The use of 6-aminoquinolyl-N-hydroxysuccinimidyl carbamate (AQC), a previously known amino acids derivatizing reagent (Sun *et al.*, 1999) has also been reported. Its main advantage is that any excess reacts with water and forms products that do not interfere with the domoic acid derivate elution. The new structure formed by AQC and domoic acid is stated to be stable and this protocol was successfully applied to phytoplankton analysis. Fluorescence detector conditions are: λ ex = 250 nm and λ em = 395 nm.

Another derivatizing reagent studied in seafood and phytoplankton analysis was 4-fluoro-7-nitro-2,1,3-benzoxadiazole (NBD-F) because it is very reactive with amines under mild conditions and produces highly fluorescent products (Pleasance *et al.*, 1990b). No extraction of derivative reagent excess is needed. Fluorescence detector conditions are: λ ex = 470 nm and λ em = 530 nm.

NBD-F reagent was also applied to seawater samples after being pre-concentrated using a sol-gel amorphous titania as a solid-phase sorbent. (Chan *et al.*, 2007)

MS Detection

Confirmation of domoic acid by spectroscopic methods has been in use since the beginning of the Canadian toxic outbreak investigation. (Blanco *et al.*, 2002). Toxic fractions isolated from LC were analysed by fast atom bombardment mass spectrometry (FAB-MS).

The development of the more accurate LC-MS method is currently drawing more efforts from researchers despite the GC-MS being the most widely used method to date for analysing shellfish samples. The LC-MS has also undergone several clean-up protocols and derivatizations.

Tandem mass spectrometry experiments for domoic acid determination in shellfish (Furey *et al.*, 2001; Hess *et al.*, 2001; Holland *et al.*, 2003; Pardo *et al.*, 2007; Piñeiro *et al.*, 2001) as well as in seawater and phytoplankton samples (Wang *et al.*, 2007) are also available. A recently published study for analysing domoic acid in seawater is based on rapid resolution liquid chromatography tandem mass spectrometry with head-column trapping (De la Iglesia *et al.*, 2008). Mass spectrometry has proved to be a more convenient alternative to the FMOC derivatization LC-FD method (Mafra *et al.*, 2009) at trace domoic acid levels in seawater and phytoplankton.

1.2. Lipophilic Toxins

1.2.1. General Extraction Procedure

Most extraction and purification processes for biological and chemical determinations of lipophilic toxins follow a general procedure with slight modifications except in case of brevetoxins and ciguatoxins (extraction procedure for these toxins is detailed in their correspondent section).

Methanol or 80% Methanol with a wet weight:volume ratio of 1:3 are the most common solvents used to extract the toxins. The following steps can include partitions with hexane for elimination of non-polar materials such as pigments, triglycerides, etc. And lastly, an optional purification step using solid phase extraction cartridges may be required depending on the matrix (Alcala *et al.*, 1988; Beress *et al.*, 1983; Fukui *et al.*, 1987; Hirata *et al.*, 1988; Kimura *et al.*, 1973; Lau *et al.*, 1993; Lau *et al.*, 1995; Mahnir *et al.*, 1992; Moore *et al.*, 1971; Oku *et al.*, 2004; Onuma *et al.*, 1999; Teh *et al.*, 1974; Uemura *et al.*, 1985; Yasumoto *et al.*, 1986). Polyps, molluscs, crabs and fish have the most complex matrixes. Extraction method can be simplified when dinoflagellates samples are obtained from cultures or from seawater since the matrix is less complex and the amount of lipophilic compounds is smaller than in molluscs and fish.

1.2.2. DSP Toxins: Okadaic Acid Group

Three groups of polyether toxins, namely; okadaic acid group (OAs), yessotoxins (YTXs) and pectenotoxins (PTXs) have been traditionally called DSP toxins (Yasumoto *et al.*, 1987) because they often co-occur and their toxins are co-extracted in the same lipophilic fraction from plankton samples and shellfish. Furthermore, the three groups are detected together by mouse bioassays. Nowadays, it is known that these three groups of toxins have different biological effects, wherein only the OAs are diarrhoegenic. They can be analyzed with distinct analytical methods, a fact, that recently led European health authorities to regulate them separately (Commission, 2002).

OAs are mainly produced by planktonic marine dinoflagellates of the genus *Dinophysis*, but there are also a few references about their presence in bivalves associated with the epibenthic *Prorocentrum* spp (Lawrence *et al.*, 2000).

Toxins causing DSP were first reported in Japan in 1978 (Yasumoto *et al.*, 1978). Until then, DSP events were probably often mistaken for gastroenteritis of bacterial origin. These severe gastrointestinal outbreaks occurred in Japan after consumption of mussels (*Mytilus edulis*) and scallops (*Patinopecten yessoensis*). One person affected by such event was the eminent Japanese chemist, Prof. Takeshi Yasumoto. His natural curiosity drove him to research into the causes of intoxication after eating cooked bivalves. He was finally able to isolate two fat-soluble thermostable toxins and to implement a MBA that allowed quantification of this kind of toxicity. Two years later, Yasumoto´s group identified the dinoflagellate *D. fortii* as the causative agent of the intoxications (Yasumoto *et al.*, 1980). OA, a polyether previously isolated and described from the sponge *Halichondria okadai* (Tachibana *et al.*, 1981), was finally identified by Murata and co-workers (Murata *et al.*, 1982) as the main bioactive compound responsible for DSP. Since then, occurrences of OA-group toxins in shellfish have been reported from almost all regions of the world, where Europe and Japan appear to be the most affected areas. Species of *Dinophysis* became target organisms in new phytoplankton monitoring programs established in the 1980s. New phytoplankton sampling and counting protocols were established so as to be able to detect *Dinophysis* spp in the water column, even at very low concentrations (<10^2 cell·l^{-1}), and alert health and fisheries authorities of the imminent arrival of toxins to shellfish resources (Lindahl, 1986).

In the mid 1980s, the list of published polyether toxins associated with *Dinophysis spp* events started to grow. Besides the dinophysis toxins (DTX1, DTX3), derivatives of OA: two new groups of polyether toxins, namely PTXs (Yasumoto *et al.*, 1984a) and YTXs (Yasumoto *et al.*, 1987) were described. Both of these new groups were obtained from lipophilic extracts by the same extraction procedure used for OA, and could be detected in the standard mouse bioassay applied for control of DSP toxins in shellfish. Consequently, PTXs and YTXs, together with OAs, were all included in the DSP toxin complex. In 1995 some intoxication with the typical DSP symptoms were detected in the Netherlands after consumption of mussels from Ireland. Since these mussels contained only traces of okadaic acid and DTX2, presence of unknown diarrhoeic toxins was suspected. Subsequent research discovered a new group of toxins called the azaspiracids, which were found to be the causative compounds (Satake *et al.*, 1998b). In 2003 heterotrophic dinoflagellates of the genus *Protoperidinium* were identified as the causative agent of Azaspiracid Shellfish Poisoning (AZP) in samples picked from natural populations (James *et al.*, 2003a). Although azaspiracids cause diarrhoea in humans they have never been included in the DSP group.

Chemistry

Okadaic acid (OA) and its congeners dinophysistoxin-1(DTX-1), dinophysistoxin-2 (DTX-2) and dinophysistoxin-3 (DTX-3) are the main toxins responsible for the DSP syndrome (Yasumoto *et al.*, 1985). They are heat-stable polyether compounds (FAO, 2004; McCarron et al., 2007; Yasumoto *et al.*, 1990) that are highly soluble in organic solvents such as methanol, acetone, chloroform or dichloromethane. OA derivatives have different polarities and consequently their solubilities in organic solvents are extremely variable. Acyl-derivatives (DTX3) and diol-esters are less polar than their original toxins. On the other hand, oxidised derivatives of diol-esters, DTX4 and DTX5, are more polar than their corresponding toxins and their solubility in some organic solvents such as dichloromethane or hexane is reduced, especially in the case of the two latter types, DTX4 and DTX5, which are water-soluble.

These lipophilic compounds are accumulated in the digestive gland (hepatopancreas) of shellfish and have been isolated from mussels and scallops. They are one of the active principles causing DSP syndrome (Murata *et al.*, 1982).

However, the real producers of these metabolites found in molluscs are planktonic dinoflagellates of the genus *Dinophysis spp.* (Yasumoto, 1980) such as *Dinophysis acuminata, Dinophysis acuta, Dinophysis rotundata, Dinophysis norvergica*; and benthic species of the genus *Prorocentrum spp.* (Murakami *et al.*, 1982) such as *Prorocentrum lima, Prorocentrum belizeanum, Prorocentrum concavum, Prorocentrum maculosum*.

OA and DTXs are polyketide compounds containing furane and pyrane-type ether rings and an alpha-hydroxycarboxyl function, the difference between them being only the number or position of the methyl groups. Their molecular masses are around 810 Da. These toxins can be classified into two different groups: a) OA esters and okadaates including the OA diol-esters (Hu *et al.*, 1992; Norte *et al.*, 1994; Paz *et al.*, 2007a; Suárez-Gómez *et al.*, 2005; Suárez-Gómez *et al.*, 2001). b) Other recently reported related compounds with changes in the OA backbone or the absence of the hydroxyl group at C-2 or C-7 (Blanco et al., 2005) are norokadanone (Fernández et al., 2003), 19-epi-okadaic acid (Cruz et al., 2007; Paz et al., 2008a) and belizeanic acid (Cruz et al., 2008). The first group of OA derivatives differ from the original toxin mainly in: i) the esterification of the hydroxyl group at the C7 position with fatty acids of a different chain length but typically C14 to C18, saturated or unsaturated, to produce DTX3 compounds. They were originally described as a group of DTX-1 derivatives, but it was later shown that OA and DTX-2 may be similarly acylated. The most predominant fatty acid identified in acylated derivatives is reported to be hexadecanoic (palmitic, C16:0) acid (EU/SANCO, 21 to 23rd, May 2001; Wright, 1995). ii) The oxidation of the diol part of the molecules of the diol-esters. iii) The esterification of the diol-esters with sulphated chains, which may or may not include an amide function in the polar side chain, to produce DTX4 and DTX5, respectively; and iv) the lack of the hydroxyl group in C2 or C7. A significant portion of the total toxin content found in bivalves consists of acylated forms of the toxins. This fact is also dependent on the bivalve species involved; for instance oysters accumulate most of the DTXs while only about a half of the total might be present in mussels. Since acyl derivatives have only been detected in the digestive gland of contaminated shellfish, it has been suggested that they are a product of animal metabolism and are not produced *de novo* by the microalgae (Wright, 1995). In fact, transformation of DTX-1 to 7-O-acyl-DTX-1 has only been demonstrated to occur in the scallop *Patinopecten yessoensis* (Suzuki et al., 1999). These acylated derivatives show an increased liposolubility as compared to their parent (unesterified) compounds and possess toxic activity upon hydrolysis in the human gastrointestinal tract.

Chemical Methods

Several published methods exist for detecting the OA-group of toxins in both plankton and bivalves. These methods can be divided into biological assays and chemical analytical methods. In assay methods, the measured signal is either a specific response to a single toxin structure or an integration of responses to several structures in a group. In order to use the assay result for evaluating seafood safety, it is most useful if an integrated response correlates with overall toxicity. In vivo assays are still applied widely for public health protection purposes, despite growing concern related to the use of such methods for reasons of animal welfare, and their inherent variability and interference with other biotoxins which may co-exist in a sample (EFSA, 2008). A chemical analytical method measures signals that correspond to a single toxin structure. Most analytical methods are based on separation methods and permit determination of several toxins in just the one analysis. Calculations of individual toxin concentrations require accurate standards to calibrate responses (which are either expensive or unavailable many times) and furthermore evaluation of seafood safety requires specific toxicity factors. The most common analytical method used for toxins is LC combined with different detection methods (Fluorescence, Mass Spectrometry) and Capillary Electrophoresis. These methods are mostly applied for research or confirmatory purposes. At present, two LC methods have been formally validated, one of them with fluorescent detection (LC-FLD) (CEN, 2004) and the other coupled with mass spectrometry (LC-MS)(EC., 2011).

Liquid Chromatography-Fluorescence Detection (LC-FLD)

Several fluorometric LC methods have been developed for the detection of lipophilic toxins (Nogueiras *et al.*, 2003). Among chemical methods, Liquid Chromatography with fluorimetric detection (LC-FD) has been widely used for many years in the determination of OA and DTXs. Due to the lack of chromophores exhibited by these molecules, most methods are based on pre-column chemical derivatization of the toxins using fluorescent reagents and further separating and detecting the fluorescent ester derivatives. A number of derivatization compounds that react with the carboxyl group of OA and certain DTXs (DTX1, DTX2 and isomers) have been described in literature. Among these, the most widely used, because of its selectivity and sensitivity, is 9-anthryldiazomethane (ADAM) (Lee *et al.*, 1987). Lee´s method is unable to detect acyl-derivatives such as DTX-3, diolesters, DTX-4 and DTX-5 due to their high molecular weight and lipophilicity. But their presence can be indirectly determined after alkaline hydrolysis of the extracts by converting DTX3 to its corresponding parent compound (Fernández *et al.*, 1996; Yasumoto *et al.*, 1990). The ADAM procedure and its critical steps have been extensively researched into and some modifications have been introduced to improve the reliability of the method (Quilliam, 1995). Use of in situ produced ADAM has been proposed by Quilliam, *et al.*, 1998. A thorough evaluation of LC methods showed that the ADAM method was the most reliable (Ramstad *et al.*, 2001a; Ramstad *et al.*, 2001b) despite the poor stability of the ADAM reagent and the possibility of toxin losses during the silica column clean-up step required after derivatization. At present, this approach is the only inter-laboratory validated method for the OA-group of toxins. Other reagents studied include: 1-bromoacetylpyrene (BAP) (Dickey *et al.*, 1993), that showed better stability but lower selectivity and sensitivity than ADAM (James *et al.*, 1997); 3-bromomethyl-6,7-dimethoxy-1-methyl-2(1H)-quinoxalinone (BrDMEQ); 9 chloromethylanthracene (CA) (Nogueiras *et al.*, 2003); several cumarin derivatives (Luckas, 1992; Marr *et al.*, 1994; Shen *et al.*, 1991); 2-(anthracene 2,3-dicarboximido); ethyl-trifluoromethanesulfonate (AE-Otf) (Akasaka *et al.*, 1996b), which can also be applied for DTX3 determinations (Akasaka, *et al.*, 1996a) and luminarin-3 (James *et al.*, 1998). The latest fluorophore used is 1-pyrenyldiazomethane (PDAM) (García *et al.*, 2003).

Liquid Chromatography-Mass Spectrometry (LC/MS)

The development and application of liquid chromatography-mass spectrometry (LC/MS) for the determination of marine toxins over the last few years together with recent developments in multi-toxin analytical methods have led to great progress in the advancement of knowledge and management of the complex phenomenon of shellfish toxicity associated with lipophilic toxins. LC/MS is currently the most powerful analytical tool to identify and determine multiple toxins. Furthermore, collision experiments (MS/MS) can provide valuable structural information needed for confirmation of known toxin identities as well as for identification of new toxins. They do not require the complex derivatization and purification steps needed for LC-FD methods. However, calibration standards are required for method development and quantitation. Furthermore, LC/MS methods can provide relevant information in the presence of closely related compounds of a known structure, even if the toxin standard for calibration is available only for one relevant toxin of the group. There are different types of commercially available mass spectrometers (single quadruple, triple quadruple, ion traps, time-of-flight, etc) that differ mainly in sensitivity and capability of providing ion fragmentation for structural information and selective detection. Two basic types of API sources are available: electrospray ionization (ESI) and atmospheric pressure chemical ionization (APCI). The ESI method has been frequently used for marine toxins because it is suitable for ionization of all known toxins. Although APCI has proved useful for some toxins, it is limited in that it requires an analyte to have some volatility and a fair amount of thermal stability. Several specific LC-MS methods differing in mobile phase, type of buffer, pH, ionic strength, stationary phase, electrospray mode (positive or negative), have been developed for the detection of OA, DTX-1 and DTX-2 (Draisci *et al.*, 1998; MacKenzie *et al.*, 2005; Pleasance *et al.*, 1990a; Quilliam, 1995; Suzuki *et al.*, 2000), DTX-3 and diol esters of OA and DTX-1 (Hu *et al.*, 1992; Marr *et al.*, 1992; Paz *et al.*, 2007a; Suzuki *et al.*, 2004). Moreover, some multi-toxin detection methods have been developed for a range of toxins including OAs, YTXs, AZAs and several other lipophilic toxins (Aasen *et al.*, 2005b; Draisci *et al.*, 1999b; Stobo *et al.*, 2005; Suzuki *et al.*, 2005). Current existing LC/MS methods for the determination of lipophilic toxins and the influence of different features and parameters on the performance of LC/MS have been comprehensively reviewed by Quilliam (Quilliam, 2003b). An important issue to be taken into account when applying LC-MS is ionization efficiency of the analytes which may be significantly affected due to matrix components accumulated on the LC column after repeated injections. The inclusion of cleanup stages by solid phase extraction was suggested by Suzuki *et al.*, 2000, to remove matrix effects. Interference with ionization may also vary from matrix to matrix. This effect was described previously by Ito and Tsukada (2001) (Ito *et al.*, 2001), who concluded that the suppression of the ionization efficiency due to co-eluting substances may cause variation in the responses, making quantification difficult by using pure standard solution, thus suggesting the use of standard addition methods for LC-MS quantification of DSP toxins. A LC-MS method for the detection of marine lipophilic toxins under alkaline conditions was recently developed by Gerssen (Gerssen *et al.*, 2009b) in order to reduce the matrix effect. The use of a mobile phase containing ammonium hydroxide instead of formic acid substantially improved the separation of marine lipophilic toxins in this method. LODs for a number of toxins improved two to three-fold, partly due to better peak shapes and partly due to improved ionisation efficiency. As for OA, the LOD improved significantly from 22 pg (under acidic conditions) to 9 pg (under basic conditions).

LC-MS is being used in routine testing of DSP and PTX toxins in New Zealand to manage commercial shellfish harvesting (McNabb *et al.*, 2003) and is currently applied to complement the mouse assay in monitoring programs (Hess *et al.*, 2003). Moreover, an ultra-performance liquid chromatography mass spectrometry (UPLC) method was developed by Fux *et al.*, 2007.

Recently, a liquid chromatography-mass spectrometry (LC-MS/MS) method for the determination of OA, PTX, AZA and YTX was validated under the coordination of the European Union Reference Laboratory on marine biotoxins (EU-RL) in an inter-laboratory validation study carried out by Member States. On 10thJanuary 2011, this validated technique was recognised as a reference method for detecting lipophilic toxins (EC., 2011). Commission Regulation states that this method must be used routinely both for official controls at any stage in the food chain and for own-checks by food business operators. This Regulation will become applicable from first July 2011 onwards.

Capillary Electrophoresis (CE)

Capillary electrophoresis is a technique that provides high resolution separation and requires a very small amount of sample. Several CE methods have been developed by Gago Martínez and collaborators (Gago Martínez *et al.*, 2003a; Gago Martínez *et al.*, 2003b) .

The capillary electrophoresis method is particularly attractive for plankton research where sample size may be limited. It is relatively inexpensive, minute amounts of sample are required and the technique provides high resolution for lipophilic toxins separations. The preferred CE technique for lipophilic toxins with a limited aqueous solubility is micellar electrokinetic capillary chromatography (MECC), in which a surfactant is added to the running buffer. The surfactant helps to dissolve low polarity substances, minimises wall interactions and achieves separations of non-ionic compounds. The only published application of CE to OA is an MECC-UVD method (Bouaïcha *et al.*, 1997). Detection limit and specificity are limited in this case due to the poor UV extinction coefficient of the analyte.

1.2.3. AZP: Azaspiracids

AZP is the most recently discovered human toxic syndrome associated with shellfish. (Ito *et al.*, 1997; James *et al.*, 2000). This syndrome emerged in the Netherlands as the cause of severe gastrointestinal illness in humans after consumption of mussels (*Mytilus edulis*) from Ireland. These mussels had previously been passed as safe for consumption by the DSP mouse bioassay. As a result of subsequent research, a new group of toxins named AZPs were found to be the causative compounds (Satake *et al.*, 1998b). Originally, *Protoperidinium crassipes* was identified as the causative agent (James *et al.*, 2000; James *et al.*, 2003a), however, subsequent studies conducted by Tillman and colleagues indicated that a small photosynthetic thecate dinoflagellate named *Azadinium spinosum* is the progenitor of AZA toxins and *P. crassipes* is only a vector (Tillman *et al.*, 2009).

Since its first discovery in Irish mussels, the development of facile sensitive and selective LC-MS/MS methods has resulted in the discovery of AZA in other countries and in other species. Thus, AZAs have been found in shellfish harvested throughout Europe (De Schrijver *et al.*, 2002; James *et al.*, 2004; James *et al.*, 2000; James *et al.*, 2002; James *et al.*, 2008; Magdalena *et al.*, 2003a; Twiner *et al.*, 2008; Vale *et al.*, 2008; Vale, 2004), North-west Africa (Taleb *et al.*, 2006) and Chile (López-Rivera *et al.*, 2010). In relation with seafood, presence of AZAs has been recorded in crabs (Torgersen *et al.*, 2008), oysters, scallops and clams (Furey *et al.*, 2002; Furey *et al.*, 2003; Magdalena *et al.*, 2003b). Mice studies indicate that this toxin class can cause serious tissue injury, especially to the small intestine, and chronic exposure may lead to the development of lung tumours. A reasonable deduction from such data is that AZA may present a greater risk during times when DSP toxins are co-occurring in shellfish because DSP has been shown to be a tumour promoter and AZA is a tumour initiator. Studies also show that tissue recovery is very slow following exposure. These observations suggest that AZA is more dangerous than the other known classes of shellfish toxins (James *et al.*, 2002).

Chemistry

AZAs are nitrogen-containing polyether toxins that are comprised of a unique spiral ring assembly containing a heterocyclic amine (piperidine) and an aliphatic carboxylic acid moiety. These toxins were called azaspiracids (AZAs) because of their spiro ring assemblies

(Satake *et al.*, 1998b). Their molecular masses are around 840 Da. The differences between them are due to the methylation of C8 and C22, and to the hydroxylation of C3 and C23. AZAs are less polar than what would be expected from the presence of an amine and a carboxylic acid, because these two functions appear to form an intramolecular ion pair. They are soluble in organic solvents but unstable in some of them, such as chloroform, under slightly alkaline conditions, and or during chromatography on silica based supports (James *et al.*, 2000). In 1997, a second poisoning incident happened in Arranmore Island (County Donegal, Ireland) and two AZA analogues were identified, namely AZA2 (8-methylazaspiracid) and AZA3 (22-demethylazaspiracid) (Ofuji *et al.*, 1999). Subsequently, two further analogues were found in mussels, namely AZA4 (3-hydroxy-22-demethylazaspiracid) and AZA5 (23-hydroxy-22-demethyla-zaspiracid) (Ofuji *et al.*, 2001). James and co-workers discovered five new hydroxyl analogues (AZA6–AZA11) (James *et al.*, 2003b). During an instrumental validation study for rapid quantification of AZA1 in complex biological matrix, another new analogue (AZA7c) was discovered (Blay *et al.*, 2003).

Recently, Rehmann and co-workers used ultra-performance liquid chromatography (UPLC) hyphenated with a range of mass spectrometry technologies to uncover the existence of five further groups of AZA, namely dihydroxy, carboxy, carboxy–hydroxy- and dehydro AZAs, as well as methyl esters of the toxins. The potential number of AZAs increased to 32 with their findings (AZA12-AZA32) (Rehmann *et al.*, 2008).

Chemical Methods

Mouse and rat bioassays were the first methods used to determine AZA. First, it was reported that these methods are non-selective and lack specificity (James *et al.*, 2001). However, a recent study to evaluate the ability of the mouse bioassay to detect presence of AZA toxins was conducted by Hess and co-workers concluding that the mouse bioassay delivered a high probability of producing a positive response around and below the regulatory limit. This indicates that bioassay, although not quantitative, is adequate for protection of human health (Hess *et al.*, 2009).

Nowadays, mouse bioassays will be replaced by LC-MS for analysis of lipophilic toxins. Their use is only allowed during periodic monitoring of production areas for detecting new or unknown marine toxins.

A multi-toxin LC-MS method for OA, PTX AZA and YTX has recently been validated by inter-laboratory studies and established as a reference method (EC., 2011) This method is able to detect AZAs at concentrations well below the current regulatory limit of 160 μg AZA1 equivalents/Kg shellfish meat.

AZAs lack a chromophore for detection by liquid chromatography with ultra violet absorption (LC-UV) and conditions for fluorescence derivatization are not yet established. However, sensitive LC-MS/MS techniques have been developed that are routinely used in some laboratories in Ireland, Norway, Germany and New Zealand. A problem for the analytical community is the limited availability of reliable calibrants for AZAs. Currently a certified standard is only available for AZA1. Quantification of other analogues is undertaken through cross-calibration by assuming equal response factors.

At present, liquid chromatography hyphenated with mass spectrometry is the best means for quantitative determination of azaspiracids in the complex matrix of shellfish (Draisci *et al.*, 2000; Quilliam *et al.*, 2001). Multiple tandem MS methods for the determination of AZAs have been developed using triple quadrupole instruments (Draisci *et al.*, 2000; Volmer, *et al.*, 2002) and ion-trap instruments (Blay *et al.*, 2003; Furey *et al.*, 2002; James *et al.*, 2003b; Lehane *et al.*, 2002). Eleven AZAs were separated on a C18 column prior to MS^n identification *(James et al., 2003b; Lehane et al.*, 2004). However, chromatographic resolution is not necessary in LC-MSn methods, if the structural differences between analogues are understood; AZA1–AZA10 were determined without the need for complete chromatographic separation. Using $LC\text{-}MS^3$ where the isomers were resolved using prescribed mass selection it was possible to select unique productions for each azaspiracid (James *et al.*, 2003b; Lehane *et al.*, 2002; Lehane *et al.*, 2004). Fux and co-workers (2007) developed an ultra-performance liquid chromatography (UPLC)–MS/MS method to determine 21 lipophilic marine toxins including AZA1, using pre-selected precursor–production m/z combinations an analysis time of 6.6 min was reported (Fux *et al.*, 2007). Rehmann and collaborators (2008) used UPLC combined with Q-TOF MS/MS for the successful determination of known and new azaspiracid analogues in blue mussels, the group also employed an API 4000 QTrap system in their investigations (Rehmann *et al.*, 2008).

1.2.4. Palytoxins

Although Palytoxins (PLTXs) were first associated with ciguateric fish poisoning (CFP), recent evidence suggests that this group of toxins should be excluded from CFP toxins (Ballantine *et al.*, 1988; Kodama *et al.*, 1989).

PLTX is one of the most poisonous non-protein substances known to date. It was first isolated and purified from corals belonging to the family Zoanthidae, order Zoantharia and phylum Coelenterata (Moore *et al.*, 1971). This zoanthid was subsequently identified as *Palythoa toxica* (Walsh *et al.*, 1971). It was also found in other zoanthids such as *Palythoa tuberculosa*, *Palythoa mammilosa* and *Palythoa caribeaorum* (Attaway *et al.*, 1974; Beress *et al.*, 1983; Kimura *et al.*, 1973) in different locations. The toxin was subsequently identified worldwide in many marine organisms ranging from fish to microalgae (Munday, 2008).

This toxin and its analogues have become a global concern due to their effects in animals and especially in humans. In this sense, human fatalities arising from consumption of seafood suspected to be contaminated with PLTX have been reported after consumption of crab (Alcala *et al.*, 1988), sardine (Onuma *et al.*, 1999), smoked fish (Kodama *et al.*, 1989), groupers (Taniyama *et al.*, 2002) and parrotfish (Okano *et al.*, 1998). One of the most commonly reported complications arising from consumption of these fishes appears to be rhabdomyolysis (Deeds *et al.*, 2010). Symptoms associated with palytoxin exposure vary greatly depending upon the exposure route. Mortalities in humans have only occurred due to ingestion but a variety of additional non-lethal symptoms have been observed upon ingestion: dermal, ocular and inhalational exposure in humans. Systemic PLTX effects have been reported through dermal contact with marine aquarium zoanthids (Deeds *et al.*, 2010; Hoffmann *et al.*, 2008), including an unusual case of inhalation exposure (Majlesi *et al.*, 2008). Many other anecdotal evidences of intoxications related with aquarium zoanthids can be found in online marine aquarium forums (Deeds *et al.*, 2010).

In recent years, *Ostreopsis ovata* blooms in the Mediterranean region were found to cause respiratory illnesses due to inhalation of aerosols released during such blooms (Bottalico *et al.*, 2002; Ciminiello *et al.*, 2006a; Sansoni *et al.*, 2003; Simoni *et al.*, 2004; Simoni *et al.*,

2003). The most alarming phenomenon occurred in summer 2005 along the Genoa coasts (Italy) when, concurrently with a massive bloom of *O. ovata*, hundreds of people showed symptoms of serious respiratory distress after exposure to marine aerosols. Putative palytoxin and ovatoxin-a were identified as toxins produced by *O. ovata* (Ciminiello *et al.*, 2006a; Ciminiello *et al.*, 2008).

There is no recognized official method to date for the determination of this toxin. Each laboratory, according to its ability, develops its particular methodology, through a combination of methods, analytical and/or biological testing in order to confirm the presence/absence of PLTX in a sample. Moreover there are no regulations on PLTX-group toxins in shellfish, either in the EU, or in other regions of the world. Recently, a provisional limit of 30 μg/Kg shellfish for the sum of PLTX and its analogue ostreocin-D was proposed by the European Food Safety authority (EFSA, 2009). Several published methods exist for the determination of PLTX, but none of them have been formally validated in inter-laboratory validation studies, probably due to the lack of certified standards and certified reference materials for this group of toxins. Little is known about the real consequences that these toxins may have on coastal communities despite the obvious acute biological impact of PLTXs. There are ongoing attempts to develop a validated assay for rapid, sensitive and specific detection of PLTXs because of human health risks. None of the PLTX determination methods are able to meet all requirements per se, and so a combination of fast and confirmatory methods still seems to be the most appropriate approach for monitoring purposes.

Chemistry

The PLTX-group toxins are complex polyhydroxylated compounds with both lipophilic and hydrophilic areas. They are white, amorphous, hygroscopic solids, which have not yet been crystallized. They are insoluble in non-polar solvents such as chloroform, ether, and acetone, are sparingly soluble in methanol and ethanol and soluble in pyridine, dimethyl sulfoxide and water. Their structure was elucidated by two independent groups, one led by Professor Hirata in Japan (Uemura *et al.*, 1985) and the other by Professor Moore at Honolulu in the United States (Moore, 1985).

Due to their two chromophores, PLTX-group toxins exhibit an ultraviolet (UV) absorption spectrum with a λmax at 233 and 263 nm. This characteristic UV profile is used to verify the presence of the toxin by LC separation with Diode Array Detection (Noguchi *et al.*, 1987).

At least 8 different PLTX analogues are known: PLTX, ostreocin D, ovatoxin-A, homopalytoxin, bishomopalyloxin, neopalytoxin, deoxypalytoxin and 42-hydroxypalytoxin (Uemura *et al.*, 1985), (Ciminiello *et al.*, 2009). Chemical structure has only been characterised for PLTX (Cha *et al.*, 1982), ostreocin-D (Ukena *et al.*, 2001) and 42-hydroxypalytoxin (Ciminiello *et al.*, 2009).

Some palytoxin-like compounds have been identified from various *Ostreopsis spp.* especially, ostreocin-D, which was isolated from *O. siamensis* and structurally elucidated by NMR (Ukena *et al.*, 2002; Ukena *et al.*, 2001; Usami *et al.*, 1995). Others such as mascarenotoxins and ovatoxin-a were identified based only on MS evidence, as palytoxin-like compounds, from *Ostreopsis mascarenensis* (Lenoir *et al.*, 2004) and *Ostreopsis ovata* (Ciminiello *et al.*, 2008) respectively.

Chemical Methods

Chemical determinations are mainly based on a prior separation and later detection of analytes. Liquid chromatography (LC) with ultraviolet (UV) or fluorescence (FLD) detection is widely used in laboratories. However, thin layer chromatography (TLC) is not widely used at present, and is restricted to isolation and purification processes.

HPCE-UV and LC-UV

UV detection has been applied in separations by means of capillary electrophoresis (CE) (Mereish *et al.*, 1991) and LC (Riobó *et al.*, 2006). There are no reports on LC-UV methods for quantitative determination of PLTX and analogues in shellfish samples probably due to interferences presented in the biological matrix that notably diminish method sensitivity (Noguchi *et al.*, 1987). It is unlikely that they could be routinely used for regulatory monitoring of PLTX and its analogues in shellfish tissues. On the other hand Lenoir *et al.*, 2004 analyzed the toxically active n-butanol soluble extract from cells of *Ostreopsis mascarenensis* from a natural bloom in the South-western Indian Ocean. Comparison of both, retention time and UV-spectra, obtained with sample and reference PLTX by LC-UV coupled with DAD, enabled verification of the presence of two different PLTX analogues called mascarenotoxin-A and -B. However, the methodology based on UV detection is unsuitable for shellfish analysis because of the strong matrix effect.

It is widely recognized that the LC-FLD methods are much more sensitive than the LC-UV methods. Therefore, Riobó and collaborators (Riobó *et al.*, 2006) took advantage of the presence of the one amino terminal group in the PLTX molecule and established a pre-column derivatization method for the separation and quantification of PLTX. They derivatized the molecule using the derivatization reagent 6-aminoquinolyl-N-hyroxysuccinimidylcarbamate (AccQ) followed by reverse-phase LC analysis with FLD detection (λex 250 nm and λem 395 nm). This method is suitable for determining and quantifying PLTX toxins in samples of benthic dinoflagellates of the genus *Ostreopsis* (Riobó *et al.*, 2006) and also in shellfish samples (Riobó unpublished). Results correlated well with those obtained through haemolysis assay. Instrumental LOD for derivatized PLTX was 0.75 ng standard injected. The principal advantages of the LC-FLD method are: that it is simple, low cost and it can be automated. However, LOD and limit of quantification (LOQ) information in shellfish tissues is unavailable.

LC–MS

Different LC–MS methods have been used for identifying and quantifying the PlTX-group of toxins in seawater and phytoplankton (Ciminiello *et al.*, 2006a; Ciminiello *et al.*, 2008; Lenoir *et al.*, 2004; Penna *et al.*, 2005; Riobó *et al.*, 2006). Presence of the PLTX-group of toxins in samples is confirmed by the presence of the m/z 327 fragment ion, the bi-charged ion m/z 1340 and/or the tri-charged ion m/z 912. Another characteristic in the MS spectrum of PLTXs is the cluster generated by multiple losses of water molecules.

The main advantages of the LC–MS methods are: they are fast, sensitive–if high resolution MS instruments are used, capable of screening and measuring PLTX group toxins individually, provide information on the profile of PLTX-group toxins and can be automated. However, they require costly equipment and highly trained personnel. Further research into the development of such methods is essential for future application in routine testing of shellfish tissues for the PLTX-group of toxins. In the last few years, the LC-MS method developed by Ciminiello *et al.* for combined detection of palytoxin and ovatoxin-a (Ciminiello *et al.*, 2006a; Ciminiello *et al.*, 2008) has been employed within monitoring programs of *O. ovata* for determining toxin content of seafood samples collected along the Italian coasts (ARPAC, 2008). LC-MS analyses were carried out on a triple quadrupole MS in MRM mode. The obtained results indicated the presence of ovatoxin-a in most mussel samples is in the range of 303–625 mg/kg, which is well above the provisional limit for palytoxins of 250 mg/kg, proposed by the Community Reference Laboratory for Marine Biotoxins in the course of the 1st Meeting of Working Group on toxicology of the national reference laboratories (NRLs) for Marine Biotoxins (Cesenatico, Italy, 24–25 October, 2005), or to the tolerance limit of 30 mg/kg suggested by the European Food Safety Authority (EFSA, 2009).

LC-MS may also become the preferred technique to be used within monitoring programs carried out in a regulatory setting because it is fast and sensitive, and furthermore can be automated and permits screening and quantitation of palytoxins that can be individually measured in various matrices (plankton, seawater and seafood). However, routine LC-MS analysis of palytoxins in real samples is still at a preliminary stage. An important drawback is the lack of certified reference standards of palytoxin and its analogues, which are difficult to obtain in acceptable amounts. This will have to wait until a large scale culture of *Ostreopsis* spp. can be performed and efficient isolation procedures of the produced toxins are developed. Quantitation is presently carried out based on the tentative assumption that palytoxin-like compounds show the same molar response as palytoxin itself. However, even limited structural features in large molecules such as palytoxins could significantly impact their ionization efficiency. Additionally, none of the proposed LC-MS methods (Ciminiello *et al.*, 2006a; Ciminiello *et al.*, 2008; Lenoir *et al.*, 2004; Penna *et al.*, 2005; Riobó *et al.*, 2006) has been currently validated and, in some cases, fully developed.

1.2.5. Ciguatoxins

Ciguatera (CFP) is considered to be the most common type of marine biotoxin food poisoning worldwide. According to De Fouw (2001), as many as 10,000-50,000 individuals are affected annually, while Fleming (1998) estimates the global number of poisonings among 50,000-500,000 per year (De Fouw *et al.*, 2001; Fleming *et al.*, 1998). This syndrome is rarely fatal but some neurological effects can last long (for months). Clinical manifestations appear 2-30 hours after consumption of ciguateric fish. Intoxication symptoms are the principal means of diagnosis and distinction of CFP from other types of ichtyosarcotoxaemias. The disease is characterised by gastrointestinal, neurological and cardiovascular disturbances.

Some characteristic neurological symptoms of ciguatera are dysesthesia (inversion of the sensation between cold and hot) and paresthesia (trouble of sensitivity in the extremities). In severe cases, fatalities occur due to cardio-respiratory failure.

Although several treatments for ciguatera intoxication have been implemented with more or less success, there is no clear therapy available yet. Administration of manitol at the earliest possibility after CFP intoxication seems to be the most efficient therapy (Palafox *et al.*, 1988; Blythe *et al.*, 1994).

Dinoflagellates of the genus *Gambierdiscus* were identified as the producers of toxins responsible for CFP. These dinoflagellates grow epiphytically on macroalgae in coral reef environments of tropical and subtropical areas. It mainly produces maitotoxins (MTXs) whose role in CFP appears to be minor, and to a lesser extent CTXs which are really responsible for CFP intoxications. CTXs arise via biotransformation in fish of less polar ciguatoxins (gambiertoxins) which are produced by these dinoflagellates. This microbial biomass is ingested by herbivorous fish, which accumulate the toxin in their musculature, liver, and viscera and in turn are preyed upon by carnivorous fish that concentrate, amplify and modify the toxin. Carnivorous reef fish pass it up the food chain and the toxin ultimately reaches humans.

More than 425 species of fish are associated with CFP in humans, but generally only a relatively small number of species are regularly associated with ciguatera. The most common fish involved are barracuda, red snapper, grouper, amberjack, sea bass, surgeonfish, and moray (eel). Ciguateric fish look, taste and smell normal, and detection of toxins in fish therefore continues to be a problem. More than 20 precursor gambiertoxins and ciguatoxins have been identified in *G. toxicus* and in herbivorous and carnivorous fish. The toxins become more polar as they undergo oxidative metabolism and pass up the food chain.

Chemistry

CTXs are lipophylic polyethers with a molecular weight of about 1110 Da. The molecule consists of 13-14 rings fused by ether linkages into a rigid ladder-like structure. They are odourless and tasteless. CTX1B ($C_{60}H_{86}O_{19}$) was the first CTX fully described by Murata *et al.* (1989) (Murata *et al.*, 1989). The total synthesis of CTX3C has been reported by Hirama and co-workers in 2001 (Hirama *et al.*, 2001).

At least 20 congeners from the Pacific (P-CTXs), having slight modifications in their lipophilicity have been described. The main CTXs in the Pacific areas are P-CTX-1, P-CTX-2 and P-CTX-3, and they are present in fish in different relative amounts (Lehane *et al.*, 2000; Lewis, 2001; Lewis *et al.*, 1991). Structural modifications are mainly seen in both termini of the toxin molecules and mostly by oxidation (Lehane et al., 2000; Lewis *et al.* 1991; Murata *et al.*, 1990; Satake *et al.*, 1998a; Satake *et al.*, 1997a; Satake *et al.*, 1993b; Satake *et al.*, 1993a).

Legrand *et al.* (1998) (Legrand *et al.*, 1998), have described two families of CTXs (types 1 and 2) based on number of carbons and structure of the E ring. Type-1 CTXs include CTX1B, CTX4A and CTX4B and Type-2 CTXs include CTX3C and CTX2A1.

Other ciguatoxins described with different structure and biological activity are CTXs from the Caribbean (C-CTXs) (Lewis, 2001; Sauviat *et al.*, 2002) and multiple I-CTXs, from the Indian Ocean (Hamilton *et al.*, 2002b).

Extraction Procedures

Several methodologies have been used for the isolation of CTX from fish extracts and *G. toxicus* cultures. In case of cultures, procedures are quite simple and fast. For example, in the protocol developed by Murata *et al.*, (1990) (Murata *et al.*, 1990) extraction is performed with acetone followed by chromatographic purification steps which involves the use of five different columns: fluorisil, ODS, Toyopearl HW-40, RP Develosil ODS-7 and RP Asahipak ODP-50. Chinain and co-workers (Chinain *et al.*, 2010) used a protocol for isolation and extraction of CTX from cultures based on the previous procedure recommended by Satake *et al.* (1993) (Satake *et al.*, 1993a). In this case extraction is first performed with methanol and then in 50% aqueous methanol. The extract is later evaporated and partitioned between dichloromethane and 60% aqueous methanol. CTXs are recovered in the dichloromethane soluble fraction.

The process of toxin recovery and purification from fish is quite complex and slow. It is estimated that a minimum of 2-3 days may be required. There are several protocols published to date for CTX extraction and liquid-liquid partition steps. The suitability of each protocol may be considered according to the type of sample to be analyzed or the objective of the research.

Yasumoto *et al.* (1984) (Yasumoto *et al.*, 1984b) developed an extraction procedure followed by several liquid-liquid partition steps. This involves extraction of CTXs from 240 g of fish flesh using 700 mL of acetone. The extract is evaporated and then purified by two liquid: liquid partition steps involving water: diethyl-ether 90%. Diethyl-ether is evaporated and then partitioned between 90% aqueous methanol and hexane. Aqueous methanol is collected and then evaporated to near dryness. The final extract can then be re-suspended in acetone and preserved at –20°C.

Pauillac *et al.* (1995) (Pauillac *et al.*, 1995) present a simple protocol using a double extraction of 100 g of fish flesh with acetone, that gives an extract of approx. 2 g (after evaporation) followed by a liquid: liquid partition with dichloromethane (CH_2-Cl_2) and methanol in water (80%). The methanol-water phase is discarded, the dichloromethane phase containing CTXs is evaporated and the extract is used.

Legrand *et al.* (1998) (Legrand *et al.*, 1998) present a simplified protocol that involves acetone extraction followed by two liquid-liquid partitions. The first one between water:diethyl ether and the second between cyclohexane:methanol 80%. CTXs are recovered from the methanolic phase. This extract is subsequently pre-purified by Low Pressure Chromatography (using Florisil, Sephadex LH-20 and Sep-Pak C_{18}) and finally purified by High Performance Liquid Chromatography.

Chemical Methods

LC-FLD Methods

CTXs do not possess a suitable chromophore for selective spectroscopic detection but some analogues contain a relatively reactive primary hydroxyl group through which (after appropriate clean-up) a fluorescent label can be attached prior to detection. In a preliminary study Yasumoto and co-workers (1993) used fluorescent 1-anthroylnitrile for derivatising CTX prior to LC separation and fluorescence detection. The authors concluded that further work is required to develop efficient cleanup procedures (Yasumoto *et al.*, 1993). This method was subsequently used to detect CTXs in fish after derivatization with fluorescent reagents (Yasumoto *et al.*, 1995). Dickey *et al.*(1992) derivatized CTX-1 using a coumarin-based fluorescent reagent but the yield of derivatized CTX-1 did not provide a LOD at μg/Kg levels (Dickey *et al.*, 1992). There are no reports of more recent development of LC-fluorescence methods.

LC-MS/MS Methods

LC-MS/MS allows specific detection of individual analogues of P-, C- and I- CTXs (Hamilton *et al.*, 2002a; Hamilton *et al.*, 2002b; Lewis *et al.*, 1999; Pottier *et al.*, 2002b; Pottier *et al.*, 2002a). Lewis, 2001 was able to determine toxins from crude fish extract but for quantification of clinically relevant levels, the LOD had to be lowered towards a concentration of 0.1µg/Kg or less. The capability to reach this LOD could be demonstrated for P-CTX-1 by using an enrichment step like solid phase extraction (SPE) prior to MS detection (Lewis et al., 2009). This method was slightly modified by (Stewart, *et al.*) and is established as a reference analysis method by the public health laboratory for the State of Queensland, Australia. LOD was determined to be 0.03 µg/Kg fish flesh with an average recovery of 53% (range 27-75%) for various kinds of spiked fish species (n=10).

The validation status of the method is quite restricted due to the lack of certified standards and reference materials and availability of limited amounts of contaminated material for method development.

1.2.6. NSP: Brevetoxin

Brevetoxins (BTXs) are potent polyether neurotoxins produced by the marine dinoflagellate *Karenia brevis*. This dinoflagellate is responsible for the most prevalent harmful algal blooms (HAB) in the Gulf of Mexico and has caused periodic blooms along the US Atlantic Coast, as far north as North Carolina. Blooms of other BTX-producing algal species (including *Karenia* and *Chattonella* spp.) have been reported throughout the world but are not as well characterized as those of *K. brevis* in the Gulf of Mexico. A unique characteristic of *K. brevis* blooms is the associated airborne (aerosolized) toxin component. Toxins are released into the water as cells rupture increasing the amount of extra-cellular toxins as a bloom progresses (Pierce *et al.,* 2005). Bubbles from breaking waves transport toxins to the sea surface where they are ejected into the air as jet and film drops from the bursting bubbles (Pierce *et al.*, 2005). Onshore winds carry the toxin-containing aerosols on to the beach causing respiratory irritation in humans and other mammals along the shore (Cheng *et al.*, 2005; Kirkpatrick et al., 2004; Pierce *et al.*, 2005). BTXs can cause significant mortalities of fish and other aquatic animals, including marine mammals, through direct exposure during harmful algal blooms, or indirectly in the food web (Baden *et al.*, 1995; Poli *et al.*, 1986). Many filter-feeding molluscs are known to accumulate BTXs with no obvious adverse effects. However, BTXs in shellfish do pose a significant human health risk. Consumption of BTX-contaminated molluscan shellfish causes neurotoxic shellfish poisoning (NSP). Symptoms and signs of NSP are mainly gastrointestinal (e.g., abdominal pain, nausea, vomiting, diarrhoea) and neurological (e.g., ataxia, myalgia, paresthesia, and reversal of temperature sensation). Symptoms develop within minutes to hours after consumption of contaminated shellfish, and resolve within a few days. In severe cases, urgent medical care is required, while no human fatalities of NSP have been recorded. Treatment is mainly palliative.

Currently there are no regulations for BTX-group toxins in shellfish or fish in Europe. However, some countries in other regions of the world have set action levels or maximum levels. The action level in the USA is 20 mouse units (MUs)/100 g (0.8 mg BTX-2 equivalents/Kg shellfish). In New Zealand and Australia the maximum level for BTX-group toxins is 20 MUs/100 g, but the BTX analogue is not specified. Shellfish and fish analysis for this group of toxins is complicated due to the fact that BTX-2 and BTX-3 are the only commercially available certified reference standards.

None of the current methods of analysis to determine BTXs in shellfish and fish have been formally validated in inter-laboratory studies.

Chemistry

BTXs constitute a suite of lipid-soluble, cyclic polyether compounds that are tasteless, odourless, and heat-stable. They are stable compounds, resistant to heat and steam autoclaving. Both acid and alkaline hydrolyses can lead to reversible lactone ring opening of the BTX-molecule and alkaline hydrolysis proceeds faster than acid hydrolyses (Hua *et al.*, 1999). Structural characterization of BTXs as originally isolated from *K. brevis* showed two distinct backbone structural types (Lin *et al.*, 1981; Shimizu *et al.*, 1986). Backbone structure of A-type BTXs consisted of ten trans-fused ether rings, while that of B-type showed eleven.

BTXs are further distinguished by their side chains (tail regions). PbTx-1(A-type) and PbTx-2 (B-type) possess identical side chains, terminating in a highly reactive a,b-unsaturated aldehyde group.

PbTx-1 and PbTx-2 are believed to be the parent algal toxins from which others are derived (Baden *et al.*, 2005). Numerous analogues and derivatives have been identified in *K. brevis* blooms and cultures, including products of reduction, oxidation, and hydrolysis.

PbTx-2 is by far the most abundant BTX produced y *K. brevis*, while PbTx-1 is the most potent. Evidence suggests that polar BTX derivatives of PbTx-1 and PbTx-2 may be formed extra-cellularly in *K. brevis* (Abraham *et al.*, 2006; Pierce *et al.*, 2008). PbTx-1 and PbTx-2 dominate the BTX profile in cell-rich fractions of *K. brevis* blooms and cultures, while their polar derivatives abound in bloom water filtrate. Filter- feeding molluscan shellfish are thus exposed to complex mixtures of intra-and extra-cellular BTXs in *K. brevis* blooms.

Extraction Procedures

Extraction of BTXs from molluscan shellfish is challenged by matrix complexity and varied physico-chemical properties of metabolites. PbTx-2, which is extensively metabolized in several shellfish species, is also poorly recoverable from spiked shellfish homogenates (Ishida *et al.*, 2004; Plakas *et al.*, 2002). Data suggest that PbTx-2 rapidly and irreversibly binds to, or reacts with, constituents in shellfish tissues, in a concentration dependent manner (Plakas *et al.*, 2002). Metabolites formed in vivo were not found in vitro when homogenates were extracted within minutes of spiking with PbTx-2 (Plakas *et al.*, 2002). With longer incubation times, polar derivatives of PbTx-2 were detected in spiked oyster homogenates by ELISA, but were not characterized structurally (Naar *et al.*, 2004).

Acetone and methanol are useful solvents for extraction of BTX metabolites in shellfish. As verified by LC–MS and ELISA analyses, acetone extracts best reflect the full spectrum of BTXs present in shellfish. Methanolic solutions may be more suitable for analysis of polar BTX metabolites. PbTx-3, BTX-B5, and BTX-B1 were generally well extracted in 80% methanol from New Zealand shellfish (Ishida *et al.*, 2006; Ishida *et al.*, 2004; Nozawa *et al.*, 2003). Wang and collaborators (2004) discussed various sample preparation methods to improve recovery and reduce matrix interferences for BTXs in shellfish (Wang *et al.*, 2004).

Stability of incurred BTX residues in shellfish and their analytical standards is an important consideration in methods development and validation. Naturally incurred residues of PbTx-2, PbTx-3, BTX-B5, and BTX-B1 were stable in shellfish tissues under freezer storage for one year (Ishida *et al.*, 2004).

On the other hand, marine aerosol samples can be collected with high volume air samplers fitted with glass fibre filters. BTXs in marine aerosol are recovered from filters by extraction for 12 h in acetone using a Soxhlet apparatus. Extracts are then evaporated and transferred to vials in methanol for LC-MS analysis (Pierce *et al.*, 2005).

Extraction of BTXs from seawater is achieved by passing the seawater though a C-18 solid phase extraction disk under vacuum according to the procedure described by Pierce *et al* (2003) (Pierce *et al.*, 2003). The C18 disks are then rinsed with reverse osmosis water and eluted with methanol for their analysis.

Chemical Methods

Mass spectrometry provides unambiguous identification, based on structural features of individual BTX congeners. LC-MS/MS methods have been extensively used for characterization and determination of the BTX-group of toxins in shellfish, fish and algae (Dickey *et al.*, 1999; Hua *et al.*, 2000; Hua *et al.*, 1995; Ishida *et al.*, 2006; Ishida *et al.*, 2004; Nozawa *et al.*, 2003; Plakas *et al.*, 2002; Wang *et al.*, 2009; Wang *et al.*, 2004).

Differences observed in ionization, fragmentation, and matrix effects warrant the need for having analytical standards for each toxin in order to apply LC–MS methods for quantifying individual BTXs. Only a few metabolite standards are currently available commercially. Most promising applications of LC–MS are for confirmation of BTXs in shellfish, and for monitoring specific BTXs as markers of composite toxin burden and toxicity.

Hua and Cole (2000) (Hua *et al.*, 2000) described diagnostic ions for the A- and B-type BTX backbone structures, by using LC–MS/MS with electrospray ionization (ESI). Wang *et al.* (2004) (Wang *et al.*, 2004) presented mass spectra for the more abundant brevetoxin metabolites in Eastern oyster. Ishida *et al.* (2004b, 2006) (Ishida *et al.*, 2006; Ishida *et al.*, 2004) described an LC–MS/MS selected reaction monitoring (SRM) method, ESI positive and negative ion modes, for determination of major BTX metabolites in New Zealand shellfish.

At present, none of these methods have been formally validated in inter-laboratory studies. Another important disadvantage for their use is the lack of standards and reference materials (EFSA, 2010a).

1.3. Lipophilic Toxins whose Effects have not been Demonstrated in Humans

1.3.1. YTX Group

Yessotoxins (YTXs) are disulphated polycyclic polyether toxins. The lead compound, YTX was isolated for the first time in Mutsu Bay, Japan, in 1986 from the digestive glands of the scallop *Patinopecten yessoensis*, from which the toxin acquired its name (Murata *et al.*, 1987). DTX-1 and DTX-3 were also isolated and the three toxins were therefore tentatively included in the DSP family. However, YTX was subsequently found in mussels collected from different parts of the world. The real producers of these toxins are the algal dinoflagellates *Protoceratium reticulatum* (Satake *et al.*, 1997b), *Lingulodinium polyedrum* (Armstrong *et al.*, 2006; Paz *et al.*, 2004) and *Gonyaulax spinifera* (Rhodes *et al.*, 2006). YTXs accumulate in shellfish and are toxic to mice only through peritoneal injection (Aune *et al.*, 2002; Tubaro *et al.*, 2003). This fact led the EU to remove the YTX group from the DSP complex even though YTXs had traditionally been included within the DSP (EC, 2002). It is now regulated separately since they neither cause diarrhoea, (Ogino *et al.*, 1997; Tubaro *et al.*, 1998; Tubaro *et al.*, 2003) nor inhibit protein phosphatases, (Ogino *et al.*, 1997; Paz et al., 2008b) implying that they are not tumour promoters.

Chemistry

The planar structure of YTX was determined using nuclear magnetic resonance spectroscopy (NMR) and mass spectrometry (MS) which revealed a molecular formula of $C_{55}H_{82}O_{21}S_2Na_2$ (Murata *et al.*, 1987). Besides YTX, about 100 analogues have been reported to date from bivalves or dinoflagellates (Miles *et al.*, 2005), although only the structure of about forty of them have been identified by NMR and/or LC-MS/MS techniques (Ciminiello *et al.*, 2007a; Miles *et al.*, 2006). Typically their molecular weight ranged between 955 and 1551 mu (Ciminiello *et al.*, 2000; Miles *et al.*, 2005; Miles *et al.*, 2006; Satake *et al.*, 1996). As is usually the case for marine toxins, some YTXs are directly produced by dinoflagellates while others are produced by shellfish metabolism. Most YTXs found in dinoflagellates are derived from *P. reticulatum*. The first YTX analogue identified in this algae was 45,46,47-trinor-YTX (=norYTX), isolated from a strain of *P. reticulatum* collected in Yamada Bay,Japan (Satake *et al.,* 1999). It was also found later in Norway (Samdal *et al.*, 2004).

In spite of the high number of analogues found in molluscs, the most abundant one is typically 45-hydroxy-YTX, followed by carboxy-YTX. Initial results from the metabolism of YTXs in molluscs indicated that both YTX and homo-YTX are probably oxidized to 45-OH-YTX and 45-OHhomo-YTX respectively (Yasumoto *et al.*, 1997). More recent studies indicate that molluscs rapidly oxidize YTX to 45-OH-YTX and more slowly to carboxy-YTX. Recent studies did not find 45-OH- YTX, or carboxy-YTX in dinoflagellates (Miles *et al.*, 2005; Paz *et al.*, 2007b; Samdal *et al.*, 2004).

Chemical Methods

The chemical methods used to determine YTXs are essentially liquid chromatography with fluorescence detection (LC-FLD) or coupled with mass spectrometry (LC-MS).

Liquid Chromatography with Fluorescence Detection (LC-FLD)

This method is extensively used for the qualitative and quantitative analysis of YTXs both in molluscs and microalgae (Ramstad *et al.*, 2001b). It is based on determination of a fluorescent derivative of the toxin obtained by pre-column derivatization using dienophile reagent 4-(2-(6,7-dimethoxy-4-methyl-3-oxo-3,4dihydroquinoxalimylethyl)-1,2,4- triazoline-3,5-dione (DMEQ-TAD). A clean-up of the sample using solid phase extraction (SPE) cartridges is required prior to analysis, in order to eliminate interferences. The fluorogenic reagent reacts with the conjugated diene in the side-chain and, as a consequence, this method can only detect YTX, 45-OH-YTX, trinor-YTX (Yasumoto *et al.*, 1997), desulfo-YTX (Daiguji *et al.,* 1998), homo-YTX, 45-OHhomo-YTX and G-YTXs (Paz *et al.*, 2006). Other analogues such as carboxy-YTX, diOH-YTX, and noroxo-YTX or adriatoxin without the diene moiety cannot be detected by LC-FLD. The fluorescent adduct gives two diasteroisomeric peaks due to the formation of both C-42 epimers in a 3:1 ratio for YTX (Yasumoto *et al.*, 1997). YTX analogues elute at different retention times and therefore the use of toxin standards is essential for toxin identification. Weaknesses of this method are that the detection of new analogues are limited to the presence of a conjugated diene, and that it is difficult to distinguish between analogues with similar retention times, such as YTX and homo-YTX (Paz *et al.*, 2007b). On the other hand, advantages are that it is a very sensitive method with a linear response (0.4-6 mg/Kg), requires relatively low-cost equipment, is affordable to many labs and is faster than the mouse bioassay.

Liquid Chromatography Coupled with Mass Spectrometry (LC-MS)

Currently this is the most powerful analytical tool used to identify multiple toxins. This technique has several advantages, such as its high sensitivity and selectivity. Additionally, collision experiments (MS/MS) can provide valuable structural information for confirming toxin identities, as well as for identifying new toxins. It does not require complex derivatization and purification steps needed for LC-FLD methods. It is possible to analyze all YTXs, even those derivatized with DMEQ-TAD (Paz *et al.*, 2006), and also different algal toxins. In fact, several multi-toxin detection methods are available (Stobo *et al.*, 2005). However, just like in the case of other chemical methods, calibration standards are needed for method development and toxin quantification. LC-MS methods can provide relevant information on the presence of compounds related to a known structure, even if the toxin standard is available only for one relevant toxin of the group. The main disadvantage of this technique is that the equipment is expensive. Several specific LC-MS methods have been developed for YTX detection, which basically differ in the mobile or stationary phase used or the electrospray mode selected (positive or negative) (Ciminiello *et al.*, 2002; Draisci *et al.*, 1999a; Draisci *et al.*, 1998; Fernandez_Amandi *et al.*, 2002; Goto *et al.*, 2001). Most methods use negative ion mode detection because YTXs easily lose a proton from the sulphate group. The fragmentation of YTX is achieved easily even when low collision energies (C.E.) are used and the loss of the sulphate group and subsequent fragmentation of the unsaturated side chain are characteristic (Ciminiello *et al.*, 2002). An ultra-performance liquid chromatography mass spectrometry (UPLC) method has been recently developed by Fux *et al.*, 2007.

Analyses of YTXs must currently be carried out by means of LC-MS multi-toxins method, since this is the recognized official method for lipophilic toxins (EC., 2011).

Capillary Electrophoresis with UV/MS Detection

This is an analytical method designed as an alternative to LC-FLD for the determination of YTXs. It shows high resolution and small amounts of sample are required. When coupled with MS instrumentation, it allows confirmation of the presence of YTXs in samples (De la Iglesia *et al.*, 2007).

1.3.2. PTX Group

Pectenotoxins (PTXs) are a family of polyether macrolide toxins, first isolated from the digestive glands of the scallop *Patinopecten yessoensis*, cultivated in Japan (Yasumoto *et al.*, 1985). However, the dinoflagellate *Dinophysis fortii* (Draisci *et al.*, 1996; Lee, 1989) was initially identified as the real producer of these toxins. More recently PTXs have also been found in *D. acuminata* (MacKenzie *et al.*, 2005), *D. acuta* (Suzuki *et al.*, 2003), *D. caudata* (Fernández *et al.*, 2006), *D. rotundata* and *D. norvegica*. Moreover, PTXs were detected in samples of heterotrophic dinoflagellates *Protoperidinium divergens*, *Protoperidinium depressum*, and *Protoperidinium crassipes* (Miles *et al.*, 2004a). PTXs have been included within the DSP group, as they are often found in combination with OA in shellfish (Creppy et al., 2002; Miles *et al.*, 2004b). However, just like in the case for YTXs there is a great debate about whether these toxins should be classified as DSP toxins or not. Some research groups found diarrhoeic effects (Ishige *et al.*, 1988) while others did not find evidence of that (Edebo *et al.*, 1988; Terao *et al.*, 1986). In the light of the above, there are authors who have proposed to remove PTXs from the DSP group (Burgess *et al.*, 2001).

Chemistry

PTXs group toxins are heat stable polyether macrolides, isolated from shellfish and from dinoflagellates of the genus *Dinophysis*. To date, 15 analogues have been isolated and characterised. Several such PTXs are derived from a parent PTX, which is metabolized within shellfish (Yasumoto and Murata, 1993). Their common structural features include a spiroketal group, threeoxolanes, a bicyclic ketal and a six- member cyclic hemiketal (Allingham *et al.*, 2007). The most commonly found PTX in algae is PTX-2 (Draisci *et al.*, 1996) and it is also suspected to be the main precursor, from which many PTXs are derived through biotransformation during metabolism in the gut of bivalves.

Only four PTXs have been identified as actual biosynthetic products of the algae, these are PTX-2, PTX-12, PTX-11 and PTX-13. Other PTXs seem to be either product of shellfish metabolism or artefacts (Blanco *et al.*, 2005).

PTXs are easily destroyed under strong basic conditions and they are also labile under acidic conditions which transform them into the seco acid forms (PTX-SA) in shellfish. Wilkins, et al., 2006, recently reported the identification of fatty acid esters of PTX- 2-SA. These fatty acid derivatives resemble what had been reported for OA/DTXs (Marr *et al.*, 1992), BTXs (Morohashi *et al.*, 1995), and SPXs (Aasen *et al.*, 2006). All these esters appear to be metabolites produced by bivalves because they have never been observed in phytoplankton (Aasen *et al.*, 2006).

Chemical Methods

Some chemical methods developed for analysing PTXs involved fluorescence detection and mass spectrometry.

Liquid Chromatography with Fluorescence Detection (LC-FLD)

This is an alternative inexpensive method to LC/MS, however its use for PTXs detection is not widespread among researchers.

Lee and co-workers developed some methods for determining DSP toxins by LC using fluorescence labelling. They derivatized toxins having a primary hydroxyl group to fluorescent esters by using 1-anthroyl nitrile (AN). This work sheds light on anthroyl nitrile derivatives of PTX1 (AN-PTX1) and PTX4 (AN-PTX4) (Lee *et al.*, 1989). Yasumoto and colleagues took advantage of a conjugated diene in the macrolide skeleton of PTX2, and developed an analytical method that uses pre-column reaction fluorescent LC by means of the dienophile reagent DMEQ-TAD (Sasaki *et al.*, 1999). The method was successfully applied to microalgae samples.

Liquid Chromatography Coupled with Mass Spectrometry (LC-MS/MS)

At present, the method that appears to be most promising is high-performance liquid chromatography in combination with tandem mass spectrometry (LC–MS/MS) (These *et al.*, 2009). Mass spectrometry has been applied for lipophilic toxins for almost 20 years (Fux *et al.*, 2007; Pleasance *et al.*, 1990a; Quilliam, 1995; Stobo *et al.*, 2005; Suzuki *et al.*, 2005; These *et al.*, 2009).

A multi-toxin-method covering the same analyte range would be required, in order to replace bioassays. In this sense, several multi-toxin methods, including PTXs, have been published (Gerssen *et al.*, 2009a; Gerssen *et al.*, 2009b; These *et al.*, 2009). A multi-toxin

LC-MS method for detecting marine biotoxins in live bivalve molluscs was recently accepted as the reference method for detecting lipophilic toxins (See 1.2.2 section in this Chapter) (EC., 2011).

1.3.3. Cyclic Imines Group: Spirolids, Gymnodimines, Pinnatoxins and Pteriatoxins

Cyclic imines (CIs) were discovered in shellfish during the last few decades. This group includes spirolids (SPXs), gymnodimines (GYMs), pinnatoxins (PnTXs) and pteriatoxins (PtTXs). Although no reports of human illness due to CIs have been identified, they nevertheless must be taken into account since they show acute toxicity in mice upon intraperitoneal injection of lipophilic extracts. Mice die rapidly when they are present in shellfish at elevated levels. Therefore, they may interfere with the mouse bioassay of other lipophilic toxins such OA, BTX, YTX and AZAs.

SPXs are pharmacologically active macrocyclic imines that were first isolated and characterized from lipophilic extracts of scallop and mussel viscera harvested from aquaculture sites in Nova Scotia, Canada (Hu *et al.*, 1995; Hu *et al.*, 1996). Since 1994, SPXs have been found in high concentrations in bulk plankton samples on an annually recurring basis, during spring and early summer, at these sites (Cembella, 1998).

GYMs are the smallest molecules of the group. They were first isolated and characterized from lipophilic extracts of oysters harvested in New Zealand. Their name comes from the dinoflagellate *Gymnodinium* which was originally supposed to be the toxin producer (Seki *et al.*, 1995). Later, it was demonstrated that *Gymnodinium* identification was wrong and the really producer is the dinoflagellate *Karenia selliformis* (Haywood *et al.*, 2004).

PnTXs present a chemical structure which most closely resembles that of the SPXs. In 1995 a finding of the first PnTX analogue (PnTX A) was reported in Japanese shellfish (Uemura *et al.*, 1995). The organism producing PnTXs has not yet been identified but has been described as a peridinoid dinoflagellate (Rhodes *et al.*, 2010).

PtTXs are polyether macrocycles which are almost structurally identical to PnTXs. These toxins have only been detected in molluscs and it has been suggested that they are biotransformation products of PnTXs in shellfish (Selwood *et al.*, 2010).

In all cases, the biological activity of undescribed cyclic imine toxins was first recognized via their 'fast acting toxicity' that caused rapid death to mice upon intraperitoneal injection, and also for having a high oral potency with apparent neurotoxic symptomatology, but the mode of action is currently unknown. No regulatory limits have been established for any of the macrocyclic imines and they are assumed to be "safe for human consumption" until proven otherwise since there are no evidences of human intoxication cases (EFSA, 2010b).

Chemistry

SPXs, GYMs, PnTXs and PtTXs are macrocyclic compounds with imine (carbon-nitrogen double bond) and spiro-linked ether moieties. All of them are soluble in organic solvents such as methanol, acetone, chloroform and ethyl acetate.

SPXs are the largest of the known cyclic imine subgroups, comprising twelve structurally distinct variants found among various strains of marine dinoflagellates (Aasen *et al.*, 2005a; Ciminiello *et al.*, 2007b; Hu *et al.*, 2001; Hu *et al.*, 1995; Hu *et al.*, 1996; Krock *et al.*, 2007; Mackinnon *et al.*, 2006; Sleno *et al.*, 2004). Based on their chemical structure they can be divided into three types.

The first type includes those SPXs that have 6-5-5-polyether ring system in addition to the heptacyclic imine ring. They are SPXs A, B, C, D,13-desmethyl SPX C (=SPX 1), 13-19-didesmethyl SPX C,13-desmethyl SPX D and 27-hydroxy 13-19-didesmethyl SPX C.

The second type has the same polyether ring system as the first type but where the heptacyclic imine ring has transformed into a keto amine. This type comprises SPX E and F which are shellfish metabolites of SPX A and SPX B, respectively.

The third type includes SPX G and its analogue 20-methyl SPX G. Both of these have a heptacyclic imine ring and an unusual 5-5-6-trispiroketal ring system unseen in other marine biotoxin groups.

GYMs, with a molecular weight of around 500 Da, are the smallest molecules among the CIs group. The molecule has a six-member imino ring and their macrocycle contains only 16 carbon units and one ether bridge. They are soluble in acetone and ethyl acetate. At present, there are three analogues described GYM A, B and C (Miles *et al.*, 2000; Miles *et al.*, 2003; Seki *et al.*, 1995; Stewart *et al.*, 1997).

PnTXs are the CIs that are most closely related in structure to SPXs. These compounds are soluble in ethanol and ethyl acetate at neutral pH. The different analogues vary slightly in the polyether ring system (6-5-6) and in an additional bicyclic ether moiety in the macrocycle, which is absent in SPXs. At present there are seven analogues described with molecular weights ranging from 693 to 783 Da, which differ in the length of their cyclohexenyl side chains. They are named PnTX A, B, C, D, E, F y G. The latest results suggest that the novel PnTX F and G are progenitors of all the known PnTXs and PtTXs that are produced via metabolic transformation in shellfish (Selwood *et al.*, 2010).

PtTXs are almost structurally identical to PnTXs, except for their cyclohexenyl side chain, which ends in a cysteine moiety. Like Pn TXs, they have only been detected in shellfish. A distinguishing characteristic of this group of toxins is the presence of one free carboxylic group at the cyclohexenyl side chain, whereas all other CIs form lactones. At present there are three analogues described: PtTX A (MW 830.5) and the other two analogues are epimers at the C-34 named PtTX B and C (MW 844.5) (Takada *et al.*, 2001). Like PnTXs, they are soluble in ethanol and ethyl acetate at neutral pH and moreover are water soluble at both low and high pH.

Chemical Methods

CIs lack chromophores and therefore detection of these compounds by optical methods is hindered due to their very unspecific UV spectra and very slow sensitivity. Recently, a reversed-phase LC-UV method was developed for routine analysis of GYMs in shellfish (Marrouchi *et al.*, 2010). UV detection at wavelength 210 nm was used to detect GYM A and B in clam samples. No other LC method with optical detection techniques for CIs was identified in the literature. In contrast, mass spectrometric detection is both highly sensitive and specific, especially because the imino function is a very good proton acceptor, and results in very high ion yields and provides the corresponding high sensitivity. Consequently, all detection methods for CIs are based on mass spectrometry, where the SPXs are the most intensively studied for developing analytical methods (Aasen *et al.*, 2005a; Álvarez *et al.*, 2010; Cembella *et al.*, 1999; Ciminiello *et al.*, 2006b; Fux *et al.*, 2007; Gerssen *et al.*, 2009a; Gerssen *et al.*, 2009b; MacKenzie et al., 2002; Selwood *et al.*, 2010; Takada *et al.*, 2001; Villar González, *et al.*, 2006).

Conclusion

Adequate method validation is essential to maintain and improve the quality of testing for marine biotoxins in seafood. Although formal collaborative studies should be conducted where possible, only a few of such studies for new marine biotoxin methods have been completed to international standards.

The main objective of the marine biotoxins monitoring programs is to ensure food safety. Bearing this fact in mind, decisions about the analyzed samples should be supported by both chemical and biological methods. Both of these methods are complementary since the chemical methods provide structural information from the toxins and biological methods provide information on the toxic potency of the sample.

The imperative to reduce reliance on mouse bioassays is a strong argument for use of validated instrumental methods. However certified reference materials are not available for all regulated toxins. This is a suitable long-term goal, but its achievement is a huge and never-ending task due to the wide and increasing range of toxins and analogues. On the other hand new toxins may go unnoticed by chemical methods while the bioassays have the potential to determine the presence of new toxins.

Acknowledgments

This work was funded by the Spanish national project EBITOX (Study of the biological and toxinologic aspects of benthic dinoflagellates associated with risks to human health) CTQ2008-06754-C04-04.

References

Aasen, J., Mackinnon, S. L., LeBlanc, P., Walter, J. A., Hovgaard, P., Aune, T., and Quilliam, M. A. (2005a). Detection and identification of spirolides in Norwegian shellfish and plankton. *Chemical Research in Toxicology, 18*, 509-515.

Aasen, J., Samdal, I. A., Miles, C. O., Dahl, E., Briggs, L., and Aune, T. (2005b). Yessotoxins in Norwegian blue mussels (Mytilus edulis): uptake from *Protoceratium reticulatum*, metabolism and depuration. *Toxicon, 45*, 265-272.

Aasen, J., Hardstaff, W. R., Aune, T., and Quilliam, M. A. (2006). Discovery of fatty acid ester metabolites of spirolide toxins in mussels from Norway using liquid chromatography / tandem mass spectrometry. *Rapid Communication in Mass Spectrometry, 20*, 1531-1537.

Abraham, A., Plakas, S. M., Wang, Z., Jester, E. L. E., ElSaid, K. R., Granade, H. R., Henry, M. S., Blum, P. C., Pierce, R. H., and Dickey, R. W. (2006). Characterization of polar brevetoxin derivatives isolated from Karenia brevis cultures and natural blooms. *Toxicon, 48*, 104-115.

Akasaka, K., Ohrui, H., Meguro, H., and Yasumoto, T. (1996a). Determination of dinophysistoxin-3 by the LC method with fluorimetric detection. *Analitycal Science, 12*, 557-560.

Akasaka, K., Ohrui, H., Meguro, H., and Yasumoto, T. (1996b). Fluorimetric determination of diarrhetic shellfish toxins in scallops and mussels by high-performance liquid chromatography. *Journal of chromatography, 729*, 381-86.

Alcala, A. C., Alcala, L. C., Garth, J. S., Yasumura, D., and Yasumoto, T. (1988). Human fatality due to ingestion of the crab *Demania reynaudii* that contained a palytoxin-like toxin. *Toxicon, 26*, 105-107.

Álvarez, G., Uribe, E., Avalos, P., Marino, C., and Blanco, J. (2010). First identification of azaspiracid and spirolides in *Mesodesma donacium* and *Mulina edulis* from Northern Chile. *Toxicon, 55*, 638-641.

Allingham, J. S., Miles, C. O., and Rayment, I. (2007). A structural basis for regulation of actin polymerization by pectenotoxins. *Journal of molecular biology, 371*, 959-970.

AOAC. (2000). *Official method 959.08. Paralytic shellfish poison, biological method. Final action. AOAC official methods of analysis, 17th ed.* Gaithersburg, MD,USA.

Armstrong, M., and Kudela, R. (2006). Evaluation of California isolates of *Lingulodinium poliedrum* for the production of yessotoxin. *African Journal of Marine Science, 25*, 399-401.

ARPAC (2008). Il monitoraggio dell´Ostreopsis ovata lungo il litorale della Campania (giugno-agosto 2007). http:// www.arpacampania.it/

Attaway, D. H., and Ciereszko, L. S. (1974). "Isolation and partial characterization of Caribbean palytoxin." Paper presented at the Proceedings of 2nd International Coral Reef Symposium I, Great Barrier Reef Comunity, Brisbane.

Aune, T., Sørby, R., Yasumoto, T., Ramstad, H., and Landsverk, T. (2002). Comparison of oral and intraperitoneal toxicity of yessotoxin towards mice. *40*, 77-82.

Baden, D., Fleming, L. E., and Bean, J. A. Marine toxins. In: De Wolff, F. A., (ed.) *Handbook of clinical neurology: Intoxication of the nervous system, Part II* Elsevier; (1995); pp. 141-175.

Baden, D. G., Bourdelais, A. J., Jacocks, H., Michelliza, S., and Naar, J. (2005). Natural and derivative brevetoxins: historical background,multiplicity,and effects. *Environmental Health Perspectives, 113*, 621-625.

Ballantine, D.L., Tosteson, T.R. and Bardales, T. (1988). Population dynamics and toxicity of natural populations of benthic dinoflagellates in southwestern Puerto Rico. *Journal of Experimental Marine Biology and Ecology, 119,*201-212.

Bargu, S., Powell, C. L., Wang, Z., Doucette, G. J., and Silver, M. W. (2008). Note on the occurrence of *Pseudo-nitzschia australis* and domoic acid in squid from Monterey Bay, CA (USA). *Harmful Algae, 7*, 45-51.

Bates, H. A., and Rapaport, H. (1975). A chemical assay for saxitoxin, the paralytic shellfish poison. *J. Agric Food Chem, 23*, 237-239.

Bates, S. S., Bird, C. J., De Freitas, A. S. W., Foxall, R., Gilgan, M., Hanic, L. A., Jonhson, G. R., McCulloch, A. W., and Odense, P. (1989). Pennate diatom *Nitzschia pungens* as the primary source of domoic acid, a toxin in shellfish from eastern Prince Edward Island, Canada. *Can J Fish Aquat Sci, 46*, 1203-1215.

Beress, L., Zwick, J., Kolkenbrock, H. J., Kaul, P. N., and Wasserrmann, O. (1983). A method for the isolation of the caribbean palytoxin (C-PTX) from the coelenterate (zoanthid) Palythoa caribaeorum. *Toxicon, 21*, 285-290.

Blanco, J., Acosta, C. P., Bermúdez de la Puente, M., and Salgado, C. (2002). Depuration and anatomical distribution of the amnesic shellfish poisoning (ASP) toxin domoic acid in the king scallop *Pecten maximus. Aquatic Toxicology, 60*, 111-121.

Blanco, J., Moroño, A., and Fernández, M. L. (2005). Toxic episodes in shellfish produced by lipophilic phycotoxins: an overview. *Revista Galega de Recursos Mariños. Monografías, 1*, 1-70.

Blay, P. K. S., Brombacher, S., and Volmer, D. A. (2003). Studies on azaspiracid biotoxins. III. Instrumental validation for rapid quantification of AZA1 in complex biological matrices. *Rapid Communications in Mass Spectrometry, 17*, 2153-2159.

Botana, L. M., Alfonso, A., Louzao, M. C., Vieytes, M. R., and Velasco, M. R. Marine toxins analysis. In: Nollet, L., (ed.) *Handbook of water analysis*. London: CRC Press; (2007); pp. 135-156.

Bottalico, A., Milella, P., and Felicini, G. p. Fioritura di *Ostreopsis* spl (Dynophyta) nel porto di Otranto. In: *Abstract Book of the Riunione Scientifica Annuale del Gruppo di lavoro per l´Algologia*. Chioggia (VE), Italy, 8-9 November: Società Botanica Italiana; (2002); pp. 21.

Bouaïcha, N., Hennion, M. C., and Sandra, P. (1997). Determination of okadaic acid by micellar electrokinetic chromatography with ultraviolet detection. *Toxicon 35*, 273-281.

Burgess, V., and Shaw, G. (2001). Pectenotoxins, an issue for public health: a review of their comparative toxicology and metabolism. *Environment international, 27*, 275-283.

Cembella, A. D. Ecophysiology and metabolism of paralytic shellfish toxins in marine microalgae. In: Anderson, D. M., Cembella, A. D., and Hallegraeff, G. M., (eds.) *Physiological Ecology of Harmful Algal Blooms. NATO-Advanced Study Institute Series*. Heidelberg, Germany: Springer-Verlag; (1998); pp. 381-404.

Cembella, A. D., Lewis, N. I., and Quilliam, M. A. (1999). Spirolide composition of micro-extracted pooled cells isolated from natural plankton assemblages and from cultures of the dinoflagellate *Alexandrium ostenfeldii. Natural toxins, 7*, 197-206.

CEN. (2003). European standard, TC 275 WI 002750104, Foodstuffs-determination of saxitoxin and dc-saxitoxin in mussels-HPLC method using pre-column derivatization with peroxide or peiodate oxidation, European Committee for Standardization (CEN).

CEN. Food stuffs Determination of Okadaic Acid and Dinophysis toxin in Mussels HPLC Method with Solid Phase Extraction Clean-up after Derivatization and Fluorimetric Detection. In: *European Standardization Committee (CEN)*. Brussels, Belgium: (2004); pp. 14524.

Cha, J. K., Christ, W. J., Finan, J. M., Fujioka, H., Kishi, Y., Klein, L. L., Ko, S. S., Leder, J., McWhorter, W. W., Pfaff, P., Yonaga, M., Uemura, D., and Hirata, Y. (1982). Stereochemistry of palytoxin. 4. Complete structure. *Journal of the American Chemical Society, 104*, 7369-7371.

Chan, I. O. M., Tsang, V. W. H., Chu, K. K., Leung, S. K., Lam, M. H. W., Lau, T. C., Lam, P. K. S., and Wo, R. S. S. (2007). Solid-phase extraction-fluorimetric high performance liquid chromatographic determination of domoic acid in natural seawater mediated by an amorphous titania sorbent. *583*, 111-117.

Cheng, Y. S., Zhou, Y., Irvin, C. M., Pierce, R. H., Naar, J., Backer, L. C., Fleming, L. E., Kirkpatrick, B., and Baden, D. G. (2005). *Environmental Health Perspectives, 113*, 638-643.

Chinain, M., Darius, T. H., Ung, A., Cruchet, P., Wang, Z., Ponton, D., Laurent, D., and Pauillac, S. (2010). Growth and toxin production in the ciguatera-causing dinoflagellate *Gambierdiscus polynesiensis* (Dinophyceae) in culture. *Toxicon, 56*, 739-750.

Ciminiello, P., Fattorusso, E., Forino, M., Poletti, R., and Viviani, R. (2000). A New Analogue of Yessotoxin, Carboxyyessotoxin, Isolated from Adriatic Sea Mussels. *Eur. J. Org. Chem.,* 291-295.

Ciminiello, P., Dell_Aversano, C., E., F., Forino, M., Magno, S., and Poletti, R. (2002). Direct detection of yessotoxin and its analogues by liquid chromatography coupled with electrospray ion trap mass spectrometry. *Journal of Chromatography A, 968*, 61-69.

Ciminiello, P., Dell_Aversano, C., Fattorusso, E., Forino, M., Magno, G. S., Tartaglione, L., Quilliam, M. A., Tubaro, A., and Poletti, R. (2005). Hydrophilic interaction liquid chromatography/mass spectrometry for determination of domoic acid in Adriatic shellfish. *Rapid communications in mass spectrometry, 19*, 2030-2038.

Ciminiello, P., Dell'Aversano, C., Fattorusso, E., Forino, M., Magno, G. S., Tartaglione, L., Grillo, C., and Melchiorre, N. (2006a). The Genoa 2005 outbreak. Determination of putative palytoxin in Mediterranean Ostreopsis ovata by a new liquid chromatography tandem mass spectrometry method. *Analytical Chemistry, 78*, 6153-6159.

Ciminiello, P., Dell´Aversano, C., Fattorusso, E., Magno, S., Tartaglione, L., Cangini, M., Pompei, M., Guerrini, F., Boni, L., and Pistocchi, R. (2006b). Toxin profile of Alexandrium ostenfeldii (Dinophyceae) from the Northern Adriatic Sea revealed by liquid chromatography-mass spectrometry. *Toxicon, 47*, 597-604.

Ciminiello, P., Dell-Aversano, C., Fattorusso, E., Forino, M., Grauso, L., Magno, S., Poletti, R., and Tartaglione, L. (2007a). Desulfoyessotoxins from Adriatic Mussels: a new problem for seafood safety control. *Chemical Research in Toxicology, 20*, 95-98.

Ciminiello, P., Dell_Aversano, C., Fattorusso, E., Forino, M., Grauso, L., Tartaglione, L., Guerrini, F., and Pistocchi, R. (2007b). Spirolide toxin profile of Adriatic *Alexandrium ostenfeldii* cultures and structure elucidation of 27-hydroxy-13,19-didesmethyl spirolide C. *Journal of Natural Products, 70*, 1878-1883.

Ciminiello, P., Dell_Aversano, C., Fattorusso, E., Forino, M., Tartaglione, L., Grillo, C., and Melchiorre, N. (2008). Putative palytoxin and its new analogue, ovatoxin-a, in *Ostreopsis ovata* collected along the Ligurian coasts during the 2006 toxic outbreak. *Journal of American Society for Mass Spectrometry, 19*, 11-120.

Ciminiello, P., Dell´Aversano, C., Dello lacovo, E., Fattorusso, E., Forino, M., Grauso, L., Tartaglione, L., Florio, C., Lorenzon, P., De Bortpmo, M., Tubaro, A., Poli, M., and Bignami, G. S. (2009). Stereostructure and biological activity of 42-hydroxy-palytoxin: A new palytoxin analogue from Hawaiian *Palythoa* Subspecies. *Chemical Research in Toxicology, 22*, 1851-1859.

Commission, E. (2002). Decision 2002/225/EC of 15 March, implementing Council Directive 91/492/EC as regards the maximum levels and the methods of analysis of certain marine biotoxins in bivalve molluscs, echinoderms, tunicates and marine gastropods. *Official J. Eur. Communities, L 75*, 62.

Costa, P. R., Rosa, R., Duarte-Silva, A., Brotas, V., and Sampayo, M. A. (2005). Accumulation, transformation and tissue distribution of domoic acid, the amnesic shellfish poisoning toxin, in the common cuttlefish, Sepia officinalis. *Aquatic Toxicology, 74*, 82-91.

Creppy, E. E., Traore, A., Baudrimont, I., Cascante, M., and Carratu, M. R. (2002). Recent advances in the study of epigenetic effects induced by the phycotoxin okadaic acid. *Toxicology,* 433-439.

Cruz, P. G., Daranas, A. H., Fernández, J. J., and Norte, M. (2007). 19-epi-Okadaicacid, a novel protein phosphatase inhibitor with enhanced selectiviy. *Organic Letters, 9*, 3045-3048.

Cruz, P. G., Fernández, J. J., Norte, M., and Daranas, A. H. (2008). Belizeanic acid: a potent protein phosphatase 1 inhibitor belonging to the okadaic acid class, with an unusual skeloton. *Chemical European Journal, 14*, 6948-6956.

Daiguji, M., Satake, M., Ramstad, H., Aune, T., Naoki, H., and Yasumoto, T. (1998). Structure and fluorometric HPLC determination of 1-desulfoyessotoxin, a new yessotoxin analog isolated from mussels from Norway. *Natural Toxins, 6*, 235-239.

De Fouw, J. C., Van Egmond, H. P., and Speijers, G. J. A. (2001). *Ciguatera fish poisoning: a review. RIVM Report Nº 388802021.*

De la Iglesia, P., Gago-Martinez, A., and Yasumoto, T. (2007). Advanced studies for the application of high performance capillary electrophoresis for the analysis of yessotoxin and 45-hydroxy yessotoxin. *Journal of chromatography A, 1156*, 160-166.

De la Iglesia, P., Giménez, G., and Diogene, J. (2008). Determination of dissolved domoic acid in seawater with reversed-phase extraction disks and rapid resolution liquid chromatography tandem mass spectrometry with head-column trapping. *Journal of chromatography, 1215*, 116-124.

De Schrijver, K., Maes, I., De Man, L., and Michelet, J. (2002). An outbreak of diarrhoeic poisoning in Antwerp, Belgium. *Eurosurveillance monthly, 7*, 138-141.

Deeds, J. R., and Schwartz, M. D. (2010). Human risk associated with palytoxin exposure. . *Toxicon, 56*, 150-162.

Dell_Aversano, C., Hess, P., and Quilliam, M. A. (2005). Hydrophilic interaction liquid chromatography-mass spectrometry for the analysis of paralytic shellfish poisoning (PSP) toxins. *Journal of chromatography A, 1081*, 190-201.

Dickey, R. W., Bencsath, F. A., Granade, H. R., and Lewis, R. J. (1992). Liquid chromatographic mass spectrometric methods for the determination of marine polyether toxins. *Bulletin of the exotic pathology society, 85*, 514-515.

Dickey, R. W., Granade, H. R., and Bencsath, F. A. Improved analytical methodology for the derivatization and HPLC-fluorometric determination of okadaic acid in phytoplankton and shellfish. In: Smayda, T. J., and Shimizu, Y., (eds.) *Toxic phytoplankton blooms in the sea,*. New York, USA: Elsevier Science; (1993); pp. 495-499.

Dickey, R. W., Jester, E. L. E., Granade, H. R., Mowdy, D., Moncreiff, C., Rebarchik, D., Robl, M., Musser, S. M., and Poli, M. (1999). Monitoring brevetoxins during a *Gymnodinium breve* red tide: comparison of sodium channel specific cytotoxicity assay and mouse bioassay for determination of neurotoxic shellfish toxins in shellfish extracts. *Natural toxins, 7*, 157-165.

Diener, M., Erler, K., Hiller, S., Christian, B., and Luckas, B. (2006). Determination of paralytic shellfish poisoning (PSP) toxins in dietary supplements by application of a new HPLC/FD method. *Eur Food Res Technol, 224*, 147-151.

Diener, M., Erler, K., Christian, B., and Luckas, B. (2007). Application of a new zwitterionic hydrophilic interaction chromatography column for determination of paralytic shellfish poisoning toxins. *J Sep Sci, 30*, 1821-1826.

Draisci, R., Lucentini, L., Giannetti, L., Boria, P., and Poletti, R. (1996). First report of pectenotoxin-2 (PTX-2) in algae (*Dinophysis fortii*) related to seafood poisoning in Europe. *Toxicon, 34*, 923-935.

Draisci, R., Giannetti, L., Lucentini, L., Ferretti, E., Palleschi, L., and Marchiafava, C. (1998). Direct identification of yessotoxin in shellfish by liquid chromatography coupled with mass spectrometry and tandem mass spectrometry. *Rapid communications in mass spectrometry, 12*, 1291-1296.

Draisci, R., Ferretti, E., Palleschi, L., Marchiafava, C., Poletti, R., Milandri, A., Ceredi, A., and Pompei, M. (1999a). High levels of yessotoxin in mussels and presence of yessotoxin and homoyessotoxin in dinoflagellates of the Adriatic Sea. *Toxicon, 37*, 1187-1193.

Draisci, R., Palleschia, L., Giannettia, L., Lucentinia, L., Jamesb, K. J., Bishopb, A. G., Satakec, M., and Yasumoto, T. (1999b). New approach to the direct detection of known and new diarrhoeic shellfish toxins in mussels and phytoplankton by liquid chromatography-mass spectrometry. *Journal of Chromatography A, 847*, 213-221.

Draisci, R., Palleschi, L., Ferretti, E., Furey, A., James, K. J., Satake, M., and Yasumoto, T. (2000). Development of a liquid chromatography tandem mass spectrometry method for the identification of azaspiracid in shellfish. *Journal of chromatography, 871*, 13-21.

EC. (2002). Commission Decision 2002/225/EC of 15 laying down detailed rules for the implementation of Council Directive 91/492/EEC as regards the maximum levels and the methods of analysis of certain marine biotoxins in bivalve molluscs,echinoderms,tunicates and marine gastropods. *Official Journal European Community, 62*.

EC. (2006). Commision regulation (EU) N° 1664/2006. Amending Regulation (EC) N° 2074/2005 as regards implementing measures for certain products of animal origin intended for human consumption and repealing certain implementing measures. *Official Journal of the European Union,* L320/13.

EC. (2011). Commision regulation (EU) No 15/201. Amending Regulation (EC) No 2074/2005 as regards recognised testing methods for detecting marine biotoxins in live bivalve molluscs. *Official Journal of the European Union,* L 6/3.

Edebo, L., Lange, S., Li, X. P., and Allenmark, S. (1988). Toxic mussels and okadaic acid induce rapid hypersecretion in the rat small intestine. *Acta pathologica, microbiologica et inmunologica scandinava, 96*, 1029-1035.

EFSA. (2008). Opinion of the Scientific Panel on Contaminants in the Food chain on a request from the European Commission on marine biotoxins in shellfish okadaic acid and analogues. *EFSA journal, 589*, 1-62.

EFSA. (2009). Panel on contaminants in the food chain . Scientific opinion on marine biotoxins in shellfish-palytoxin group. *EFSA Journal, 7*, 1-38.

EFSA (2010a). Scientific Opinion on marine biotoxins in shellfish – Emerging toxins: Brevetoxin group *EFSA Journal, 8(7)*,1677.

EFSA (2010b). Scientific Opinion on marine biotoxins in shellfish – Cyclic imines (spirolides, gymnodimines, pinnatoxins and pteriatoxins) *EFSA Journal, 8(6)*,1628.

EU/SANCO. (21 to 23rd, May 2001). "Report of the meeting of the working group on toxicology of DSP and AZP." Paper presented at the European Union / Directorate General for Health and Consumer Affairs (EU/SANCO), Brussels.

FAO. (2004). Marine Biotoxins. *Food and Agriculture organization of the United Nations (FAO). Food and Nutrition, Food and Agriculture Organization. Rome, Italy., 80*, 53.

Fernández, J. J., Suárez-Gómez, B., Souto, M. L., and Norte, M. (2003). Identification of new okadaic acid derivatives from laboratory cultures of *Prorocentrum lima*. *Journal of Natural Products, 66*, 1294-1296.

Fernández, M. L., Míguez, A., Cacho, E., and Martínez, A. (1996). Detection of okadaic acid esters in the hexane extracts of spanish mussels. *Toxicon, 34*, 381-387.

Fernández, M. L., Reguera, B., González-Gil, S., and Míguez, A. (2006). Pectenotoxin-2 in single cell isolates of *Dinophysis caudata* and *Dinophysis acuta* from the Galician Rías (NW Spain). *Toxicon, 48*,

Fernandez_Amandi, M., Furey, A., Lehane, M., Ramstad, H., and James, K. J. (2002). Liquid chromatography with electrospray ion-trap mass spectrometry for the determination of yessotoxins in shellfish. *976*, 329-34.

Fleming, L.E., Baden, D.G., Bean, J.A., Weisman, R., Blythe, D.G. (1998). Seafood toxin diseases: issues in epidemiology and community outreach. In: Reguera, B., Blanco, J., Fernandez, M.L., Wyatt, T. (Eds.), Harmful Algae. Xunta de Galicia and Intergovernmental Oceanographic Commission of UNESCO, pp. 245–248.

Franco, J. M., and Fernández-Vila, P. (1993). Separation of paralytic shellfish toxins by reversed phase high performance liquid chromatography, with postcolumn reaction and fluorimetric detection. *Chromatographia, 35*, 613-620.

Fritz, L., Quilliam, M. A., Wright, J. L. C., Beal, A. M., and Work, M. An outbreak of domoic acid poisoning attributed to the pinnate diatom *Pseudonitzsschia australis*. *Journal of Phycology, 28*, 439-442.

Fukui, M., Murata, M., Inoue, A., Gawel, M., and Yasumoto, T. (1987). Ocurrence of palytoxin in the Trigger fish *Melichtys vidua*. *Toxicon, 25*, 1121-1124.

Furey, A., Lehane, M., Gillman, M., Fernandez-Puente, P., and James, K. J. (2001). Determination of domoic acid in shellfish by liquid chromatography with electrospray ionization and multiple tandem mass spectrometry. *Journal of chromatography, 938*, 167-174.

Furey, A., Braña-Magdalena, A., Lehane, M., Moroney, C., James, K., Satake, M., and Yasumoto, T. (2002). Determination of azaspiracids in shellfish using liquid chromatography-tandem electrospray mass spectrometry. *Rapid communications in mass spectrometry, 16*, 238-242.

Furey, A., Moroney, C., Magdalena, A. B., Saez, M. J. F., Lehane, M., and James, K. J. (2003). Geographical, temporal, and species variation of the polyether toxins, azaspiracids, in shellfish. *Environmental Science and Technology, 37*, 3078-3084.

Fux, E., McMillan, D., Bire, R., and Hess, P. (2007). Development of an ultra-performance liquid chromatography mass spectrometry method for the detection of lipophilic marine toxins. *Journal of chromatography A, 1157*, 273-280.

Gago Martínez, A., Leao, J. M., Piñeiro, N., Carballal, E., Vaquero, E., Nogueiras, M., and Rodríguez Vazquez, J. A. (2003a). An application of capillary electrophoresis for the analysis of algal toxins from the aquatic environment. *Journal of Environmental Analytical Chemistry, 83*, 443-456.

Gago Martínez, A., Piñeiro, N., Aguete, E. C., Vaquero, E., Nogueiras, M., Leao, J. M., Rodríguez Vazquez, J. A., and Dabek Zlotorzynska, E. (2003b). Further improvements in the application of high-performance liquid chromatography, capillary electrophoresis and capillary electrochromatography to the analysis of algal toxins in the aquatic environment. *Journal of chromatography A, 992*, 159-168.

García, C., Pereira, P., Valle, L., and Lagos, N. (2003). Quantitation of diarrhetic shellfish poisoning toxins in Chilean Mussel using pyrenyldiazomethane as fluorescent labeling reagent. *Biological Research, 36*, 171-183.

Gerssen, A., McElhinney, M. A., Mulder, P., Bire, R., Hess, P., and Boer, J. (2009a). Solid phase extraction for removal of matrix effects in lipophilic marine toxin analysis by liquid chromatography-tandem mass spectrometry. *Analytical and Bioanalytical Chemistry, 394*, 1213-1226.

Gerssen, A., Muldera, P. J., McElhinneyb, M. A., and Boerc, J. (2009b). Liquid chromatography–tandem mass spectrometry method for the detection of marine lipophilic toxins under alkaline conditions. *Journal of Chromatography A, 1216*, 1421–1430.

Goto, H., Igarashi, T., Yamamoto, M., Yasuda, M., Sekiguchi, R., Watai, M., Tanno, K., and Yasumoto, T. (2001). Quantitative determination of marine toxins associated with diarrhetic shellfish poisoning by liquid chromatography coupled with mass spectrometry. *Journal of Chromatography A, 907*, 181-189.

Hadley, S. W., Braun, S. K., and Wekell, M. M. Confirmatinon of domoic acid as an N-formyl-O-methyl derivative in shellfish tissues by gass chromatography /mass spectrometry. In: Shahidi F, J. Y., Kitts DD,, (ed.) *Seafood safety, process and biotechnology*. Lancaster: Technomic Pub; (1990); pp. 25-32.

Hallegraeff, G. M. (1993). A review of harmful algae blooms and their apparent global increase. *Phycologia, 32*, 79-99.

Hamilton, B., Hurbungs, M., Jones, A., and Lewis, R. J. (2002a). Multiple ciguatoxins present in Indian Ocean reef fish. *Toxicon, 40*, 1347-1353.

Hamilton, B., Hurbungs, M., Vernoux, J. P., Jones, A., and Lewis, R. J. (2002b). Isolation and characterisation of Indian Ocean ciguatoxin. *Toxicon, 40*, 685-693.

Hatfield, C. L., Wekell, J. C., Gauglitz, E. J., and Barnett, H. J. (1994). Salt clean-up procedure for the determination of domoic acid by HPLC. *Natural Toxins, 2*, 206-211.

Haywood, A.J., Steidinger, K.A., Truby, E.W., Bergquist P.R., Bergquist P.L., Adamson, J. and Mackenzie, L. (2004). Comparative morphology and molecular phylogenetic analysis of three new species of the genus *Karenia* (Dinophyceae) from New Zealand. *Journal of phycology,40,* 165-179.

Hess, P., Gallagher, S., Bates, L. A., Brown, N., and Quilliam, M. A. (2001). Determination and confirmation of the amnesic shellfish poisoning toxin, domoic acid, in shellfish from Scotland by liquid chromatography and mass spectrometry. *Journal Association of Official Analytical Chemists, 84*, 1657-1667.

Hess, P., McMahon, T., Slattery, D., Swords, D., Dowling, G., McCarron, M., Clarke, D., Gibbons, W., Silke, W., and OCinneide, M. Use of LC-MS testing to identify lipophilic toxins, to establish local trends and interspecies differences and to test the comparability of LC-MS testing with mouse bioassay an example from the Irish Biotoxin Monitoring Programme. In: Villalba, A., Reguera, B., Romalde, J. L., and Beiras, R., (eds.) *Molluscan Shellfish Safety* Xunta de Galicia and IOC of UNESCO; (2003); pp. 57-66.

Hess, P., Butter, T., Petersen, A., Silke, J., and McMohan, T. (2009). Performance of the EU-harmonised mouse bioassay for lipophilic toxins for the detection of azaspiracids in naturally contaminated mussel(*Mytilus edulis*) hepatopancreas tissue homogenates characterised by liquid chromatography coupled to tandem mass spectrometry. *Toxicon, 53*, 713-722.

Hirama, M., Oishi, T., Uehara, H., Inoue, M., Maruyama, M., Oguri, H., and Satake, M. (2001). Total synthesis of Ciguatoxin CTX3C. *Science 294*, 1904-1907.

Hirata, Y., Uemura, D., and Ohizumi, Y. Chemistry and pharmacology of palytoxin. In: Dekker, M., (ed.) *Handbook of natural toxins*. New York: (1988); pp. 241-258.

Hoffmann, K., Hermanns-Clausen, M., Buhl, C., Buechler, M. W., Schemmer, P., Mebs, D., and Kauferstein, S. (2008). A case of palytoxin poisoning due to contact with zoanthid corals through a skin injury. *Toxicon, 51*, 1535-1537.

Holland, P. T., McNabb, P., Selwood, A. I., and Neil, T. (2003). Amnesic shellfish poisoning toxins in shellfish: estimation of uncertainty of measurement for a liquid chromatography/tandem mass spectrometry method. *Journal Association of Official Analytical Chemists, 86*, 1095-1100.

Hu, T., Marr, J., Defreitas, A. S. W., Quilliam, M. A., Walter, J. A., and Wright, J. L. C. (1992). New diolester (ofokadaic acid) isolated from cultures of *Prorocentrum lima* and *Prorocentrum concavum*. *Journal of Natural Products, 55*, 1631-1637.

Hu, T., Curtis, J. M., Oshima, Y., Quilliam, M. A., Walter, J. A., Watson-Wright, W. M., and Wright, J. L. C. (1995). Spirolides B and D, two novel macrocycles isolated from the digestive glands of shellfish. *Journal of the chemical socciety, chemical communications.*, 2159-2161.

Hu, T., Curtis, J. M., Walter, J. A., and Wright, J. L. C. (1996). Characterization of biologically inactive spirolides E and F: identification of the spirolide pharmacophore. *Tetrahedron Letters, 37*, 7671-7674.

Hu, T., Burton, I. W., Cembella, A. D., Curtis, J. M., Quilliam, M. A., Walter, J. A., and Wright, J. L. C. (2001). Characterization of spirolides a, c and 13-desmethyl c, new marine toxins isolated from toxic plankton and contaminated shellfish. *Journal of Natural Products, 64*, 308-312.

Hua, Y., Lu, W., Henry, M., Pierce, R., and Cole, R. B. (1995). On-line high-performance liquid chromatography-electrospray ionization mass spectrometry for the determination of brevetoxins in "red tide" algae. *Analytical Chemistry, 67*, 1815-1823.

Hua, Y., and Cole, R. B. (1999). Solution reactivity of brevetoxins as monitored by electrospray ionization mass spectrometry and implications for detoxification. *Chemical Research in Toxicology, 12*, 1268-1277.

Hua, Y., and Cole, R. B. (2000). Electrospray ionization tandem mass spectrometry for structural elucidation of protonated brevetoxins in red tide algae. *Analytical Chemistry, 72*, 376-383.

Ishida, H., Nozawa, A., Nukaya, H., and Tsuji, K. (2004). Comparative concentrations of brevetoxins PbTx-2, PbTx-3, BTX-B1 and BTX-B5 in cockle, *Austrovenus stutchburyi*, green shellmussel, *Perna canaliculus*, and Pacific oyster, *Crassostrea gigas*, involved in neurotoxic shellfish poisoning in New Zealand. *Toxicon, 43*, 779-789.

Ishida, H., Nozawa, A., Nukaya, H., Rhodes, L., McNabb, P., Holland, P. T., and Tsuji, K. Brevetoxin metabolism in shellfish associated with neurotoxic shellfish poisoning. In: Njapau, H., Trujillo, S., Van Egmond, H. P., and Park, D. L., (eds.) *Mycotoxins and phycotoxins: advances in determination, toxicology and exposure management* Wageningen Academic; (2006); pp. 297-307.

Ishige, M., Satoh, N., and Yasumoto, T. (1988). Pathological studies on the mice administered with the causative agent of diarrhetic shellfish poisoning (okadaic acid and pectenotoxin-2). *Report from the Hokkaido Institute of Health, 38*, 15-19.

Ito, E., Terao, K., McMahon, T., Silke, J., and Yasumoto, T. (1997). Acute pathological changes in mice caused by crude extracts of novel toxins isolated from Irish mussels. *Harmful Algae,* 588-589.

Ito, S., and Tsukada, K. (2001). Matrix effect and correction by standard addition in quantitative liquid chromatographic–mass spectrometric analysis of diarrhetic shellfish poisoning toxins. *Journal of Chromatography A, 943*, 39-46.

James, K., Bishop, A. G., Gillman, M., Kelly, S. S., Roden, C., Draisci, R., Lucentini, L., Giannetti, L., and Boria, P. (1997). Liquid chromatography with fluorimetric mass spectrometric and tandem mass spectrometric detection for the investigation of the seafood toxin producing phytoplankton, *Dinophysis acuta*. . *Journal of Chromatography A, 777*, 213-221.

James, K., Furey, A., Satake, M., and Yasumoto, T. Azaspiracid poisoning (AZP): A new shellfish toxic syndrome in Europe. In: *Ninth International Conference on Harmful Algae Blooms* (2000); pp. 250-253.

James, K., Lehane, M., Moroney, C., Fernández-Puente, P., Satake, M., Yasumoto, T., and Furey, A. (2002). Azaspiracid Shellfish Poisoning: unusual toxin dynamics in shellfish and the increased risk of acute human intoxications. *Food Additives and Contaminants, 19*, 555-561.

James, K., Moroney, C., Roden, C., Satake, M., Yasumoto, T., Lehane, M., and Furey, A. (2003a). Ubiquitous benign alga emerges as the cause of shellfish contamination responsible for the human toxic syndrome, azaspiracid poisoning. *Toxicon, 41*, 145-151.

James, K., Fidalgo, M. J., Furey, A., and Lehane, L. (2004). Azaspiracid poisoning, the food-borne illness associated with shellfish consumption: a review. *Food Additives and Contaminants, 21*, 879-892.

James, K., O´Driscoll, D., Fernández, J. G., and Furey, A. Azaspiracids: chemistry, bioconversion and determination. In: Botana, L. M., (ed.) *Seafood and Freshwater Toxins: Pharmacology, Physiology and Detection*. New York: CRC Press, Taylor and Francis; (2008); pp. 763-773.

James, K. J., Furey, A., Healy, B., Kelly, S. S., and Twohig, M. Luminarin-3, a new fluorimetric reagent for the HPLC determination of diarrhoeic shellfish poisoning toxins. In: Reguera, B., Blanco, J., Fernandez, M. L., and Wyatt, T., (eds.) *Harmful Algae*. Santiago de Compostela, Spain: IOC of UNESCO and Xunta de Galicia; (1998); pp. 519-20.

James, K. J., Gillman, M., Lehane, M., and Gago-Martinez, A. (2000). New flourimetric method of liquid chromatography for the determination of the neurotoxin domoic acid in seafood and marine phytoplanckton. *Journal of chromatography, 871*, 1-6.

James, K. J., Furey, A., Lehane, M., Moroney, C., Satake, M., and Yasumoto, T. (2001). LC-MS methods for the investigation of a new shellfish toxic syndrome Azaspiracid Poisoning (AZP). *Mycotoxins and Phycotoxins in Perspective at theTurn of the Century,* 401-408.

James, K. J., Sierra, M. D., Lehane, M., Magdalena, A. B., and Furey, A. (2003b). Detection of five new hydroxyl analogues of azaspiracids in shellfish using multiple tandem mass spectrometry. *Toxicon, 41*, 277-283.

Kao, C. Y. Paralytic shellfish poisoning. In: Falconer, I. R., (ed.) *Algal toxins in seafood and drinking water*. London: Academic Press; (1993); pp. 75-86.

Kimura, S., and Hashimoto, K. (1973). Purification of the toxin in a zoanthid *Palythoa tuberculosa. Publ. Seto. Mar. Biol. Lab., 20*, 713-718.

Kirkpatrick, B., Fleming, L. E., Squicciarini, D., Backer, L. C., Clark, R., Abraham, W., Benson, B., Cheng, Y. S., Johnson, D., Pierce, R., Zaias, J., Bossart, G. D., and Baden, D. G. (2004). Literature review of Florida red tide: implications for human health effects. *Harmful Algae, 3*, 99-115.

Kodama, A. M., Hokama, Y., Yasumoto, T., Fukui, M., Manea, S. J., and Sutherland, N. (1989). Clinical and laboratory findings implicating palytoxin as cause of ciguatera poisoning due to Decapterus macrosoma (mackerel). *Toxicon, 27*, 1051-1053.

Krock, B., and Cembella, A. D. (2007). "Mass spectral characterisation of undescribed spirolides in an isolate of the dinoflagellate Alexandrium ostenfeldii from Atlantic Canada." Paper presented at the Sixth International Conference on Molluscan Shellfish Safety.

Lau, C. O., Khoo, H. E., Yuen, R., Wan, M., and Tan, C. H. (1993). Isolation of a novel fluorescent toxin from the coral reef crab, *Lophozozymus pictor. Toxicon, 31*, 1341-1345.

Lau, C. O., Tan, C. H., Khoo, H. E., Yuen, R., Lewis, R. J., Corpuz, G. P., and Bignami, G. S. (1995). Lophozozymus pictor toxin: A fluorescent structural isomer of palytoxin. *Toxicon, 33*, 1373-1377.

Lawrence, J. A., Charbonneau, C. F., Ménard, C., Quilliam, M. A., and Sim, P. G. (1989a). Liquid chromatographic determination of domoic acid in shellfish products using the paralytic shellfish poison extraction procedure of the association of official analytical chemists. *Journal of chromatography, 462*, 349-356.

Lawrence, J. A., Charbonneau, C. F., Page, B. D., and Lacroix, G. M. A. (1989b). Confirmation of domoic acid in molluscan shellfish by chemical derivatization and reversed-hase liquid chromatography. *Journal of chromatography, 462*, 419-425.

Lawrence, J. E., Grant, J., Quilliam, M. A., Bauder, A. G., and Cembella, A. D. (2000). Colonization and growth of the toxic dinoflagellate *Prorocentrum lima* and associated fouling macroalgae on mussels in suspended culture. *Marine Ecology Progress Series, 201*, 147-154.

Lawrence, J. F., and Menard, C. (1991). Liquid chromatographic determination of paralytic shellfish poisons in shellfish after prechromatographic oxidation. . *J Assoc Off Anal Chem, 74*, 1006-1012.

Lawrence, J. F., Niedzwiadek, B., and Menard, C. (2005). Quantitative determination of paralytic shellfish poisoning toxins in shellfish using prechromatographic oxidation and liquid chromatography with fluorescence detection: collaborative study. *J. AOAC. Int, 88*, 1714-1732.

Lee, J.-S., Igarashi, T., Fraga, S., Darhl, E., Hovgaard, P., Yasumoto, T. (1989). Determination of diarrhetic shellfish toxins in various dinoflagellate species. *J. Appl. Phycol., 1*, 147-152.

Lee, J. S., Yanagi, T., Kenma, R., and Yasumoto, T. (1987). Fluorometric determination of diarrhetic shellfish toxins by high-performance liquid chomatography. *Agriculturan and Biological Chemistry, 51*, 877-881.

Lee, J. S., Murata, M., and Yasumoto, T. Analytical methods for determination of diarretic shellfish toxins. In: Natori, S., Hashimoto, K., and Ueno, Y., (eds.) *Bioactive molecules: Mycotoxins and phycotoxins´88*. Amsterdam: Elsevier; (1989); pp. 327-335.

Legrand, A. M., Teai, T., Cruchet, P., Satake, M., Murata, K., and Yasumoto, T. Two structural types of ciguatoxin involved in ciguatera fish poisoning in French Polynesia. In: Reguera, B., Blanco, J., M.L., F., and Wyatt, T., (eds.) *Proceedings of the VIII International Conference on Harmful Algae* Xunta de Galicia and International Oceanographic Commission of UNESCO; (1998b); pp. 473-475.

Lehane, L., and Lewis, R. J. (2000). Ciguatera: recent advances but the risk remains. *International Journal of Food Microbiology, 61*, 91–125.

Lehane, M., Brana-Magdalena, A., Moroney, C., Furey, A., and James, K. J. (2002). Liquid chromatography with electrospray ion trap mass spectrometry for the determination of five azaspiracids in shellfish. *Journal of Chromatography A, 950*, 139-147.

Lehane, M., Saez, M. J. F., Magdalena, A. B., Canas, I. R., Sierra, M. D., Hamilton, B., Furey, A., and James, K. J. (2004). Liquid chromatography multiple tandem mass spectrometry for the determination of ten azaspiracids, including hydroxyl analogues in shellfish. *Journal of Chromatography A, 1024*, 63-70.

Lenoir, S., Ten-Hage, L., Turquet, J., Quod, J. P., Bernard, C., and Hennion, M. C. (2004). First evidence of palytoxin analogues from an *Ostreopsis mascarenensis* (Dinophyceae) benthic bloom in southwestern Indian Ocean. *Journal of Phycology, 40*, 1042-1051.

Lewis, R. J., Sellin, M., Poli, M., Norton, R. S., MacLeod, J. K., and Sheil, M. M. (1991). Purification and characterization of ciguatoxins from moray eel (*Lycodontis javanicus*, Muraenidae). *Toxicon, 29*, 1115-1127.

Lewis, R. J., Jones, A., and Vernoux, J. P. (1999). HPLC/Mass spectrometry and HPLC/MS/MS methods for the detection of multiple ciguatoxins. *Proceedings of the 5th Indo-Pacific Fish conference, Noumea - New Caledonia, 3-8 November 1997., Societe Francaise d' Ichtyologie, Paris (France),*, 739-744.

Lewis, R. J. (2001). The changing face of ciguatera. *Toxicon, 39*, 97-106.

Lewis, R. J., Yang, A. J., and Jones, A. (2009). Rapid extraction combined with LC-tandem mass spectrometry (CREM-LC/MS/MS) for the determination of ciguatoxins in ciguateric fish flesh. *Toxicon, 54*, 62-66.

Lin, Y. Y., Risk, M., Ray, S. M., VanEngen, D., Clardy, J., Golik, J., James, J., and Nakanishi, K. (1981). Isolation and structure o fbrevetoxin B from the ''red tide'' dinoflagellate *Ptychodiscus brevis* (*Gymnodinium breve*). *Journal of the American Chemical Society, 103*, 6773-6775.

Lindahl, O. (1986). A dividable hose for phytoplankton sampling *International Council for the Exploration of the Sea, C.M.1982/L:27,* 8.

López-Rivera, A., O´Callaghan, K., Moriarty, M., O´Driscoll, D., Hamilton, B., Lehane, M., James, K. J., and Furey, A. (2010). First evidence of azaspiracids (AZAs): a family of lipophilic polyether marine toxins in scallops (*Argopecten purpuratus*) and mussels (*Mytilus chilensis*) collected in two regions of Chile. *Toxicon, 55*, 692-701.

Luckas, B. (1992). Phycotoxins in seafood: toxicological and chromatographic aspects. *Journal of chromatography 624*, 439-56.

MacKenzie, L., Holland, P., McNabb, P., Beuzenberg, V., Selwood, A., and Suzuki, T. (2002). Complex toxin profiles in phytoplankton and Greenshell mussels (Perna canaliculus), revealed by LC-MS/MS analysis. *Toxicon : Official Journal of the International Society On Toxinology, 40*, 1321-30.

MacKenzie, L., Beuzenberg, V., Holland, P., McNabb, P., Suzuki, T., and Selwood, A. (2005). Pectenotoxin and okadaic acid based toxin profiles in *Dinophysis acuta* and *Dinophysis acuminata* from New Zealand. . *Harmful Algae, 4*, 75-85.

Mackinnon, S. L., Walter, J. A., Quilliam, M. A., Cembella, A. D., LeBlanc, P., Burton, I. W., Hardstaff, W. R., and Lewis, N. I. (2006). Spirolides isolated from Danish strains of the toxigenic dinoflagellate *Alexandrium ostenfeldii. Journal of Natural Products, 69*, 983-987.

Mafra, J. L. L., Leger, C., Bates, S. S., and Quilliam, M. A. (2009). Analysis of trace levels of domoic acid in seawater and plankton by liquid chromatography without derivatization, using UV or mass spectrometry detection. *Journal of chromatography A, 1216*, 6003-6011.

Magdalena, A. B., Lehane, M., Krys, S., Fernández, M. L., Furey, A., and James, K. J. (2003a). The first identification of azaspiracids in shellfish from France and Spain. *Toxicon, 42*, 105-108.

Magdalena, A. B., Lehane, M., Moroney, C., Furey, A., and James, K. J. (2003b). Food safety implications of the distribution of azaspiracids in the tissue compartments of scallops (*Pecten maximus*). *Food Additives and Contaminants, 20*, 154-160.

Mahnir, V. M., and Kozlovskaja, E. J. (1992). Isolation of palytoxin from the sea anemone *Radianthus macrodactylus*. *Bioorg. Chim., 18*, 751-752.

Majlesi, N., Su, M. K., Chang, G. M., Lee, D. C., and reller, H. A. (2008). A case of inhalational exposure to palytoxin. *Clinical Toxicology, 46*, 637.

Marr, J., Hu, T., Pleasance, S., Quilliam, M. A., and Wright, J. L. C. (1992). Detection of a new 7-O- acyl derivates of diarrhetic shellfish poisoning toxins by liquid chromatography-mass spectrometry. *Toxicon, 30*, 1621-1630.

Marr, J. C., Mc Dowell, L. M., and Quilliam, M. A. (1994). Investigation of derivatization reagents for the analysis of DSP toxins by liquid chromatography with fluorescence detection. *Natural Toxins, 2*, 302-311.

Marrouchi, R., Dziri, F., Belayouni, N., Hamza, A., Benoit, E., Molgo, J., and Kharrat, R. (2010). Quantitative determination of Gymnodimine-A by high performance liquid chromatography in contaminated clams from Tunisia Coastline. *Marine Biotechnology, 12*, 579-585.

McCarron, P., Emteborg, H., and Hess, P. (2007). Freeze-drying for the stabilisation of shellfish toxins in mussel tissue (Mytilus edulis) reference materials. *Analytical and Bioanalytical Chemistry, 387*, 2475-2486.

McNabb, P., and Holland, P. Using liquid chromatography mass spectrometry to manage shellfish harvesting and protect public health. In: Villalba, A., Reguera, B., Romalde, J. L., and Beiras, R., (eds.) *Proceedings of the 4th Internacional Conference on Molluscan Shellfish Safety* Consellería de Pesca e Asuntos Marítimos da Xunta de Galicia and Intergovernmental Oceanographic Commission of Unesco; (2003); pp. 179-86.

Meldrum, B. Excitatory amino acid antagonists as potencial therapeutic agents. In: Jenner, P., (ed.) *Neurotoxins and their pharmacological implications*. New York: Raven Press; (1987); pp. 33-53.

Mereish, K. A., Morris, S., McCullers, G., Taylor, T. J., and Bunner, D. L. (1991). Analysis of palytoxin by liquid chromatography and capillary electrophoresis. *Journal of Liquid Chromatography, 14*, 1025-1031.

Miles, C. O., Wilkins, A. L., Stirling, D. J., and MacKenzie, A. L. (2000). New analogue of gymnodimine from a *Gymnodinium* species. *Journal of Agricultural and Food Chemistry, 48*, 1373-1376.

Miles, C. O., Wilkins, A. L., Stirling, D. J., and MacKenzie, A. L. (2003). Gymnodimine C, an isomer of gymnodimine B, from *Karenia selliformis. Journal of Agricultural and Food Chemistry, 51*, 4838-48-40.

Miles, C. O., Wilkins, A. L., Jensen, D. J., Cooney, J. M., Quilliam, M. A., Aasen, J., and MacKenzie, A. L. (2004a). Isolation of 41a-homoyessotoxin and the identification of 9-Methyl-41a-homoyessotoxin and Nor-ring-A-yessotoxin from *Protoceratium reticulatum. Chemical Research in Toxicology, 17*, 1414-1422.

Miles, C. O., Wilkins, A. L., Munday, R., Dines, M. H., Hawkes, A. D., Briggs, L. R., Sandvik, M., Jensen, D. J., Cooney, J. M., Holland, P. T., Quilliam, M. A., MacKenzie, A. L., and Beuzenberg, V. (2004b). Isolation of pectenotoxin-2 from *Dinophysis acuta* and its conversion to pectenotoxin-2 seco acid and preliminary assessment of their acute toxicities. *Toxicon, 43*, 1-9.

Miles, C. O., Samdal, I. A., Aasen, J. A. G., Jensen, D. J., Quilliam, M. A., Petersen, D., Briggs, L. R., Wilkins, A. L., Rise, F., Cooney, J. M., and MacKenzie, L. (2005). Evidence for numerous analogs of yessotoxin in *Protoceratium reticulatum. Harmful Algae, 4*, 1075-1091.

Miles, C. O., Wilkins, A. L., Allan, D. H., Selwood, A. I., Jensen, D. J., Cooney, J. M., Beuzenberg, V., and MacKenzie, A. L. (2006). Identification of 45-hydroxy-46,47-dinoryessotoxin, 44-oxo-45,46,47-trinoryessotoxin and 9-methyl-42,43,44,45,46,47,55-heptanor-38-en-41oxoyesso-toxin and partial characterization of some minor yessotoxins from *Protoceratium reticulatum Toxicon, 47*, 229-240.

Moore, R. E., and Scheuer, P. J. (1971). Palytoxin: A new marine toxin from a coelenterate. *Science, 172*, 495-498.

Moore, R. E. (1985). Structure of palytoxin. *Fortschr Chem Org Naturst, 48*, 81-202.

Morohashi, A., Satake, M., Murata, K., Naoki, H., Kaspar, H. F., and Yasumoto, T. (1995). Brevetoxin B3, a new brevetoxin analog isolated from the green shell mussel *Perna canaliculus* involved in neurotoxic shellfish poisoning in New Zealand. *Tetrahedrom Letters, 36*, 8995-8998.

Munday, R. Occurrence and toxicology of palytoxins. In: Botana, L. M., (ed.) *Seafood and Freshwater Toxins* CRC Press Taylor & Francis Group, Boca Raton FL; (2008); pp. 693-713.

Murakami, Y., Oshima, Y., and Yasumoto, T. (1982). Identification of okadaic acid as a toxic component of a marine dinoflagellate *Prorocentrum lima Bulletin of the Japanese Society of Scientific Fisheries, 48*, 69-72.

Murata, M., Shimatani, M., Sugitani, H., Oshima, Y., and Yasumoto, T. (1982). Isolation and structural elucidation of the causative toxin of diarrhetic shellfish poisoning. *Bulletin of the Japanese Society of Scientific Fisheries, 48*, 549-552.

Murata, M., Masanori, K., Lee, J.-S., and Yasumoto, T. (1987). Isolation and structure of Yessotoxin, a novel polyether compound implicated in diarrhetic shellfish poisoning. *Tetraedron Letters, 28*, 5869-5872.

Murata, M., Legrand, A. M., Ishibashi, Y., and Yasumoto, T. (1989). Structures of ciguatoxin and its congener. *Journal of the American Chemical Society, 111*, 8929-8931.

Murata, M., Legrand, A. M., Ishibashi, Y., Fukui, M., and Yasumoto, T. (1990). Structures and configurations of ciguatoxin from the moray eel *gymnothorax-javanicus* and its likely precursor from the dinoflagellate *gambierdiscus-toxicus*. *Journal of the American Chemical Society, 112*, 4380-4386.

Naar, J., Kubanek, J., Weidner, A., Flewelling, L., Bourdelais, A., Steidinger, K., and Baden, D. G. Brevetoxin depuration in shellfish via production of non-toxic metabolites: consequences for seafood safety and the environmental fate of biotoxins. In: Steidinger, K. A., Landsberg, J. H., Tomas, C. R., and Vargo, G. A., (eds.) *HarmfulAlgae*. St.Petersburg, Florida: Florida Fish and Wildlife Conservation Commission, Florida Institute of Oceanography and Intergovernmental Oceanographic Commission of UNESCO; (2004); pp. 488-549.

Noguchi, T., Hwang, D., Arakawa, O., Daigo, K., Sato, S., Ozaki, H., Kwoi, N., Ito, M., and Hashimoto, K. Palytoxin as the causative agent in the parrotfish poisoning. In: Tan, P. G. a. C. K., (ed.) *Progress in Venom and Toxin Research*. Kent Ridge, Singapore: National University of Singapore; (1987); pp. 325-335.

Nogueiras, M. J., Gago-Martinez, A., Paniello, A. I., Twohig, M., James, K., and Lawrence, J. F. (2003). Comparison of different fluorometric HPLC method for analysis of acidic polyether toxins in marine phytoplankton. *Analitycal Bioanalitycal Chemistry, 377*, 1202-1206.

Norte, M., Padilla, A., Fernández, J. J., and Souto, M. L. (1994). Structural determination and biosynthetic origin of two ester derivatives of okadaic acid isolated from *Prorocentrum lima*. *Tetrahedron, 50*, 9175-9180.

Nozawa, A., Tsuji, K., and Ishida, H. (2003). Implication of brevetoxin B1 and PbTx-3 in neurotoxic shellfish oisoning in New Zealand by isolation and quantitative determination with liquid chromatography tandem mass spectrometry. *Toxicon, 42*, 91-103.

Ofuji, K., Satake, M., McMahon, T., Silke, J., James, K. J., Naoki, H., Oshima, Y., and Yasumoto, T. (1999). Two analogs of azaspiracid isolated from mussels, *Mytilus edulis*, involved in human intoxication in Ireland. *NaturalToxins, 7*, 99-102.

Ofuji, K., Satake, M., McMahon, T., James, K. J., Naoki, H., Oshima, Y., and Yasumoto, T. (2001). Structures of azaspiracid analogs, azaspiracid-4 and azaspiracid-5, causative toxins of azaspiracid poisoning in Europe. *Bioscience Biotechnology and Biochemistry, 65*, 740-742.

Ogino, K., Kumagai, M., and Yasumoto, T. (1997). Toxicologic evaluation of yessotoxin. *Natural Toxins, 5*, 255-259.

Ohfune, Y., and Tomita, M. (1982). Total synthesis of domoic acid. A revision of the original structure. *J Am Chem Soc, 104*, 3511-3513.

Okano, H., Masuoka, H., Kamei, S., Seko, T., Koyabu, S., Tsunemoto, K., Tamiya, T., Ueda, K., Nakazawa, S., Sugawa, M., Suzuki, H., Watanabe, M., Yatani, R., and Nakano, T. (1998). Rhabdomyolysis and myocardial damage induced by palytoxin, a toxin of blue humphead parrotfish. *Internal Med., 37*, 330-333.

Oku, N., Sata, N. U., Matsunaga, S., Uchida, H., and Fusetani, N. (2004). Identification of palytoxin as a principle which causes morphological changes in rat 3Y1 cells in the zoanthid Palythoa aff. margaritae. *Toxicon, 43*, 21-25.

Onuma, Y., Satake, M., Ukena, T., Roux, J., Chanteau, S., Rasolofonirina, N., Ratsimaloto, N., Naoki, H., and Yasumoto, T. (1999). Identification of putative palytoxin as the cause of clupeotoxism. *Toxicon, 37*, 55-65.

Oshima, Y., Sugino, K., and Yasumoto, T. Latest advances in HPLC analysis of paralytic shellfish toxins. In: Natori, S., Hashimoto, K., and Ueno, Y., (eds.) *Mycotoxins and phycotoxins ´88*. Amsterdam: Elsevier; (1989); pp. 319-328.

Pardo, O., Yusa, V., Leon, N., and Pastor, A. (2007). Development of a pressurised liquid extraction and liquid chromatography with electrospray ionization-tandem mass spectrometry method for the determination of domoic acid in shellfish. *Journal of chromatography, 1154*, 287-294.

Pauillac, S., Blehaut, J., Cruchet, P., Lotte, C., and Legrand, A.-M. Recent advances in detection of ciguatoxins in French Polynesia. In: Lassus, P., Arzul, G., Erard, E., Gentien, P., and Mercaillou, C., (eds.) *Harmful Marine Algae Blooms* (1995); pp. 801-808.

Paz, B., Riobó, P., Franco, J. M., and Fraga, S. (2004). "Extracción y purificación de YTX a partir de cultivos de Protoceratium reticulatum." Paper presented at the VIII reunión ibérica sobre fitoplancton tóxico y biotoxinas, La Laguna, Tenerife.

Paz, B., Riobó, P., Souto, M. L., Gil, L. V., Norte, M., Fernandez, J. J., and Franco, J. M. (2006). Detection and identification of glycoyessotoxin A in a culture of the dinoflagellate *Protoceratium reticulatum. Toxicon 48*, 611-619.

Paz, B., Daranas, A. H., Cruz, P. G., Franco, J. M., Pizarro, G., Souto, M. L., Norte, M., and Fernández, J. J. (2007a). Characterisation of okadaic acid related toxins by liquid chromatography coupled with mass spectrometry. *Toxicon, 50*, 225-235.

Paz, B., Riobó, P., Ramilo, I., and Franco, J. M. (2007b). Yessotoxins profile in strains of *Protoceratium reticulatum* from Spain and USA. *Toxicon, 50*, 1-17.

Paz, B., Daranas, A. H., Cruz, P. G., Franco, J. M., Norte, M., and Fernández, J. J. (2008a). Identification of 19- epi-okadaic acid, a new diarrhetic shellfish poisoning toxin, by liquid chromatography with mass spectrometry detection. *Marine Drugs, 6*, 489-495.

Paz, B., Daranas, A. H., Norte, M., Riobó, P., Franco, J. M., and Fernández, J. J. (2008b). Yessotoxin, a group of marine polyether toxins: an overview. *Marine Drugs, 6*, 73-102.

Penna, A., Vila, M., Fraga, S., Giacobbe, M. G., Andreoni, F., Riobó, P., and Vernesi, C. (2005). Characterization of *Ostreopsis* and *Coolia* (Dinophyceae) isolates in the Western Mediterranean Sea based on morphology, toxicity and internal transcribed spacer 5.8s rDNA sequences. *Journal of Phycology, 41*, 212-245.

Pierce, R., Henry, M., and Blum, P. (2008). Brevetoxin abundance and composition during ECOHAB Florida field monitoring cruises in the Gulf of Mexico. *Continental Shellfish Research, 28*, 45-58.

Pierce, R. H., Henry, M. S., Blum, P. C., Lyons, J., Cheng, T. S., and Yazzie, D. (2003). Brevetoxin concentrations in marine aerosol: Human exposure levels during a *Karenia brevis* harmful algal bloom. *Bulletin of the environmental contaminants and toxicology, 70*, 161-165.

Pierce, R. H., Henry, M. S., Blum, P. C., Hamel, S. L., Kirkpatrick, G. J., Cheng, Y. S., Zhou, Y., Irvin, C. M., Naar, J., Weidner, A., Fleming, L. E., Backer, L. C., and Baden, D. G. (2005). Brevetoxin composition in water and marine aerosol along a Florida beach assessing potential human exposure to marine biotoxins. *Harmful Algae, 4*, 965-972.

Piñeiro, N., Vaquero, E., Leao, J. M., Gago-Martinez, A., and Rodriguez-Vazquez, J. A. (2001). Optimization of conditions for the liquid chromatography-electrospray ionization-mass spectrometric analysis of amnesic shellfish poisoning toxins. *Chromatographia supplement, 53*, S-231-S235.

Plakas, S. M., ElSaid, K. R., Jester, E. L. E., Granade, H. R., Musser, S. M., and Dickey, R. W. (2002). Confirmation of brevetoxin metabolism in the Eastern oyster (*Crassostrea virginica*) by controlled exposures to pure toxins and to *Karenia brevis* cultures. *Toxicon, 40*, 721-729.

Pleasance, S., Quilliam, M. A., de Freitas, A. S. W., Marr, J. C., and Cembella, A. D. (1990a). Ion-spray mass spectrometry of marine toxins. II. Analysis of diarrhetic shellfish toxins in plankton by liquidchromatrography / mass spectrometry. . *Rapid communications in mass spectrometry, 4*, 206-213.

Pleasance, S., Xie, M., LeBlanc, Y., and Quilliam, M. A. (1990b). Analysis of domoic acid and related compounds by mass spectroscopy as N-trifluoroacetyl-O-silyl derivates. *Biomedical and Environmental Mass Spectrometry, 19*, 420-427.

Pocklington, R., Milley, J. E., Bates, S. S., Bird, C. J., De Freitas, A. S. W., and Quilliam, M. A. (1990). Trace determination of domoic acid in seawater and phytoplankton by HPLC of the fluorenyl-metoxhycarbonyl (FMOC) derivative. *Journal of Environmental Analytical Chemistry, 38*, 211-214.

Poli, M., Mende, T. J., and Baden, D. (1986). Brevetoxins, unique activators of voltage sensitive sodium channels bind to specific sites in rat synaptosomes. *Molecular pharmacology, 30*, 129-135.

Pottier, I., Vernoux, J. P., Jones, A., and Lewis, R. J. (2002a). Characterisation of multiple Caribbean ciguatoxins and congeners in individual specimens of horse-eye jack (*Caranx latus*) by high-performance liquid chromatography/mass spectrometry. *Toxicon, 40*, 929-939.

Pottier, I., Vernoux, J. P., Jones, A., and Lewis, R. J. (2002b). Analysis of toxin profiles in three different fish species causing ciguatera fish poisoning in Guadeloupe, French West Indies. *Food Additives and Contaminants, 19*, 1034-1042.

Quilliam, M. A., Sim, P. G., McCulloch, A. W., and McInnes, A. G. (1989a). High-performance liquid chromatography of domoic acid, a marine neurotoxin, with application to shellfish and plankton. *International Journal of Environmental Analytical Chemistry, 36*, 139-154.

Quilliam, M. A., and Wright, J. L. C. (1989b). The amnesic shellfish poisoning mystery. *Anal Chem, 61*, 1053-1060.

Quilliam, M. A., Xie, M., and Harstaff, W. R. (1991). A rapid extraction and clean-up procedure for the determination of domoic acid in tissue samples. *National Research Council Canada Technical Report No 64. (NRCC No 33001),*

Quilliam, M. A. (1995). Analysis of Diarrhetic Shellfish Poisoning Toxins in shellfih tissue by LC with fluorometric and Mass Spectrometric detection. *Journal of AOAC International, 78*, 555-569.

Quilliam, M. A., Xie, M., and Hardstaff, W. R. (1995). Rapid extraction and cleanup for liquid chromatographic determination of domoic acid in unsalted seafoof. *Journal Association of Official Analytical Chemists, 78*, 543-554.

Quilliam, M. A., Gago-Martínez, A., and Rodríguez-Vázquez, J. A. (1998). Improved method for preparation and use of 9-anthryldiazomethane for derivatization of hydroxycarboxylic acids. Application to diarrhetic shellfish poisoning toxins. *Journal of Chromatography, A, 807*, 229-239.

Quilliam, M. A., Hess, P., and Dell'Aversano, C. (2001). Recent developments in the analysis of phycotoxins by liquid chromatography mass spectrometry. *Mycotoxins and Phycotoxins in Perspective at theTurn of the Century,* 383-391.

Quilliam, M. A. (2003a). The role of chromatography in the hunt for red tide toxins. *Journal of Chromatography A, 1000,* 527-548.

Quilliam, M. A. Chemical methods for lipophilic shellfish toxins. In: Hallegraeff, G. M., Anderson, D. M., and Cembella, A. D., (eds.) *Manual on harmful marine microalgae.* París: Intergovernmental Oceanographic Commission (UNESCO); (2003b); pp. 211-246.

Ramstad, H., Hovgaard, P., Yasumoto, T., Larsen, S., and Aune, T. (2001a). Monthly variations in diarrhetic toxins and yessotoxin in shellfish from coast to the inner part of the Sognefjord, Norway. *Toxicon, 39,* 1035-1043.

Ramstad, H., Larsen, S., and Aune, T. (2001b). Repeatability and validity of a fluorimetric HPLC method in the quantification of yessotoxin in blue mussels (*Mytilus edulis*) related to the mouse bioassay. *Toxicon, 39,* 1393-1397.

Rehmann, N., Hess, P., and Quilliam, M. A. (2008). Discovery of new analogs of the marine biotoxin azaspiracid in blue mussels Mytilus edulis by ultra-performance liquid chromatography / tandem mass spectrometry. *Rapid Communications in Mass Spectrometry, 22,* 549-558.

Rhodes, L., McNabb, P., DeSalas, M., Briggs, L., Beuzenberg, V., and Gladstone, M. (2006). Yessotoxin production by Gonyaulax spinifera. *Harmful Algae, 5,* 148-155.

Rhodes, L., Smith, K., Selwood, A., McNabb, P., Van Ginkel, R., Holland, P., and Munday, R. (2010). Production of pinnatoxins by a peridinoid dinoflagellate isolated from Northland, New Zealand. *Harmful Algae, 9,* 384-389.

Riobó, P., Paz, B., and Franco, J. M. (2006). Analysis of palytoxin-like in *Ostreopsis* cultures by liquid chromatography with precolumn derivatization and fluorescence detection. *Analytica Chimica Acta, 566,* 217-223.

Riobó, P., Franco, J. M., Blanco, J., Arévalo, F., Correa, J., Muñiz, S., Cacho, E., Ocaña, M. A., and Gordillo, M. (2010). "Comparison of the influence of different extraction procedures and analytical methods on the qualitative and quantitative results for PSP toxins determination " Paper presented at the 14th International Conference on Harmful Algae, Hersonissos, Crete.

Rourke, W. A., Murphy, C. J., Van de Riet, J. M., Burns, B. G., Thomas, K. M., and Quilliam, M. A. (2008). Rapid postcolumn methodology for determination of paralytic shellfish toxins in shellfish tissue. *J. AOAC. Int, 91,* 589-597.

Samdal, I. A., Naustvoll, L. J., Olseng, C. D., Briggs, L. R., and Miles, C. O. (2004). Use of ELISA to identify Protoceratium reticulatum as a source of yessotoxin in Norway. *Toxicon, 44,* 75-82.

Sansoni, G., Borghini, B., Camici, G., Casotti, M., Righini, P., and Rustighi, C. (2003). Fioriture algali di *Ostreopsis ovata* (Gonyaucales: Dinophyceae): un problema emergente. *Biol. Anim., 17,* 17-23.

Sasaki, K., Takizawa, A., Tubaro, A., Sidari, L., Della Loggia, R., and Yasumoto, T. (1999). Fluorometric Analysis of Pectenotoxin-2 in Microalgal Samples by High Performance Liquid Chromatography. *Natural Toxins, 7,* 241-246.

Satake, M., Murata, M., and Yasumoto, T. (1993a). The structure of CTX3C, a ciguatoxin congener isolated from culture Gambierdiscus toxicus. *Tetrahedron Letters, 34,* 1975-1978.

Satake, M., Murata, M., and Yasumoto, T. (1993b). Gambierol: a new toxic polyether compound isolated from the marine dinoflagellate *Gambierdiscus toxicus*. *Journal of the American Chemical Society, 115*, 361-362.

Satake, M., Terasawa, K., Kadowaki, Y., and Yasumoto, T. (1996). Relative configuration of YTX and isolation of two new analogs from toxic scallops. *Tetrahedron Letters, 37*, 5955-5958.

Satake, M., Ishibashi, Y., Legrand, A. M., and Yasumoto, T. (1997a). Isolation and structure of ciguatoxin-4A, a new ciguatoxin precursor from cultures of dinoflagellate Gambierdiscus toxicus and parrotfish Scarus gibbus. *Bioscience, Biotechnology, and Biochemistry, 60*, 2103-2105.

Satake, M., McKenzie, L., and Yasumoto, T. (1997b). Identification of Protoceratium reticulatum as the biogenic origin of yessotoxin. *Nat Toxins, 5*, 164-167.

Satake, M., Fukui, M., Legrand, A. M., Cruchet, P., and Yasumoto, T. (1998a). Isolation and structures of new ciguatoxin analogs, 2,3-DihydroxyCTX3C and 51-HydroxyCTX3C, accumulated in tropical reef fish. *Tetrahedron Letters, 39*, 1197-1198.

Satake, M., Ofuji, K., Naoki, H., James, K. J., Furey, A., Mahon, T. M., Silke, J., and Yasumoto, T. (1998b). Azaspiracid, a new marin toxin having unique spiro ring assemblies, isolated from Irish mussels *Mytilus edulis*. *Journal American Chemistry Society, 120*, 9967-9968.

Satake, M., Ichimura, T., Sekiguchi, K., Yoshimatsu, S., and Oshima, Y. (1999). Confirmation of yessotoxin and 45,46,47-trinoryessotoxin production by Protoceratium reticulatum collected in Japan. *Natural Toxins, 7*, 147-50.

Sauviat, M., Marquais, M., and Vernoux, J. (2002). Muscarinic effects of the Caribbean ciguatoxin C-CTX-1 on frog atrial heart muscle. *Toxicon, 40*, 1155-1163.

Seki, T., Satake, M., Mackenzie, L., Kaspar, H. F., and Yasumoto, T. (1995). Gymnodimine, a New Marine Toxin of Unprecedented Structure Isolated from New Zealand Oysters and the Dinoflagellate, *Gymnodinium sp*. *Tetrahedron Letters, 36*, 7093-7096.

Selwood, A., Miles, C. O., Wilkins, A. L., Van Ginkel, R., Munday, R., Rise, F., and McNabb, P. (2010). Isolation, structural determination and acute toxicity of pinnatoxins E, F and G. *Journal of Agricultural and Food Chemistry, 50*, 6532-6542.

Shen, J. L., Ganzlin, G., and Luckas, B. (1991). "HPLC determination of DSP toxins." Paper presented at the Proceedings of Symposium on Marine Biotoxins, Paris, France.

Shimizu, Y., Chou, H. N., and Bando, H. (1986). Structure of brevetoxin A (GB-1 toxin), the most potent toxin in the Florida red tide organism *Gymnodinium breve* (*Ptychodiscus brevis*). *Journal of the American Chemical Society, 108*, 514-515.

Simoni, F., Gaddi, A., Di Paolo, C., and Lepri, L. (2003). Harmful epiphytic dinoflagellates on Tyrrhenian Sea. *Harmful Algae News, 24*, 13-14.

Simoni, F., Di Paolo, C., Gori, L., and Lepri, L. (2004). Further investigation on blooms of *Ostreopsis ovata*, *Coolia monotis*, *Prorocentrum lima*, on the macroalgae of artificial and natural reefs in the Northern Tyrrhenian Sea. *Harmful Algae News, 26*, 5-7.

Sleno, L., Chalmers, M. J., and Volmer, D. A. (2004). Structural study of spirolide marine toxins by mass spectrometry. Part II. Mass spectrometric characterization of unknown spirolides and related compounds in a cultured phytoplankton extract. *Analitycal Bioanalitycal Chemistry, 378*, 977-986.

Stewart, I., Eaglesham, G. K., Poole, S., Graham, G., Paulo, C., Wickramasinghe, W., Sadler, R., and Shaw, G. R. Establishing a public health analytical service based on chemical methods for detecting and quantifying Pacific ciguatoxin in fish samples. *Toxicon, 56*, 804-812.

Stewart, M., Blunt, J. W., Munro, M. H. G., Robinson, W. T., and Hannah, D. J. (1997). The absolute stereochemistry of the New Zeland shellfish toxin Gymnodimine. *Tetrahedron Letters, 38*, 4889-4890.

Stobo, L. A., Lacaze, J. C. L., Scott, A. C., Gallacher, s., Smith, E. A., and Quilliam, M. A. (2005). Liquid chromatography with mass spectrometry detection of lipophilic shellfish toxins. *J AOAC Int 88, 5*, 1371-1382.

Suárez-Gómez, B., Souto, M. L., Norte, M., and Fernández, J. J. (2001). Isolation and structural determination of DTX-6, a new okadaic acid derivative. *Journal of Natural Products, 64*, 1363-1364.

Suárez-Gómez, B., Souto, M. L., Cruz, P. G., Fernández, J. J., and Norte, M. (2005). New target in diarrhetic shelfish poisoning control. *Journal of Natural Products, 68*, 596-599.

Sullivan, J. J., and Wekell, M. M. Determination of paralytic shellfish poisoning toxins by high pressure liquid chromatography. In: Ragelis, E. P., (ed.) *Seafood toxins, ACS symposium series 262.* Washington DC: American Chemical Society; (1984); pp. 197-205.

Sun, T., and H., W. W. (1999). Determination of domoic acid in phytoplankton by HPLC of the 6-aminoquinolyl-N-hydroxysuccinimidyl carbamate derivative. *J. Agric. Food Chem., 47*, 4678-4681.

Suzuki, H., Beuzenberg, V., MacKenzie, A. L., and Quilliam, M. A. (2004). Discovery of okadaic acid esters in the toxic dinoflagellate Dinophysis acuta from New Zealand using liquid chromatography/tandem mass spectrometry. *Rapid communications in mass spectrometry, 18*, 1131-1138.

Suzuki, T., Ota, H., and Yamasaki, M. (1999). Direct evidence of transformation of dinophysistoxin-1 to 7- O-acyl-dinophysistoxin-1 (dinophysistoxin-3) in the scallop Patinopecten yessoensis. *Toxicon, 37*, 187-198.

Suzuki, T., and Yasumoto, T. (2000). Liquid chromatography electrospray ionization mass spectrometry of the diarrhetic shellfish poisoning toxins okadaic acid, dinophysistoxin-1 and pectenotoxin-6 in bivalves. *Journal of chromatography A, 874*, 199-206.

Suzuki, T., Beuzenberg, V., Mackenzie, L., and Quilliam, M. A. (2003). Liquid chromatography mass spectrometry of spiro ketal stereoisomers of pectenotoxins and the analysis of novel pectenotoxin isomers in the toxic dinoflagellate *Dinophysis acuta* from New Zealand. *Journal of chromatography A, 992*, 141-150.

Suzuki, T., Jin, T., Shirota, Y., Mitsuya, T., Okumura, Y., and Kamiyama, T. (2005). Quantification of lipophilic toxins associated with diarrhetic shellfish poisoning in Japanese bivalves by liquid chromatography–mass spectrometry and comparison with mouse bioassay. *Fisheries Science, 71*, 1370-1378.

Tachibana, K., Scheuer, P. J., Tsukitani, Y., Kikuchi, H., Enden, V., Clardy, J., Gopichand, Y., and Schmitz, F. (1981). Okadaic acid, a cytotoxic polyether from two marine sponges of the genus *Halichondria. Journal of the American Chemical Society, 103*, 2469-2471.

Takada, N., Uemura, N., Suenaga, K., and Uemura, D. (2001). Structural determination of pteriatoxins A, B and C, extremely potent toxins from the bivalve *Pteria penguin. Tetrahedron Letters, 42*, 3495-3497.

Takemoto, T., and Daigo, K. (1958). Constituentes of *Chondria armata. Chem Pharm Bull, 6*, 578-580.

Takemoto, T. Isolation and structural identification of naturally occurring excitatory amino acids. In: McGeer, E. G., Olney, J. W., and McGeer, P. L., (eds.) *Kainic acid as a tool in neurobiology*. New York: Raven Press; (1978); pp. 1-15.

Taleb, H., Vale, P., Amanhir, R., Benhadouch, A., Sagou, R., and Chafik, A. (2006). First detection of azaspiracids in mussels in north xest Africa. *Journal of Shellfish Research, 25*, 1067-1070.

Taniyama, S., Mahmud, Y., Terada, m., and Takatani, T. (2002). Ocurrence of a food poisoning incident by palytoxin form a serranid *Epinephelus sp.* in Japan. *Journal of Natural Toxins, 11*, 277-282.

Teh, Y. F., and Gardiner, J. E. (1974). Partial purification of *Lophozozymus pictor* toxin. *Toxicon, 12*, 603-610.

Terao, K., Ito, E., Yanagi, T., and Yasumoto, T. (1986). Histopathological studies on experimental marine toxin poisoning. I. Ultrastructural changes in the small intestine and liver of suckling mice induced by dinophysistoxin-1 and pectenotoxin-1. *Toxicon, 24*, 1141-1151.

These, A., Scholz, J., and Preiss-Weigert, A. (2009). Sensitive method for the determination of lipophilic marine biotoxins in extracts of mussels and processed shellfish by high-performance liquid chromatography-tandem mass spectrometry based on enrichment by solid-phase extraction. *Journal of Chromatography A, 1216*, 4529-4538.

Tillman, U., Elbrachter, M., Krock, B., Jonh, U., and Cembella, A. D. (2009). *Azadinium spinosum gen. et sp. nov.* (Dinophycae) identified as primary producer of azaspiracid toxins. *European journal of phycology, 44*, 63-79.

Torgersen, T., Bremnes, N. B., Rundberget, T., and Aune, T. (2008). Stuctural confirmation and occurrence of azaspiracids in Scandinavian brown crabs (*Cancer pagurus*). *Toxicon, 51*, 93-101.

Tubaro, A., Sidari, L., Della-Loggia, R., and Yasumoto, T. (1998). Occurrence of yessotoxin-like toxins in phytoplankton and mussels from northern Adriatic Sea. *Harmful Algae*, 470-472.

Tubaro, A., Sosa, S., Carbonatto, M., Altinier, G., Vita, F., Melato, M., Satake, M., and Yasumoto, T. (2003). Oral and intraperitoneal acute toxicity studies of yessotoxin and homoyessotoxins in mice. *Toxicon, 41*, 783-792.

Twiner, M. J., Rehmann, N., Hess, P., and Doucette, G. J. (2008). Azaspiracid Shellfish Poisoning: a review on the chemistry, ecology and toxicology with an emphasis on human health impact. *Marine Drugs, 6*, 39-72.

Uemura, D., Ueda, K., Hirata, Y., Iwashita, T., and Naoki, H. (1985). Further studies on palytoxin II. Structure of palytoxin. *Tetrahedron Letters, 22*, 1007-1017.

Uemura, D., Chou, T., Haino, T., Nagatsu, A., Fukuzawa, S., Zeng, S., and Chen, H. (1995). Pinnatoxin A: A toxic amphoteric macrocycle from the Okinawa bivalve *Pinna muricata*. *Journal of the American Chemical Society, 117*, 1155-1156.

Ukena, T., Satake, M., Usami, M., Oshima, Y., Naoki, H., Fujita, T., Kan, Y., and Yasumoto, T. (2001). Structure elucidation of Ostreocin D, a Palytoxin analog isolated from the Dinoflagellate *Ostreopsis siamensis*. *Bioscience, Biotechnology, and Biochemistry, 65*, 2585-2588.

Ukena, T., Satake, M., Usami, M., Oshima, Y., Fujita, T., Naoki, H., and Yasumoto, T. (2002). Structural confirmation of ostreocin-D by application of negative-ion fast-atom bombardment collision-induced dissociation tandem mass spectrometric methods. *Rapid communications in mass spectrometry, 16*, 2387-2393.

Usami, M., Satake, M., Ishida, S., Inoue, A., Kan, Y., and Yasumoto, T. (1995). Palytoxin analogs from the dinoflagellate Ostreopsis siamensis. *Journal of the American Chemical Society, 117*, 5389-5390.

Vale, C., Wandscheer, C., Nicolaou, K. C., Frederick, M. O., Alfonso, C., Vieites, M. R., and Botana, L. M. (2008). Cytotoxic effect of azaspiracid-2 and azaspiracid-2- methyl ester in cultured neurons: involvement of the c-jun N-terminal kinase. *Journal of Neuroscience Research, 86*, 2952-2962.

Vale, P. (2004). Is there a risk of human poisoning by azaspiracids from shellfish harvested at the Portuguese coast? *Toxicon, 44*, 943-947.

Van Egmond, H. P., Van den Top, H. J., Paulsch, W. E., Goenaga, X., and Vieytes, M. R. (1994). Paralytic shellfish poison reference materials: an intercomparison of methods for the determination of saxitoxin. *Food Addit Contam, 11*, 39-56.

Villar González, A., Rodríguez-Velasco, M. L., Ben-Gigirey, B., and Botana, L. M. (2006). First evidence of spirolides in Spanish shellfish. *Toxicon, 48*, 1068-1074.

Volmer, D. A., Brombacher, S., and Whitehead, B. (2002). Studies on azaspiracid biotoxins. I. Ultra fast high resolution liquid chromatography / mass spectrometry separations using monolithic columns. *Rapid Communications in Mass Spectrometry, 16*, 2298-2305.

Walsh, G. E., and Bowers, R. L. (1971). A review of Hawaiian zoanthids with descriptions of three new species. *Zool. J. Linn. Soc., 50*, 161-180.

Walter, J. A., Falk, M., ., and Wright, J. L. C. (1994). Chemistry of the shellfish toxin domoic acid: characterization of related compounds. *Canadian Journal of Chemistry, 72*, 430-436.

Wang, Z., Plakas, S. M., ElSaid, K. R., Jester, E. L. E., Granade, H. R., and Dickey, R. W. (2004). LC/MS analysis of brevetoxin metabolites in the Eastern oyster (*Crassostrea virginica*). *Toxicon, 43*, 455-465.

Wang, Z., King, K. L., Ramsdell, J. S., and Doucette, G. (2007). Determination of domoic acid in seawater and phytoplankton by liquid chromatography-tandem mass spectrometry. *Journal of chromatography, 1163*, 169-176.

Wang, Z., and Cole, R. B. (2009). Enhanced collision-induced decomposition efficiency and unraveling of fragmentation pathways for anionic adducts of brevetoxins in negative ion electrospray mass spectrometry. *Analytical Chemistry, 81*, 8826-8838.

Wekell, J., Gauglitz, E., Barnett, H., Hatfield, C., and Eklund, M. (1994). The occurrence of domoic acid in razor clams (*Siliqua patula*), Dungeness crab (*Cancer magister*), and anchovies (*Engraulis mordax*). *Journal of Shellfish Research, 13*, 587-593.

Wiese, M., D´Agostino, P.M., Mihali, T.K., Moffitt, M.C., Neilan, B.A. (2010). Neurotoxic alkaloids: saxitoxin and its analogs. *Marine Drugs, 8* (7), 2185-2211.

Wilkins, A. L., Rehmann, N., Torgersen, T., Rundberget, T., Keogh, M., Petersen, D., Hess, P., Rise, F., and Miles, C. O. (2006). Identification of fatty acid esters of pectenotoxin-2 seco acid in Blue mussels (*Mytilus edulis*) from Ireland. *Journal of agricultural and food chemistry, 54*, 5672-5678.

Wright, J. L. C., Falk, M., McInnes, A. G., and Walter, J. A. (1990). Identification of isodomoic acid D and two new geometrical isomers of domoic acid in toxic mussels. *Canadian Journal of Chemistry, 68*, 22-25.

Wright, J. L. C. (1995). Dealing with seafood toxins: present approaches and future options. *Food Research International, 28*, 347-358.

Yasumoto, T., Oshima, Y., and Yamaguchi, M. (1978). Ocurrence of a new type of shellfish poisoning in the Tohoku district. *Bulletin of the Japanese Society of Scientific Fisheries, 44*, 1249-1255.

Yasumoto, T., Sugawara, W., Fukuyo, Y., Oguri, H., Igarashi, T., and Fujita, T. (1980). Identification of *Dinophysis fortii* as the causative organism of diarrhetic shellfish poisoning in the Tohoku district. *Bulletin of the Japanese Society of Scientific Fisheries, 46*, 1405-1411.

Yasumoto, T., Murata, M., Oshima, Y., Matsumoto, G. K., and Clardy, J. Diarrhetic Shellfish Toxins. In: Ragelis, E. P., (ed.) *Seafood Toxins* Am. Chem. Soc.; (1984a); pp. 207-214.

Yasumoto, T., Raj, U., and Bagnis, R. (1984b). *Seafood Poisonings in Tropical Regions*, Kagoshima.

Yasumoto, T., Murata, M., Oshima, Y., Sano, M., Matsumoto, G. K., and Clardy, J. (1985). *Diarrhetic Shellfish Toxins*. Am. Chem. Soc.

Yasumoto, T., Yasumura, D., Ohizumi, Y., Takahashi, M., Alcala, A. C., and Alcala, L. C. (1986). Palytoxin in two species of xanthid crab from the Philippines. *Agric. Biol. Chem., 50*, 163-167.

Yasumoto, T., Seino, N., Murakami, Y., and Murata, M. (1987). Toxins produced by benthic dinoflagellates. *Biol. Bull., 172*, 128-131.

Yasumoto, T., and Murata, M. Polyether toxin involved in seafood poisoning In: Hall, S., and Strichartz, G. P., (eds.) *ACS Symposium Series*. Washington D.C.: American Chemical Society; (1990); pp. 120-132.

Yasumoto, T., Satake, M., Fukui, M., Nagai, H., Murata, M., and Legrand, A. A turning point in ciguatera study. In: Smayda, T. J., and Shimizu, Y., (eds.) *Toxic Phytoplankton blooms in the sea*. New York: Elsevier; (1993); pp. 455-461.

Yasumoto, T., Fukui, M., Sasaki, K., and Sugiyama, K. (1995). Determinations of marine toxins in foods. *Journal of AOAC International, 78*, 574-582.

Yasumoto, T., and Takizawa, A. (1997). Fluorometric measurement of yessotoxins in shellfish by high-pressure liquid chromatography. *Bioscience, Biotechnology, and Biochemistry, 61*, 1775-7.

Yasumoto, T., Oshima, Y., Sugawara, W., Fukuyo, Y., Oguri, H., Igarashi, T., Fujita, N.,. (1980). Identification of Dinophysis fortii as the causative organism of diarrhetic shellfish poisoning. *Bulletin of the Japanese Society of Scientific Fisheries, 46*, 1405-1411.

In: New Trends in Marine and Freshwater Toxins
Editors: A. G. Cabado and J. M. Vieites
ISBN: 978-1-62257-787-3

Chapter 9

Immunological Methods for Detection of Toxic Algae and Phycotoxins: Immunofluorescence, ELISAs and other Innovative Techniques

África González-Fernández[1], Elina Garet Fernández[1], Hans Kleivdal[2], Christopher T. Elliott[3] and Katrina Campbell[3]
[1]Immunology, University of Vigo, Campus Lagoas Marcosende, Vigo, Pontevedra, Spain
[2]Uni Environment, Uni Research AS, Thormohlensgate, Bergen, Norway
[3]Institute of Agri-Food & Land Use, Queen's University, David Keir Building, Belfast, UK

Abstract

In recent years, the application of polyclonal and monoclonal antibodies in several areas of the marine field has shown a great potential. In this chapter you will find some immunological methods for the specific detection of *toxic algae* and *phycotoxins*. Antibodies can be used in the monitoring of toxic algae blooms and in the identification and discrimination of species by Immunofluorescence techniques. In the case of marine toxin controls, several alternative methods to biological tests are described such as ELISA, Phage display libraries, Biosensors or Lateral flow immunochromatography. In the case of ELISAs, one example is the Amnesic Shellfish Poisoning (ASP) ELISA for the determination of domoic acid, which has been approved by the EU as an alternative method to the biological method. The use of phage display libraries carrying antibody fragments opens enormous possibilities as well. Phages carrying specific antibody fragments offer the required flexibility for the identification of new toxins or complex toxin groups whilst avoiding the need for animal immunization to obtain conventional antibodies. Here, it is shown the example of palytoxin, a very potent toxin that could be regulated in the near future. Recently, different types of biosensors using antibodies have been developed. Many of them employ optical-based technology, particularly Surface Plasmon Resonance (SPR), and offer rapid and reliable methods for detection and quantification of marine toxins. Other immunological tool that offers a portable, low cost, and fast method of detection, is the lateral Flow Immunochromatography, that could be the method of election for qualitative rapid screening on shellfish. The application of antibodies in marine field can offer these and others potential applications such as the study of the immune system of various marine organisms, immune response to vaccines, aquaculture management, ecological studies and many others.

1. INTRODUCTION

1.1. Polyclonal and Monoclonal Antibodies

1.1.1. Introduction

Antibodies are one of the main effector arms of the specific immune system in vertebrate animals. They are a group of glycoproteins, also called immunoglobulins (Igs), which are produced by specialized B lymphocytes. Vertebrate animals can generate many different B cells, each one expressing a different type of antibody receptor on its surface. The entry of a pathogen or a foreign element (carbohydrate, protein, DNA, etc...) usually called "antigen", selectively activates some of these B cells. The specific B-cell clones will undergo activation and maturation with the generation of plasma cells, which secrete antibodies to plasma and extracellular fluids (Figure 1.1). In these locations, soluble antibodies will specifically recognize the antigens, in order to selectively neutralize, block or eliminate them. The binding to the antigen is highly specific (like a key with its specific lock), which is why antibodies are being widely used in several fields and in different techniques for the identification of components (Pyo & Hahn 2009, Spier et al. 2009, Zhou et al. 2009), diagnostic assays (Aguirre et al. 2007, Xu et al. 2010) and even in human therapy (Majeti 2010, Southwell & Patterson 2010).

Most antibodies have a similar structure made up of four chains, two identical heavy chains of 55-70kD (in mammals called γ, ε, δ, α, or μ, which give the name to the Ig classes: IgG, E, D, A or M, respectively) and two identical light chains of 24-25kD (either, κ or, λ), all of them bound by disulphide bonds (intra and inter-chains). Although almost all antibodies share this common structure, Igs without light chains have been described in camels (Hamers-Casterman et al. 1993), different Ig classes have been described in several vertebrates (such as Ig Y, X or P) (Leslie & Clem 1969, Hsu et al. 1985, Schaerlinger et al. 2008), and there are also some Ig subclasses (such as G1, G2, G3, G4, A1 and A2 in humans). The conventional antibody shows two identical domains for antigen recognition (known as Fab, fragment of "antigen binding") that are quite unique for each antibody thanks to their variable regions, and two identical domains with effector functions (known as Fc, crystallizable fragment), shared by different antibodies of the same class.

1.1.2. Polyclonal Antibodies

As vertebrates are able to produce hundreds of millions of different B cells, the entry of a complex antigen (a protein, bacteria or parasite carrying different antigenic regions or epitopes) will induce the activation of several B cells on the lymphoid organs. Each one will specifically recognize its target, and each lymphocyte will become activated, starting its proliferation (Figure 1.1).

This cell division originates a B-cell clone with identical daughter B-cells, all carrying the same Ig receptor. The activated B cells can undergo differentiation into plasma cells, which will secrete antibodies in large quantities. Thus, after the entry of a complex antigen and further activation of different B-cell clones in the immune response, a mixture of specific antibodies (also called *polyclonal antibodies, PAbs*) against the antigen will be present in the serum of the vertebrate animal.

Since the discovery of antibodies and their structure and specificity, PAbs have been produced from different sources by the immunization of domestic animals, preferably large ones, such as horses, goats, rabbits, or even by the immunization of humans. Although the initial use of polyclonal antibodies was in therapy (for the treatment of diphtheria and tetanus, using anti-diphtheria and anti-tetanic gammaglobulins, respectively), their use in continuous therapy is now rather restricted due to their secondary effects. Some examples are the administration to Rh-negative mothers giving birth to Rh-positive children to avoid mother immunization, in acute toxicity by poisons or in some cases of immunodeficiency. However, they are mainly used in several immunological techniques, generally as secondary antibodies, and in this chapter the focus will be on their application in the marine field.

PAbs are being extensively used in the recognition of antigens from different sources (for example, in marine studies for the identification of fish, algae, toxins, proteins, carbohydrates, etc.). In addition, they can be used in different identification techniques (immunofluorescence, flow cytometry, Western blot, ELISA, dot blot, lateral flow, immunohistochemistry, biosensors, etc.) or in the purification of antigens (Desai & Dworecki 2004, Carey et al. 2010). They offer great versatility and a wide range of possible applications, as PAbs can be used either un-conjugated or conjugated to molecules such as enzymes, fluorescent elements or toxins.

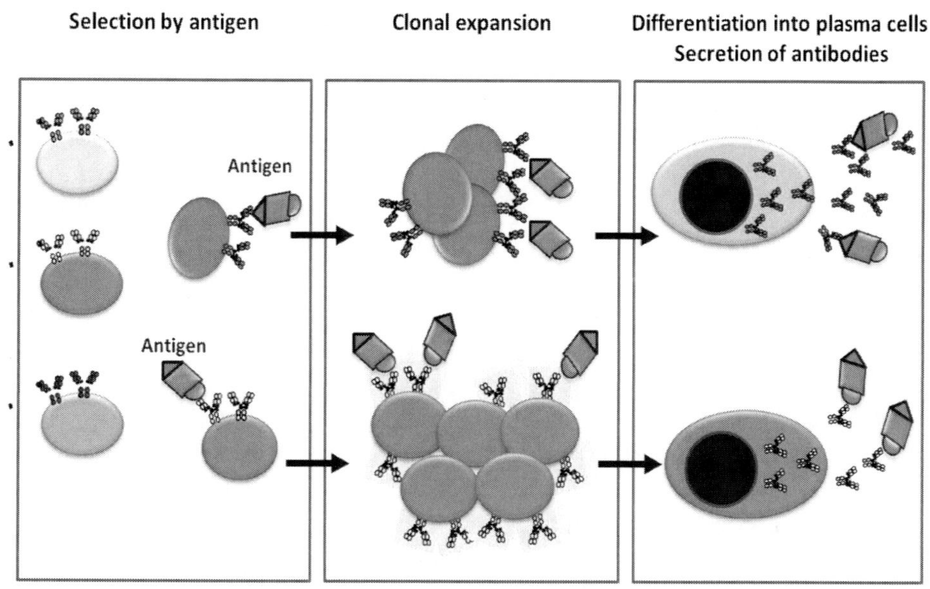

Figure 1.1. B cells are selected by one antigen, which can show different epitopes (in the figure, two epitopes are shown as a triangle and a semicircle). After B-cell activation, clonal expansion and differentiation into plasma cells, they will produce soluble antibodies directed against the specific epitope.

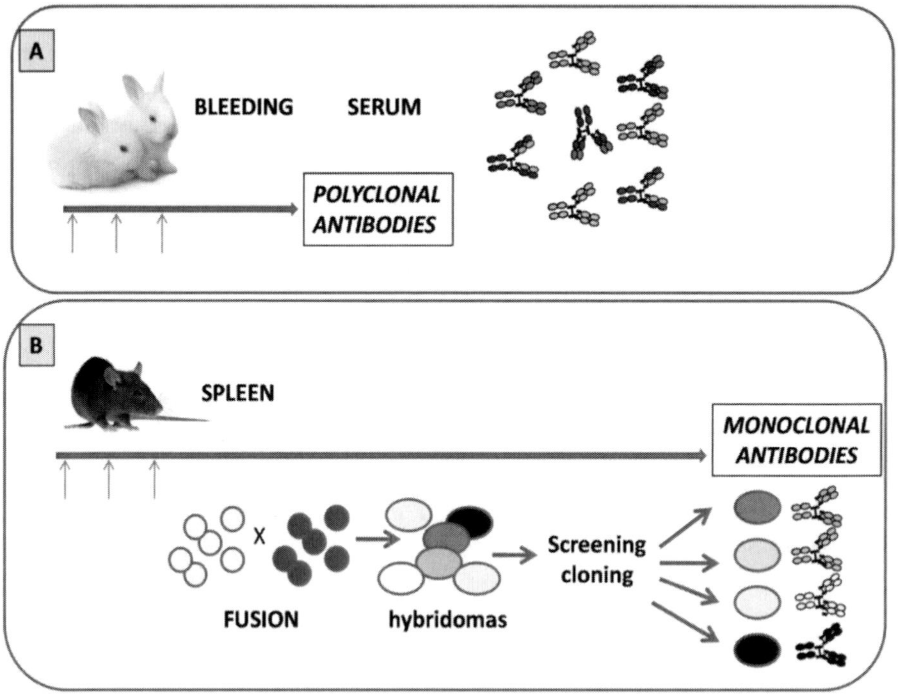

Figure 1.2. Differences between the steps required for the production of polyclonal (A) and monoclonal antibodies (B). Polyclonal production is simple and only requires repeated immunization with the antigen (marked with arrows) followed by serum collection. However, the generation of monoclonal antibodies is much more complex. Following immunization, spleen cells are collected and fused to myeloma cells to generate hybrid cells. After screening of the secretory cells recognizing the specific antigen, the hybridomas are cloned and one single antibody is produced for each hybridoma (4 hybridoma cells are shown in the figure).

To target PAbs against fish cells, microalgae or mussel larvae proteins, the selected animal (rabbit, goat, horse...) must be immunized at least 2-3 times with the antigen of interest, in the presence of adjuvants (such as alum), to increase its immune response and to induce the production of serum antibodies. After certain time periods, serum is collected from the peripheral blood of immunized animals, where specific polyclonal antibodies are present (Figure 1.2A).

The use of PAbs in immunological techniques has several advantages. The first one is that their production is not an expensive process. They are quite easy to obtain in a short time and the only requirement for obtaining more anti-serum is repeated bleeding of the immunized animals. Moreover, PAbs are very useful as secondary antibodies (for example goat anti-mouse Igs), because several secondary antibodies can bind to the primary ones, thereby amplifying the signal in the indirect immunological reactions. For many research groups, the use of polyclonal antibodies is the best choice, because neither special equipment nor a tissue culture laboratory are required for their production.

Yet, PAbs do have several disadvantages, including differences between batches, a mixture of Ig specificities and functionalities, polyreactivity when different antigens are used in the immunization, loss of immune animals (by infectious processes, aging or death), and the presence in the immune sera of many other proteins in addition to antibodies, which can affect the immunological techniques.

1.1.3. Monoclonal Antibody

The concept of *monoclonal antibody (MAb)* refers to the antibody secreted by a single B-cell clone. However, it is not easy to obtain it from conventional cells, because isolated B and plasma cells die quickly in *in vitro* cultures and little antibody is obtained. In 1975, César Milstein and George Köhler were able to immortalize specific antibody-producing B cells from immunized mice by fusion with tumoral B cells (myeloma) to obtain hybrid cells or "hybridomas" (Figure 1.3). A hybridoma can now be divided into identical secreting cells able to secrete a single type of antibody *(monoclonal)* of a predefined specificity in large quantities, and these cells can be maintained either in culture or in cycles of freezing and thawing for long periods of time (De Leij et al. 1983, Fuller et al. 2001). Milstein and Köhler were awarded the Nobel Prize in Medicine and Physiology in 1984 for the development of this technique.

The generation of MAbs begins in a similar way to the production of polyclonal antibodies: with repeated immunization (usually two or three times) of laboratory animals (in this case mice or rats), which are injected at different time intervals (monthly) with the antigen in the presence of adjuvants (Figure 1.2B). After several immunizations, the activated lymphoid organ is removed, generally the spleen, which has become a primary source of antigen-specific B lymphocytes. Subsequently, splenocytes are fused with mouse B tumoral cells (myelomas), which are capable of proliferating indefinitely when they are cultivated *in vitro* under specific conditions. This results in the generation of several hybridomas, cells that combine the capacity of the B lymphocyte to produce a specific antibody, with that of the myeloma cell to reproduce indefinitely. At this point, the hybrid cells must be individualized in wells (called "cloning") to assure the monoclonality of the secreted antibody. Once the selected hybridoma has been cloned several times and remains stable, the cells can be frozen making it possible to store them for an indefinite time. In addition to other attributes, this implies the availability of an inexhaustible source of specific and homogenous antibodies. As the production of MAbs requires a specialized laboratory and the maintenance of the hybridoma cells (either growing on culture or frozen on liquid nitrogen for long periods of time, their generation is much more expensive than that for polyclonal antibodies. However, they have many advantages over PAbs, such as specificity, single epitope targeting and homogenicity.

MAbs have meant a revolution in several fields, especially in biomedicine, and they are now also essential in most Immunology, Microbiology, Physiology, Virology, Pathology, Ecology, and Cell Biology laboratories, for the recognition of structures, identification of species, isolation of proteins, and quantification of different products. They are mainly used as primary reagents in techniques, such as ELISA and RIA, immunofluorescence, inmunohistochemistry, Western blot, Dot blot and lateral flow immunochromatography.

In some cases it is difficult to obtain specific MAbs following the traditional methods described above, especially against small molecules (haptens) or toxic compounds, such as marine toxins. Haptens are non-immunogenic or poorly immunogenic, requiring to be chemically coupled to a large "carrier" protein to increase their immunogenicity. Animals immunized with such hapten-carrier complexes, besides producing some antibodies directed against the hapten, mainly produce antibodies against different epitopes on the carrier, or against epitopes on the hapten-protein complex. Although this is clearly a problem, somehow it is possible to improve the amount of anti-hapten antibodies, by modifying the proportion of hapten copies/protein in the complexes and/or by designing a good external location of the hapten on the protein surface. Antibodies generated against epitopes derived from the hapten-protein complex will not recognize the hapten in isolated form, because it is usually found in natural samples. Furthermore, during the process of conjugation, chemical modification of the small molecules (e.g. a marine toxin) could alter their structure and, therefore, the generated antibodies may no longer recognize the molecules in their native or natural state. The fact that hapten can adopt a variety of different structures causes further problems. For marine toxins, it is difficult to develop antibodies using conventional techniques against the myriad of possibilities that may arise, because, besides the large variety of different toxins, there are also several isomers, which can change their structures depending on pH conditions or temperature.

In addition to their small size, another significant disadvantage for obtaining antibodies to marine toxins is that immunogenic doses are usually lethal in mice. In an effort to avoid these problems using conventional antibodies, alternative techniques such as phage display libraries containing Ig fragments (described in section 1.4 of this chapter) could provide solutions.

1.1.4. Applications of Antibodies

In the marine field, there are a number of areas in which the use of antibodies has considerable potential as a powerful tool for:

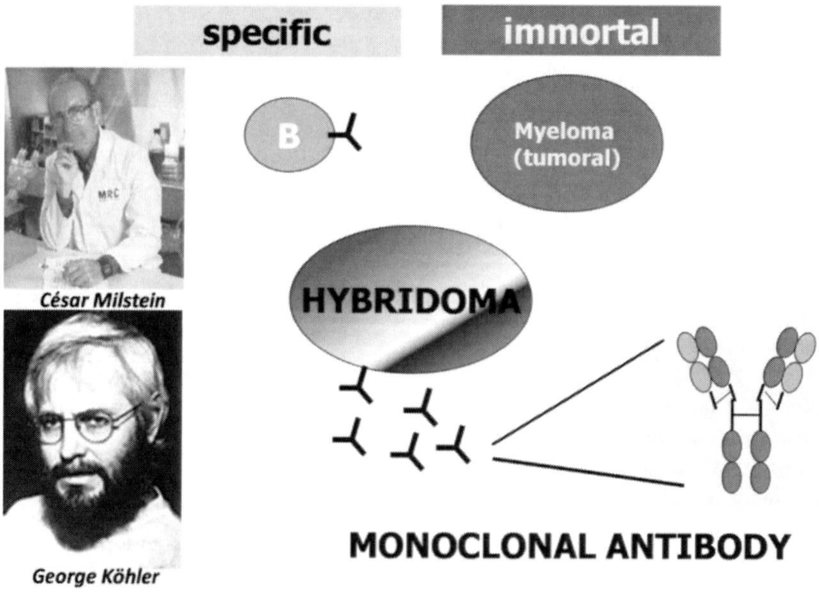

Figure 1.3. Principle of hybridoma production. A B cell carrying the specific antibody is fused to an immortal B cell (myeloma) that is tumoral. The new cell shows the properties of both parental cells: specificity and immortality. The hybrid cell or hybridoma now secretes one antibody of predefined specificity in large quantities. Photographs of César Milstein and George Köhler, both Nobel Prize winners in 1984 for the development of this technique.

- Monitoring of toxic algae blooms.
- Identification and discrimination of species (algae, larvae).
- Toxin controls: detection of toxins by immunology assays.
- Study of the immune system of various marine organisms.
- Immune response to vaccines
- Fishing control.
- Aquaculture management.
- Ecological studies.

Some examples of immunological methods applied to marine samples are described in the following sections of this chapter. Polyclonal and monoclonal antibodies can be used in the specific detection of toxic algae by immunofluorescence techniques (described by Elina Garet in *section 1.2*); in a similar way, Hans Kleivdal illustrates the ELISA method (see *section 1.3*) for the identification and quantification of marine toxins, such as domoic acid. In *section 1.4* Elina Garet shows how the use of phage display libraries carrying antibody domains can be a useful tool in the quantification of marine toxins, whilst avoiding the need for animal immunization. In *section 1.5* Christopher T Elliott describes the development of biosensors for the detection of algal toxins and, finally, in *section 1.6* Katrina Campbell describes the use of antibodies in the lateral flow immunochromatography method.

1.2. Immunofluorescence: Microscopy, Flow Cytometry and Flow Cam with Toxic Algae

1.2.1. Introduction

The application of antibodies has revolutionized the marine field, allowing the specific identification of a target molecule through a wide variety of techniques. In 1941, Albert Coons showed that antibodies could be labelled with fluorescent molecules (Coons et al. 1941), and opened a new era in physiological and ecological studies. Fluorescent dyes or fluorochromes can be coupled to the Fc region of an antibody molecule without affecting its specificity or binding properties. Fluorochrome absorbs light at one wavelength (excitation) and emits light at a longer wavelength (emission). This allows the detection of a target of interest by coloured light emission, when excited by light of the appropriate wavelength, a technique known as immunofluorescence. The light emitted by a labelled antibody can be viewed with a fluorescence microscope (equipped with a UV light source), a flow cytometer, a confocal microscope or a flow cam, allowing the detection and enumeration of phytoplankton species, or quantification and localization of cell constituents, such as enzymes, proteins or toxins. A wide variety of fluorochromes are now available, but fluorescein (FITC), phycoerythrin (PE), and rhodamine, are still the most frequently used. A very important aspect of immunofluorescence detection is choosing the appropriate fluorochrome for carrying out the required studies. This choice will depend on the number of antigens to be detected. Simultaneous identification of various antigens on the same cell is made possible by the use of antibodies specific to several antigens, conjugated with fluorochromes of different emission wavelengths.

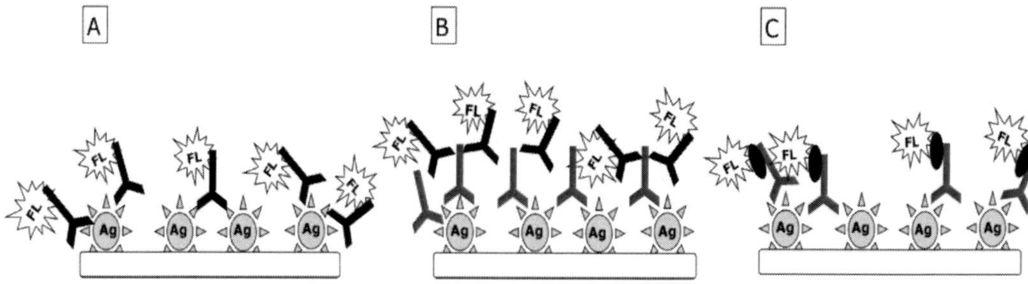

Figure 2.1. Direct and indirect fluorescence staining of an immobilized antigen. In direct labelling (A), antigen (Ag) is stained with an anti-Ag antibody coupled to a fluorescent dye (FL). In the indirect method (B & C), antigen is first incubated with an unlabelled anti-Ag antibody, and then stained with a fluorochrome-labelled secondary reagent, such as an antibody against the primary antibody (B) or streptavidin against a biotinilated primary antibody (C).

Immunofluorescence detection can be carried out directly or indirectly (Figure 2.1). Direct staining involves the detection of fluorescence from a fluorochrome conjugated directly to a specific antibody (the primary antibody, usually a MAb). Indirect immunofluorescence is characterized by detection of the primary antibody using a secondary reagent (usually PAbs), labelled with a fluorochrome that binds to the primary antibody or to another molecule coupled to it. A number of reagents have been developed for indirect staining. The most common ones are fluorochrome-labelled secondary antibodies raised in one species and directed against antibodies of different species, such as fluorescein-labelled goat anti-mouse Igs. However, other reagents such as protein A, which binds to the Fc regions of some Ig classes especially IgG, or streptavidin, which binds very strongly to biotin, could also be used to detect unlabelled or biotinilated primary antibodies, respectively.

Although direct staining reduces the background and involves a smaller number of reagents and washing steps than indirect immunofluorescence, it has some inconveniences such as a low signal if only a small number of epitopes are present. Moreover, direct staining may not be possible, for example in intracellular staining, due to the charge or size constraints of the antibody-fluorochrome complex. Indirect staining is normally employed to avoid these problems, using an unlabelled primary antibody followed by labelled secondary antibodies. The indirect process increases the signal of staining, because several secondary antibodies can bind to each primary antibody, but it also increases the background signal.

1.2.2. Fluorescence Microscopy

The ability of immunofluorescence methods to recognize labelled cells depends on the difference in fluorescence intensities between the positive label and background (autofluorescence). In phytoplankton species, due to chlorophylls, autofluorescence covers a wide range of wavelength and may interfere with many fluorophores (Vrieling et al. 1993). Although autofluorescence can be reduced by elimination of chlorophylls using a suitable fixative agent, such as methanol, the fixative agent can dramatically affect the affinity of the antibody for its antigen (Vrieling & Anderson 1996). Other solutions include the selection of fluorophores and filter sets with the optimal wavelength that could maximize the probe fluorescence, while minimizing the autofluorescence.

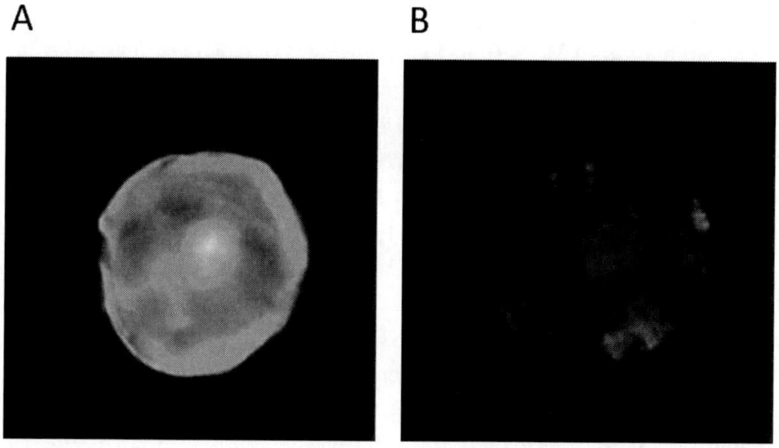

Figure 2.2. Microscopy images of indirect immunofluorescence of the dinoflagellate *Alexandrium minutum* stained with (A) and without (B) a specific primary antibody followed by Goat-anti mouse PAb-FITC conjugated.

Immunofluorescence is a helpful tool for the specific identification of phytoplankton species. Since the first efforts in 1969 to obtain antibodies that would make it possible to distinguish between species of marine microalgae (Bernhard et al. 1969), several PAbs have been generated (Anderson et al. 1990b, Bates et al. 1993, Mendoza et al. 1995, Lopez-Rodas & Costas 1999, Lin et al. 2003, Cho & Costas 2004). However, PAbs present cross-reaction problems between species sharing common antigens and, as mentioned in section 1, their production is limited. These problems could be solved by the generation of MAbs, which are reactive to only a single epitope and are, therefore, more specific. Several MAbs directed against different species of phytoplankton have been described, mostly to harmful algal bloom species (Hiroishi et al. 1988, Uchida et al. 1989, Adachi et al. 1993, Bates et al. 1993, Sako et al. 1993, Córdova et al. 2002, Stauffer et al. 2008, Carrera et al. 2010). For example, our group has recently developed MAbs directed against the dinoflagellate *Alexandrium minutum*, which can be used for the specific identification of this species (Figure 2.2). The identification of *A. minutum* has, until now, been based on morphological characteristics determined by microscopic examination of plankton samples. This process is time-consuming and requires considerable taxonomic expertise. Immunofluorescence using MAb could help to identify this species in natural samples during coastal monitoring (Carrera et al. 2010).

1.2.3. Flow Cytometry

The immunofluorescence technique with a fluorescence microscope is better than conventional optical microscopy for the identification of species, but in some cases it is not easy to discriminate and count the small particles, particularly for cells smaller than 4 μm. Moreover, new methods of detection and analysis are required for monitoring programs or in other circumstances where a large number of samples have to be analysed. The flow cytometer was designed to automate the rapid analysis and identification of cells, when studying size, complexity and fluorescence. Hence, flow cytometry has become the method of choice for the analysis of a large variety of cells and organisms (algae, cells from invertebrate and vertebrate animals, sperm, mussel larvae), allowing their rapid identification even in complex samples. For the detection of algae, flow cytometry could be a reliable method for the routine monitoring of the abundance of phytoplankton species in coastal ecosystems, leading to an early detection of Harmful Algal Blooms (HABs) caused by these microalgae.

The flow cytometer uses a laser beam and light detector to count single intact cells in suspension. A suspension of immunolabelled cells is brought in a sheath flow, resulting in a hydrodynamically focused sample stream. The cells in this stream are transferred as a single cell file with a constant velocity to one or two laser beams. Each time a cell passes the laser beam, light is deflected from the detector, and this interruption of the laser signal is recorded. Under very narrow forward angles, forward scattered (FS) light correlates with cell size and shape (Cunningham & Buonnacorsi 1992). Perpendicular light scatter represents cell morphology and structure (Dubelaar et al. 1987). Cells with a fluorescently tagged antibody bound to their antigens are excited by the laser, and emit light that is recorded by a second detector system. Several dichroic mirrors are displayed to partially reflect the light of a defined wavelength, which is collected in different detectors. Thus, in the same sample, different fluorophores can be detected simultaneously (green, yellow, orange, red, etc.). The simplest form of the instrument counts each cell as it passes the laser beam, and records the level of fluorescence that the cell emits; an attached computer generates plots of the number of cells (e.g. in the y axis) and fluorescence intensity (e.g. in the x axis). More sophisticated versions of the instrument are those that can detect many fluorophores at the same time (including combinations of lasers, more dichroic mirrors and filters), and those capable of sorting populations of cells into different containers according to their fluorescence profile (sorter capacity).

The potential of flow cytometry in the marine field is enormous. This technology may allow the detection of toxic marine phytoplankton before a bloom has started, providing not only a helpful tool in monitoring harmful algae (Zubkov & Burkill 2006, Carrera et al. 2010), but also in the isolation of microalgae and cyanobacteria species by cell sorting (Sensen et al. 1993, Reckermann 2000, Crosbie et al. 2003). Flow cytometry is an optimal instrument for analysing cells ranging from 0.5 to 20 μm in diameter that exist at concentrations between 10^6 to 10^9 cells per litre. In this context, a modified flow cytometer, known as FlowCam, has been developed to handle larger cells, including many species present in the phytoplankton.

1.2.4. Flow Cam

FlowCam is an integrated system combining the capabilities of Flow Cytometry, Microscopy, Imaging and Fluorescence. FlowCam automatically counts, takes images, and analyses the particles or cells that range in size from 20 to 200 μm in a discrete sample (fewer than 10^4 cells per litre) or a continuous flow, providing significantly increased rapid data collection (Sieracki et al. 1998). Microplankton samples are pumped through a thin glass chamber illuminated by a green laser. A video camera captures an image of each object that passes through the field of view of an objective microscope lens. Images of cells can be further magnified on the computer screen for visual identification. Information on the relative fluorescence of the cell is captured by a photomultiplier tube (PMT), and information on the dimensions of the image (equivalent spherical diameter [ESD], maximum and minimum dimension, among others) is collected in Excel format.

Potential applications include studies of plankton community structure, ocean optics, and monitoring of harmful species of algae. According to See et al. (See et al. 2005), FlowCAM estimates overall nanoplankton and microplankton cell abundance in a similar way to epifluorescence microscopy. Hence, when used in tandem, Fluoroprobe and FlowCAM could be important additions to biological observation systems of the coastal waters of the ocean. Specifically, FlowCAM is less tedious and time-consuming than microscopy for the enumeration of the red tide species *Karenia brevis*, and allows the examination of more cells for greater counting precision (Buskey & Hyatt 2006). This technology is not only applicable to the identification of harmful algae, but is also an efficient tool for the rapid analysis of zooplankton feeding on different types and sizes of food particles (Ide et al. 2008).

1.3. ELISAs for Natural Biotoxin Control

1.3.1. Introduction

The use of antibodies to detect and quantify the presence of certain molecules can be very specific and accurate, and has been used in human diagnosis for several decades. The beginnings of modern immunodiagnostic assays, as practiced today, can be attributed to discoveries such as the use of coupled labels to provide an indirect signal of antigen-antibody interaction (Coons et al. 1941), and to the refinement for improving the separation of the bound and unbound antigen, by absorbing one of the assay components in polystyrene test tubes (Catt et al. 1967). The independent description of the non-isotopic enzyme labels for the immunoassay by two research groups (Engvall & Perlman 1971, van Weemen & Schuurs 1971) led to the introduction of the enzyme-labelled immunosorbent assay (ELISA). The following rapid developments and the agreed use of the 96-well microtitre polystyrene plate as the standardized ELISA format paved the way for automation and the extensive use of ELISA as a powerful analytical tool in various disciplines, including as diagnostic tests in medicine, and as quantitative tools in environmental monitoring, food safety and quality control.

1.3.2. The ELISA Principle

The process of production and the large-scale availability of antibodies with the required affinity, specificity and cross-reactivity were fundamental for the development of sensitive, precise, and specific immunoassay products. In ELISA, toxin-specific antibodies are used in a solid support assay format on a 96-well polystyrene plate. The solid phase acts as a support for the immobilization of either the antigen (the toxin) or the anti-toxin antibody, to allow washing and removal of unbound components during the assay procedure. The basis for ELISA technology is the detection and quantification of the binding event between the antibody and toxin components in the assay, and the subsequent estimation of the amount of toxin in the original sample solution. The toxin-antibody bound fraction is quantified by labelling the free component with a reporter enzyme, acting as an amplifier to produce many signal molecules. Thus, the high performance of toxin-specific antibodies is only one important aspect of the assay. The other main parameter is the detection of the binding event, finally defining the sensitivity and robustness of the system. The most commonly used enzyme detection and signal amplification system is the conjugation of horseradish peroxidase (HRP) to either the antibody or the toxin, leading to an observable colour change upon the addition of appropriate chromogenic substrates. Although there is a constant development of new immunoassay techniques utilizing fluorescent or electrochemiluminescent reporters, the classical enzyme-based ELISA is still the most common commercialized format, because of its robustness to various biological matrices.

1.3.3. ELISA Designs and Formats

Although immunoassays can be carried out in a wide variety of designs and formats, the most commonly used commercial ELISAs are in the competitive (reagent-limited) or capture (sometimes called sandwich) formats. The *antigen-capture ELISA* format refers to the binding of the specific antigen to immobilized "capture" antibodies in the microwell. In some cases, capture ELISAs can be designed as a *sandwich ELISA,* in which the ligand or toxin is sandwiched between antibodies coated on the microplate well and an enzyme-labelled antibody. This can only be used for large antigens (normally a protein) with two antibodies capable of binding to discrete, non-overlapping epitopes of the antigen. Therefore, it is rarely used for small toxin molecules.

Most commercial immunoassays for natural toxins are based on the *competitive ELISA* principle (Figure 3.1), where the toxin is coated to the bottom of the microplate. The toxin in the sample is then added into the well together with the labelled antibody. The subsequent incubation allows the competitive binding of the toxin-specific labelled antibody to either the immobilized toxin coated on the plate or the free toxin in the sample. The antibodies will be split in two fractions: one fraction of antibodies bound to the toxin molecules in solution, and another fraction of antibodies bound to the surface-bound toxin molecules. If there is toxin present in the sample (or standard), the amount of labelled antibody bound to the well will be reduced. When the soluble fraction of toxin-bound antibodies is washed away, only the surface-bound antibody fraction is left in the microwell. Chromogenic substrates are added to activate the reporter enzyme to give a strong signal from the remaining enzyme-labelled antibody fraction (Figure 3.1). The colour intensity produced is directly proportional to the amount of enzyme-labelled antibody present. The correlation between an increasing toxin present in the sample and colour intensity is used to form calibration curves (Figure 3.2) with toxin standard solution, which is used to determine the toxin present in unknown samples. Calibration curves, showing the correlation between an increasing toxin content and the decreasing reporter signal/colour reaction in competitive assays, are used to calculate the toxin content in an unknown sample by extrapolation. The reporter signal is usually measured as the intensity of the enzymatic colour reaction by using a microplate reader for absorbance analysis at given wavelengths. The typical ELISA procedure involves adding reagents and samples in a series of steps before the incubation step, followed by the development of the colourimetric reaction. The typical time for the analysis of up to 40 samples in one 96-well plate is one to two hours.

1.3.4. Quantitative ELISAs and Qualitative Screening ELISA Tests

All ELISAs with multipoint calibration curves are in principle regarded *as quantitative ELISAs*, but the quantitative property is strongly dependent on the ability to provide an accurate and reproducible determination of toxins in a sample set. An ELISA should only be regarded as a quantitative test, if appropriate analytical properties are confirmed through comprehensive validation studies, demonstrating an adequate correlation to the given reference method. Some ELISAs may not be fit for accurate toxin determination, because of poor analytical performance, even when multipoint calibration curves are used. Such ELISAs may have a lower level of accuracy and precision, but their high sensitivity and selectivity can make them useful as qualitative *screening tests.* In this context, it is important to assess if the sensitivity and specificity of the screening test is sufficient for the intended purpose. Reliable screening tests are widely used to confirm negatives related to a regulatory toxin level, where positive samples are usually subject to re-analysis by the reference method or a suitable confirmatory method. In particular, such qualitative screening tests, based on the specific risk assessment, may be suitable for pre-harvest monitoring in the shellfish industry, if a "yes/no" result is considered to be sufficient. However, while food business operators can cope with a certain level of false positives samples when using screening tests, false negatives cannot be tolerated in quality control systems designed to protect the health of consumers.

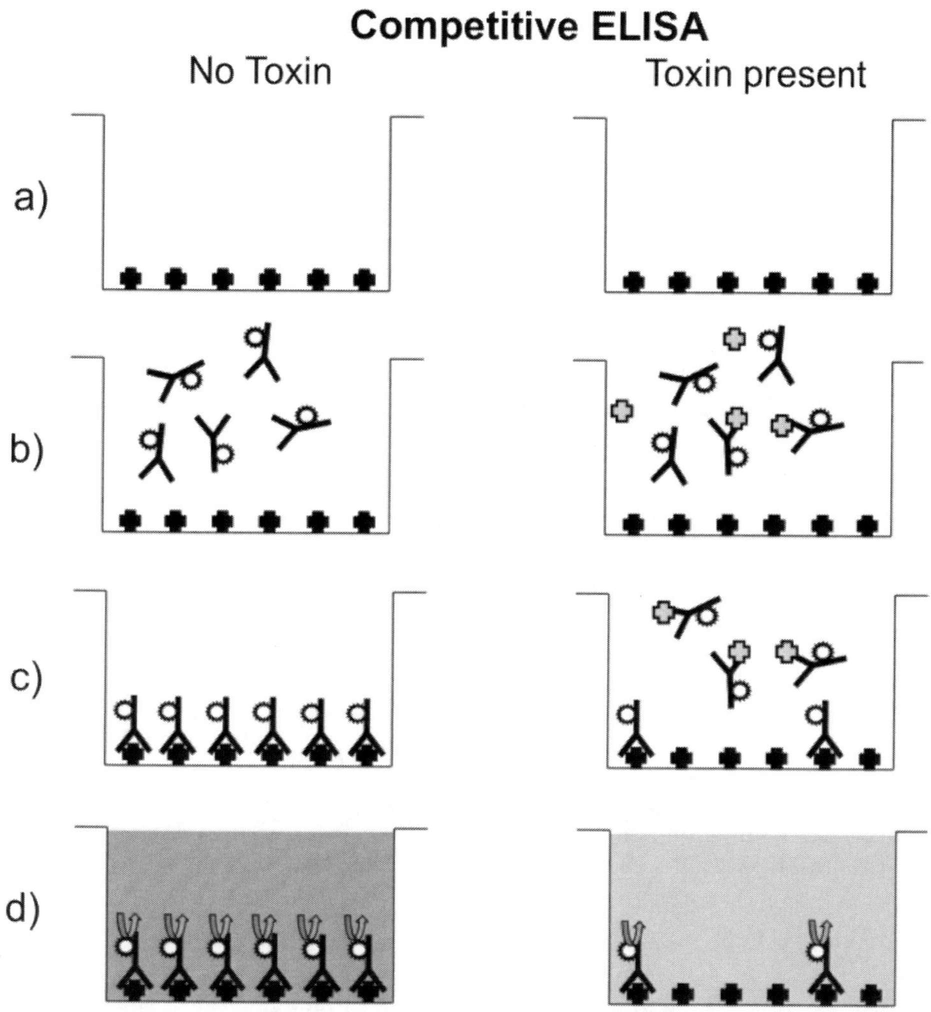

Figure 3.1. The major steps of a competitive toxin ELISA; a) coating of toxin conjugate; b) addition of toxin sample and enzyme-labelled antibody; c) competitive binding of antibody to either soluble toxin in sample or immobilized toxin; d) after washing away soluble components, the antibody enzyme label is activated by addition of substrate. *Left panel with no toxin present in the sample:* All labelled antibodies are bound to the surface-coated toxin, resulting in a strong enzymatic reporter signal. Right *panel with toxin present in the sample:* The toxin molecules in the sample will compete with the surface-bound toxins in binding to the antibodies. As antibodies bound to the soluble toxin molecules are washed away, only the few enzyme labels from the remaining surface-bound antibodies will be present to produce a weak colour signal.

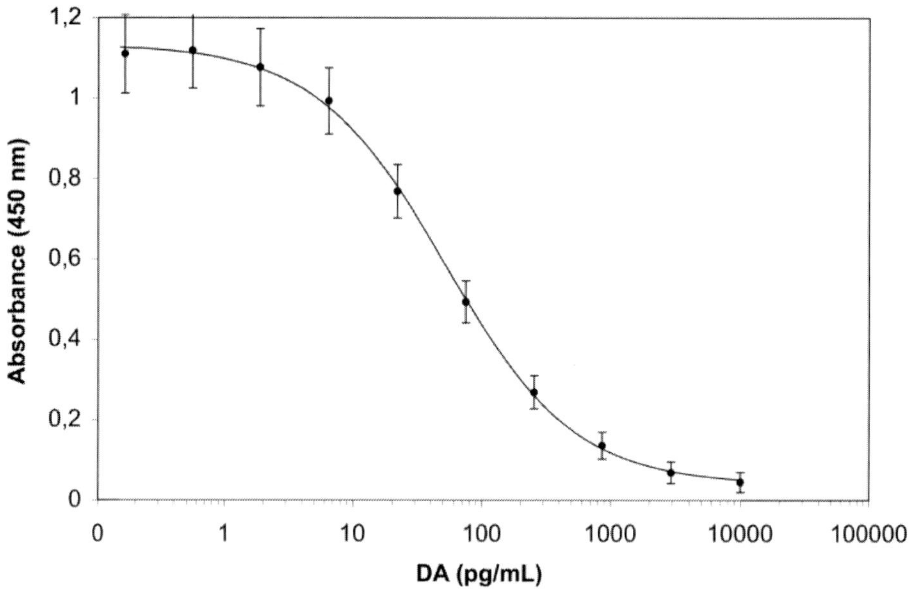

Figure 3.2. Multipoint calibration curve from an ELISA for the quantification of domoic acid (Kleivdal et al. 2007a). The absorbance read from the 10 calibration standard solutions, presented as a sigmoid calibration curve, shows how the increasing amount of toxin leads to decreasing absorbance values. For this particular ELISA, the calibration working range is defined along the linear range in the middle of the curve (Kleivdal et al. 2007a).

1.3.5. Factors Affecting the Analytical Performance of ELISA Tests

ELISA-based tests designed for a single toxin analyte can be very sensitive, accurate and precise in comparison to chemical analysis using instrumental techniques, such as high-performance liquid-chromatography (HPLC) (Kleivdal et al. 2007b, Garet et al. 2010a), but ELISA *specificity* can be challenged if the analyte group is both structurally and toxicologically diverse. The most important characteristic of a typical confirmatory method for a given toxin group is the ability to both identify and accurately quantify the main structural analogues of the toxin group. In contrast, for ELISA, the most important features for a specific toxin are to be able to integrate the signal response created by all the different structures of the toxin group, following a simple procedure without the use of highly sophisticated instrumentation, and to be able to provide an estimate of the total amount of that toxin group present in the sample. However, as even a slightly different structure can have a major impact on the toxicity, not all analogues are toxicologically relevant to a toxin group. The specific antibody in an ELISA will bind strongly to the particular toxin structure for which it was designed, but may have variable cross-reactivity to other analogues, and may show a weaker or stronger binding to other structures, irrespective of their toxicity. Thus, an ELISA with a good correlation to the actual toxicity should be based on a well-designed antibody with strong binding ability (high cross-reactivity) to the toxic analogues, and low binding ability (low cross-reactivity) to less toxic or /non-toxic analogues. The design of the immunizing toxin conjugate can direct antibody production towards the more conserved parts of a toxin structure to obtain a broad-spectrum ELISA, as in the case of an ELISA directed towards the lipophilic yessotoxin group (Briggs et al. 2004) or the ADDA ELISA for the detection of microcysteins (Fischer et al. 2001). The cross-reactivity of the acting antibody in an ELISA is an important property that, to a considerable extent, affects the method's performance (Chu et al. 1996, Fremy & Usleber 2003). Nevertheless, it should be stressed that the *specificity* challenge, described here for the ELISA, has to be met by all antibody-based test technologies, regardless of the format.

Selectivity is another factor that may affect the performance of the ELISA test, as the assay antibodies can bind in a non-specific manner to components in complex sample matrices, leading to unwanted masking effects. In competitive ELISAs, such a "matrix effect" will produce false positive signals, as a result of the competitive non-specific binding of the labelled antibodies to the soluble proteins, which will in turn be removed during the washing step. ELISAs with high selectivity values demonstrate low cross-reactivity to potentially interfering compounds, or to non-toxic compounds with structural similarities to the toxin group. It is important to characterize the ELISA response to the toxin-free sample matrix of interest, before using an ELISA to analyse a new sample matrix, as such non-specific interference effects can be overcome by extract dilution (Kleivdal et al. 2007a). The required minimal dilution factor of the sample extract is determined to eliminate the unwanted matrix effects. Once the minimal dilution has been established, the lower limit of detection for a given biological sample matrix can be estimated for the specific ELISA.

1.3.6. Validation of ELISAs for Toxin Analysis

For the different groups of marine biotoxins, there are several analytical methods available, but EU legislation (EC 2074/2005) requires a specific reference method for each toxin group for the purposes of official controls. However, when the legislation does not specify analytical methods, Food Business Operators (FBOs) may use appropriate methods laid down in EU legislation or national legislation, or in the absence of such methods, techniques that offer equivalent results to those obtained using the reference method, providing that they are scientifically validated in accordance with internationally recognized rules or protocols.

The validation of a method involves its comprehensive testing against predetermined parameters for measuring its ability to give an acceptable result, which correlates well with the results of the reference method. Following a basic method characterisation, a thorough evaluation is required to demonstrate the suitability of the method to measure a specific toxin group in a specified matrix. This is achieved by validating the assay performance in accordance with international guidelines (Thompson et al. 2002), by investigating the assay performance characteristics, including specificity, limit of detection, limit of quantification, working range, accuracy/trueness, recovery, precision (repeatability and reproducibility), and robustness. In addition, the novel methods should demonstrate a good correlation to the reference ones, by a method comparison study. Once a novel method is characterized in a single laboratory validation study, an inter-laboratory study may be performed to confirm the precision and accuracy of performance parameters between laboratories, and that the method is transferable without significant loss of performance.

One example of this is the Amnesic Shellfish Poisoning (ASP) ELISA from Biosense, developed for the determination of the domoic acid in shellfish. This ELISA was originally described as an indirect ELISA by Garthwaite and co-workers (Garthwaite et al. 1998), but was subject to comprehensive validation studies to convert the research-grade in-house ELISA into a flexible standardized ELISA format suitable for routine use. The ELISA performance was characterized in the initial single laboratory study (Kleivdal et al. 2007a), followed by a comprehensive inter-laboratory validation study, involving 15 laboratories in ten countries (Kleivdal et al. 2007b). The comprehensive validation studies demonstrated that the ASP ELISA sample results were in good agreement with the expected domoic acid values spiked into the study samples (Figure 3.3), with an accuracy (104% recovery) and precision (23% reproducibility) comparable to instrumental methods (Kleivdal et al. 2007b). The same study results from ASP ELISA compared well with the reference method using liquid chromatography, revealing a Pearson correlation coefficient of 0.984. The data was, therefore, submitted to the Association of Analytical Communities (AOAC), rewarded an Official Method Accreditation (AOAC OMA 2006.02) and, subsequently, approved by the EU as an alternative method to the reference one (EC 1244/2007).

The strong analytical performance and validation data presented for the ASP ELISA, as a food safety tool for shellfish samples, has also led to further exploration into food safety (Reboreda et al. 2010, Garet et al. 2010a) and the monitoring of other biological matrices (Maucher & Ramsdell 2005, Lefebvre et al. 2007).

In general, very few rapid tests have been subject to a similar full validation process because of the prohibitive costs involved, which have to be covered by the developers' of the test. However, the validation and method accreditation of more rapid ELISA tests would be beneficial to all users, so that each end-user would be able to choose the most suitable method for the intended analytical purpose from a range of validated rapid methods.

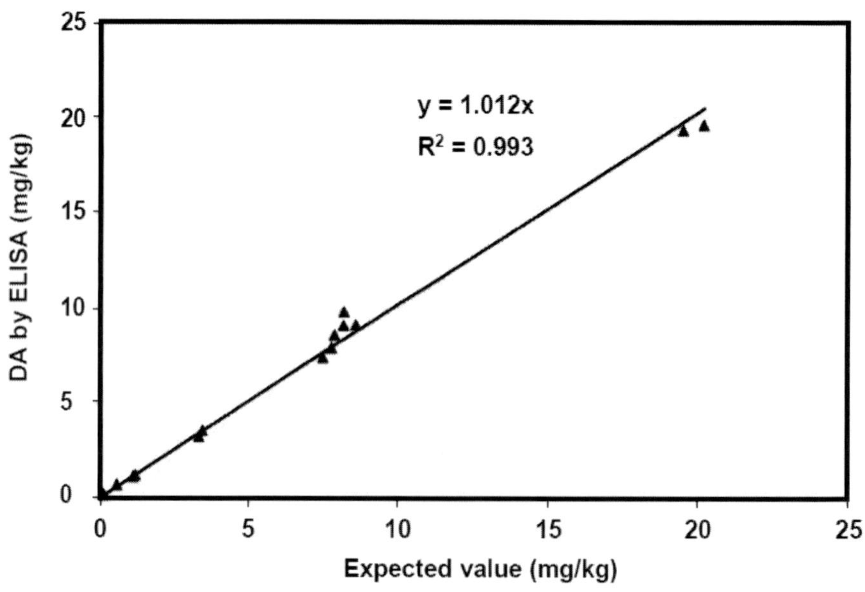

Figure 3.3. A correlation plot showing good agreement between the analytical ASP ELISA data and the expected domoic acid spike values from an inter-laboratory validation study (from Kleivdal et al., 2007b). A total of 194 shellfish samples were prepared by spiking blank shellfish with known amounts of domoic acid reference material, and distributed to 15 participant laboratories for analysis by ASP ELISA. The reported sample results were compiled and compared directly with the known spike values of the corresponding study samples.

1.3.7. The Future Development of ELISA Testing

Following the international focus on food safety control systems and the increased effort to bring the monitoring systems closer to critical control points, there has been an increased effort to develop rapid and reliable ELISA tests for marine biotoxins. Quantitative methods able to detect variations and trends in toxin levels can be suitable for pre-harvest monitoring, production line monitoring and end product testing (Reboreda et al. 2010). In general, screening methods are dominated by ELISAs, which offer greater benefits in terms of speed and sensitivity than the reference methods.

In the future development of classical ELISAs, more effort should be focussed on the challenge to match toxicity with toxin levels, as quantified by a single antibody (Chu et al. 1996, Dubois et al. 2010). The challenge of designing antibodies with the optimal specificity will be further complicated by the structural and toxicological diversity within a toxin group, together with potential chemical conversion and biotransformation. However, as recombinant antibodies have been successfully constructed and produced against the potent palytoxin (Garet et al. 2010b) (*see section 1.4*), such strategies may offer the required flexibility for developing antibody solutions that can contribute to resolving the specificity challenge for complex toxin groups. A trend in classical ELISA development is the efforts to further simplify the test and to reduce analysis time, as seen with the Instant ELISA used in biomedical research (www.bendermedsystems.com).

In the past, the most significant challenge has been the lack of standardization and transferability of in-house ELISA tests between laboratories. This should be overcome by increasing the focus on method validation for ELISA toxin tests, based on new strategies for the design and production of high-quality antibodies.

1.4. Phage Display Libraries for Identification of Marine Toxins

1.4.1. Introduction

Recent advances in antibody engineering and molecular biology have made possible the development of recombinant antibodies with high specificity and affinity for, theoretically, any antigen. This has allowed the generation of a broad repertoire of antibodies expressed in various systems, such as bacteria (Skerra & Pluckthun 1988), insects (Bei et al. 1995), yeast (Ridder et al. 1995) and plants (Whitelam et al. 1994). Among several strategies for *in vitro* expression of peptides or proteins, one of the most useful technologies is the exposure of antibody fragments on the surface of the viral particles or *Phage display* (Hust & Dubel 2004), which mimics the B cell. Phage display was described by George P. Smith in 1985 (Smith 1985), and refers to the expression of peptides, proteins or functional antibodies on the surface of filamentous bacteriophage viruses (Smith 1985, Clackson et al. 1991). In recent decades, several antibody libraries have been developed (Bass et al. 1990, McCafferty et al. 1990, Smith 1991, Hoogenboom 2002) of antibodies with a single-chain variable fragment (scFv), or Fab fragment, expressed as a fusion to the phage pIII capsid protein. The expression of Fab fragments (Barbas et al. 1991, Hoogenboom et al. 1991), heterodimers formed by a light chain (VL-CL) connected by a disulphide bond to the variable domain of a heavy chain (VH) and to the CH1 domain, are typically in univalent format (i.e. a single copy of fragment per viral particle). Yet, these libraries may be inefficient for recovering antibodies with low affinities.

1.4.2. Antibody Libraries

Most of the antibody libraries that have been obtained are based on single-chain variable fragments (*scFv*), one smaller unit with the ability to recognize and bind to a specific antigen (Figure 4.1). These libraries were created using monomers of light-chain (VL) and heavy-chain (VH) variable fragments, linked by a flexible peptide chain (linker). Due to their small size, scFv monomers are presented in the phage more efficiently than the Fab fragments (McCafferty et al. 1990).

These antibody libraries have been constructed by PCR amplification of naturally un-immunized rearranged V genes (naïve Libraries) (Marks et al. 1991, Vaughan et al. 1996, Sheets et al. 1998, de Haard et al. 1999, Sblattero & Bradbury 2000), by PCR assembly of synthetic human V genes (Synthetic Libraries) (Griffiths et al. 1994, Nissim et al. 1994, de Kruif et al. 1995, Knappik et al. 2000), or created from V genes from immunized humans (Immune Libraries) (Barbas et al. 1991, Amersdorfer & Marks 2000, Amersdorfer et al. 2002), or mice (Orum et al. 1993, Ames et al. 1994).

Generally, antibody libraries consist of billions of different scFv genes fused at the N-terminus of the capsid proteins (pVIII or pIII) of a filamentous bacteriophage, such as Fd, F1 or M13 (known as Ff phages) (Gao et al. 1999). Filamentous phages, a large family of viruses that infect many Gram-negative bacteria, are an excellent tool for use as cloning vector and for display proteins. The viral particles tolerate insertions into non-essential regions of their genome, retain their infectivity even after the modification of proteins, can accumulate to high titres without killing the host cell, and are stable in a wide range of conditions.

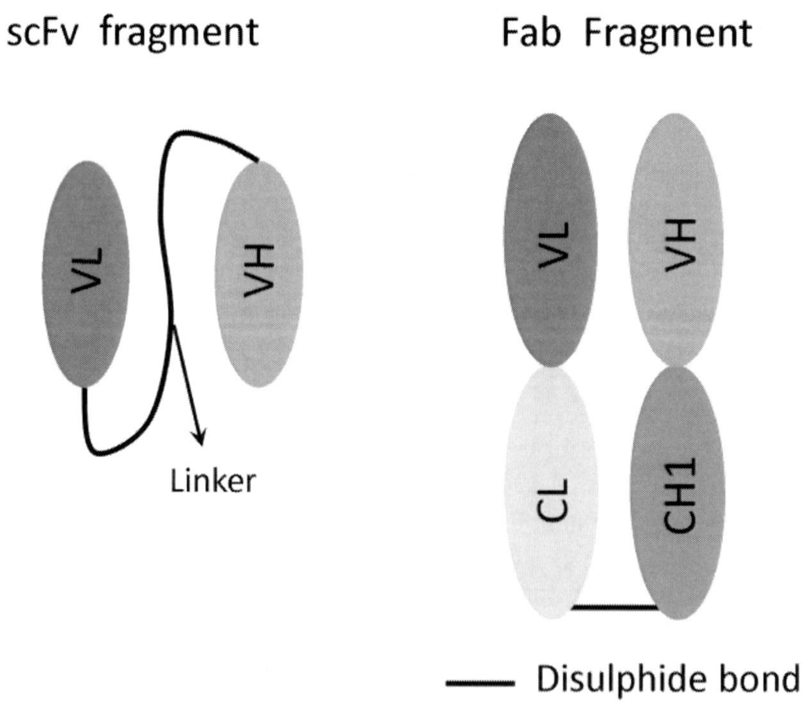

Figure 4.1. Comparison of a single-chain variable fragment (scFv) with an antigen-binding fragment (Fab).

Recombinant viral proteins encoded on a plasmid (phagemid vector), which also contains the packaging signal of the Ff virus and an antibiotic resistance gene to allow propagation as plasmid in *E. coli*, are transformed into bacterial cells to create the library. This phagemid vector produces large amounts of the recombinant protein, but it requires the co-infection of the bacteria carrying the phagemid with a helper phage, which supplies all the other proteins necessary to make functional viral particles that display scFv fragments. Helper phages are basically normal Ff phages with some modifications (i.e. a disabled packaging signal or a modified pIII) (Bradbury & Marks 2004). They usually carry antibiotic resistance genes that make it possible to select bacteria that contain the helper phage. Thus, the genetic material encoding different scFvs is integrated into the genome of different phages, providing a direct linkage between phenotype and genotype. This technology has enabled the construction of phage libraries of up to 10^{12} different scFv genes, allowing selection of one or more viral particles displaying antibody fragments specific against an antigen of interest.

1.4.3. Selection of Specific scFvs

The random sequences of the displayed antibody fragments provide sufficient diversity to allow the selection of high affinity ligands of targets of interest (Devlin et al. 1990). The selection of specific scFvs from phage display libraries includes an *Affinity Selection or Panning* (Figure 4.2), a process where phage populations are exposed to a target molecule to capture those phage-carrying scFvs that bind specifically to the antigen (Hoogenboom et al. 1998).

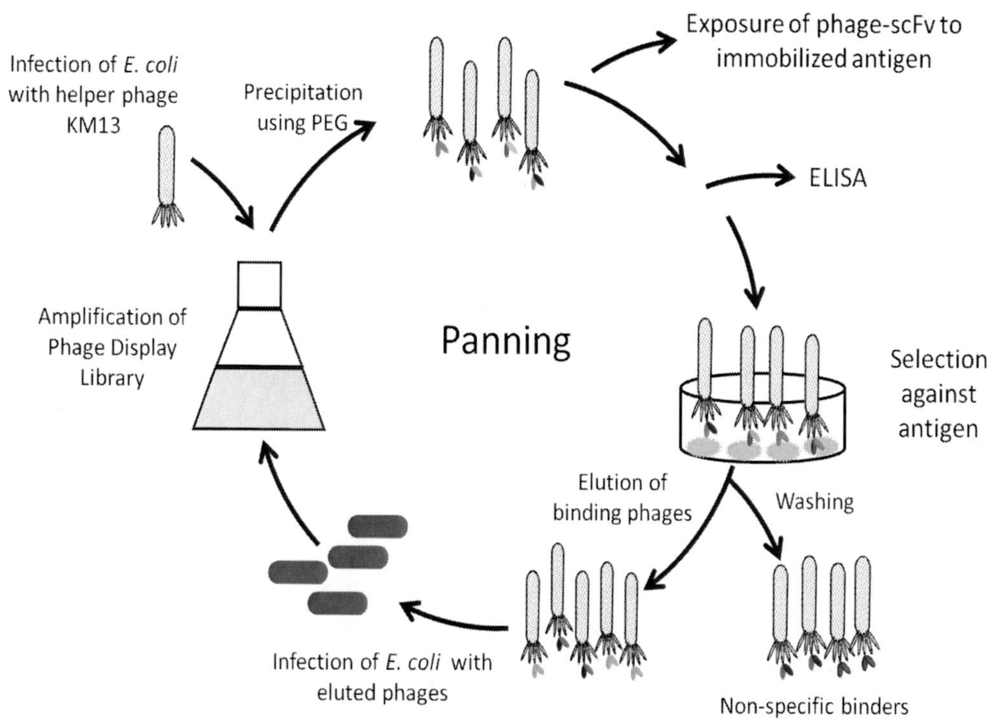

Figure 4.2. Illustration of the Selection or Panning process.

Through successive rounds of binding, washing, elution and amplification, the original population of phages, widely diverse, is enriched especially in those particles that are specific against the antigen. The number of cycles of binding selection depends on the suitability of the variants obtained. In general, the proportion of phages bound to the specific antigen increases from 1% in the first round, to 100% after 2-4 rounds.

An important aspect to take into account is the antigen, because selection is most efficient when a purified antigen is used. Once the pure antigen is obtained, it must be immobilized on a solid support, such as immunotubes or plates, before starting the panning process. However, selection using immobilized antigen may be difficult, since some molecules are too small or do not have the chemical groups that allow their attachment to the support, or even change their conformation when they are directly coated on a surface. In such cases, phage binding to the target can also be carried out in solution by using, for example, a biotinylated antigen and capturing the phage-antigen complex via avidin or streptavidin.

Non-specific scFv-phages are washed away, while those viral particles bound to the antigen are recovered by elution. One of the most effective methods for elution of specific scFv-phages is proteolytic selection (Kristensen & Winter 1998), which ensures the recovery of only infective phages.

1.4.4. Screening Process

After the selection process, a *screening assay* is necessary**,** using the specific antigen as target. Usually, it is accomplished by a simple ELISA method employing 96-well plates coated with the antigen. However, immunohistochemistry, immunofluorescence or Western blot may also be used for this purpose. The selected clones are amplified in *E. coli* to obtain scFv-phages or soluble antibody fragments, as phagemid vectors incorporate an amber stop codon between the scFv and pIII genes (Hoogenboom et al. 1991), together with a tag that allows the detection of the soluble fragment. The scFv fragments are displayed on phage produced from infected amber suppressor strains of bacteria, or secreted into the periplasm of infected non-suppressor strains. Antibody fragments isolated from phage display libraries have now been used in similar applications to those of traditional antibodies, such as ELISA, immunohistochemistry, immunofluorescence, inhibition studies, or Western blots.

As each phage carries the genetic information that results in scFv selected, once the antibody of interest has been isolated, the coding sequence can be determined and/or altered to increase affinity, by the introduction of mutations either randomly or at specific sites. Mutant residues could produce changes in the side chains of the contacting antigen that improve their affinity (Russell et al. 1993) or could provide a new paratope. Different methods are used to introduce diversity in the antibody genes, including mutator strains (Low et al. 1996), error-prone PCR (Hawkins et al. 1992), chain shuffling (Clackson et al. 1991, Marks et al. 1992) and DNA shuffling (Crameri et al. 1996).

The phage display library is a widely used technology, which has been employed successfully to find specific antibody fragments against a wide variety of antigens, such as foreign or "self" human antigens (Matsumoto-Takasaki et al. 2009, Guettinger et al. 2010, Wolf et al. 2010), freshwater toxins (McElhiney et al. 2000, Zhao et al. 2005) or low molecular weight environmental contaminants, such as marine toxins (Nagumo et al. 2004, Shaw et al. 2008). In this context, our group has recently obtained a specific phage-scFv directed against one of the most powerful marine toxins, called palytoxin (Garet et al. 2010b). Phage display technology has allowed us to isolate a specific antibody fragment using free palytoxin adsorbed immunotubes, thereby avoiding the coupling to a carrier protein. High affinity shown by the phage-scFv, against both free and immobilized palytoxin, made it possible to develop a competitive ELISA with a low detection limit and a wide working range. The assay was compatible with the methanol extraction method and was successfully applied to spiked shellfish samples, suggesting that this ELISA could be a powerful aid for the detection of palytoxin in routine controls of shellfish samples. Our results, and those from other authors, indicate that phage display technology could be a cheaper, more convenient and more efficient method for obtaining specific antibodies against complex antigens, such as small molecules or poorly immunogenic molecules.

1.5. Biosensor Detection of Algal Toxins

1.5.1. Introduction

Biosensors are generally defined as any measuring system that relies upon a biological recognition element. The broad principles of signal transduction and an overview of the numerous emerging applications of biosensors in medicine, life science research, environmental and food analysis, together with other potential areas of application, such as defence, have been extensively reviewed by Buerk (Buerk 1993). The numerous areas of interest in biosensor technology relating to food quality applications have been detailed by Wagner and Guilbault (Wagner & Guilbault 1994). The range of technology platforms and uses has expanded considerably since the publication of these two pioneering studies, nearly twenty years ago.

In general terms, biosensors consist of three main functioning parts:

- The biological element, such as human and animal cells, microorganisms, cell receptors, enzymes, antibodies and nucleic acids.
- The transducer that is capable of measuring a biomolecular interaction and converting it into a signal that can be read by some form of instrumentation (Table 5.1)
- The electronics responsible for the display of the results in a manner that can be easily interpreted by the user.

There were more than six thousand publications relating to the development and applications of biosensors between 1984 and 1997 (Luong et al. 2008). The estimated annual worldwide investment in biosensor research and development was around $300 US million at the end of the 20th century and, as new technologies and applications are developed, it is likely that it will continue to increase (Weetall 1999). The optical biosensor is the most widely used in the large majority of reported biosensor platforms. By the end of the 20th century, more than 75% of the studies published about applications related to this biosensor technology used platforms with Surface Plasmon Resonance (SPR) based optical biosensors (Homola et al. 1999).

Table 5.1. Range of types of transducers used for converting biomolecular interactions into measurable signals

Type of Transducer	Measurement Principle
Amperometric	H_2O_2 or O_2 measurements
potentiometric	pH or pIon measurements
photometric	Optical measurements
Piezoelectric	Change in mass
Conductive	conductivity change
Capacitive	dielectric constant
Thermometric	Temperature

Many SPR-based optical biosensors employ evanescent wave techniques. These techniques use a property of gold, where a thin layer of gold on a high refractive index glass surface absorbs laser light, producing electron waves referred to as surface plasmons on the gold surface.

This phenomenon occurs only at a specific angle and wavelength of incident light, and is highly dependent on the surface of the gold, such that binding of the target analyte to a biological receptor on the gold surface produces a signal that can be measured. SPR sensors generally employ some form of sensor chip, which consists of a cartridge encompassing a glass plate coated with gold on one side that comes into contact with the optical detection unit. A microfluidic flow through system delivers reagents across the chip on the reverse side. In most cases, the gold layer is modified with a range of molecules that permit the straightforward attachment of receptors or analytes. The literature reports carboxymethyl dextran as the most widely used linker layer.

A light source at a fixed wavelength is reflected off the gold layer at the angle of total internal reflection. This induces the evanescent wave to penetrate through the glass plate into the reagent stream being pumped across the surface. Biomolecular interactions occurring at the chip surface have an effect on the refractive index and these can be measured with a very high degree of sensitivity. SPR has been incorporated into a wide range of devices aimed at measuring the concentrations of analytes present in a sample and the rates of binding and dissociation of a pair of biomolecules. While the main commercial products based on SPR are undoubtedly the Biacore range of biosensors (produced by GE Healthcare), other manufacturers, such as ICX Technologies, Biorad, Ibis and Horiba, have commercialized SPR-based measuring systems capable of fulfilling a very wide range of applications.

1.5.2. Measurement of Small Molecular Weight Targets

For most biomolecules, the change in response is proportional to the mass of the material that has bound to the surface of the sensor chip. A typical 'sensorgram', a measure of the binding response against time, is illustrated in Figure 5.1 (Abery 2001). For binding molecules to cause a measureable change in response, they must be greater than ~1000 Daltons. For small molecular weight compounds, including many vitamins, drugs and toxins, it is very difficult to measure the change in mass, when binding events occur, using many SPR sensors. Therefore, an alternate strategy had to be developed to measure the molecular interactions between drugs/toxins and antibody/receptors. This approach, often referred to as the inhibition assay format, is based on the immobilization of the low molecular weight analytes to the chip surface, and the injection of the associated binding partner over the chip surface. Using this assay format, the level of binding to the surface is inversely proportional to the concentration of target analyte present in the sample (Figure 5.2).

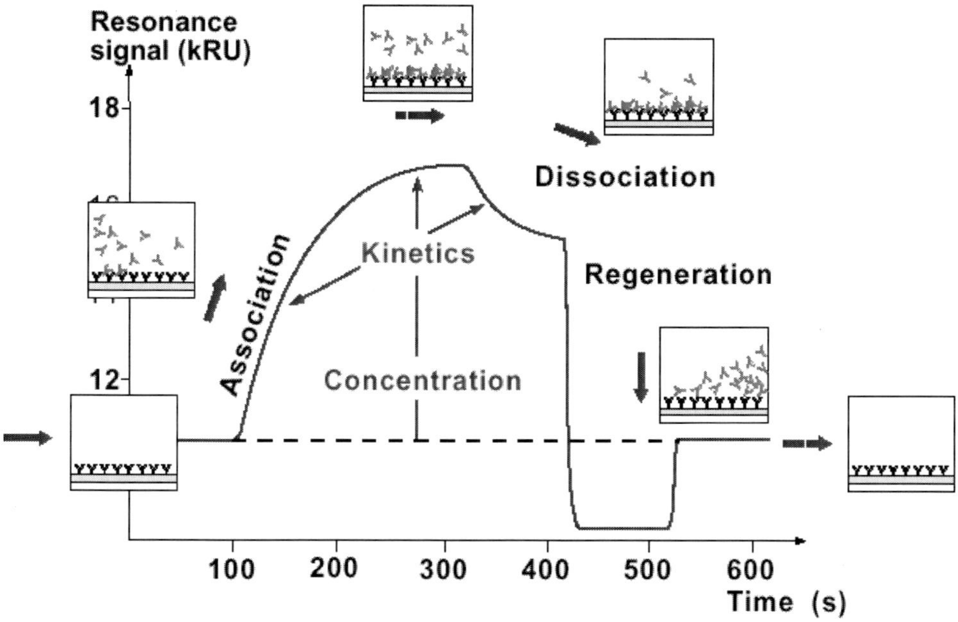

Figure 5.1. An SPR optical biosensor sensorgram of the molecular interaction between a specific binder and immobilized ligand (Abery 2001) (Courtesy of Biacore International AB).

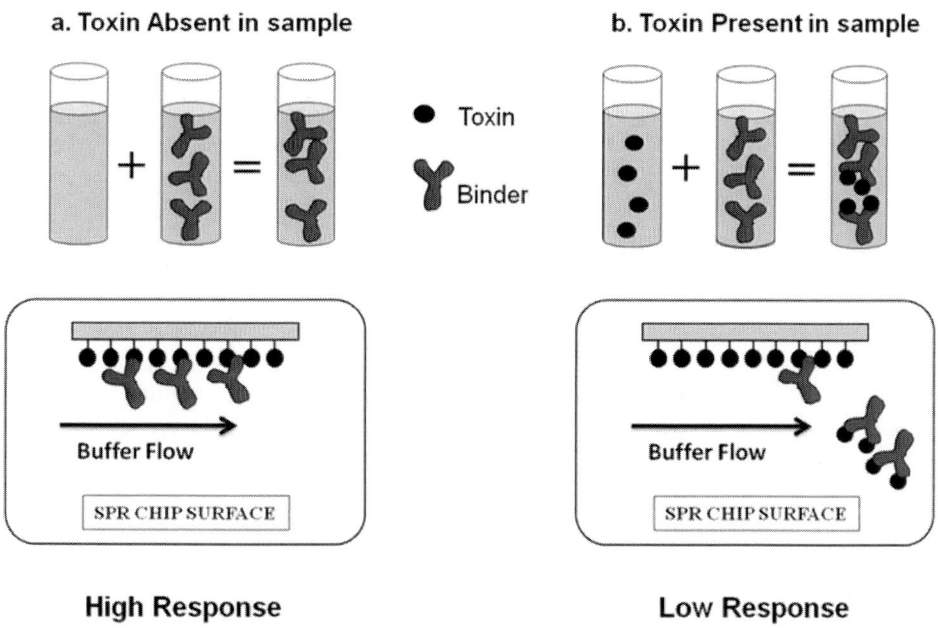

Figure 5.2. The inhibition assay format designed for SPR-based devices for measuring the concentrations of low molecular weight analytical targets.

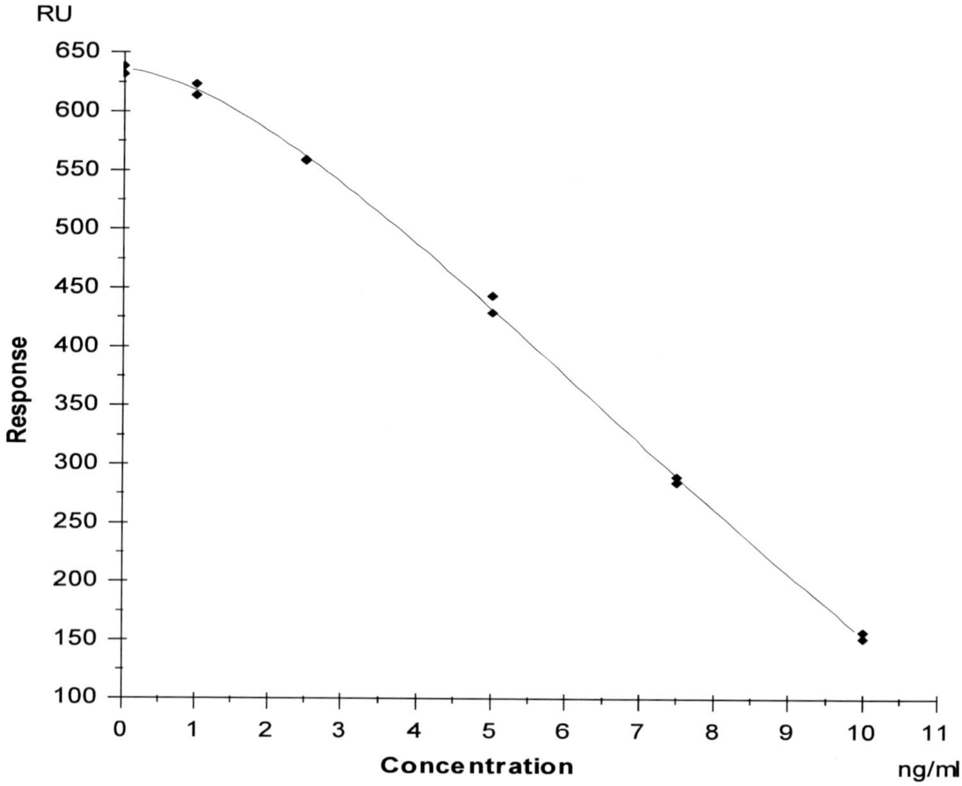

Figure 5.3. Typical inhibition assay calibration curve obtained for the measurement and quantification of toxins in complex biological samples.

A large number of toxin SPR assays have now been successfully developed using the inhibition approach. Traynor et al. (Traynor et al. 2006) developed an assay for domoic acid, Llamas et al. (Llamas et al. 2007) for okadaic acid and Fonfria et al. (Fonfria et al. 2007) for the class of Paralytic Shellfish Poisoning (PSP) toxins. In each case, the quantification of the amount of toxin present could be calculated by the construction of a calibration curve (Figure 5.3).

1.5.3. The Components of a High Performing SPR Assay

The phycotoxin assays mentioned using SPR are examples of very high quality detection and measurement tests, as has been proven by the validation data published to support the methods (Traynor et al. 2006, Stewart et al. 2009, Campbell et al. 2010). To achieve such reliable biosensor assays, an examination of the key attributes of the methods is revealing. The SPR system itself is a vital component of the method and it is clear that the Biacore instruments, and in particular the Biacore Q, dominate the literature in terms of delivering excellent performing methods for small molecular weight molecules and, in particular, drugs, vitamins and biotoxins. However, three key elements of the actual assay determine the way in which the methods perform.

- The binding protein (often an antibody) is essential in terms of delivering a specific molecular interaction with the target analyte or analytes. This is no different to any other form of immunoassay. The need to characterise different binders for sensitivity and specificity has been well described by Campbell et al. (Campbell et al. 2007) and Stewart et al. (Stewart et al. 2009). Yet, in terms of on-rates, the affinity constant of the binder plays a much more important role in SPR assays. Inhibition assays are non-equilibrium based procedures and rely on the ability of the antibody to bind to the immobilized target in a flow through system. Antibodies with fast on-rates tend to perform better than those with slower rates, as with such binders a faster flow rate can be used. This ability greatly hinders the interference caused by low affinity non-specific interactions between matrix components and the chip surface.

- The surface chemistry is a very important parameter in the development of an SPR assay. Small molecular weight components can be immobilized via the production of an analyte-protein conjugate and this, in turn, is immobilized onto the surface on a sensor chip via an amine or carboxyl coupling strategy. While this approach can be relatively straightforward, it has two main disadvantages. Firstly, during the regeneration of the chip surface, the immobilized conjugate can alter its three-dimensional structure and, thereby, bring about a different level of interaction with antibodies during subsequent interaction measurements. Secondly, the lifespan of an immobilized conjugate on a chip surface tends to be very short and, generally, only between 10 and 50 assay cycles are possible. As the sensor chips themselves tend to be the most expensive part of the assay, this negative attribute results in a high cost per test. A much better way to produce chip surfaces is through the direct immobilization of the analyte onto the sensor surface. Such surfaces tend to be much more robust and stable than the conjugate approach, with surfaces lasting for hundreds of analytical cycles. In some cases, however, the chemistry involved can be complex and requires a specialist organic knowledge.
- The sample preparation protocol, as with any (bio) analytical technique, is crucial for obtaining reliable data. SPR methods are renowned for being rapid and very user-friendly and, therefore, the type of method used for sample preparation has to be compatible with this analytical approach. In general, due to the advantages of SPR over conventional immunoassays, i.e. as they are non-equilibrium based and flow through based, in many cases the requirement for a highly purified sample extract can be avoided. The examples of the high quality phycotoxin assays outlined in this chapter are a testament to these attributes. In all cases, methods that are relatively simple to use have been developed, tested and validated in trials.

1.5.4. Conclusions

Biosensor-based assays for biotoxins, especially phycotoxins, have been shown to be extremely rapid and reliable methods for detection and quantification. The most widely used procedures employ optical-based technology, particularly SPR. The next challenges for biosensor methods are to develop and validate systems capable of remote analysis, and a recent publication has pointed towards the importance of this requirement (Doucette et al. 2009). Another challenge is the ability to simultaneously measure multiple biotoxins in a single bioanalytical system. Such an effort is currently being undertaken in a Framework 7 Project entitled CONFFIDENCE (www.Conffidence.eu). Within this project, the SPR methods developed for three different classes of phycotoxins (DSP, ASP, PSP) are being combined onto a single sensor chip using a multichannel biosensor developed as part of another European Framework project Biocop (www.Biocop.org) (Campbell et al. 2011).

1.6. Lateral Flow Immunochromatography

16.1. Introduction

An alternative antibody-based assay is lateral flow immunochromatography (LFI), more commonly referred to as the lateral flow test. This type of assay is an adaptation of the latex agglutination tests that were developed by Plotz and Singer in 1956 (Plotz & Singer 1956). The first commercial and most widely recognized application was for a home pregnancy test launched as Clearview by Unipath in 1988. This technology has since been used to develop an ever-expanding range of tests for clinical, veterinary, agricultural, food industry, bio-defence and environmental applications.

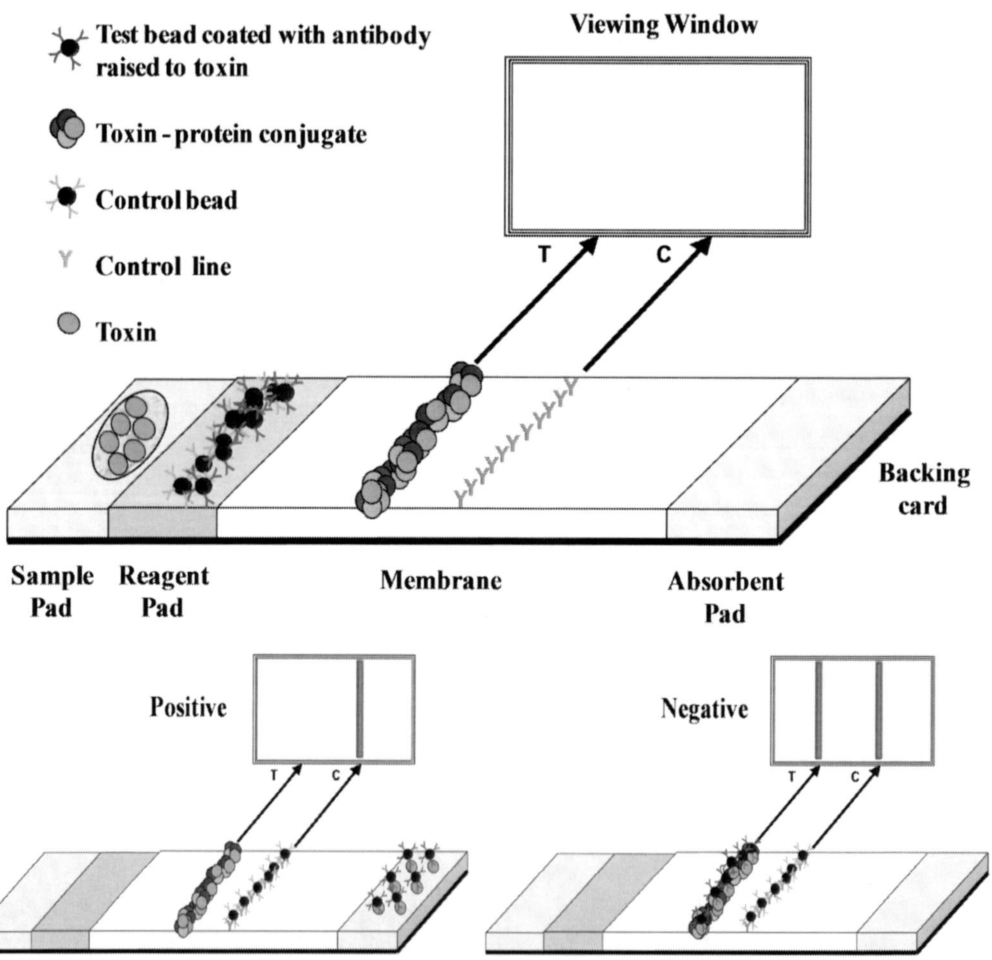

Figure 6.1. Diagrammatic view of competitive lateral flow immunoassay.

Lateral flow tests are designed as robust, user-friendly, low cost, portable immunoassays that offer a significant reduction in assay time compared to ELISA and other chemical detection methods. This immunoassay has been tailored to operate along a single axis or strip format to detect a specific substance in an aqueous sample. The key components of a typical test strip are the sample pad, reagent pad, membrane, absorbing or wicking pad and backing membrane (Figure 6.1). The sample pad is an absorbent pad onto which the test sample is applied. The reagent pad contains antibodies specific to the target analyte conjugated to coloured particles (usually colloidal gold particles, or latex microspheres). The membrane is typically a hydrophobic nitrocellulose or cellulose acetate membrane onto which anti-target analyte antibodies or protein conjugates are immobilized in a line across the membrane as a capture zone or test line (a control zone may also be present, containing antibodies specific for the conjugate antibodies). The absorbing or wicking pad is designed to draw the sample across the reaction membrane by capillary action and collect it.

These components of the strip are fixed to an inert backing material and may be presented in a simple dipstick format or within a plastic casing with a sample port and viewing window. This window displays the results in the form of a test and control line. The absence or presence of the test line indicates the result for qualitative analysis (yes/no). However, electronic readers can now be used to quantify results.

The individual assay layouts are adapted to the particular application or analyte being detected, but there are two principal formats: non-competitive (sandwich) and competitive. In the non-competitive (sandwich) assay, the sample migrates from the sample pad through the reagent pad where any target analyte present will bind to the labelled antibody. The sample then continues to migrate across the membrane until it reaches the test line, where the target/antibody complex will bind to immobilized antibodies producing a visible line on the membrane. Subsequently, the sample migrates further along the strip until it reaches the control line, where labelled antibody will bind and produce a second visible line on the membrane indicating that the strip is functioning correctly. A single line in the control zone is a negative result. Two lines on the membrane indicate a positive result and the more intense the test line the higher the concentration of the target analyte. Double antibody sandwich assays are most suitable for larger analytes, such as bacteria and viruses, with multiple antigenic sites.

As competitive assays are primarily used for testing small molecular weight molecules, they are more applicable for the analysis of marine biotoxins. The difference between competitive and non-competitive formats is that in the competitive format the test line contains a protein conjugated to the target analyte or its derivative (Figure 6.1). If the target analyte is not present in the sample, the labelled antibody in the reagent pad will migrate along the membrane and bind to the target analyte conjugated to the protein at the test line, and a line will be detected. If the target analyte is present in the sample, it will therefore bind to labelled antibody in the reagent pad, and this complex will migrate past the protein conjugate test line. The intensity of the test line is greatest in the complete absence of the target analyte and decreases with increasing concentration of this target. The tests are generally designed by manufacturers, so that two visible lines indicate a negative result and a single control line a positive result.

Competitive lateral flow assays have been developed for three key regulated toxins covered by European legislation: paralytic shellfish poisoning (PSP) toxins, domoic acid and okadaic acid. Maritime *in vitro* Shellfish Tests (MIST) for PSP and domoic acid were first developed in 2002 by Jellett Biotek Limited (JBL) in Nova Scotia, Canada. The PSP test demonstrated a detection limit of 0.4mg STXeq/kg in less than 20 minutes, whereas the domoic acid test is sensitive well below the regulatory guideline of 20-mg/kg shellfish tissue. Each one of the tests has been evaluated (MacKintosh et al. 2002, MacKintosh & Smith 2002, Inami et al. 2004, Oshiro et al. 2006). The general conclusion is that this approach has considerable potential for qualitative rapid screening of shellfish. An improved MIST test for PSP has been developed and reported in 2010 (Laycock et al. 2010), with a detection limit of 0.25mg STXeq/kg. In 2006, the first report of a new rapid test for diarrhetic shellfish poisoning toxins was published (Laycock et al. 2006). This test was able to detect okadaic acid and dinophysistoxin 1 and 2, with 50% reduction in test line colour intensity at 4.02 ng/mL for the solutions applied to test strips. These MIST Alert™ tests are qualitative (yes/no) tests that provide a visual determination of the presence of toxicity in shellfish samples in 10 to 20 minutes. They were designed as a screening method for regulatory laboratories to eliminate negative samples, thereby leaving a smaller number of positive samples to be tested with more sophisticated and time-consuming quantitative methods. MIST Alert™ tests have been demonstrated to be sufficiently simple for field use as a harvest management tool for determining a safe time to harvest, and as a quality control tool in processing plants to ensure that incoming shellfish are free of toxins. These tests have been evaluated for laboratory and field studies in shellfish (Jellett et al. 2002), phytoplankton (Rafuse et al. 2002) and bioterrorism defence (Komano et al. 2010) without the need for investment in laboratory equipment.

As antibody-based assays for the analysis of marine biotoxins, lateral flow tests offer similar successful attributes in terms of sensitivity and specificity to other types of immunoassay. However, they also demonstrate similar problems associated with cross-reactivity not correlating to the toxicity of varying analogues within a toxin group, relative to the mouse bioassay. The most important difference is that lateral flow tests measure concentration relative to the cross-reactivity of the antibody, whereas the mouse bioassay measures toxicity.

For this reason, Usleber et al. (Usleber et al. 2001) stated that, because of quantitative uncertainties, antibody-based methods are only suitable for screening. Yet, the design of multiplex lateral flow devices with multiple test lines and multichannel nano-microfluidic devices (Assadollahi et al. 2009) may provide new tools for overcoming cross-reactivity matching toxicity issues, using multiple antibodies with variable affinities to the family of toxins.

The key feature of this assay type is that, to date, it is the most suitable immunoassay for field or rapid on-site testing, due to its relatively low cost and portability. It can be readily incorporated into a test kit, because sample preparation is minimal and training and laboratory facilities are not required. After sample preparation, an answer can be obtained in a few minutes. These advantages must be weighed up against quantitative limitations characteristic of all antibody assays in cases where there is a family of related compounds, such as the PSP toxins, to which the antibodies bind with different affinities. In contrast to the situation for PSP toxins, for those shellfish toxins where cross-reactivities are similar, as for example those for ASP and DSP, the lateral flow format is relatively simple to apply. For lipophilic Diarrhetic Shellfish Poisoning (DSP) toxins, there is a problem in selecting suitable extraction solvents that will not interfere with the components of the strip.

The limit of sensitivity for all qualitative (yes/no) tests is calibrated and built-in by the manufacturer, and quality control measures determine confidence limits for the test and ensure reproducibility. A controversial issue that arose concerning the implementation of MIST tests as screening tools to replace or reduce the use of the mouse bioassay was the risk of depending on a single supplier for the test. In recent years, other companies are producing lateral flow tests for biotoxins, with Neogen Corporation releasing onto the market in 2010 their test for ASP for detecting domoic acid in shellfish and in 2011 their test for a range of DSP toxins. A PSP device is due for commercialisation in 2012.

Conclusion

Since the discovery of antibodies, they are being used in several immunological techniques with application in many areas, especially in the biomedical field. In recent years, antibodies are being incorporated in routine techniques in many marine laboratories and also new applications are positively coming up. This chapter shows the recent application of antibodies or antibody fragments in several immunological methods for the identification of algae and phycotoxins. However, antibodies can offer many other possibilities in the marine field, and more applications will sure appear in the near future. Immunological methods provide specificity, sensitivity and rapid identification, offering interesting alternatives to conventional methods.

References

Abery J (2001) Detecting the molecular ties that bind. *Modern Drug Discovery* 4:34-40.

Adachi M, Sako Y, Ishida Y (1993) The identification of conspecific dinoflagellates *Alexandrium tamarense* from Japan and Thailand by monoclonal antibodies. *Nipp Suis Gakk* 59:327-332.

Aguirre AA, Keefe TJ, Reif JS, Kashinsky L, Yochem PK, Saliki JT, Stott JL, Goldstein T, Dubey JP, Braun R, Antonelis G (2007) Infectious Disease Monitoring of the endangered Hawaiian Monk Seal. *J Wildl Dis* 43:229-241.

Amersdorfer P, Marks JD (2000) Phage libraries for generation of anti-botulinum scFv antibodies. *Methods Mol Biol* 145:219-240.

Amersdorfer P, Wong C, Smith T, Chen S, Deshpande S, Sheridan R, Marks JD (2002) Genetic and immunological comparison of anti-botulinum type A antibodies from immune and non-immune human phage libraries. *Vaccine* 20:1640-1648.

Ames RS, Tornetta MA, Jones CS, Tsui P (1994) Isolation of neutralizing anti-C5a monoclonal antibodies from a filamentous phage monovalent Fab display library [published erratum appears in J Immunol 1994 Jul 15;153(2):910] *J Immunol* 152:4572-4581.

Anderson DM, Kulis DM, Cosper EM (1990b) Immunofluorescent detection of the brown tide organism *Aureococcus anophagefferens*. In: Cosper EM, Bricelj, V.M., Carpenter, E.J. (ed) *Phytoplankton Blooms: Causes and Impacts of Recurrent Brown Tides and Other Unusual Blooms*. Springer-Verlag, Berlin, p 213-228.

Assadollahi S, Reininger C, Palkovits R, Pointl P, Schalkhammer T (2009) From Lateral Flow Devices to a Novel Nano-Color Microfluidic Assay. *Sensors* 9:6084-6100.

Barbas CF, 3rd, Kang AS, Lerner RA, Benkovic SJ (1991) Assembly of combinatorial antibody libraries on phage surfaces: the gene III site. *Proc Natl Acad Sci U S A* 88:7978-7982.

Bass HW, Sisco PH, Murray DL, Boston RS (1990) Probes for the b-32 protein hybridize to loci on 7L and 8L. *Maize Genet Coop Newslett* 64:97.

Bates M, Léger C, Keafer BA, Anderson DM (1993) Discrimination between domoic-acid-producing and non-toxic forms of the diatom *Pseudonitzchia pungens* using immunofluorescence. *Mar Ecol Progr Ser* 100:185-195.

Bei R, Schlom J, Kashmiri SVS (1995) Baculovirus expression of a functional single-chain immunoglobulin and its IL-2 fusion protein. *J Immunol Methods* 186:245-255.

Bernhard M, Lomi G, Riparbelli G, Saletti M, Zaterra A (1969) Un metodo immunologico per la caratterizzazzione del fitoplancton. *Publ Staz Zool Napoli* 37:64-72.

Bradbury AR, Marks JD (2004) Antibodies from phage antibody libraries. *J Immunol Methods* 290:29-49.

Briggs LR, Miles CO, Fitzgerald JM, Ross KM, Garthwaite I, Towers NR (2004) Enzyme-linked immunosorbent assay for the detection of yessotoxin and its analogues. *J Agric Food Chem* 52:5836-5842.

Buerk DG (1993) Biosensors: Theory and Applications, Vol. *Technomic Publishing Company,* Lancaster, Pennsylvania.

Buskey EJ, Hyatt CJ (2006) Use of the FlowCAM for semi-automated recognition and enumeration of red tide cells (Karenia brevis) in natural plankton samples. *Harmful Algae* 5:685-692.

Campbell K, Haughey SA, van den Top H, van Egmond H, Vilarino N, Botana LM, Elliott CT (2010) Single Laboratory Validation of a Surface Plasmon Resonance Biosensor Screening method for Paralytic Shellfish Poisoning Toxins. *Anal Chem* 82:2977-2988.

Campbell K, McGrath T, Sjölander S, Hanson T, Tidare M, Janssonb Ö, Moberg A, Mooney M, Elliott C, Buijs J (2011) Use of a novel micro-fluidic device to create arrays for multiplex analysis of large and small molecular weight compounds by surface plasmon resonance *Biosens Bioelectron* 26, 3029-3036

Campbell K, Stewart LD, Doucette GJ, Fodey TL, Haughey SA, Vilarino N, Kawatsu K, Elliott CT (2007) Assessment of specific binding proteins suitable for the detection of paralytic shellfish poisons using optical biosensor technology SO. *Anal Chem* 79:5906-5914.

Carey MF, Peterson CL, Smale ST (2010) *Purification of Epitope-Tagged Transcription Factor IID.* Cold Spring Harb Protoc:doi:10.1101/pdb.prot5450.

Carrera M, Garet E, Barreiro A, Garcés E, Pérez D, Guisande C, González-Fernández Á (2010) Generation of monoclonal antibodies for the specific immunodetection of the toxic dinoflagellate *Alexandrium minutum* Halim from Spanish waters. *Harmful Algae* 9:272-280.

Catt K, Niall HD, Tregear GW (1967) Solid phase radioimmunoassay. Nature 213:825-827.

Cho ES, Costas E (2004) Rapid monitoring for the potentially ichthyotoxic dinoflagellate *Cochlodinium polykrikoides* in Korean coastal waters using fluorescent probe tools. *J Plankton Res:*175-180.

Chu FS, Hsu KH, Huang X, Barrett R, Allison C (1996) Screening of paralytic shellfish poisoning toxins in naturally occurring samples with three different direct competitive enzyme-linked immunosorbent assays. *J Agric Food Chem* 44:4043-4047.

Clackson T, Hoogenboom HR, Griffiths AD, Winter G (1991) Making antibody fragments using phage display libraries. *Nature* 352:624-628.

Coons AH, Creech HJ, Jones R (1941) Immunological properties of an antibody containing a fluorescent group. *Proc Soc Exp Biol Med* 47:200-202.

Córdova JL, Jamett A, Lembeye GS (2002) Immunolocalization of theca antigens on *Alexandrium catenella* and their use to isolate cells from fixed phytoplankton samples. *Harmful Algae* 1:351-360.

Crameri A, Cwirla S, Stemmer WPC (1996) Construction and evolution of antibody-phage libraries by DMA shuffling. *Nat Med* 2:100-102.

Crosbie ND, Pöckl M, Weisse T (2003) Rapid establishment of clonal isolates of freshwater autotrophic picoplankton by single-cell and single-colony sorting. *J Microbiol Methods* 55:361-370.

Cunningham A, Buonnacorsi GA (1992) Narrow-angle forward light scattering from individual algal cells: implications for size and shape discrimination in flow cytometry. *J Plankton Res* 14:223-234.

de Haard HJ, van Neer N, Reurs A, Hufton SE, Roovers RC, Henderikx P, de Bruine AP, Arends JW, Hoogenboom HR (1999) A large non-immunized human Fab fragment phage library that permits rapid isolation and kinetic analysis of high affinity antibodies. *J Biol Chem* 274:18218-18230.

de Kruif J, Boel E, Logtenberg T (1995) Selection and Application of Human Single Chain Fv Antibody Fragments from a Semi-synthetic Phage Antibody Display Library with Designed CDR3 Regions. *Journal of Molecular Biology* 248:97-105.

De Leij L, Poppema S, The TH (1983) Cryopreservation of newly formed hybridomas. *J Immunol Methods* 62:69-72.

Desai S, Dworecki BR (2004) Coated microwell plate-based affinity purification of antigens. *Analytical Biochemistry* 328:162-165.

Devlin JJ, Panganiban LC, Devlin PE (1990) Random peptide libraries: A source of specific protein binding molecules. *Science* 249:404-406.

Doucette GJ, Mikulski CM, Jones K, King K, Greenfield DI, Marin R, Jensen S, Roman B, Elliott CT, Scholin CA (2009) Remote, subsurface detection of the algal toxin domoic acid onboard the Environmental Sample Processor: Assay development and field trials. *Harmful Algae* 8:880-888.

Dubelaar GB, Visser JW, Donze M (1987) Anomalous behaviour of forward and perpendicular light scattering of a cyanobacterium owing to intracellular gas vacuoles. *Cytometry* 8:405-412.

Dubois M, Demoulin L, Charlier C, Singh G, Godefroy SB, Campbell K, Elliot CT, Delahaut P (2010) Development of ELISAs for detecting domoic acid, okadaic acid, and saxitoxin and their applicability for the detection of marine toxins in samples collected in Belgium. *Food Additives and Contaminants* 27:859-868.

EC 1244/2007 Commission Regulation (EC) 1244/2007 of 24 October 2007 amending Regulation (EC) Nº 2074/2005 as regards implementing measures for certain products of animal origin intended for human consumption and laying down specific rules on official controls for the inspection of meat. In: Union OJE (ed). *Off. J. Eur. Union,* p 12-18.

EC 2074/2005 Commission Regulation (EC) Nº 2074/2005 of 5 December 2005 laying down implementing measures for certain products under Regulation (EC) 853/2004 of the European Parliament and of the Council and for the organization of official controls under Regulation (EC) 854/2004 of the European Parliament and of the Council and Regulation (EC) 882/2004 of the European Parliament and of the Council, derogating from Regulation (EC) 852/2004 of the European Parliament and of the Council and amending Regulations (EC) 853/2004 and (EC) 854/2004. In: Union OJE (ed). *Off. J. Eur. Union,* p 27-59.

Engvall E, Perlman P (1971) Enzyme-linked immunosorbent assay (ELISA). Quantitative assay of immunoglobulin G. *Immunochemistry* 8:871-874.

Fischer WJ, Garthwaite I, Miles CO, Ross KM, Aggen JB, Chamberlin AR, Towers NR, Dietrich DR (2001) Congener-independent immunoassay for microcystins and nodularins. *Environ Sci Technol* 35:4849-4856.

Fonfria ES, Vilarino N, Campbell K, Elliott C, Haughey SA, Ben-Gigirey B, Vieites JM, Kawatsu K, Botana LM (2007) Paralytic shellfish poisoning detection by surface Plasmon resonance-based Biosensors in shellfish matrixes. *Anal Chem* 79:6303-6311.

Fremy JM, Usleber E (2003) Policy on characterization of antibodies used in immunochemical methods of analysis for mycotoxins and phycotoxins. *J AOAC Int* 86:868-871.

Fuller SA, Takahashi M, Hurrell JGR (2001) *Freezing and Recovery of Hybridoma Cell Lines,* Vol. John Wiley & Sons, Inc.

Gao C, Mao S, Lo CH, Wirsching P, Lerner RA, Janda KD (1999) Making artificial antibodies: a format for phage display of combinatorial heterodimeric arrays. *Proc Natl Acad Sci* U S A 96:6025-6030.

Garet E, González-Fernández Á, Lago J, Vieites JM, Cabado AG (2010a) Comparative Evaluation of Enzyme-Linked Immunoassay and Reference Methods for the Detection of Shellfish Hydrophilic Toxins in Several Presentations of Seafood. *J Agric Food Chem* 58:1410-1415.

Garet E, Cabado AG, Vieites JM, González-Fernández Á (2010b) Rapid isolation of single-chain antibodies by phage display technology directed against one of the most potent marine toxins: Palytoxin. *Toxicon* 55:1519-1526.

Garthwaite I, Ross KM, Miles CO, Hansen RP, Foster D, Wilkins AL, Towers NR (1998) Polyclonal antibodies to domoic acid, and their use in immunoassays for domoic acid in sea water and shellfish. *Natural Toxins* 6:93-104.

Griffiths AD, Williams SC, Hartley O, Tomlinson IM, Waterhouse P, Crosby WL, Kontermann RE, Jones PT, Low NM, Allison TJ, et al. (1994) Isolation of high affinity human antibodies directly from large synthetic repertoires. *EMBO J* 13:3245-3260.

Guettinger Y, Barbin K, Peipp M, Bruenke J, Dechant M, Horner H, Thierschmidt D, Valerius T, Repp R, Fey GH, Stockmeyer B (2010) A Recombinant Bispecific Single-Chain Fragment Variable Specific for HLA Class II and FcαRI (CD89) Recruits Polymorphonuclear Neutrophils for Efficient Lysis of Malignant B Lymphoid Cells. *J Immunol* 184:1210-1217.

Hamers-Casterman C, Atarhouch T, Muyldermans S, Robinson G, Hammers C, Bajyana Songa E, Bendahman N, Hammers R (1993) Naturally occurring antibodies devoid of light chains. *Nature* 363:446-448.

Hawkins RE, Russell SJ, Winter G (1992) Selection of phage antibodies by binding affinity. Mimicking affinity maturation. *J Mol Biol* 226:889-896.

Hiroishi S, Uchida A, Nagasaki K, Ishida Y (1988) A new method for identification of inter- and intra-species of the red tide algae *Chattonella antiqua* and *Chattonella marina* (Raphidophyceae) by means of monoclonal antibodies. *Phycol* 24:442-444.

Homola J, Yee SS, Gauglitz G (1999) Surface plasmon resonance sensors: review. *Sens Actuators B* 54:3-15.

Hoogenboom HR (2002) Overview of antibody phage-display technology and its applications. *Methods Mol Biol* 178:1-37.

Hoogenboom HR, de Bruïne AP, Hufton SE, Hoet RM, Arends J-W, Roovers RC (1998) Antibody phage display technology and its applications. *Immunotechnology* 4:1-20.

Hoogenboom HR, Griffiths AD, Johnson KS, Chiswell DJ, Hudson P, Winter G (1991) Multi-subunit proteins on the surface of filamentous phage: methodologies for displaying antibody (Fab) heavy and light chains. *Nucleic Acids Res* 19:4133-4137.

Hsu E, Flajnik MF, Du Pasquier L (1985) A third immunoglobulin class in amphibians. *J Immunol* 135 1998-2004.

Hust M, Dubel S (2004) Mating antibody phage display with proteomics. *Trends in Biotechnology* 22:8-14.

Ide K, Takahashi K, Kuwata A, Nakamachi M, Saito H (2008) A rapid analysis of copepod feeding using FlowCAM. *J Plankton Res* 30:275-281.

Inami GB, Crandall C, Csuti D, Oshiro M, Brendens RA (2004) Feasibility of Reduction in Use of the Mouse Bioassay: Presence/Absence Screening for Saxitoxin in Frozen Acidified Mussel and Oyster Extracts from the Coast of California with *In Vitro* Methods *J AOAC Int* 87:1133-1142.

Jellett JF, Roberts RL, Laycock MV, Quilliam MA, Barrett RE (2002) Detection of paralytic shellfish poisoning (PSP) toxins in shellfish tissues using the MIST AlertTM, a new rapid test, in parallel with the regulatory AOAC bioassay. *Toxicon* 40:1407-1425.

Kleivdal H, Kristiansen SI, Nilsen MV, Briggs L (2007a) Single-laboratory validation of the biosense direct competitive enzyme-linked immunosorbent assay (ELISA) for determination of domoic acid toxins in shellfish. *J AOAC Int* 90:1000-1010.

Kleivdal H, Kristiansen SI, Nilsen MV, Goksoyr A, Briggs L, Holland P, McNabb P (2007b) Determination of domoic acid toxins in shellfish by Biosense ASP ELISA - A direct competitive enzyme-linked immunosorbent assay: Collaborative study. *J AOAC Int* 90:1011-1027.

Knappik A, Ge L, Honegger A, Pack P, Fischer M, Wellnhofer G, Hoess A, Wölle J, Plückthun A, Virnekäs B (2000) Fully synthetic human combinatorial antibody libraries (HuCAL) based on modular consensus frameworks and CDRs randomized with trinucleotides. *J Mol Biol* 296:57-86.

Komano A, Maruko H, Sekiguchi H, Seto Y (2010) Detection of saxitoxin in counterterrorism using a commercial lateral flow immunoassay kit. *Forensic Toxicol* doiI:10.1007/s11419-010-0102-1.

Kristensen P, Winter G (1998) Proteolytic selection for protein folding using filamentous bacteriophages. *Folding and Design* 3:321-328.

Laycock MV, Donovan MA, Easy DJ (2010) Sensitivity of lateral flow tests to mixtures of saxitoxins and applications to shellfish and phytoplankton monitoring. *Toxicon* 55:597-605.

Laycock MV, Jellett JF, Easy DJ, Donovan MA (2006) First report of a rapid assay for diarrhetic shellfish poisoning toxins. *Harmful Algae* 5:74-78.

Lefebvre KA, Noren DP, Schultz IR, Bogard SM, Wilson J, Eberhart BT (2007) Uptake, tissue distribution and excretion of domoic acid after oral exposure in coho salmon (*Oncorhynchus kisutch*). *Aquat Toxicol* 81:266-274.

Leslie GA, Clem LW (1969) Phylogen of immunoglobulin structure and function. 3. Immunoglobulins of the chicken. *J Exp Med* 130:1337-1352.

Lin S, Feinstein TN, Zhang H, Carpenter EJ (2003) Development of an immunofluorescence technique for detecting *Pfiesteria piscicida*. *Harmful Algae* 2:223-231.

Lopez-Rodas V, Costas E (1999) Immunochemical characterization of morphospecies and strains of *Prorocentrum* (dinophyceae). *J Exp Mar Biol Ecol* 238:293-308.

Low NM, Holliger P, Winter G (1996) Mimicking somatic hypermuation: affinity maturation of antibodies displayed on bacteriophage using a bacterial mutator strain. *J Mol Biol* 260:359-368

Luong JHT, Male KB, Glennon JD (2008) *Biosensor technology: Technology push versus market pull Biotech Advances* 26:492-500

Llamas N, Stewart L, Fodey T, Higgins HC, Velasco M, Botana L, Elliott CT (2007) Development of a novel immunobiosensor method for the rapid detection of okadaic acid contamination in shellfish extracts. *Anal Bioanal* 389:581-585.

MacKintosh FH, Gallagher S, Shanks A, Smith EA (2002) Assessment of MIST Alert, a commercial qualitative assay for detection of paralytic shellfish poisoning toxins in bivalve molluscs. *Food Chemical Contaminants* 85:632-641.

MacKintosh FH, Smith EA (2002) Evaluation of Mist Alert Rapid Test Kits for the Detection of Paralytic and Amnesic Shellfish Poisoning Toxins in Shellfish. *J Shellfish Research* 21:455-460.

Majeti R (2010) Monoclonal antibody therapy directed against human acute myeloid leukemia stem cells. *Oncogene* 30, 1009-1019.

Marks JD, Griffiths AD, Malmqvist M, Clackson TP, Bye JM, Winter G (1992) By-Passing Immunization: Building High Affinity Human Antibodies by Chain Shuffling. *Nat Biotech* 10:779-783.

Marks JD, Hoogenboom HR, Bonnert TP, McCafferty J, Griffiths AD, Winter G (1991) By-passing immunization. Human antibodies from V-gene libraries displayed on phage. *J Mol Biol* 222:581-597.

Matsumoto-Takasaki A, Horie J, Sakai K, Furui Y, Sato R, Kawakami H, Toma K, Takayanagi A, Fujita-Yamaguchi Y (2009) Isolation and characterization of anti-T-antigen single chain antibodies from a phage library. *BioScience Trends* 3:87-95.

Maucher JM, Ramsdell JS (2005) Ultrasensitive detection of domoic acid in mouse blood by competitive ELISA using blood collection cards. *Toxicon* 45:607-613.

McCafferty J, Griffiths AD, Winter G, Chiswell DJ (1990) Phage antibodies: filamentous phage displaying antibody variable domains. *Nature* 348:552-554.

McElhiney J, Lawton LA, Porter AJR (2000) Detection and quantification of microcystins (cyanobacterial hepatotoxins) with recombinant antibody fragments isolated from a naïve human phage display library. *FEMS Microbiology Letters* 193:83-88.

Mendoza H, López-Rodas V, González-Gil S, Aguilera A, Costas E (1995) The use of polyclonal antisera and blocking of antibodies in the identification of marine dinoflagellates: species-specific and clone-specific antisera against *Gymnodinium* and *Alexandrium*. *J Exp Mar Biol Ecol* 186:103-115.

Nagumo Y, Oguri H, Tsumoto K, Shindo Y, Hirama M, Tsumuraya T, Fujii I, Tomioka Y, Mizugaki M, Kumagai I (2004) Phage-display selection of antibodies to the left end of CTX3C using synthetic fragments. *J Immunol Methods* 289:137-146.

Nissim A, Hoogenboom HR, Tomlinson IM, Flynn G, Midgley C, Lane D, Winter G (1994) Antibody fragments from a 'single pot' phage display library as immunochemical reagents. *EMBO J* 13:692-698.

Orum H, Andersen PS, Oster A, Johansen LK, Riise E, Bjornvad M, Svendsen I, Engberg J (1993) Efficient method for constructing comprehensive murine Fab antibody libraries displayed on phage. *Nucleic Acids Research* 21:4491-4498.

Oshiro M, Pham L, Csuti D, Dodd M, Inami GB, Brendens RA (2006) Paralytic shellfish poisoning surveillance in California using the Jellett Rapid PSP test. *Harmful Algae* 5:69-73.

Plotz CM, Singer JM (1956) The latex fixation test. I. Application to the serologic diagnosis of rheumatoid arthritis. *Am J Med* 21:888-892.

Pyo D, Hahn JH (2009) Determination of Trace Amount of Cyanobacterial Toxin in Water by Microchip Based Enzyme-Linked Immunosorbent Assay. *J Immunoassay Immunochem* 30:97-105.

Rafuse C, Cembella AD, Laycock MV, Jellett JF (2002) Rapid monitoring of toxic phytoplankton and zooplankton with a lateral-flow immunochromatographic assay for ASP and PSP toxins. In: Steidinger KAL, J.H.; Tomas, C.R.; Vargo, G.A. (ed) *Florida Fish and Wildlife Conservation. Intergovernmental Oceanographic Commission of UNESCO*, St. Petersburg, Florida, USA, p 309-311.

Reboreda A, Lago J, Chapela MJ, Vieites JM, Botana LM, Alfonso A, Cabado AG (2010) Decrease of marine toxin content in bivalves by industrial processes. *Toxicon* 55:235-243.

Reckermann M (2000) Flow sorting in aquatic ecology. *Sci Mar* 64:235-246.

Ridder R, Schmitz R, Legay F, Gram H (1995) Generation of Rabbit Monoclonal Antibody Fragments from a Combinatorial Phage Display Library and Their Production in the Yeast *Pichia pastoris*. *Nat Biotech* 13:255-260.

Russell SJ, Hawkins RE, Winter G (1993) Retroviral vectors displaying functional antibody fragments. *Nucleic Acids Research* 21:1081-1085.

Sako Y, Adachi M, Ishida Y (1993) Preparation and characterization of monoclonal antibodies to *Alexandrium* species. In: Smayda TY, Shimizu, Y. (ed) *Toxic Phytoplankton Blooms in the Sea Elsevier Publishers* B.V., Amsterdam, p 87-93

Sblattero D, Bradbury A (2000) Exploiting recombination in single bacteria to make large phage antibody libraries. *Nat Biotechnol* 18:75-80.

Schaerlinger B, Bascove M, Frippiat J-P (2008) A new isotype of immunoglobulin heavy chain in the urodele amphibian Pleurodeles waltl predominantly expressed in larvae. *Mol Immunol* 45:776-786.

See JH, Campbell L, Richardson TL, Pinckney JL, Shen R, Guinasso NL (2005) Combing new technologies for determination of phytoplankton community structure in the northern gulf of Mexico 1. *J Phycol* 41:305-310.

Sensen CW, Heimann K, Melkonian M (1993) The production of clonal and axenic cultures of microalgae using fluorescence-activated cell sorting. *Eur J Phycol* 28:93-97.

Shaw I, O'Reilly A, Charleton M, Kane M (2008) Development of a High-Affinity Anti-Domoic Acid Sheep scFv and its Use in Detection of the Toxin in Shellfish. *Anal Chem* 80:3205-3212.

Sheets MD, Amersdorfer P, Finnern R, al. e (1998) Efficient construction of a large nonimmune phage antibody library: the production of high-affinity human single-chain antibodies to protein antigens. *Proc Natl Acad Sci USA* 95:6157-6162.

Sieracki CK, Sieracki ME, Yentsch CS (1998) An imaging-in-flow system for automated analysis of marine microplankton. *Mar Ecol Prog Ser* 168:285-296.

Skerra A, Pluckthun A (1988) Assembly of a functional immunoglobulin Fv fragment in *Escherichia coli*. *Science* 240:1038-1041.

Smith GP (1985) Filamentous fusion phage: novel expression vectors that display cloned antigens on the virion surface. *Science* 228:1315-1317.

Smith GP (1991) Surface presentation of protein epitopes using bacteriophage expression systems. *Curr Opin Biotechnol* 2:668-673.

Southwell AL, Patterson PH (2010) Antibody therapy in neurodegenerative disease. *Rev Neurosci* 21:273-287.

Spier CR, Bromage ES, Harris TM, Unger MA, Kaattari SL (2009) The development and evaluation of monoclonal antibodies for the detection of polycyclic aromatic hydrocarbons. *Anal Biochem* 387:287-293.

Stauffer BA, Schaffner RA, Wazniak C, Caron DA (2008) Immunofluorescence Flow Cytometry Technique for Enumeration of the Brown-Tide Alga, *Aureococcus anophagefferens*. *Appl Environ Microbiol* 74:6931-6940.

Stewart LD, Elliott CT, Walker AD, Curran RM, Connolly L (2009) Development of a monoclonal antibody binding okadaic acid and dinophysistoxins-1,-2 in proportion to their toxicity equivalence factors. *Toxicon* 54:491-498.

Thompson M, Ellison SLR, Wood R (2002) Harmonized guidelines for single-laboratory validation of methods of analysis (IUPAC Technical Report). *Pure Appl Chem* 74:835-855.

Traynor I, Plumpton L, Fodey TL, Higgins C, Elliott CT (2006) Immunobiosensor detection of domoic acid as a screening test in bivalve molluscs: comparison with LC based analysis. *JAOAC* 89:868-872.

Uchida A, Nagasaki K, Hirioshi S, Ishida Y (1989) The application of monoclonal antibodies to and identification of *Chattonella marina* and *Chattonella antiqua*. *Nippon Suisan Gakk* 55:721-725.

Usleber E, Dietrich R, Burk C, Schneider E, Märtlebauer E (2001) Immunoassay methods for paralytic shellfish poisoning toxins. *J AOAC Int* 84:1649-1656.

van Weemen BK, Schuurs AH (1971) Immunoassay using antigen-enzyme conjugates. *FEBS Letters* 15:232-236.

Vaughan TJ, Williams AJ, Pritchard K, Osbourn JK, Pope AR, Earnshaw JC, McCafferty J, Hodits RA, Wilton J, Johnson KS (1996) Human antibodies with sub-nanomolar affinities isolated from a large non-immunized phage display library. *Nat Biotechnol* 14:309-314.

Vrieling EG, Anderson DM (1996) Immunofluorescence in phytoplankton research: applications and potential. *J Phycol* 32:1-16.

Vrieling EG, Draaijer A, van Zeijl WJM, Peperzak L, Gieskes WWC, Veenhuis M (1993) The effect of labeling intensity, estimate by real-time confocal laser scanning microscopy, on flow cytometric appearance and identification of immunochemically labeled marine dinoflagellates. *J Phycol* 29:180-188.

Wagner G, Guilbault GG (1994) *Food Biosensor Analysis,* Vol 60. Marcel Dekker, Inc., New York.

Weetall HH (1999) Chemical sensors and biosensors, update, what, where, when and how. *Biosens Bioelectron* 14:237-242.

Whitelam GC, Cockburn W, Owen MR (1994) Antibody production in transgenic plants. *Biochem Soc Trans* 22:940-944.

Wolf P, Alt K, Wetterauer D, Bühler P, Gierschner D, Katzenwadel A, Wetterauer U, Elsässer-Beile U (2010) Preclinical evaluation of a recombinant anti-prostate specific membrane antigen single-chain immunotoxin against prostate cancer. *J Immunother* 33:262-271.

Xu X, Sui J, Cao L, Lin H (2010) Direct competitive enzyme-linked immunosorbent assay (ELISA) for rapid screening of anisakid larvae in seafood. *J Sci Food Agric* 90:877-881

Zhao SW, Shen PP, Zhou Y, Wei Y, Xin XB, Hua ZC (2005) Selecting peptide ligands of microcystin-LR from phage displayed random libraries. *Environment International* 31:535-541.

Zhou Y, Pan F-G, Li Y-S, Zhang Y-Y, Zhang J-H, Lu S-Y, Ren H-L, Liu Z-S (2009) Colloidal gold probe-based immunochromatographic assay for the rapid detection of brevetoxins in fishery product samples. *Biosensors and Bioelectronics* 24:2744-2747.

Zubkov MV, Burkill PH (2006) Syringe pumped high speed flow cytometry of oceanic phytoplankton. *Cytometry Part* A 69A:1010-1019.

In: New Trends in Marine and Freshwater Toxins
Editors: A. G. Cabado and J. M. Vieites
ISBN: 978-1-62257-787-3

Chapter 10

SHELLFISH TOXIN MONITORING SYSTEM IN JAPAN AND SOME ASIAN COUNTRIES

Toshiyuki Suzuki and Ryuichi Watanabe
National Research Institute of Fisheries Science, Fukuura, Kanazawa, Yokohama, Kanagawa, Japan

ABSTRACT

Diarrhetic shellfish poisoning (DSP) and paralytic shellfish poisoning (PSP) cause serious quality assurance problems for bivalve industries in Japan. The toxicities of cultured bivalves are periodically monitored by mouse bioassay (MBA) at selected monitoring stations. When the toxicity of the bivalves exceeds the quarantine levels (0.05 MU/g wet weight for DSP, i.e. 0.2 μg okadaic acid equivalents/g shellfish flesh), 4.0 MU/g wet weight for PSP, i.e. 80 μg saxitoxin equivalents/100g shellfish flesh), harvesting ceases. Shellfish harvesting is not resumed until testing indicates that the toxicity of the bivalves are below the quarantine levels on three successive weeks. Since the establishment of the monitoring system in 1979, no human poisonings due to contaminated bivalves distributed on commercial markets have been reported. In addition to the direct monitoring of bivalve toxicity, the occurrence and abundance of the toxic dinoflagellates which cause DSP (*Dinophysis* spp.) and PSP (*Alexandrium* spp., *Gymnodinium catenatum*) are also monitored. This provides an early warning and risk assessment of the chances of toxicity developing in the bivalves. Several analytical methods have been developed as the alternative to MBA in Japan. DSP and PSP toxin analytical calibration standards are also produced in Japan by funding of Japanese government. These calibration standards are distributed to laboratories participating in shellfish toxin monitoring in Japan to promote the use of high-performance liquid chromatography (HPLC) and liquid chromatography mass spectrometry (LC-MS) for monitoring and research.

INTRODUCTION

Scallops, *Patinopecten yessoensis*, and oysters, *Crassostrea gigas*, are respectively ranked the second and third most important aquaculture species in Japan. The total production of scallops and oysters in Japan in 2008 was 226,000 and 190,000 tons, respectively (Ministry of Agriculture, Forestry and Fisheries in Japan, 2008). Scallops are mainly produced in the northern part of Japan whereas oysters are produced mainly in Hiroshima Prefecture in the western part of Japan. Another important bivalve species is the Japanese short-neck clam, *Tapes philippinarum*, which is mainly produced in central regions.

Diarrhetic shellfish poisoning (DSP) and paralytic shellfish poisoning (PSP) cause serious quality assurance problems for bivalve industries in Japan (Yasumoto et al., 1978; Oshima et al., 1982), however no human poisonings due to contaminated bivalves with DSP or PSP toxins distributed on commercial markets have been reported since the establishment of the monitoring system for shellfish toxins in 1979. This indicates that the monitoring system is successful. In this chapter, the present Japanese shellfish toxin monitoring system and the brief situation in some of Asian countries are described.

In addition to DSP and PSP, other marine biotoxins in Japan are Ciguatera fish poisoning (CFP) and Puffer fish poisoning caused by tetrodotoxin (TTX). Selling and consumption of the ciguatera fish are not allowed in Japan. Puffer fish is also banned to sell and consume in Japan, however some parts of some puffer fish (Table 1) are allowed to eat if the puffer fish is cooked by the licensed chef. Because of no monitoring system for CFP and TTX in Japan, these issues are not described in this chapter.

Table 1. Edible puffer fish and the edible parts controlled by Ministry of Health, Labour and Welfare in Japan

		Parts		
English name	Species	Muscles	Skins	Testes
Grass puffer	*Takifugu niphobles*	○	-	-
Finepatterned puffer	*Takifugu poecilonotus*	○	-	-
Panther puffer	*Takifugu pardalis*	○	-	-
Vermiculated puffer	*Takifugu snyderi*	○	-	○
Genuine puffer or Purple puffer	*Takifugu porphyreus*	○	-	○
*1	*Takifugu obscurus*	○	-	○
Red-eyed puffer	*Takifugu chrysops*	○	-	○
Ocella puffer or Tiger puffer	*Takifugu rubripes*	○	○	○
Eyespot puffer	*Takifugu chinensis*	○	○	○
Striped puffer or Yellowfin puffer	*Takifugu xanthopterum*	○	○	○
Spottyback puffer	*Takifugu stictonotus*	○	-	○
Smooth-backed blowfish	*Lagocephalus inermis*	○	○	○
Brown-backed toadfish	*Laocephalus wheeleri*	○	○	○
Brown-backed toadfish	*Laocephalus gloveri*	○	○	○
Slackskinned puffer	*Sphoeroides pachygaster*	○	○	○
*1	*Takifugu flavidus*	○	-	-

Notes: ○: edible. -: not edible. *1: Corresponding English name is not found.

CONCLUSION

Monitoring System for Shellfish Toxins

For the prevention of shellfish poisoning and to maintain the safe supply of major shellfish products, the guidelines on monitoring and inspection of shellfish toxins were reinforced in 1979. Since then the monitoring system for shellfish toxins has been revised several times. The latest version of the monitoring system is basically carried out as follows.

Monitoring stations are established in the areas where shellfish are harvested. The amount of toxins contained in bivalves are monitored by the official mouse bioassay method, at least monthly in the harvest season, and weekly when the probability of toxicity appearing in bivalves is expected to be high.

Table 2 shows the mouse bioassay method for DSP and PSP in Japan. In the case of DSP, the frequency of toxicity testing of bivalves is increased when the toxicity in the midgut gland of scallops, *P. yessoensis*, and the other bivalves, such as *Mytilus galloprovincialis* and *Chlamys farreri akazara*, exceeds 0.5 and 0.3 MU/g, respectively. When the toxicity of bivalves exceeds the quarantine levels (0.05 MU/g for wet weight of whole meat), harvesting ceases. In the case of scallops, products are allowed to be distributed to commercial markets after removal of the midgut gland at authorized factories if the toxicity of whole meat removed the midgut gland is below the quarantine levels. In the case of PSP, the toxicity testing of bivalves is increased in frequency when the toxicity in the midgut gland exceeds 20 MU/g which corresponds to approximately 2 MU/g whole meat. When the PSP toxin level in bivalves exceeds the quarantine level (4 MU/g for wet weight of whole meat), harvesting ceases. Shellfish harvesting is not resumed until testing indicates that the toxicity of the bivalves is below the quarantine levels on three successive weeks. In addition to DSP and PSP, amnesic shellfish poisoning (ASP) in scallops *P. yessoensis* that are exported from Japan to member countries of European Union is monitored by the HPLC-UV method (Quilliam et al., 1995). As far as the authors know, there are no reports that the scallops exported to European Union are contaminated with domoic acid (DA). Because of no historical evidence of contamination of bivalves with DA, monitoring of ASP in bivalves for domestic consumption in Japan has not been established.

Table 2. The official testing methods of shellfish toxins in Japan

Type of poisonings	Toxins	Methods	End point of the method	Closure levels (MU/g)
DSP[*1]	Okadaic acid analogues	i.p. Mouse bioassay	Death time	0.05 MU/g for whole meat[*2]
PSP	Saxitoxin analogues	i.p. Mouse bioassay	Death time	4 MU/g for whole meat[*3]

[*1] Pectenotoxins and Yessotoxins are detectable as false positive compounds.
[*2] MU = Quantity of toxins which can kill a 16-20 g ddY or ICR strain male mouse in 24 hours.
[*3] MU = Quantity of toxins which can kill a 19-21 g ddY strain male mouse in 15 minutes.

The results of shellfish testing are immediately reported to the authorities of each prefectural government as shown in Figure 1. The information that toxicities of shellfish exceed the quarantine levels is distributed quickly through the network and the Association of Fishery Cooperatives orders the unit cooperatives to cease harvesting. The marine product department also reports the situation to the Food Safety and Consumer Affairs Bureau in Ministry of Agriculture, forestry and fisheries and the other prefectures, as well as to the wholesale markets and mass media. The Food Safety and Consumer Affairs Bureau transmits the information to the Ministry of Health, Labour and Welfare and the General Food Policy Bureau to reinforce necessary surveillance in the markets.

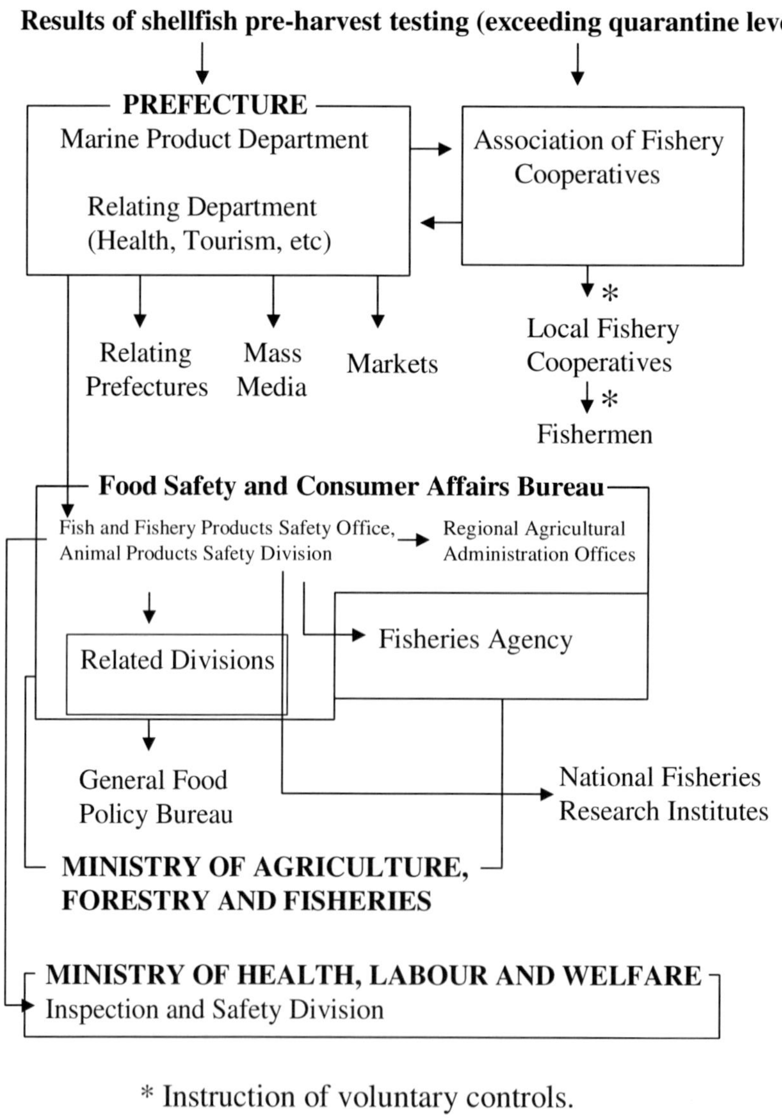

Figure 1. Flow of information of shellfish toxins in Japan. * instruction of voluntary controls.

Figure 2 shows the areas where toxicity levels of bivalve exceeded the quarantine levels in 2009. This is a typical map of the distribution of bivalves contaminated with DSP and PSP toxins in Japan. Contamination of bivalves with DSP toxins exceeding the quarantine level has been observed in the northern coastal areas of Japan, Hokkaido and the Tohoku district. There are few recorded DSP events in the western areas of central Japan. Ascidian, *Ascidiacea Nielsen*, is consumed in Japan and Korea. Although human intoxication cases by consumption of the ascidian has not been reported, positive results in the DSP testing by the mouse bioassay has recently confirmed. In contrast to DSP, PSP toxins exceeding the quarantine level have been detected in bivalves throughout the country.

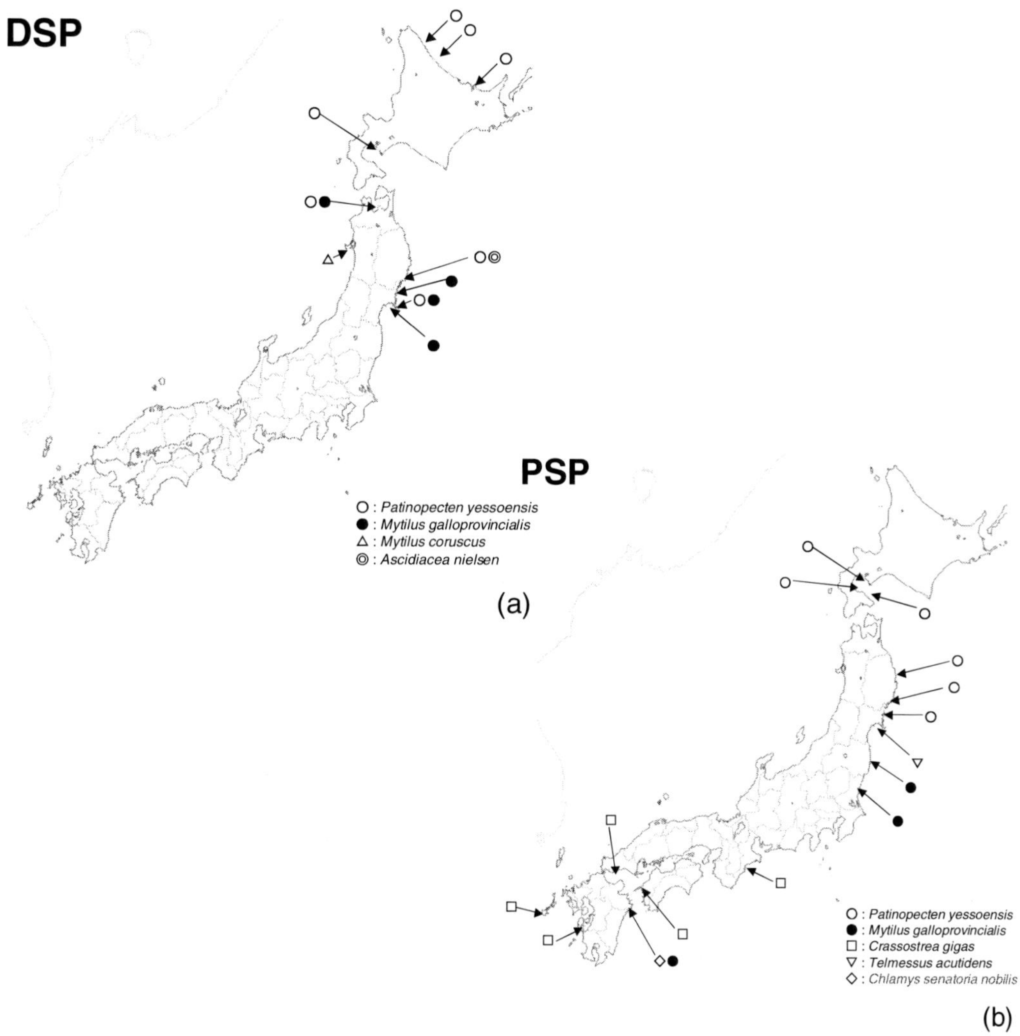

Figure 2. Locations of contamination of shellfish with DSP (a) and PSP (b) in Japan in 2009. Only those exceeding the quarantine limits are shown with the affected shellfish species.

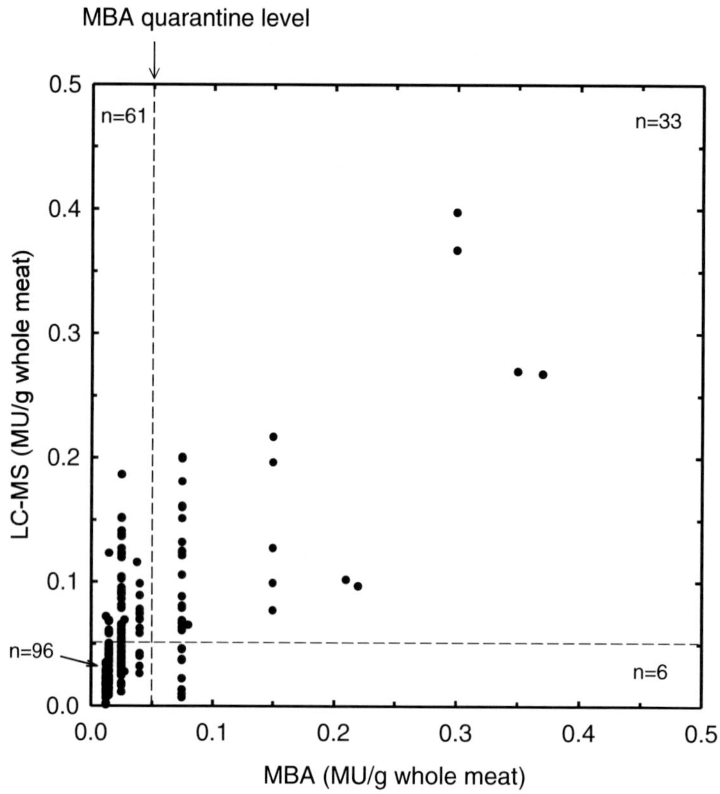

Figure 3. Comparison of total mouse toxicities obtained by LC-MS and MBA for the same scallop and mussel samples (n=196). n; sample numbers in the square.

In addition to the official monitoring system for shellfish toxins, fisheries experimental stations of each prefecture regularly monitor the toxic plankton in their areas. The main targets of the investigations are *Dinophysis fortii, D. acuminata* and some of other *Dinophysis* species as the causes of DSP and *Alexandrium tamarense, A. catenella, A. tamiyavanichii* and *Gymnodinium catenatum* as the causes of PSP. This provides an early warning and risk assessment of the chances of toxicity developing in the bivalves.

In conclusion, no human poisonings due to contaminated bivalves with DSP or PSP toxins distributed on commercial markets have been reported since the establishment of the monitoring system for shellfish toxins in 1979. This indicates that the monitoring system is successful.

Development of Alternative Testing Method to Mouse Bioassay in Japan

Although the mouse bioassay is a practical monitoring method for the detection of toxins, it has the following drawbacks: 1) it provides little information on toxin composition, 2) it gives variable results depending on mouse strains, 3) it requires the maintenance of a large stock of mice, and 4) it poses an ethical problem of killing a large number of mice.

Fluorometric high-performance liquid chromatography (FL-HPLC) and liquid chromatography mass spectrometry (LC-MS) methods provide highly sensitive, specific and reproducible alternatives to the use of animal assays (Oshima, 1995; Suzuki et al., 2005). Protein phosphatase 2A (PP2A) inhibition assay kit for okadaic acid analogues (DSP rapid kit) was recently developed and this kit is commercially available from Tropical Technology Center LTD in Japan.

To demonstrate the applicability of FL-HPLC and LC-MS, necessary data have been accumulated. Figure 3 shows comparison of the mouse toxicities quantified by LC-MS and MBA for the 196 scallop and mussel samples in DSP testing (Suzuki et al., 2005). Because the MBA can detect other lipophilic toxins, pectenotoxins (PTXs) and yessotoxins (YTXs), total mouse toxicities of DSP and other lipophilic toxins quantified by LC-MS were compared to the MBA results. The conversion factors to the mouse toxicities of okadaic acid (OA), dinophysistoxin-1 (DTX1), dinophysistoxin-3 (DTX3) (7-O-palDTX1), pectenotoxin-1 (PTX1), pectenotoxin-2 (PTX2), pectenotoxin-6 (PTX6), yessotoxin (YTX), and 45-hydroxy-yesstoxin (45-OHYTX) are 4.0, 3.2, 5.0, 5.0 4.6, 10, 2.0, and 10 μg/MU, respectively (Yasumoto et al., 1995). Mouse toxicities obtained by MBA are shown between minimum and maximum toxicities according to the dilution factors of bivalve extracts, eg. 0.05 – 0.1 MU/g. Average values obtained by the MBA were used for the comparison to the LC-MS results. Ninety-six samples were quantified by both LC-MS and MBA as being below the quarantine level. Thirty-three samples were above the quarantine level by both LC-MS and MBA. Six samples tested as exceeding the quarantine level by MBA were quantified by LC-MS as being below the quarantine level. As the toxicities of these six samples were between 0.05 and 0.1 MU/g by MBA, which is the lowest level measurable by the MBA, false positives due to free fatty acids in the samples are suspected. Sixty-one samples as exceeding the quarantine level by LC-MS were quantified by MBA as being below the quarantine level. The Japanese MBA protocol includes a partition of bivalve extracts between water and diethyl ether to remove water soluble saxitoxin analogues. Because a major portion of the YTXs are partitioned into the aqueous fraction, YTXs are therefore not quantitatively measured by MBA. The discrepancies are probably caused by the YTXs that are poorly detected by MBA. It is noteworthy that numbers of samples exceeding the quarantine level increase if MBA is replaced to LC-MS in routine monitoring of toxins in Japanese bivalves. This impact will be huge to scallop industries because a dominant toxin in scallops is YTXs. The risk assessment of YTX to mammal is thought to be lower than the other toxin groups, and the European Union (EU) recently set a higher control limit for the YTXs than OA analogues (Regulation EC No. 853, 2004). When the LC-MS is applied to the toxin monitoring in Japanese bivalves, an adequate control limit for the YTXs as proposed by the EU should be considered to decrease economic impacts to scallop industries.

The limited availability of toxin standards to calibrate these analyses has been an obstacle in promoting the use of HPLC and LC-MS for monitoring and has hindered research. To solve this problem and to establish a reliable monitoring system, the Japanese Fisheries Agency started a project to produce and distribute toxin calibrants to laboratories participating in the shellfish toxin monitoring program in 1995 (Goto et al., 1998). Table 3 lists the standard toxins distributed under this project. This research program is operated by Japan Food Research Laboratories (JFRL) by 2009 and the program is now operated by National Research Institute of Fisheries Science (NRIFS) from 2010.

Table 3. List of shellfish toxins produced in Japan Food Research Laboratories under the standard toxin distribution project

	Toxins
DSP	Okadaic acid
	Dinophysistoxin-1
Lipophilic toxins	Pectenotoxin-1
	Pectenotoxin-2
	Pectenotoxin-6
	Yessotoxin
PSP	C1,2[*1]
	Gonyautoxin-1,4[*1]
	Gonyautoxin-2,3[*1]
	Decarbamoyl-gonyautoxin-2,3[*1]
	Neo-saxitoxin

[*1] prepared as equilibrium mixtures. Purities of all toxins were confirmed to be above 95%.

Monitoring System for Shellfish Toxins in other Asian Countries

Table 4 shows the latest situation on shellfish toxin monitoring system in some Asian countries. Many Asian countries have the shellfish toxin monitoring system on mouse bioassay for PSP and DSP. ASP is monitored by HPLC-UV. Korea and Singapore are using only LC-MS for DSP monitoring. Singapore is also using only ELISA and HPLC or LC-MS/MS for PSP monitoring. In Korea, the regulatory level and the analytical method of DSP and PSP are established in the Food Sanitation Law, however the regulatory level and the analytical method of ASP are not designated in Food Sanitation Law. ASP in some shellfish growing area is monitored by National Fisheries Research Development Institute (NFRDI) under the Law for the Quality Management of Fisheries Products.

ACKNOWLEDGMENTS

We express our gratitude to Dr. Ji Hoe Kim (Korea), Ms Helen Phang (Singapore), Ms Sandra Victoria R. Arcamo (Philippines), Mrs Supanoi Subsinserm (Thailand), Mr Ngo Hong Phong (Vietnam) for providing information about the shellfish toxin monitoring system in their countries. We express our gratitude to Ms. Huiyi SAW (Singapore) for her help correcting the information about the shellfish toxin monitoring system in Southeast Asian countries.

Table 4. Shellfish toxin monitoring system in some Asian countries

Country	Biotoxins analyzed and regulatory limits	Method of analysis	Frequency of Sampling
Korea (ROK)	PSP (80 μg/100g) DSP (0.16 μg /g) ASP (20 μg /g)	Mouse Bioassay LC-MS/MS HPLC-UV	Normal state (Toxin is not detect): once a month. If toxin is detected: once a week. Over action level: twice a week
Philippines	PSP (60 μg /100g) DSP (death of 2 mice of 3w/in 24 hrs) ASP (20 μg /g)	Mouse bioassay Mouse bioassay HPLC-UV	Twice a week specifically for Manila Bay For Provincial areas: once a month during non-HAB occurrences and weekly during HAB occurrences
Singapore	PSP (80 μg / 100g) DSP (0.2 - 0.4 ppm) ASP (20 ppm)	ELISA for screening Confirmation by HPLC or LC-MS/MS LC-MSMS HPLC-UV	(a) Import – varies, dependent on product type (b) Local Farms – once a month or move if alert arises
Thailand	PSP (80 μg / 100g) DSP - limit for mouse bioassay - 2 out of 3 mice die & limit for LCMSMS - 0.16 μg /g ASP (20 ppm)	Mouse Bioassay HPLC-FL Mouse Bioassay, LC-MSMS HPLC-UV	Monthly
Vietnam	PSP (80 μg / 100g) DSP (0.16 μg /g) ASP (20 ppm)	Mouse Bioassay, HPLC-FL Mouse Bioassay, LC-MSMS HPLC-FL HPLC-UV	2 weeks

REFERENCES

Goto, H., Igarashi, T., Sekiguchi, R., Tanno, K., Satake, M., Oshima, Y. & Yasumoto, T. (1998). A Japanese project for production and distribution of shellfish toxins as calibrants for HPLC analysis, p. 216-219. *In* B. Reguera, J. Blanco, M.L. Fernández, and T. Wyatt (eds), *Harmful Algae. Xunta de Galicia and Intergovernmental Oceanographic Commission of UNESCO,* Spain.

Oshima, Y., Yasumoto, T., Kodama, M., Ogata, T., Fukuyo, Y, & Matsuura, F. (1982). Features of paralytic shellfish poison occurring in Tohoku District. *Bull. Japan. Soc. Sci.* Fish., 48, 525-530.

Oshima, Y., Postcolumn derivatization liquid chromatographic method for paralytic shellfish toxins, *J. AOAC Int.,* 1995, 78, 528-532.

Quilliam, MA., Xie, M. & Hardstaff, W.R. (1995). Rapid extraction and cleanup for liquid chromatographic determination of domoic acid in unsalted seafood, *J. AOAC Int.,* 78, 543-554.

Regulation (EC) *No 853/2004 of the European Parliament and the Council laying down specific hygiene of foodstuffs.*

Statistics on fisheries and water culture production in 2008. p, 4. *Ministry of Agriculture, Forestry and Fisheries in Japan.*

Suzuki, T., Jin, T., Shirota, Y., Mitsuya, T., Okumura, Y. & Kamiyama, T. (2005). Quantification of lipophilic toxins associated with diarrhetic shellfish poisoning in Japanese bivalves by liquid chromatography-mass spectrometry and comparison with mouse bioassay. *Fish. Sci.,* 71, 1370-1378.

Yasumoto, T., Oshima, Y. & Yamaguchi, M. (1978). Occurrence of a new type of shellfish poisoning in the Tohoku district. *Bull. Japan. Soc. Sci. Fish.,* 44, 1249-1255.

Yasumoto, T., Fukui, M., Sasaki, K. & Sugiyama, K. (1995). Determinations of marine toxins in foods. *J. AOAC. Int.,* 78, 574-582.

In: New Trends in Marine and Freshwater Toxins
Editors: A. G. Cabado and J. M. Vieites
ISBN: 978-1-62257-787-3

Chapter 11

Biotoxin Control Programs in North, Central and South American Countries

Stacey L. DeGrasse and Karen Martinez-Diaz
U.S. FDA Center for Food Safety and Applied Nutrition,
Office of Regulatory Science, Division of Analytical Chemistry,
College Park, Maryland, US

Abstract

Global commercial and aquaculture fishery production continues to expand in order to meet consumer demand for seafood, which offers numerous benefits to our diet. However, the associated natural risks of seafood increase concomitantly with increased production. Wherever seafood is harvested or imported, natural marine biotoxins may pose a public health risk. To mitigate these biotoxin risks, most countries rely on legislation or a range of guidance and monitoring procedures to protect public health. However, some countries currently have no formal programs aimed at addressing potential seafood biotoxin concerns. This chapter serves as a review of the biotoxin control programs employed by certain North (Canada, Mexico, and the United States), Central (Belize, Costa Rica, El Salvador, Guatemala, Honduras, Nicaragua and Panama), and South (Argentina, Brazil, Chile, Colombia, Ecuador, Peru, Uruguay, and Venezuela) American countries. Ongoing international efforts, notably through the Codex Alimentarius Commission, to standardize biotoxin control programs in order to create global equivalency for ensuring seafood safety are also discussed.

1. Introduction

Seafood consumption is beneficial to human health and is a source of proteins, essential omega-3 fatty acids, and a variety of micronutrients (e.g., Kite-Powell et al., 2008; Lloret, 2010), and maternal fish intake has been associated with improved early child development (Oken et al. 2008). Despite the benefits, there are also known risks (chemical and microbial) that could result in seafood-borne illness. These risks are expected to rise with the growing production of seafood. According to the US National Marine Fisheries Service (NMFS 2010), global commercial and aquaculture fishery production reached 142.3 million metric tons in 2008, an increase of 2.5 million metric tons since 2007. While microbial contaminants can be destroyed upon cooking, most of the chemical contaminants, such as marine biotoxins, are not denatured by heat and withstand the cooking process. Therefore, the risks associated with marine biotoxins must be controlled through monitoring and control programs to prevent contaminated products from entering the market.

There are numerous known natural biotoxins, as well as emerging ones, that are associated with seafood-borne illnesses globally. These toxins are generally of algal origin and may accumulate in shellfish and/or fish. Common human syndromes caused by contaminated shellfish include paralytic shellfish poisoning (PSP; see review by Etheridge 2010), amnesic shellfish poisoning (ASP; see review by Lefebvre and Robertson 2010), neurotoxic shellfish poisoning (NSP; see review by Plakas and Dickey 2010), diarrheic shellfish poisoning (DSP; see review by Dominquez et al. 2010), and azaspiracid poisoning (AZP; see review by Furey et al. 2010). Ciguatera fish poisoning (CFP; see review by Dickey and Plakas 2010) is responsible for most seafood-borne illnesses caused by finfish and is endemic to tropical and subtropical coral reef environments.

The above poisoning events may or may not be associated with the presence of an obvious algal bloom. Colloquially, harmful algal blooms (HABs) are often referred to as red tides, or *las mareas rojas* in Spanish, due to the water discoloration that often accompanies algal blooms. However, the use of the term "red tide" is misleading for several reasons. First, not all algal blooms are caused by toxic or harmful species. Second, HABs may occur without imparting any noticeable changes in water color. Lastly, toxin levels in shellfish may persist long after a HAB, regardless of whether the water was discolored at the onset of toxin accumulation. Thus the reliance on water discoloration alone may negatively impact public health if recreational harvesters rely on water color when deciding whether to harvest seafood in a given location.

Most countries with extensive coastlines and histories of seafood production have developed biotoxin control measures to safely harvest or import seafood products. As international trade continues to expand, land-locked countries are being forced to develop biotoxin control measures to protect their consumers from contaminated imported seafood products. Some countries, such as the European Union Members States (EC 2004), establish formal legislation to ensure that biotoxin contaminated seafood product does not enter into their market. Other countries, such as the United States, provide guidance to industry and implement monitoring programs for biotoxin control (http://www.fda.gov/Food/GuidanceComplianceRegulatoryInformation/GuidanceDocuments/Seafood/FishandFisheriesProductsHazardsandControlsGuide/default.htm [accessed 27 December 2010]). Such monitoring programs vary in nature, focusing on the monitoring of (1) oceanographic conditions, (2) presence of toxic algae, and/or (3) presence of toxins in seafood. Other countries have not established any formal programs to ensure seafood products are safe from natural biotoxins.

The goal of this chapter is to provide an overview of North, Central, and South American biotoxin control programs (Figure 1). In addition to describing what is known about these 18 countries and their means of managing these risks, this chapter also highlights the need and ongoing efforts to harmonize programs to create equivalency between countries for controlling risks associated with marine biotoxins. Global harmonization of seafood safety standards benefits public health and eliminates trade barriers.

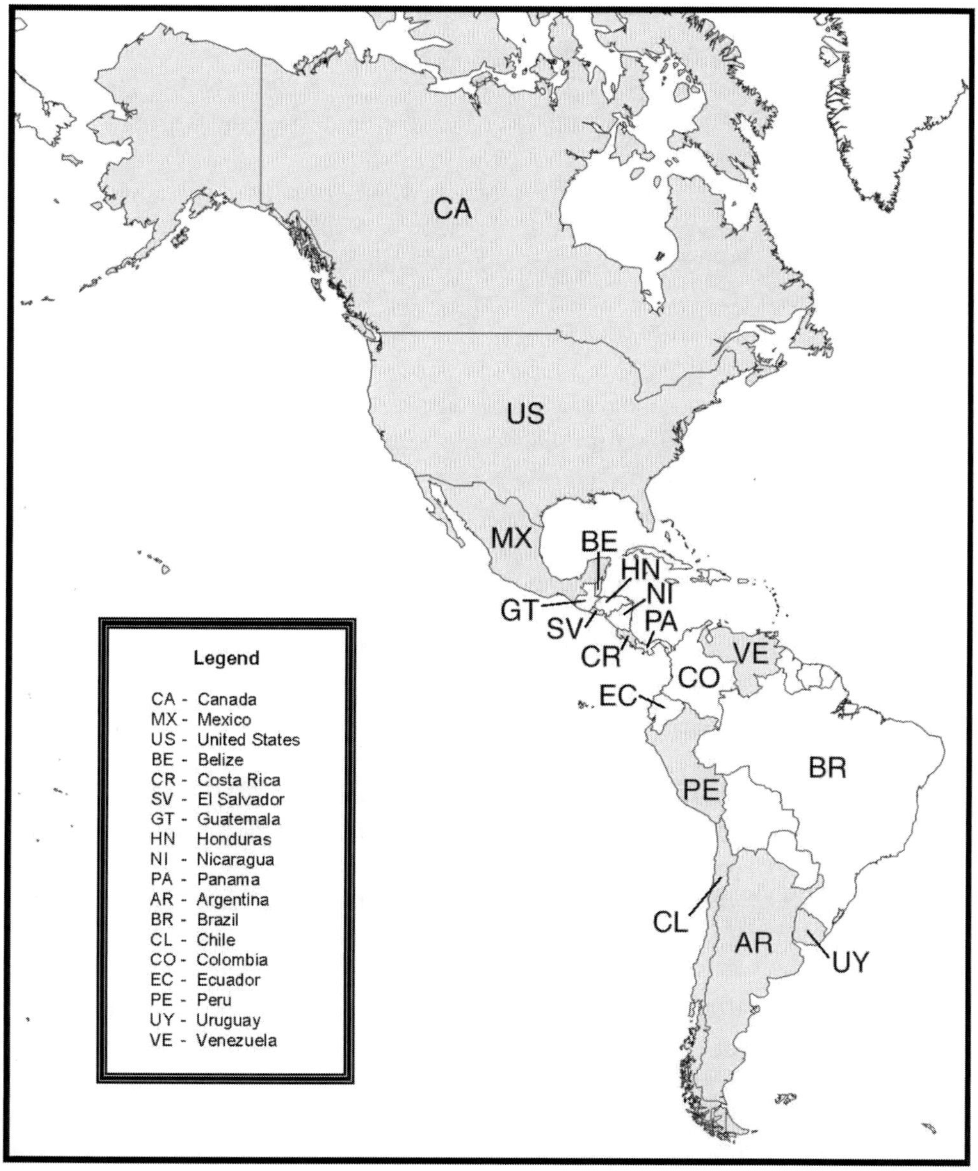

Figure 1. North, Central, and South American countries covered in this chapter. Countries in grey represent those that have established monitoring programs. Those in white either do not have formal monitoring programs or little to no information regarding programs was found.

1.1. North American Countries

Canada

Canada has a coastline of 243,042 km, which is by far the longest coastline in the world. Natural Resources Canada reported that approximately one-third of Canada's Gross National Product is generated by activities related to the coast. Per capita seafood consumption was reported to be 23.7 kg on average between the years of 2005-2007 (NMFS 2010). Canada exported 139,925 metric tons and imported 291,063 metric tons of edible fishery products in 2009 (NMFS 2010).

Of the list of known marine biotoxins, three groups are of most concern to Canadian seafood safety: toxins associated with PSP, ASP and DSP (Wright 1995). The first documented report of PSP occurred in the late 1700s during Captain George Vancouver's expedition of British Columbia when several crew members died after consuming contaminated mussels (Kao 1993). ASP was first reported in Canada in 1987 when an unexpected seafood outbreak occurred and was isolated to mussels collected from Prince Edward Island. The ASP toxin domoic acid was identified as the etiological agent (Wright et al. 1989) and has been detected in Canadian waters ever since. Owing to the established and effective monitoring programs, no such major ASP events have occurred since. Evidence of DSP toxins was found in the early 1990s (e.g., Quilliam et al. 1992; Gilgan et al. 1994), though no fatalities have been reported.

The Canadian Shellfish Sanitation Program (CSSP) was formed as a partnership between the Canadian Food Inspection Agency (CFIA), Fisheries and Oceans Canada (DFO), and Environment Canada (EC). CSSP was formed to "*classify harvesting areas and control the commercial and recreational harvesting of mollusks and processing of product for the consumer market.*" While there is no specific legislation regulating the control of seafood biotoxins in Canada, CSSP partners have legal authority to address biotoxin concerns established by the Fisheries Act, the Management of Contaminated Fisheries Regulations, the Fish Inspection Act, and the Fish Inspection Regulations (http://www.inspection.gc.ca/english/fssa/fispoi/man/csspccsm/chap1e.shtml [accessed 19 January 2011]). The CSSP is analogous to the US National Shellfish Sanitation Program (NSSP), and the two are considered equivalent in terms of providing assurance that bivalve mollusks are safe for consumption.

Within the CSSP partnership, CFIA (www.cfia-acia.gc.ca [accessed 19 January 2011]) serves as the lead agency with respect to marine biotoxin control programs. In the CSSP Manual of Operations, Chapter 11, entitled *Control of Marine Biotoxins,* describes Canada's approach to managing the risks associated with marine biotoxins (http://www.inspection.gc.ca/english/fssa/fispoi/man/csspccsm/chap11e.shtml [accessed 19 January 2011]). CFIA is responsible for collecting and testing shellfish samples for biotoxins, and they also make recommendations to DFO regarding the opening and closing of harvest areas. Appropriate sampling sites and frequencies are determined, and sampling frequency rates are designed to scale with observed toxicity.

Food safety guidelines and permissible levels of toxins in foods are set by Health Canada (http://www.hc-sc.gc.ca/fn-an/securit/chem-chim/contaminants-guidelines-directives-eng.php [accessed 19 January 2011]). Based on those guidelines, shellfish areas are closed when toxins exceed the established levels (Table 1). There are also considerations for the sampling of shellfish from processing plants with enforcement policies based on the regulatory levels stated above. Special sampling and testing requirements are specified for sea scallops (*Placopecten magellanicus*) to be marketed whole or with roe on. In addition to PSP risks in molluscan bivalves, there is also the potential for PSP toxin accumulation in lobster hepatopancreas (known as tomalley). As such, in March 2009 Health Canada advised limiting the daily consumption to two lobster tomalleys for adults and one for children (http://www.hc-sc.gc.ca/ahc-asc/media/advisories-avis/_2008/2008_70-eng.php [accessed 19 January 2011]).

According to the CSSP Manual of Operation, each CFIA area/region/district must develop an annual plan for marine biotoxin monitoring. All testing in support of CSSP is conducted by accredited federal laboratories (notably CFIA for seafood toxins). It is required that validated methods be employed and proficiency must be demonstrated (J. van de Riet, personal communication). Currently, a post-column oxidation high performance liquid chromatography with fluorescence detection (HPLC-FD) method (PCOX; van de Riet 2009) and the mouse bioassay (MBA; Official Method 959.08, AOAC 2005a) are used to test for PSP toxins. An HPLC-UV (ultra-violet) method (Official Method 991.26, AOAC 2005b)

Table 1. North American biotoxin regulatory action levels and methods

Country	Regulatory Levels	Analytical Methods
Canada		
PSP	80 µg STX eq 100 g^{-1}	MBA; HPLC-FD
ASP	20 µg g^{-1}	HPLC-UV
NSP	-	-
DSP	0.2 µg g^{-1} OA+DTX1 or PTX	LC-MS
AZP	-	-
CFP	-	-
Mexico		
PSP	80 µg STX eq 100 g^{-1}	MBA
ASP	20 mg kg^{-1}	HPLC-UV
NSP	20 MU 100 g^{-1}	MBA
DSP	160 µg OA kg^{-1}	-
AZP	160 µg AZA kg^{-1}	-
CFP	2.5 MU 100 g^{-1}	MBA
United States		
PSP	80 µg STX eq 100 g^{-1}	MBA; HPLC-FD
ASP	20 ppm DA*	HPLC-UV
NSP	0.8 ppm PbTx2 eq	MBA
DSP	0.2 ppm OA+DTX1	-
AZP	-	-
CFP	-	-

* 30 ppm in Dungeness crab viscera.

allows for the analysis of ASP toxins, and an LC-MS (mass spectrometry) method (McNabb et al. 2005) with an in-house extraction procedure is conducted for the determination of DSP toxins. When harvesting closures are made due to excessive toxin levels in the shellfish, the areas are only re-opened once three consecutive samplings during a minimum period of 14 days indicate toxin levels have decreased below the established action level. In the event of suspected poisonings or outbreaks, all consumer illness information must be submitted into the Issues Management System (IMS) for investigation.

Mexico

Mexico has coastline along the Pacific Ocean, Gulf of California, Gulf of Mexico, and Caribbean Sea totaling 9,330 km. Per capita human consumption of fish and shellfish in Mexico was reported to be 10.9 kg on average from 2005-2007 (NMFS 2010). Total aquaculture and commercial catches of fish, crustaceans, and mollusks were reported to be >1.7 million metric tons in 2008 (NMFS 2010). Mexico exported and imported 15,454 metric tons and 72,174 metric tons, respectively, of edible fishery products in 2009 (NMFS 2010). In 2005, aquaculture production of shellfish reached 1,500 metric tons, ranking Mexico fourth among Latin American countries in bivalve production at that time (Maeda-Martínez 2008).

There are several HAB species and associated toxins that pose seafood safety risks along the Mexican coasts. For example, PSP toxins associated with *Alexandrium* spp., *Pyrodinium bahamenses* var. *compressum*, and *Gymnodinium catenatum* have been reported (Sierra-Beltrán et al. 1998; Altamirano and Sierra-Beltrán 2008 and references therein). NSP toxin producing *Karenia brevis* cells are often highly abundant in the Gulf of Mexico off of the Mexican coast (Altamirano and Sierra-Beltrán 2008 and references therein), and almost yearly along the coasts of Veracruz and Tamaulipas. *Pseudo-nitzschia* spp. known to produce domoic acid have also been found in Mexican waters (e.g., Hernandez-Becerril 1998; Licea et al. 1999; Meave del Castillo and Hernandez-Becerril 1998). According to Altamirano and Sierra-Beltrán (2008), there have been no reported DSP cases; however, DSP toxin-producing organisms (e.g., *Dinophysis* spp., *Prorocentrum* spp.) have been found. Further, cyanobacterial toxins and ciguatoxin have been detected as well (Altamirano and Sierra-Beltrán 2008 and references therein).

In 2003 a HAB monitoring program was established, and as a result more than 40 HABs have been identified. In addition, from 2005 through 2010 the routine monitoring system for marine biotoxins yielded >2000 mollusk analyses (analyses were for saxitoxin, domoic acid, okadaic acid, brevetoxins, and/or yessotoxin). Greater than 90% of the samples tested have been below the maximum level permitted and safe for consumption (http://www.cofepris.gob.mx/wb/cfp/5_marea_roja/_rid/1901?page=2 [accessed 20 December 2010]). When the sample analyzed contains marine toxins (or harmful phytoplankton) that exceed the regulatory limit (Table 1; http://www.cofepris.gob.mx/work/sites/cfp/resources/LocalContent/1128/26/InstMuestreoFitoyMBIX.pdf [accessed 19 January 2011]), the accredited laboratory notifies the necessary parties (see below) (http://www.cofepris.gob.mx/wb/cfp/laboratorio_de_referencia [accessed 22 November 2010]).

To protect seafood from these biotoxin risks and other such hazards, the Ministry of Health, through the Federal Commission for Protection from Sanitary Risks (COFEPRIS), established the Mexican Shellfish Sanitation Program (MSSP). COFEPRIS maintains legal, administrative, safety, quality, and sanitary controls over molluscan shellfish. In addition, COFEPRIS (1) issues certifications related to laboratories and processing/packaging of shellfish and (2) establishes monitoring, surveillance, and sanitary control of all natural and artificial harvest areas for shellfish products that will be consumed both nationally and internationally (http://www.cofepris.gob.mx/wb/cfp/moluscos_bivalvos [accessed on 19 November 2010]).

This organization closely follows the guidelines established by the US National Shellfish Sanitation Program (NSSP), US Food and Drug Administration (FDA), and Directorate General for Health and Consumer Affairs of the European Union (SANCO). In 2008, the governments of Mexico (COFEPRIS) and the US (FDA) signed a memorandum of understanding (MOU) that covered the safety and quality of fresh and frozen aquacultured molluscan shellfish exported from Mexico to the US. The MOU established the set of guidelines to be implemented to assure seafood harvested in Mexico and transported to the US is safe and that it has been harvested, processed, transported and labeled in conformance with the NSSP and the US Food Drug and Cosmetic Act (http://www.fda.gov/InternationalPrograms/Agreements/MemorandaofUnderstanding/ucm107608.htm [accessed 21 December 2010]).

To deal with seafood related poisonings, the National Center of Epidemiological Surveillance and Disease Control (CENAVECE), COFEPRIS, and the Sanitary Authority of the State developed an outbreak contingency plan. The plan starts with an epidemiological investigation to identify the source of the possible outbreak. If the product is contaminated post-harvesting (via product handling) then the harvest area will not be closed, and the Federal, State and Local Authorities (including FDA and EU, if applicable) will be notified and the product will be removed immediately from the market. In cases where contamination occurs in the harvesting area, the area is closed immediately and the monitoring frequency increases for quantitative phytoplankton from biweekly to every 3 days and for shellfish toxins from monthly to 4 days (http://www.cofepris.gob.mx/work/sites/cfp/resources/LocalContent/1128/26/INSTCONTRSANITIX.pdf [accessed 26 January 2011]). The harvest areas are re-opened when the levels of contaminants no longer represent a threat to public health. Finally, a report is generated summarizing the findings of the investigations and actions taken by the sanitary authorities (http://www.cofepris.gob.mx/wb/cfp/moluscos_bivalvos [accessed on 19 November 2010]). Information on toxic events is provided on the COFEPRIS webpage and is available for public access (http://www.cofepris.gob.mx/wb/cfp/antecedentes_en_mexico ([accessed on 20 December 2010]).

United States

The United States coastline is 19,924 km in length. The 2005-2007 averages for human consumption place the United States as the world's third largest consumer of seafood, behind China and Japan (NMFS 2010). The total catch of fish, crustaceans, and mollusks in 2008 exceeded 4.8 million metric tons (NMFS 2010). Despite high levels of seafood production within the US, 84% of seafood consumed in the US in 2009 was imported (http://www.nmfs.noaa.gov/fishwatch/trade_and_aquaculture.htm [accessed 19 January 2011]).

All of the seafood-borne biotoxin syndromes listed above have been of concern in the US in either domestic or imported products. HABs known to cause PSP, ASP, NSP, DSP and CFP have been found in US waters (http://www.whoi.edu/redtide/page.do?pid=14898 [accessed 19 January 2011]). Harvesting closures due to the threat of PSP are common in certain regions of the US (e.g., Bean et al. 2005). NSP closures are made when *Karenia brevis* cells exceed the established threshold, and harvesting areas are not re-opened until shellfish testing by MBA demonstrates toxin levels have declined below the action level (e.g., Plakas and Dickey 2010). The first shellfish closures due to a DSP event occurred in 2008 (Deeds et al. 2010), yet no illnesses resulted due to the effective and timely closure. While azaspiracid has not been confirmed to be a hazard in seafood harvested within US waters, there have been AZP illnesses in the US due to imported mussels (Klontz et al. 2009).

The US Food and Drug Administration (FDA) has statutory authority and responsibility to ensure safety of most foods in the US, including seafood. The food safety program operates under provisions of the Food, Drug, and Cosmetic Act, the Public Health Service Act, and related regulations. There is no regulation pertaining specifically to seafood biotoxin concerns. An integral part of FDA's seafood safety program involves the publication of the *Fish and Fisheries Products Hazards and Controls Guidance* (referred to as the Guide), which is currently in its Third Edition (http://www.fda.gov/Food/GuidanceComplianceRegulatoryInformation/GuidanceDocuments/Seafood/FishandFisheriesProductsHazardsandControlsGuide/default.htm [accessed 19 January 2011]). This serves as a critical guide to industry and provides the most up-to-date science and policy on potential hazards in fish and fishery products as well as effective means for controlling their occurrence. Chapter 6 of the Guide, entitled *Natural Toxins*, relates to seafood biotoxin risks. The syndromes PSP, ASP, NSP, DSP, and CFP are covered and established regulatory action levels are provided (Table 1).

A Fourth Edition of the Guide is anticipated in the near future, and it will include updates on these action levels (particularly a modification for DSP toxins and the inclusion of CFP toxin guidance levels). If the scientific understanding of risks is revised in between editions, the FDA will post an update reflecting the most up-to-date information. In 2008, the Third Edition was updated to notify seafood processors that purchase grouper, amberjack, and other predatory reef fishes from the northern Gulf of Mexico that numerous CFP outbreaks occurred because of contaminated fish caught in the area near the Flower Garden Banks National Marine Sanctuary, an area previously thought to pose a minimum risk (http://www.fda.gov/Food/GuidanceComplianceRegulatoryInformation/GuidanceDocuments/Seafood/ucm142704.htm [accessed 19 January 2011]).

The Guide is used to formulate the seafood hazard analysis critical control point (HACCP) plans implemented by fish and fishery product processors in the US (and those that export to the US). HACCP is a systematic, science-based, management system whereby food safety is addressed through analysis and control of anticipated biological, chemical, and physical hazards from raw product, during the procurement and handling, and through processing, distribution and consumption of the product. In 1997, regulation was passed (final rule 21 CFR 123, *Procedures for the safe and sanitary processing and importing of fish and fishery products*) to require seafood processors to develop and implement HACCP plans.

The US FDA also releases advisories to the general public on food safety concerns. One such example related to seafood safety occurred in the summer of 2008 when FDA released an advisory against consuming American lobster tomalley due to the risk of PSP contamination (http://www.fda.gov/NewsEvents/Newsroom/PressAnnouncements/2008/ucm 116927.htm [accessed 19 January 2011]). An additional public resource available from the FDA is the Bad Bug Book, a handbook detailing foodborne pathogenic microorganisms and natural toxins, including marine biotoxins. (http://www.fda.gov/Food/FoodSafety/ FoodborneIllness/FoodborneIllnessFoodbornePathogensNaturalToxins/BadBugBook/default. htm [accessed 19 January 2011]).

While seafood safety falls under the purview of FDA, there is no statutory authority for the FDA to close a fishery. When seafood toxins exceed the established action levels in federal waters, the FDA submits a request to close harvesting in the contaminated area to National Marine Fisheries Services (NMFS), which has statutory authority to do so under the Magnuson-Stevens Fishery Conservation and Management Act. States within the US are responsible for seafood safety for product harvested from their own territorial waters. Most states along the coasts with known seafood toxin risks have established monitoring systems in place and make harvesting closures as necessary. For example, PSP toxin levels have been historically high in shellfish collected from waters off the coast of the state of Maine and so a marine biotoxin program was established in the Department of Marine Resources Public Health Division (http://www.maine.gov/dmr/rm/public_health/biotoxinmonitoring.htm [accessed 19 January 2011]). Sampling is conducted between March and October and closures are made accordingly, with updates and announcements of closures posted on their webpage. FDA would only become involved if contaminated products entered into interstate commerce and/or if assistance was requested by the states.

FDA, NMFS, state shellfish authorities, the seafood industry and other stakeholders work together through the Interstate Shellfish Sanitation Conference (ISSC), which was formed in 1982 to "*foster and promote shellfish sanitation through the cooperation of state and federal control agencies, the shellfish industry, and the academic community*" (http://www.issc.org/ [accessed 19 January 2011]). As such, the ISSC (1) adopts uniform procedures, (2) provides current and comprehensive guidelines for shellfish harvesting, processing, and shipping, (3) provides a forum for shellfish sanitation concerns to be addressed and resolved, and (4) informs all interested parties of shellfish sanitation concerns and recent developments. There are a number of ISSC committees tasked with a range of shellfish sanitation aspects. The two most relevant to marine biotoxins are the (1) Biotoxin Committee and (2) Laboratory Methods Review Committee. The Laboratory Methods Review Committee is responsible for reviewing methods that are proposed for regulatory use. The NSSP is the federal/state cooperative program recognized by FDA and ISSC for the sanitary control of shellfish produced and sold for human consumption. To ensure that all parties are in compliance, the NSSP provides the Guide for the Control of Molluscan Shellfish, which includes a Model Ordinance of guidelines and standards (http://www.fda.gov/Food/FoodSafety/Product-SpecificInformation/Seafood/FederalStatePrograms/NationalShellfishSanitationProgram/ucm 046353.htm [accessed 19 January 2011]). Chapter III entitled *Laboratory* outlines the NSSP approved methods for biotoxin testing (http://www.fda.gov/Food/FoodSafety/Product-SpecificInformation/Seafood/FederalStatePrograms/NationalShellfishSanitationProgram/ucm 154766.htm [accessed 19 January 2011]).

A new experimental monitoring and management tool employed to control for marine biotoxins in seafood harvested from US federal waters is known as the Onboard Screening Dockside Testing (ObSDT) Protocol (http://www.nero.noaa.gov/sfd/clams/ [accessed 19 January 2011]). The Protocol, currently in pilot study phase, was published in the Federal

Register in 2007 and was approved to allow one fishing vessel the opportunity to harvest surfclams and quahogs from a section of closed federal waters on Georges Bank in the northern Atlantic Ocean following the requirements designated in the Protocol. Briefly, the fishing vessel is granted a permit from NMFS and is required to sample and test for PSP toxins at sea in order to make the determination to harvest product. If all required samples are negative (<40 μg STX eq 100 g^{-1}, set more conservatively than the action level) using the field test kits, shellfish may be harvested and landed upon which time they will be tested using the MBA (which must not exceed 80 μg STX eq 100 g^{-1}) by an NSSP approved laboratory to assure they are safe before they enter the market (http://www.nero.noaa.gov /sfd/clams/Protocol%20for%20Onboard%20Screening%20and%20Dockside%20Testing%20 November%202007.pdf [accessed 19 January 2011]).

In 2010 New Jersey and Rhode Island allowed harvests under the Protocol, and Massachusetts has approved a slightly revised version of the Protocol allowing product to be landed in their state during the pilot study in 2011 (http://www.mass.gov/ ?pageID=eohhs2pressrelease&L=4&L0=Home&L1=Government&L2=Departments+and+Divisions&L3=Department+of+Public+Health&sid=Eeohhs2&b=pressrelease&f=101221_atlantic_surf_clams&csid=Eeohhs2 [accessed 26 January 2011]). To date, the successful pilot study has allowed safe product to be harvested and sold from resources that would have otherwise been unavailable. Such monitoring approaches may serve as a new model to manage biotoxin risks in offshore fish and fishery products.

1.2. Central American Countries

Belize

The coastline of Belize extends for 386 km along the Caribbean Sea. The country is home to the Belize Barrier Reef System, which is second in length behind Australia's Great Barrier Reef. Fishing is important to the economy of Belize as they have 169,840 km^2 of exclusive economic zone (EEZ) and a continental shelf of 9800 km^2 (Perez 2009). According to Perez, there are approximately 4000 fishermen in the country, and ~60% belong to fishermen co-operatives. Per capita human consumption of fish and shellfish in Belize was reported to be 12.8 kg on average from 2005-2007 (NMFS 2010).

The Belize Agricultural Health Authority (BAHA) is the statutory body tasked with modernizing the agricultural health services in Belize (http://www.baha.bz/ [accessed 7 January 2011]). Their mission is "*to provide optimum, competent, and professional services in food safety, quarantine, and plant and animal health in order to safeguard the health of the nation and facilitate trade and commerce*." BAHA is also the organization responsible for the certification for export and imports of aquaculture products. Currently, BAHA is not monitoring for biotoxins in fish or fishery products, nor do they have specific regulations regarding biotoxins in seafood (M. Figueroa, personal communication). A HACCP food safety system for fish and fishery products has been established in Belize, which allows for the export of their products to the US and EU. Should a seafood-borne illness occur, BAHA would, in conjunction with the Ministry of Health, investigate the event. To date, there have been no reported human poisonings from seafood toxins in Belize.

Costa Rica

Costa Rica has 1290 km of coastline along the Caribbean Sea and the Pacific Ocean. The Gulf of Nicoya (Pacific Coast) is one of the more productive areas and is an important resource for artisanal fishing, sustaining ~600 families (Vargas and Freer 2004; http://www.fao.org/fishery/countrysector/FI-CP_CR/en [accessed 21 January 2011]). The key commercial shellfish harvested in this area include *Spondylus calcifer* (thorny oyster), *Pinctada mazatlantica* (pearl oyster), *Anadara tuberculosa* (ark clam), and *Mytella guyanensis* (mangrove mussel).

HABs have been reported since at least 1989 (Mata et al. 1990). Extensive harmful algal blooms persisted from December 1999 to early 2002, alternating between the following species: *Gymnodinium catenatum*, *Trichodesmium erythraeum*, *Pyrodinium bahamense* and *Pyrodinium bahamense* var. *compressum*, and *Cochlodinium polykrikoides* (Vargas and Freer 2003). This was the first report of the co-occurrence of both *Pyrodinium* morphotypes on the coast of Costa Rica. The bloom affected 70 people, and there were reports of high PSP toxicity in commercial shellfish such as *Spondylus calcifer* (4000 μg STX eq 100 g^{-1}), *Pinctada masatlantica* (3800 μg STX eq 100 g^{-1}) and *Anadara similis* (1000 μg STX eq 100 g^{-1}) (INCOPESCA 2001). Although, several species of *Pseudonitszchia* (linked to the production ASP toxins) have been found in waters off Costa Rica, they have not shown toxicity to date. Vargas and Freer (2004) indicated that Costa Rica's phycotoxin monitoring program only addresses PSP toxins. However, the phytoplankton community in Costa Rican waters is diverse and includes other species known to be harmful/toxic. The last recorded HAB event began in November 2000 and lasted until early 2003 (http://www.una.ac.cr/biol/marearoja/Info2004.htm [accessed 29 December 2010]). According to the National University (UNA) webpage, a total of 2 deaths and 130 human poisonings, 75% of which occured during the last HAB episode, have resulted over the course of recorded HABs. PSP events have occurred primarily in the areas of Tárcoles and Guacalillo (Puntarenas) and during 2001 and 2002 in the Guanacaste area.

Table 2. Central American biotoxin regulatory action levels and methods

Country	Regulatory Levels	Analytical Methods
Belize		
-	-	-
Costa Rica		
PSP	80 μg STX eq 100 g^{-1}	MBA
El Salvador		
PSP	80 μg STX eq 100 g^{-1}	MBA
Guatemala		
PSP	80 μg STX eq 100 g^{-1}	MBA
Honduras		
-	-	-
Nicaragua		
PSP	400 MU 100 g^{-1}	MBA
Panama		
PSP	400 MU 100 g^{-1}	MBA

The Interagency Commission for Monitoring Red Tides was activated to deal with such posioning events. This committee comprises representatives of the Costa Rican Fishery and Aquaculture Institute (INCOPESCA), the Ministry of Agriculture and Livestock (MAG) through LANASEVE Laboratory, Ministry of Health, UNA, the Social Security Fund (CCSS) and the University of Costa Rica (UCR) (http://www.una.ac.cr/biol/marearoja/Info2004.htm [accessed 29 December 2010]). The authorities issued decrees establishing the legal framework for managing the harvest and marketing of bivalve mollusks in the country. The decrees facilitated the tracking of lots and batches of bivalve products in the market, which they consider essential to detect products from contaminated areas and establish selective closures in marine coastal areas accordingly. Moreover, periodic sampling of seawater was implemented by INCOPESCA at stations located in each of the administrative regions of the Pacific coast. As part of the emergency plan, seawater samples have been analyzed in the Laboratory of Marine Phytoplankton at the Marine Biological Station of the National University in Puntarenas for taxonomic identification and quantification of toxic dinoflagellates. In natural shellfish beds, numerous samples of shellfish, such as *Anadara similis*, *Spondylus calcifer*, *Pinctada mazatlanica*, *Donax navicula*, *Mytella guyanensis*, *Polymesoda inflata*, *Prototaca sp* and *Pinna rugosa*, have been collected. The LANASEVE laboratory conducted toxicology tests using mouse bioassay and following established regulatory levels (Table 2).

After the last seafood poisoning event, the Interagency Commission for Monitoring Red Tides proposed a plan to monitor HABs in Costa Rica (http://www.una.ac.cr/biol/marearoja/Info2004.htm [accessed 29 December 2010]). The plan proposes that the LANASEVE Laboratory conduct regular sampling (monthly or bimonthly) in areas of the Pacific coast where the accumulation of toxins in shellfish has been observed. The plan also mentions expanding monitoring for DSP and ASP toxins. The sampling frequency would intensify in areas with greater toxin accumulation and during HAB seasons. If algal patches are observed in the water that are suggestive of a possible toxic event, water samples are taken for phytoplankton analysis. When the samples are positive for toxic species, then periodic sampling in the impacted and surrounding areas is immediately implemented. In addition, a report is prepared and sent to INCOPESCA and the Interagency Commission for Monitoring and Prevention of Red Tide, who will inform the public of the preventive measures.

These agencies provide workshops to communities that harvest molluscs in the Gulf of Nicoya to explain HAB phenomena and the potential risks. Finally, INCOPESCA, the National University and the University of Costa Rica have on-going research programs to provide a knowledge base about physical and chemical conditions that regulate the appearance and disappearance of HABs in the Pacific coast. In addition, UNA and INCOPESCA have established international collaborations for projects involving the study of HABs, monitoring systems, and accumulation of marine toxins in shellfish. The FAO reports that INCOPESCA also provides courses, theoretical and practical in nature, on farming technologies and fishery/aquaculture product management in accordance with HACCP guidelines (http://www.fao.org/fishery/countrysector/FI-CP_CR/en [accessed 21 January 2011]).

El Salvador

El Salvador is the smallest country in Central America. While it is the only Central American country without a Caribbean coastline, it is home to 307 km of Pacific coastline. FAO reports El Salvador to be an important source of aquacultured tilapia, shrimp, and prawn (http://www.fao.org/fishery/countrysector/naso_elsalvador/en [accessed 21 January 2011]).

A PSP event was documented from August 2001 to January 2002 where toxin levels in certain shellfish species exceeded the safe level of 400 mouse units 100 g^{-1} (80 μg STX eq 100 g^{-1}) established by El Salvador, and therefore the Ministry of Health and Social Assistance (MSPAS) banned the consumption and trade of marine mollusks (Barraza et al. 2004). A total of 41 intoxications were reported to be associated with this event (Armero-Guardado 2002). Toxins were found in oysters (*Ostrea iridescens*) and rock mussels (*Modiolus* sp.). A monitoring study conducted along the Pacific Coast of El Salvador from November 2005 through March 2006 described common blooms of *Pyrodinium bahamense* (particularly during autumn and winter) that resulted in human illnesses and deaths, as well as turtle (*Lepidochelis olivacea* and *Chelone mydas*) mortalities (Licea et al. 2008). During that period, three people died, 7 others became ill, and 206 turtle deaths were reported. Turtle mortality was presumably the result of feeding on saxitoxin-contaminated jellyfish (*Stomolophus meleagris*). Saxitoxin concentrations were found to be up to 628 μg STX eq 100 g^{-1} in turtles, up to 12 μg STX eq 100 g^{-1} in live jellyfish, and >4000 μg STX eq 100 g^{-1} in shellfish collected from Salvadorian waters/beaches. At the end of this bloom season, one week after the shellfish bans were lifted four girls collected, cooked, and ate ~50 dog winkles (*Plicopurpura columellaris*) (Barraza 2009). All 4 girls experienced PSP symptoms within 1 hour of ingestion, and the eldest girl (age 11) died after 6 hours. Barraza (2009) further reported that dog winkle specimens were subsequently collected from the same location and were found to contain 7 million mouse units 100 g^{-1} using the mouse bioassay. This was the first documented human fatality due to the consumption of PSP contaminated gastropods in El Salvador.

After this bloom, in order to maintain a permanent monitoring system in the coastal water of El Salvador, a Red Tide National Commission or CONAMAR was created. This project was developed in collaboration with Salvadoran Health Authorities, the National Center for Promotion of Fisheries and Aquaculture (CENDEPESCA), Ministry of the Environment and Natural Resources (MARN), Universities, National Police (PNC), and Navy (http://www.salud.gob.sv/archivos/vigi_epide2001/marea_roja2001_archivos/frame.htm [accessed 3 January 2011]).

In April 2010, laboratory test results performed on different species of mollusks indicated the presence of toxins and the Government declared a national ban to safeguard public health (http://www.marn.gob.sv/index.php?option=com_content&view=article&id=185:autoridades-confirman-presencia-de-marea-roja-en-costa-salvadorena&catid=1:noticias-ciudadano&Itemid=227 [accessed 3 January 2011]). In so doing, health authorities indefinitely prohibited the harvest, sale and consumption of oysters, mussels, clams and snails from the impacted area. Authorities intensified the monitoring frequency and kept the public informed about the threat.

Guatemala

Guatemala has a total of 400 km of coastline on the Pacific and Caribbean. Per capita consumption of seafood in Guatemala was reported to be 2.8 kg on average between 2005-2007 (NMFS 2010). Tilapia and shrimp aquaculture is commercially important (http://www.fao.org/fishery/countrysector/naso_guatemala/en [accessed 21 January 2011]).

Paralytic shellfish poisoning toxins are of greatest concern for Guatemalan shellfish safety. A major outbreak of PSP occurred in Guatemala in 1987. The dinoflagellate *Pyrodinium bahamense* was found to be the source of the toxin, which was transferred to humans via the local clam *Amphicaena kindermani* (Rodrigue et al. 1990). This even resulted in 187 illnesses and 26 fatalities. The Guatemalan government created a surveillance program in response to this event with the intention of conducting shellfish testing using the mouse bioassay at 3 sampling stations along the Pacific coast of Guatemala (Rodrigue et al. 1990).

There is an ongoing PICES Seafood Safety Project focused on preparing and teaching country-specific training courses designed to ensure seafood safety (https://www.pices.int/publications/pices_press/volume18/v18_n2/pp36_PICES_Seafood_Safety_Guatemala_project.pdf [accessed 28 December 2010]). The Project is currently focused on Guatemala and is addressing (1) the magnitude of the HAB problem, (2) the need for training, and (3) the likelihood of sustainability. The program identified the current greatest needs to be (1) training in detection methods for PSP toxins, (2) training/review in phytoplankton identification, and (3) basic training in oceanography. As a result, training on PSP detection has been conducted.

The Fisheries and Aquaculture Management Unit (UNIPESCA) is the government entity involved with the fishery sector (http://www.fao.org/fishery/countrysector/naso_guatemala/en#tcNB00E3 [accessed 3 January 2011]). However, private investors, universities, and organizations are united to evaluate proposals for a new Guatemalan regulatory entity for fisheries and aquaculture. These institutions are: the tuna fisheries sub-sector, the Marine and Aquaculture Studies Center (CEMA), the Non-traditional Agricultural Exports Trade Association (AGEXPRONT), the Shrimp Producers Association (ACRICON), Managers of the Ministry of Agriculture, Husbandry and Food (MAGA) and the Regional Programme for the Support of Fisheries Development in the Central American Isthmus (PRADEPESCA).

The regulatory limit in Guatemala for PSP toxins is 80 STX eq 100 g^{-1}, and the method of analysis is the mouse bioassay (Table 2; Fernández 1998). No further information about monitoring and contingency plans was found.

Honduras

Honduras borders with the Pacific Ocean and Caribbean Sea, creating 820 km of coastline. Per capita seafood consumption in the country was reported to be 3.1 kg on average between 2005-2007 (NMFS 2010). Honduras imported 17,869 metric tons of edible fishery products in 2009 (NMFS 2010). In Honduras, fishing activities are conducted in the Coastal areas of the Caribbean and Pacific (Gulf of Fonseca) including the most important islands: El Tigre, Zacate Grande, Exposición, San Carlos and Inglesera. The fishing performed in these islands is mostly artisanal and white shrimp (*Litopenaeus vannamei*) is the most important species, with an average of production of 1000 tons annually (http://www.fao.org/fishery/countrysector/FI-CP_HN/es [accessed 3 January 2011]). Along the Pacific coast, the most important activity is industrial shrimp farming, the production of which in recent years actually exceeded the industrial production of shrimp in the Caribbean despite its larger spatial coverage. Along the Caribbean coast, fishing is performed in regional communities and in certain areas of the Bay Islands. Industrial fishing is one of the most important industries for exportation and the national economy of the country.

The Ministry of Agriculture and Livestock is the institution responsible for promoting and developing aquaculture in the country, including research projects for the development of artisanal fisheries. No information was found about bitoxin monitoring or control programs.

Nicaragua

Nicaragua claims 910 km of coastline, divided between the Pacific Ocean and Caribbean Sea. Per capita seafood consumption in Nicaragua was reported to be 4.8 kg on average between 2005-2007 (NMFS 2010).

The first reports of PSP in Nicaragua were made in November of 2005 and were associated with blooms of the toxin *Pyrodinium bahamense* (Chow et al. 2010). The event resulted in 50 cases of PSP illness, including one fatality. The vector responsible for the poisonings was the mangrove cockle *Anadara tuberculosa*. This event highlighted the need for phytoplankton monitoring and shellfish testing, as well as improved communication and cooperation among Central American countries that share the threat of HABs.

The Nicaraguan Fisheries Institute (INPESCA) is the regulatory body for the fisheries industry and is authorized to oversee enterprises, associations and productive cooperatives. On the other hand, the National Fisheries and Aquaculture Administration (ADPESCA) is responsible for the research activities, and the control and surveillance of all the resources. Research and educational programs for aquaculture can be found in the Central American University (UCA), the Agronomic University of Nicaragua (UNA), the Ave María University, the National Autonomous University of Leon, and the University of Bluffields (BICU); all have educational programs that include subjects and/or courses related to aquaculture.

Recently, Nicaragua has received technical and financial assistance from various bilateral and international agencies for the development of its fisheries. However, funding for the services they provide remains the major limitation and has resulted in noticeable challenges in research capacity, control and surveillance, and administration (http://www.fao.org/fishery/countrysector/FI-CP_NI/es [accessed 3 January 2011]).

Panama

Panama claims 2,490 km of coastline along the Pacific Ocean and Caribbean Sea. In terms of fishery products, aquacultured shrimp, trout, and tilapia are the main exports from Panama (http://www.fao.org/fishery/countrysector/naso_panama/en [accessed 5 January

2011]). Aquaculture practices for bivalve mollusks are still only in pilot phase. The main species of interest are scallops (*Argopecten ventricosus*) and oysters (*Crassostrea gigas* and *C. corteziensis*).

The agency responsible for the development of aquaculture in Panama is the Ministry of Agriculture and Husbandry Development (MIDA) through the National Aquaculture Directorate. In addition, the National Aquaculture Directorate (DINAAC) oversees the exploitation of aquatic resources for the benefit of the national population. While no specific information about biotoxin monitoring programs was found, PSP toxins are the only toxins for which action levels in Panama were found (Table 2).

1.3. South American Countries

Argentina

Argentina has 4989 km coastline along the Atlantic Ocean. Per capita seafood consumption in the country was reported to be 7.9 kg on average between 2005-2007 (NMFS 2010). Argentina exported 21,019 metric tons of edible fishery products in 2009 (NMFS 2010).

The first recorded incidence of PSP was in 1980 and resulted in 2 fatalities (Carreto et al. 1981; Ferrari 2004; http://www.fao.org/docrep/007/y5486e/y5486e0c.htm#TopOfPage [accessed 19 November 2010]). During the event toxin levels reached 50,000 μg STX eq 100 g^{-1}, and *Alexandrium tamaerense* was found to be the cause. Since that time *A. tamarense* has recurrently formed blooms along the Argentinian coast (Carreto et al. 1998). PSP toxicity has also been attributed to *A. catenella* and *Gymnodinium catenatum* (Ferrari 2004 and references therein). *Pseudo-nitzschia australis* was found to be the dominant ASP producing species off the coast of Argentina (Ferrari 2004). Domoic acid has also been detected in mussels and anchovy. A range of *Dinophysis* species as well as *Prorocentrum lima* have been documented in Argentinian waters (Ferrari 2004), and DSP toxins have been detected in mussels from Patagonia in 1999 when 40 persons became ill (Gayoso and Ciocco 2001).

In December 1997, the Federal Fisheries Law was passed (Law 24.922) and established the Federal Fisheries Council as the main body leading national fisheries policy. The primary responsibilities of the council are to manage national fisheries and develop policies and regulations regarding fishing activities in waters subject to national jurisdiction (http://www.cfp.gov.ar/funciones_ing.htm [accessed 19 November 2010]). In addition, the National Institute for Fisheries Research and Development (INIDEP) supports the Federal Fisheries Council mission by creating and conducting research projects for the development of fisheries and evaluating novel aquaculture technology processes (http://www.inidep.edu.ar/home.htm [accessed 19 November 2010]). This organization possesses three research boats to obtain samples and monitor changes in temperatures, phytoplankton, and zooplankton, among others. INIDEP has been collaborating with the National Direction of Aquatic Resources (DINARA) in Uruguay for the last several years studying algal blooms in Mar del Plata (Carreto et al. 2004).

The National Service for Agri-food Health and Quality (SENASA), however, is responsible for regulating, executing, overseeing and certifying processes (including safety, hygiene and quality) of food (including imports and exports) (http://www.senasa.gov.ar/indexhtml.php [accessed 19 November 2010]). SENASA is charged with the control of the marine toxins in seafood nationwide. In order to accomplish this, SENASA has developed a national monitoring program for PSP, ASP, and DSP in provinces along the coast, using regional laboratories (Ferrari 2004). The toxin limits are based on the European Economic Community (EEC) legislation (Table 3). The method of analysis for PSP and DSP in mussels is the mouse bioassay approved by AOAC, and HPLC is used for determining levels of ASP. INIDEP laboratories cooperate providing analytical assistance (Carreto et al. 2004). Although authorities have adopted the mouse bioassay as the official methods for PSP and DSP, SENASA recommends another analysis using LC-MS as a complementary method for the confirmation of PSP and DSP toxins (Carreto et al. 2004).

Brazil

Table 3. South American biotoxin regulatory action levels and methods

Country	Regulatory Levels	Analytical Methods
Argentina		
PSP	80 μg STX eq 100 g^{-1}	MBA
ASP	20 μg DA g^{-1}	HPLC-UV
DSP	16 μg DSP 100 g^{-1}	MBA
Brazil		
-	-	-
Chile		
PSP	80 μg STX eq 100 g^{-1}	MBA
ASP	20 μg DA g^{-1}	HPLC-UV
DSP	20 μg OA 100 g^{-1}	MBA
Colombia		
-	-	-
Ecuador		
-	-	-
Peru		
PSP	80 μg STX eq 100 g^{-1}	MBA
ASP	20 μg DA g^{-1}	HPLC-UV
DSP	If detected	MBA
Uruguay		
PSP	80 μg STX eq 100 g^{-1}	MBA
ASP	20 μg DA g^{-1}	HPLC-UV
DSP	If detected	MBA
Venezuela		
PSP	80 μg STX eq 100 g^{-1}	MBA
DSP	If detected	MBA

Brazil's 7491 km of coastline along the Atlantic Ocean makes it the 16th longest coastline in the world. Per capita seafood consumption was reported to be 6.5 kg on average between 2005-2007 (NMFS 2010). Brazil was responsible for exporting 2,134 metric tons of edible fishery products in 2009 (NMFS 2010) and importing 6,739 metric tons. In 2005, the national production of aquacultured shellfish reached 14,900 tons placing Brazil among the top 3 Latin America producers (Rupp et al. 2008).

PSP toxicity has been demonstrated in waters off the coast of Brazil (Ferrari 2004). DSP illnesses were observed in 1990, with MBA testing positive for DSP and *D. acuminata* confirmed in the water (Ferrari 2004).

A Brazilian National Shellfish Sanitation Program (BNSSP) has been established under the direction of EPAGRI (Agricultural Research and Rural Extension Company of Santa Catarina) (http://ciram.epagri.sc.gov.br/portal/website/ [accessed 19 November 2010]; http://pecmb.wordpress.com/about/ [accessed 19 November 2010]). This program uses NSSP and EU requirements to set the minimum quality standards for seawater and bivalve meat. BNSSP tasks include collecting samples of seawater and shellfish for analysis of bacteria and toxins produced by algae. As part of a pilot monitoring system, environmental variables, water samples, and shellfish are collected and sent to the laboratory for analysis from 40 locations every two weeks along the coast of Santa Catarina (http://www.portaldamaricultura.com.br/noticiasartigos/48-noticias/82-monitoramento-de-algas-nocivas [accessed 22 November 2010]). The monitoring results will be used to create a database with geographical and meteorological information, including the areas with potential for the safe culture of mollusks. In addition, these results are expected to be available to the general public online along with interpretation of results.

Chile

Chile claims 6,435 km of coastline, primarily along the Pacific Ocean. Per capita seafood consumption was reported to be 22.1 kg on average between 2005-2007 (NMFS 2010). Chile exported 1,170 metric tons of edible fishery products in 2009 (NMFS 2010) and imported 100,341. Chile is one of the ten most important countries in the world for aquaculture, the third country in oyster production worldwide and the leading producer of cultured salmon in the world along with Norway (http://www.fao.org/fishery/topic/13540/en [accessed 17 November 2010]). In terms of bivalve aquaculture, Chile is the principal producer in South America with more than 100,000 tons in 2005 (Maeda-Martínez 2008). In 2006, the top destination countries of mussels (*M. chilensis*) were Spain (27.3%), Portugal (22.3%), Italy (16.5%), Argentina (13.9%), and United States (6.6 %) (Uriarte 2008).

PSP has been linked to human intoxications, fatalities, and fish kills along the Chilean coast (Fuentes et al. 2008 and references therein). In 2009, there were reports of *Alexandrium catenella* expansion in Chilean waters (Mardones et al. 2010). Specifically, this observation showed a northward expansion of the species range and documented the first record of this species in SE Pacific coastal oceanic waters off Chiloé Island.

The Institute of Public Health (ISP), the Ministry of Health and the National Fisheries Service (Sernapesca) are the most important organizations that protect public health in Chile. ISP and the Ministry of Health monitor toxins, phytoplankton, and environmental variables from 151 different stations nationwide that are associated with seafood products to be consumed nationally (Seguel 2008). This is performed in coordination with the monitoring program of the Regional Health Authority (Seremi), and toxins analyses are performed at authorized Seremi laboratories. Sernapesca is responsible for the quality of seafood products for exportation (http://www.sernapesca.cl/ [accessed 10 November 2010]). In addition, it oversees the enforcement of bilateral trade agreements of shellfish susceptible to toxicity with European Economic Community (EEC) and the US NSSP. In order to minimize illegal harvest and to comply with new law established by EU in January 2010, all seafood products must be certified by this agency according to the national and international laws applicable. Sernapesca established the Chilean Shellfish Sanitation Program (CNSSP), which certifies harvesting sites and monitors their sanitary conditions (Seguel 2008). CNSSP is primarily funded by the local shellfish farmers (private industry) and laboratories performing the analyses must be authorized by Sernapesca. The University Austral de Chile in Puerto Montt and the University of Chile in Santiago and Castro are currently the only authorized laboratories for the analyses of marine toxins (Seguel 2008).

CNSSP has developed a contingency plan for marine biotoxins as well as other contamination emergencies. When laboratory analyses result in excessive values that pose a threat to public safety, the laboratories immediately notify (by telephone, fax or e-mail) the central and regional offices of Sernapesca. Quantitative and qualitative water samples are tested for the presence of HABs such as *Pseudo-nitzchia australis*, *Alexandrium catenella,* and *Dinophisis acuta*, and shellfish are tested for toxins. For marine toxins and phytoplankton, the monitoring frequency is every two days until the toxin levels return to below action values (http://www.sernapesca.cl/ [accessed 10 November 2010]). In the event that intensive sampling is not possible, the harvest areas will remained closed until a monitoring program is established.

If tissue samples contain high levels of toxins (Table 3), the harvesting zones are closed immediately. The areas adjacent to those affected are also evaluated to establish whether they will remain open or closed. In addition, all potentially affected products will be identified and may not be exported until they are determined to be safe for consumption. Each available lot of finished product, produced after levels were last considered safe and before the contigency plan is initiated, will be sampled and analyzed for marine toxins. In the event that contaminated products are shipped, the corresponding authorities (e.g., US FDA, EU) must be informed immediately with details about the exported product, including the date of shipment, the final recipient, date of arrival in the destination market, and other relevant information. The company must act according to their established quality assurance program regarding the withdrawal of potentially contaminated products. Servapesca will lift the restriction on harvests once the results show negative values after two consecutive analyses. Additionally, they will inform the fisheries that the contingency plan has expired, and routine monitoring resumes.

Regulatory limits of toxins in seafood established in Chile are provided in Table 3. According to Seguel (2008), PSP and DSP are the most abundant toxins in Chile, and the mouse bioassay is the official method for their detection. ASP toxins are detected using HPLC-UV. Other detection techniques such as post-column derivatization HPLC have been employed for research purposes in order to obtain more information about toxin profiles in phytoplankton and seafood products (Lembeye 2008).

With respect to exports to the US, the government of Chile and the US FDA signed a memorandum of understanding (MOU) in May 1989, which was renegotiated in February 2001. These parties agreed to continue the beneficial and cooperative collaboration concerning the safety of fresh and frozen oysters, clams and mussels being exported to the US. In this MOU, the FDA acknowledged that the CNSSP program meets the US NSSP guidelines and that Sernapesca retains the overall responsibility and program direction of the CNSSP. As is the case for exports to the EU, industries interested in exporting shellfish products must submit an application to CNSSP and comply with the requirements described in the Quality Assurance Program (PAC). CNSSP personnel will evaluate and classify the area, describe possible contamination sources and install monitoring programs. The industry must utilize laboratories authorized by CNSSP for sample analyses and must continuously send laboratory results to the regional office of Sernapesca. Based on the information collected from CNSSP (classification of areas, certification of harvesting areas, compliance of processing plants and certification of dealers exporting the shellfish), the central office of Sernapesca decides if the company will be approved by the CNSSP. Once the industry is approved, Sernapesca will prepare a report and send it to the US FDA with details of the new private industry. Finally, in order to ensure compliance, Sernapesca will continuously monitor and inspect the facilities.

Colombia

Colombia's 3,208 km of coastline is along the Pacific Ocean and Caribbean Sea. Per capita seafood consumption was reported to be 5.6 kg on average between 2005-2007 (NMFS 2010). Colombia was responsible for exporting 1,144 metric tons of edible fishery products in 2009 (NMFS 2010).

Although 50% of Colombian territory is near the sea, the production of bivalve is one of the lowest in Latin America and it is almost exclusively artisanal (Velasco and Barros 2008). The Colombian Institute for Rural Development (INCODER) implements policies, in coordination with communities, and public and private institutions related to the agricultural, forestation, and fishing sector, aiming to improve their quality of life and the socio-economic development (http://www.incoder.gov.co/ [accessed 16 December 2010]). However, there is no national policy for development of bivalve aquaculture. Insufficient government resources for research, development and technology transfer in aquaculture (<0.03% of the national budget), poor sanitary conditions, limited markets for bivalves, and the lack of monitoring systems hinder the development of bivalves aquaculture in Colombia. No information could be found on marine biotoxin problems or means of control for Colombia.

Ecuador

The coastline of Ecuador totals 2,237 km and extends along the Pacific Ocean. Per capita human consumption of fish and shellfish in Ecuador was reported to be 4.4 kg on average from 2005-2007 (NMFS 2010). Ecuador imported 102,816 metric tons of edible fishery products in 2009 (NMFS 2010).

The National Centre for Aquaculture and Marine Research (CENAIM) is responsible for promoting the production and diversification of aquaculture in marine fish and bivalve molluscs in Ecuador (http://www.cenaim.espol.edu.ec/organizacion/index.html [Accessed 16 December 2010]). Harvest from natural shellfish beds has not been developed due to the lack of government policies (Alvarez et al., 2008), so shellfish production has been limited to aquaculture activities. CENAIM has established laboratories for molluscs and phytoplanckton analysis. Research is concentrated on species that have commercial potential. In addition, CENAIM has a technical cooperation program with Japan (JICA Project) for the cultivation of bivalve molluscs. No further information on marine biotoxin controls was available.

Peru

Peru claims 2,414 km of coastline along the Pacific Ocean. While large-scale, industrial fishing yields most of the catch targeted mainly for fishmeal export, small-scale fisheries in Peru are important as they provide the majority of domestic seafood and employ more people than the industrial fisheries (Alfaro-Shigueto et al. 2010). Per capita seafood consumption in Peru was reported to be 19.9 kg on average between the years 2005 to 2007 (NMFS 2010). In 2006, the amount of scallops alone that were exported increased from 1,000 tons in previous years to 13,000 tons (~$40 million dollars), which represented 44% of the total aquaculture products (Mendo et al. 2008). In terms of world landings, Peru tied with India at 5% (~7.5 million metric tons) of the total world harvest of fish, crustaceans, and mollusks in 2008 (NMFS 2010).

Bivalve harvesting has increased in Peru and the dominant commercial species include scallops (*Argopecten purpuratus*), clams (*Semele* sp.), coquina (*Donax* sp.), razorclams (*Ensis* sp. and *Tagelus dombeii*) and mussels (*Aulacomya ater*) (Antinori et al. 2003). Actual red tides with water discoloration are frequent in Peruvian waters (e.g., Smith 1978).

The General Directorate of Environmental Health (DIGESA) is the sanitary agency that promotes and protects public health by establishing controls, regulations and surveillance to increase the environmental safety nationwide (http://www.digesa.sld.pe/ [accessed 16 November 2010]). DIGESA accomplishes its mission through collaborations and technical advice of institutions such as the Peruvian Marine Institute (IMARPE) and the Peruvian Institute of Fish Technology (ITP), which is a part of the Ministry of Production (known as the Ministry of Fishery prior to 2002) (www.itp.gob.pe/acerca-perfil.php [accessed 19 November 2010]). Currently, the General Fish Law (LGP) and the Promotion and Development of the Aquaculture Law (LPDA) regulate the exploitation of the fisheries resources in Peru. Responsibilities of IMARPE and ITP include monitoring of toxic phytoplankton, conducting research, and advising the State about the best use of fishery resources and conservation of the marine environment.

When high levels of potentially toxic phytoplankton are found, the monitoring is intensified (Mendo et al. 2008). The monitoring program has a total of 32 sampling stations and focuses on areas where the export products are collected. The mouse bioassay is the official method for PSP and DSP toxins, while HPLC is the method for ASP toxins (Table 3) (Mendo et al. 2008). The regulatory limit established for PSP toxins is 80 μg STX eq 100 g^{-1} and the limit for ASP toxins is 20 μg g^{-1}. Closures are in effect if the MBA for DSP is positive. The shellfish production areas and natural beds are closed by DIGESA if the toxin levels exceed regulatory limits. This prohibition can be lifted after two consecutive negative results (Antinori et al. 2003).

Uruguay

The Uruguayan coast consists of 680 km on the Atlantic Ocean and the Río de la Plata. The country is divided into 19 departments, 6 of which are located on the coast. The Río de la Plata is an extensive fluvio-marine system with high spatio-temporal variability (Hareau et al. 1999), and it represents an important fishing ground. In Uruguay, there are artisanal fishing communities that focus on the harvest of mussels (*Mytilus edulis*), clams (*Mesodesma mactroides*), and cockles (*Donax hanleyanus*) along the coast, and there are offshore fisheries of clams (*Pitar rostrata*) and scallops (*Zigoclamys patagonica*) (Medina et al. 2003). Per capita seafood consumption in Uruguay was reported to be 9.7 kg on average between the years 2005 and 2007 (NMFS 2010). It is notable that Uruguay is an exporter of seafood. They were approved to export frozen clams (*Pitar rostrata*) to the EU in accordance with EU Directive 91/492/CEE (Medina et al. 2003).

Shellfish toxins were first reported in Uruguay in 1980 when approximately 60 consumers experienced neurological symptoms following the consumption of mussels (Davison and Yentsch 1985). Since that time the toxic dinoflagellates *Alexandrium tamarense* and *Gymnodonium catenatum* have been observed and PSP toxins have been detected (Medina et al. 2003). *Dinophysis acuminata* and *D. caudata* have been associated with DSP events. And *Pseudo-nitzschia multiseries* has been reported, as has domoic acid in mussels (Medina et al. 2003). Further, cyanobacteria blooms of *Microcystis aeruginosa* are common and pose public health problems, are responsible for fish kills, and impact drinking water supplies (Ferrari and Méndez 2004).

As a result of the 1980 event, a monitoring program was formed by the National Directorate for Aquatic Resources (DINARA-Dirección Nacional de Recursos Acuáticos) to detect phycotoxins and harmful algae. DINARA, which falls under the Ministry of Animal Husbandry, Agriculture, and Fisheries, is the government agency responsible for the regulation, monitoring, and control of the national fisheries activities in Uruguay (http://www.dinara.gub.uy/ [accessed 17 November 2010]).

Reports indicate that samples are collected for cockles, clams, mussels, and scallops for testing (Medina et al. 2003; Méndez 2006). Phytoplankton samples are collected weekly throughout the year, and shellfish samples are collected weekly from December to March and biweekly the rest of the year. Phytoplankton samples are generally taken from docks, rocky points, or small boats. Shellfish are sampled from natural beds as well as from fresh market stalls (Méndez 2006). At the time of sample collection, the monitoring program also makes observations and records a range of environmental conditions. The collected samples are sent to the laboratory for analysis. According to Méndez (2006), PSP and DSP toxins are determined by mouse bioassay (AOAC 2005a; Yasumoto et al. 1984, respectively) and ASP toxins are analyzed using HPLC (AOAC 2005b).

Ferrari and Méndez (2004) indicated that bans were implemented when (1) PSP toxins exceeded 80 μg STX eq 100 g^{-1}, (2) DSP assays were positive, and (3) toxic phytoplankton species were observed in harvest areas (Table 3). Based on the results, regulatory measures are implemented and the public is informed of the findings as necessary. If shellfish products exceed the regulatory limit or if toxic algae are present in shellfish harvest areas, DINARA adopts contingency procedures in order to safeguard public health. The contingency plan includes a press release to the general public as well as an official statement to all specific institutions involved (e.g., fisherman associations, health authorities).

Venezuela

Venezuela has 2,800 km of Caribbean coastline. Per capita human consumption of fish and shellfish in Venezuela was reported to be 16.9 kg on average from 2005-2007 (NMFS 2010). Venezuela was responsible for exporting 3,943 metric tons of edible fishery products in 2009 (NMFS 2010).

Bivalve mollusk production in the 1970s and 80s reached between 200-300 tons per year, concentrating mainly on the brown mussel *Perna perna* and oysters (e.g., *Crassostrea virginica* and *rhizophorae*). Bivalve mollusks aquaculture is currently in a developmental stage, and considered a small scale farming activity (www.insopesca.gov.ve [accessed 23 December 2010]). This is mainly due to environmental factors (harmful algal blooms) and also, to a certain extent, price competition of artisanal fisheries. The main piece of legislation regulating the national aquaculture is the Fisheries and Aquaculture Law (2001), according to which the Ministry of Agriculture and Land is in charge of drafting the national fisheries and aquaculture policy, as well as the international fisheries policy. This Law provides the Ministry the power to authorize the conduct of fisheries, aquaculture and related activities. However, the granting of permits, licenses and concessions is vested in the National Institute for Fisheries and Aquaculture (INAPesca).

From 1989 to 1998 a monitoring program was established to detect toxic algae and toxicity in bivalve mollusks along the coast of Sucre state, Venezuela (Lodeiros and Freites 2008). The results of the monitoring program showed that *Gymnodinium catenatum* blooms were observed yearly along the coast of Venezuela (Barbara-Sánchez and Gamboa-Máruez 2001). In 1989, a *G. catenatum* bloom was responsible for death of mussels in natural banks and intoxication of 15 persons by consumption of bivalve mollusks (Barbara-Sánchez and Gamboa-Máruez 2001). In a different study performed near Margarita Island (northwestern coast), mussel samples were obtained from an outbreak of PSP in 1991 and analyzed using HPLC to determine toxins profile. The major toxins detected were dcSTX and dcGTX in mussels, consistent with the identification of *G. catenatum* as the causative organism for this outbreak. However, HPLC analyzes performed on green mussels near Playa Patilla (on the northern part of Venezuela) revealed high concentrations of PSP toxins (a total of 468 $\mu g\ 100\ g^{-1}$) and the widest range of toxins found in the country: STX, NeoSTX, dcSTX, GTX 1/4, GTX3/4, dc GTX 2/3 and C/2 (Barbara-Sánchez and Gamboa-Máruez 2001).

The national government, through the Institute of Agricultural Research (INIA), INSOPESCA, and the Foundation for Science and Technology (FUNDACITE) in Sucre, and in agreement with the Ministry of Popular Power for Health and Social Protection (MPPSDS), have established a continuous monitoring of marine biotoxins in seafood products, with particular emphasis on bivalve mollusks (www.insopesca.gov.ve [accessed 23 December 2010]). In addition, FUNDACITE developed a foundation to develop and promote aquaculture as a food safety function, transferring technology to coastal communities for the development of mussel farming in order to create a supplier industry for processing plants. No further details were found about the monitoring frequency and/or contingency plan in the event of HABs or toxin presence.

1.4. Global Standardization

The countries highlighted in this chapter exhibit a range of biotoxin control programs. Globally, there are even greater disparities and inconsistencies in how risks from marine biotoxins are regulated/ managed. Thus, there is a strong need to harmonize food safety standards, including those related specifically to seafood toxins (e.g., Fernández 2000; Fernández and van Egmond 2004; Toyofuku 2004; Choo 2004). In a review by Fernández (2000) on marine biotoxin regulations and efforts towards harmonization, the least consensus was found for the DSP toxins with respect to (1) included toxins, (2) acceptable levels, and (3) appropriate testing methods. Reaching an international consensus and establishing standardization not only on this group of toxins, but on all marine biotoxins, would allow for heightened food safety awareness and would greatly facilitate trade.

The Food and Agriculture Organization of the United Nations (FAO) and the World Health Organization (WHO) together established the Codex Alimentarius Commission as a single international reference point for food standards, including standards relating to seafood biotoxins. Codex Alimentarius Commission "*presents a unique opportunity for all countries to join the international community in formulating and harmonizing food standards and ensuring their global implementation*" (http://www.codexalimentarius.net/web/index_en.jsp [accessed 18 January 2011]). The Codex Alimentarius Commission is the "*body responsible for compiling the standards, codes of practice, guidelines and recommendations that constitute the Codex Alimentarius.*" All 18 countries covered in this chapter are members of Codex. In order to operate and address food standardization, Codex establishes two types of subsidiary bodies: (1) Codex Committees and (2) Coordinating Committees (Figure 2). Codex Committees prepare specific draft standards for the Commission, whereas Coordinating Committees focus on regional food standards activities.

Codex Committees consist of General Subject Committees and Commodity Committees. General Subject Committees are also known as horizontal committees because their focus is applicable to all food commodities. The General Subject Committee most related to marine biotoxin issues is the Committee on Methods of Analysis and Sampling (CCMAS). While all member countries are welcome to participate in each committee, not every country sends representatives to each committee meeting and participation is generally dependent on the topics of discussion and travel funds. Canada, the United States, Honduras, Panama, Argentina, Brazil, Chile and Venezuela have actively participated since 2008 in CCMAS (http://www.codexalimentarius.net/web/archives.jsp?lang=en [accessed 18 January 2011]). The main role of CCMAS in the area of biotoxin control is to endorse methods proposed by the Commodity Committees, which are referred to as vertical committees.

The Committee on Fish and Fishery Products (CCFFP) is the Commodity Committee that is applicable to seafood biotoxins. Since 2008 Canada, Mexico, the United States, Costa Rica, Honduras, Argentina, Brazil, Chile, Peru and Uruguay have actively participated in CCFFP (http://www.codexalimentarius.net/web/archives.jsp?lang=en [accessed 18 January 2011]).

This particular Committee seeks to elaborate world wide standards for fresh, frozen, or otherwise processed fish, crustaceans, and mollusks. An example of CCFFP involvement in biotoxin standardization was the formation of an electronic working group in 2010 that focused on the Proposed Draft List of Methods for the Determination of Biotoxins in the Standard for Raw and Live Bivalve Molluscs. The primary focus of this group was to develop performance criteria that must be met for reference methods used for biotoxin monitoring for regulatory purposes. The results and summary from this working group were reported back to CCFFP for Codex consideration of adoption.

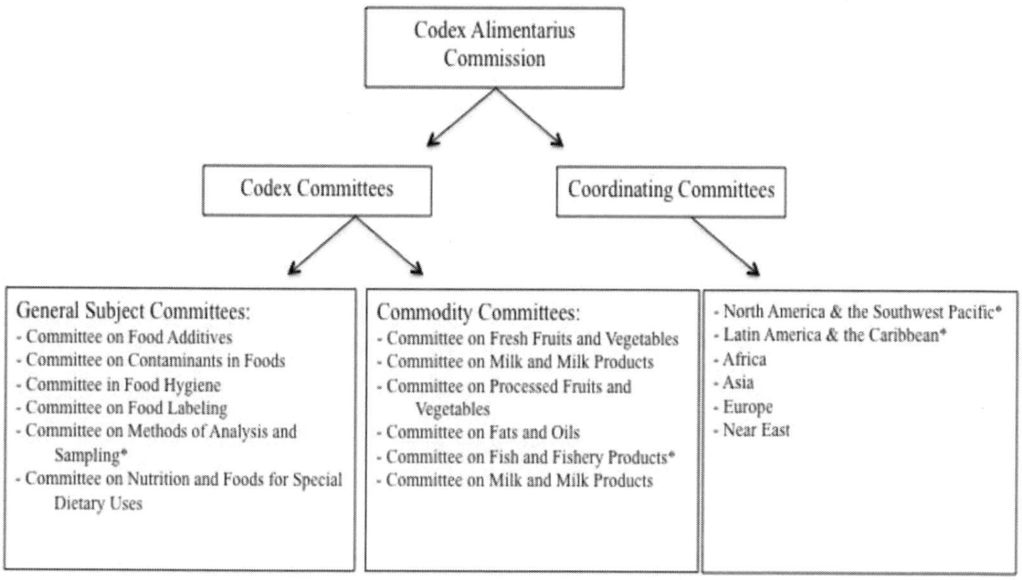

Figure 2. Representation of Codex Alimentarius Commission subsidiary bodies. An * indicates the Codex Committees most relevant to standard setting for biotoxin controls.

There are also six Coordinating Committees, including one focused on (1) North America and the Southwest and another on (2) Latin America and the Caribbean. Coordinating Committees serve to coordinate regional food standards and establish regional positions when relevant. Certainly, there are challenges that prevent some countries from adopting Codex standards, such as available resources, infrastructure, and adequate food law. Despite the numerous challenges facing implementation of global standardization efforts, such harmonization is gaining favor due to the increasing need and desire for international trade.

CONCLUSION

Countries with established fishing histories have developed control programs for managing marine biotoxin threats, with the strongest programs in North America. Programs monitoring for PSP toxins were the most widespread, owing to its extensive presence in coastal waters. In most cases, the action levels established in different countries are quite comparable. In countries where control programs are lacking, HABs and associated toxins appear to be minimal and present a low probability risk. It is likely that global standardization efforts will facilitate (1) establishing biotoxin monitoring programs in areas where they are needed and (2) further harmonization regarding official methods and regulatory action levels, thereby enhancing the capacity to protect public health while creating equivalency for international trade.

ACKNOWLEDGMENTS

We extend special thanks to Bob Dickey and Ana Cabado for the invitation to write this chapter. We would also like to thank the numerous people who contributed helpful information or made important contacts for us in the drafting of this manuscript: Paul Seligman, Mike Wehr, Camille Brewer, Melissa Ellwanger, Gisela Kopper, Gonzalo Ibanez, Benjamin Suarez, Miguel Figueroa, and Jeff van de Riet. Thanks to Steve Conrad for preparing the map for figure 1. Thanks to Jeff DeGrasse, John Callahan, Greg Diachenko, and Bill Jones for reviewing the manuscript. KMD was supported by an appointment with the Research Participation Program at CFSAN administered by the Oak Ridge Institute for Science and Education operated under an interagency agreement between the U.S. Department of Energy and the U.S. Food and Drug Administration.

REFERENCES

Alfaro-Shigueto, J., Mangel, J. C., Pajuelo, M., Dutton, P. H., Seminoff, J. A., & Godley, B. J. (2010). Where small scan have a large impact: Structure and characterization of small-scale fisheries in Peru. *Fisheries Research* 106, 8-17.

Altamirano, R. C. & Sierra-Beltrán, A. P. (2008). Biotoxins from freshwater and marine harmful algal blooms occurring in Mexico. *Toxin Reviews* 27, 27-77.

Alvarez, R., Cobo, L., Sonnenholzner, S., & Stern, S. (2008). Estado actual de la acuicultura de moluscos bivalvos en Ecuador. In Estado actual del cultivo y manejo de moluscos bivalvos y su proyección futura: factores que afectan su sustentabilidad en América Laina, Lovatelli, A., Farías, A., Uriarte, I., (Eds.). *Organización de las naciones unidas para la agricultura y la alimentación,* Roma, pp. 129-134.

Antinori, J. A., Sanchez-Ycochea, L. M., Mendoza, J. A., Flores, O., & Ascon, D. (2003). Monitoring of harmful microalgae and phycotoxins in Peru: first report of PSP toxins in shellfish. In: Villalba, A., Reguera, B., Romalde, J. L., & Beiras, R. (Eds.), Molluscan Shellfish Safety. *Intergovernmental Oceanographic Commission of UNESCO,* pp. 203-210.

AOAC Official Methods of Analysis. 2005a. 18th ed.; AOAC International, Gaithersburg, MD, Method 959.08, *Method locator* 49.10.01 (Chapter 49, Subchapter 10, 1st Method). [http://www.eoma.aoac.org/gateway/readFile.asp?id=959_08.pdf]

AOAC Official Methods of Analysis. 2005b. 18th ed.; AOAC International, Gaithersburg, MD, Method 991.26, Method locator 49.10.02 (Chapter 49, Subchapter 10, 2nd Method).

Armero-Guardado, J. A. (2002). *Brote de intoxicaciones paraliticas por mariscos,* El Salvador, Agosto 2001-Enero 2002. Ministerio de Salud Publica y Asistencia Social, Unidad Nacional de Epidemiologia. p. 6.

Barbara-Sanchez, A. L. & Gamboa-Maruez, J. F. (2001). Distribution of *Gymnodinium catenatum* graham and shellfish toxicity on the coast of Sucre state, Venezuela, from 1989 to 1998. *Journal of Shellfish Research* 20(3), 1257-1261.

Barraza, J. E., Armero-Guardado, J., & de Toledo, Z. M. V. (2004). The red tide event in El Salvador, August 2001-January 2002. *Rev. Biol. Trop.* 52 (Suppl. 1), 1-4.

Barraza, J. E. (2009). Food poisoning due to consumption of the marine gastropod *Plicopurpura columellaris* in El Salvador. *Toxicon* 54, 895-896.

Bean, L. L., McGowan, J. D., & Hurst, J. W. (2005). Annual variations of paralytic shellfish poisoning in Maine, USA 1997-2001. *Deep-Sea Research Part* II 52, 19-21.

Carreto, J. I., Lasta, M., Negri, R. M., & Benavides, H. R. (1981). Los fenomenos de marea roja y toxicidad de moluscos bivalvos en el Mar Argentino. *Contrib. Inst. Nac. Invest. Des. Pesq.* (Mar del Plata), 399, 55 pp.

Carreto, J. I., Montoya, N. G., Cucchi Colleoni, A. D., & Akselman, R. (1998). *Alexandrium tamarense* blooms and shellfish toxicity in the Argentine Sea: a retrospective view. In: Reguera, B., Blanco, J., Fernandez, M. L., & Wyatt, T. (Eds.), *Harmful Microalgae*, IOC-UNESCO, pp. 131-134.

Carreto, J. I., Montoya, N. G., & Akselman, R. (2004). *Proyecto "Proteccion Ambiental del Rio de la Plata y su Frente Maritimo.* Prevencion y Control de la Contaminacion y Restauracion de Habitats" Projecto PNUD/GEF RLA/99/G31, 20 pp. [http://www.freplata.org/documentos/archivos/Documentos_FREPLATA/Marea_roja_diagnostico_final.pdf]

Choo, P. S. (2004). Seafood production and trade in various APEC economies: the need for harmful biotoxin regulatory mechanisms. In: Hall, S., Etheridge, S., Anderson, D., Kleindinst, J., Zhu, M., & Zou, Y., (Eds.), *Harmful Algae Management and Mitigation,* pp. 88-91.

Chow, N., Vammen K., & Reguera, B. (2010). First report of PSP on Pacific coast of Nicaragua associated with *Pyrodinium bahamense*. *Harmful Algae News* 41, 6-7.

Davison, P. & Yentsch, C. (1985). Occurrence of toxic dinoflagellates and toxic shellfish along coastal Uruguay – South America. In: Anderson, D. M., White, A. W., & Badden, D. G. (Eds.*), Toxic Dinoflagellates,* Elsevier Science Publishing, New York, pp. 153-158.

Deeds, J. R., Wiles, K., Heideman, G. B., White, K. D., & Abraham, A. (2010). First US report of shellfish harvesting closures due to confirmed okadaic acid in Texas Gulf coast oysters. *Toxicon* 55(6), 1138-1146.

Dickey, R. W. & Plakas, S. M. (2010). Ciguatera: A public health perspective. *Toxicon* 56(2), 123-136.

Dominguez, H. J., Paz, B., Daranas, A. H., Norte, M., Franco, J. M., & Fernández, J. J. (2010). Dinoflagellate polyether within the yessotoxin, pectenotoxin, and okadaic acid toxin groups: Characterization, analysis and human health implications. *Toxicon* 56(2), 191-217.

EC Regulation No. 854/2004, 2004. *Off. J. Eur. Commun.* L226, 83-127.

Etheridge, S. M. (2010). Paralytic shellfish poisoning: seafood safety and human health perspectives. *Toxicon* 56, 108-122.

Fernández, M. L. (1998). Phycotoxins: regulatory limits and effects on trade. In: Miraglia, M., et al., (Eds.), *Mycotoxins and phycotoxins – developments in chemistry, toxicology, and food safety*, Fort Collins, Colorado, Alaken Inc., pp. 503-516.

Fernández, M. L. (2000). Regulations for marine microalgal toxins: towards harmonization of methods and limits. S. *Afr. J. Mar. Sci.* 22, 339-346.

Fernández, M. L., & van Egmond, H. (2004). European approaches to marine toxin control and harmonisation. In: Hall, S., Etheridge, S., Anderson, D., Kleindinst, J., Zhu, M., & Zou, Y. (Eds.), *Harmful Algae Management and Mitigation,* APEC, Singapore, pp. 77-83.

Ferrari, G. (2004). HABs in the southwestern Atlantic Ocean coast: a general review. In: Hall, S., Etheridge, S., Anderson, D., Kleindinst, J., Zhu, M., & Zou, Y. (Eds.), *Harmful Algae Management and Mitigation,* APEC, Singapore, pp. 66-73.

Ferrari, G. & Méndez, S. (2004). Harmful algal monitoring program in Uruguay. In: Hall, S., Etheridge, S., Anderson, D., Kleindinst, J., Zhu, M., & Zou, Y. (Eds.), *Harmful Algae Management and Mitigation,* APEC, Singapore, pp. 144-148.

Fuentes, C., Clement, A., & Aguilera, A. (2008). Summer *Alexandrium catenella* bloom and the impact on fish farming in the XI Aysén region, Chile. In: Moestrup, Ø. et al. (Eds.), *Proceedings of the 12th International Conference on Harmful Algae.* International Society for the Study of Harmful Algae and Intergovernmental Oceanographic Commission of UNESCO, Copenhagen, pp. 183-186.

Furey, A., O'Doherty, S., O'Callaghan, K., Lehane, M., & James, K. J. (2010). Azaspiracid poisoning (AZP) toxins in shellfish: Toxicological and health considerations. *Toxicon* 56(2), 173-190.

Gayoso, A. M. & Ciocco, N. (2001). Observations on Procentrum lima from North-Patagonian coastal waters (Argentina) associated with a human diarrhoeic disease episode. *Harmful Algae News* 22, 4.

Gilgan, M. W., Powell, C., van de Riet, J., Burns, B. G., Quilliam, M. A., Kennedy, K., & McKenzie, C. H. (1994). The occurrence of a serious DSP epidose in mussels from Newfoundland during the late autumn of 1993. *Fourth Canadian Workshop on Harmful Algae. DFO Institute for Ocean Sciences,* Sidney, BC, 3-5 May.

Hareau, A., Hofstadter, R., & Saizar, A. (1999). Vulnerability to climate change in Uruguay: potential impacts on the agricultural and coastal resource sectors and response capabilities. *Clim. Res.* 12, 185-193.

Hernandez-Becerril, D. U. (1998). Species of the planktonic diatom genus *Pseudo-nitschia* of the Pacific coasts of Mexico. *Hydrobiologia* 379, 77-84.

INCOPESCA. 2001. *Informes del Instituto Costarricense de Pesca y Acuicultura*, 2001.

Kao, C. Y. (1993). Paralytic shellfish poisoning. In: Falconer, I. R. (Ed.), *Algal Toxins in Seafood and Drinking Water,* Academic Press, San Diego, pp. 75-86.

Kite-Powell, H. L., Fleming, L. E., Backer, L. C., Faustman, E. M., Hoagland, P., Tsuchiya, A., Younglove, L. R., Wilcox, B. A., & Gast, R. J. (2008). Linking the oceans to public health: current efforts and future directions. *Environmental Health* 7(S2), S6.

Klontz, K. C., Abraham, A., Plakas, S. M., & Dickey, R. W. (2009). Mussel-associated azaspiracid intoxication in the United States. *Annals of Internal Medicine* 150(5), 361.

Lefebvre, K. A. & Robertson, A. (2010). *Domoic acid and human exposure risks: A review. Toxicon* 56(2), 218-230.

Lembeye, G. (2008). Harmful algal blooms in the austral Chilean channels and fjords. In: Silva, N., & Palma, S. (Eds.), *Progress in the oceanographic knowledge of Chilean interior waters, from Puerto Montt to Cape Horn.* 2008 Comité Oceanográfico Nacional - Pontificia Universidad Católica de Valparaíso, Valparaíso, pp. 99-103.

Licea, S., Gomez-Aguirre, S., Cortes-Altamirano, R., & Gomez, S. (1999). Notas sobre algunos florecimientos algales y la presencia de especies toxicas en cinco localidades del Pacifico Mexicano (1996-1999). In: Tresierra-Aguilar, E. A. & Culquichicón-Malpica, Z. G. (Eds.), *Libro de resúmenes ampliados. COLACMAR VIII. Universidad Nacional de Trujillo,* Editora Nuevo Norte, Peru, Vol. I, pp. 335-337.

Licea, S., Navarrete, A., Bustillos, J., & Martínez, B. (2008). Monitoring a bloom of *Pyrodinium bahamense* var. *compressum* in El Salvador and the southern coast of Mexico (November 2005 – March 2006). In: Moestrup, ∅. et al. (Eds.*), Proceedings of the 12th International Conference on Harmful Algae. International Society for the Study of Harmful Algae and Intergovernmental Oceanographic Commission of UNESCO,* Copenhagen, pp. 86-89.

Lloret, J. (2010). Human health benefits supplied by Mediterranean marine biodiversity. *Marine Pollution Bulletin* 60, 1640-1646.

Lodeiros, S. C. & Freites, V. L. (2008). Estado actual y perpectivas del cultivo de moluscos bivalvos en Venezuela. In: Lovatelli, A., Farías, A., & Uriarte, I. (Eds.), *Estado actual del cultivo y manejo de moluscos bivalvos y su proyección futura: factores que afectan su sustentabilidad en América Laina, Organización de las naciones unidas para la agricultura y la alimentación,* Roma, pp. 135-150.

Maeda-Martínez, A.N. (2008). Estado actual del cultivo de bivalvos en México. En: Lovatelli, A., Farías, A., & Uriarte, I. (Eds), Estado actual del cultivo y manejo de moluscos bivalvos y su proyección futura: factores que afectan su sustentabilidad en América Latina. Taller Técnico Regional de la FAO. 20–24 de agosto de 2007, Puerto Montt, Chile. *FAO Actas de Pesca y Acuicultura.* No. 12. Roma, FAO. pp. 91–100.

Mardones, J., Clément, A., Rojas, X., & Aparicio, C. (2010). *Alexandrium catenella* during 2009 in Chilean waters, and recent expansion to coastal ocean. *Harmful Algae News* 41, 8-9.

Mata, L., Abarca, G., Marranghello, L., & Víquez, R. (1990). Rev. Biol. Trop. 38, 129-136.

McNabb, P., Selwood, A. I., & Holland, P. T. (2005). Multiresidue method for determination of algal toxins in shellfish: single-laboratory validation and interlaboratory study. *J. AOAC Intl.* 88(3), 761-772.

Meave del Castillo, M. E. & Hernandez-Becerril, D. U. (1998). Fitoplancton. *En El golfo de Tehuantepec: el ecosistema y sus recursos,* Tapia, G., (Ed.), UAM Iztapalaps, Mexico, pp. 59-74.

Medina, D., Mendez, S., Inocente, G., Ferrari, G., Salhi, M., Giudice, H., Mendez, E., Odizzio, M., & Otero, M. D. (2003). Shellfish monitoring program in Uruguay. In: Villalba, A., Reguera, B., Romalde, J. L., & Beiras, R. (Eds.), *Molluscan Shellfish Safety.* Intergovernmental Oceanographic Commission of UNESCO, pp. 197-202.

Méndez, S. M. (2006). Impacto de las floraciones algales nocivas en Uruguay: origen, dispersión, monitoreo, control, y mitigación. In: Menafra, R., Rodríguez-Gallego, L., Scarabino, F., & Conde, D. (Eds.), *Bases para la conservación y el manejo de la costa uruguaya.* Vida Silvestre Uruguay, Montevideo, pp. 57-69.

Mendo, J., Wolff, M., Carbajal, W., Gonzáles, I., & Badjeck, M. (2008). Manejo y explotación de los principales bancos naturales de concha de abanico (*Argopecten purpuratus*) en la costa Peruana. En: Lovatelli, A., Farías, A., & Uriarte, I. (Eds), *Estado actual del cultivo y manejo de moluscos bivalvos y su proyección futura.Organización de las naciones unidas para la agricultura y la alimentación.* Roma, pp. 101-114.

NMFS (National Marine Fisheries Service) 2010. Fisheries statistics of the United States – 2009. *Current Fisheries Statistics No. 2009.* Silver Spring, MD, 104 pp. Available from http://www.st.nmfs.noaa.gov/st1/fus/fus09/fus_2009.pdf

Oken, E., Østerdal, M. L., Gillman, M. W., Knudsen, V. K., Halldorsson, T. I., Strom, M., Bellinger, D. C., Hadders-Algra, M., Michaelson, K. F., & Olsen, S. F. (2008). Associations of fish intake during pregnancy and breastfeeding duration with attainment of developmental milestones in early childhood: a study from the Danish National Birth Cohort. *Am. J. Clin. Nutr.* 88, 789-796.

Perez, A. (2009). Fisheries management at the tri-national border between Belize, Guatemala, and Honduras. *Marine Policy* 33, 195-200.

Plakas, S. M. & Dickey, R. W. (2010). Advances in monitoring and toxicity assessment of brevetoxins in molluscan shellfish. *Toxicon* 56(2), 137-149.

Quilliam, M. A., Gilgan, M. W., Pleasance, S., deFreitas, A. S. W., Douglas, D., Fritz, L., Hu, T., Marr, J. C., Smyth, C., & Wright, J. L. C. (1992). Confirmation of an incident of diarrhetic shellfish poisoning in eastern Canada. *Proceedings of the 5th International Conference on Toxic Marine Phytoplankton.* Elsevier Science Publishers B. V. Amsterdam, 547-552.

Rodrigue, D. C., Etzel, R. S., Hall, S., De Porras, E., Velasquez, O. H., Tauxe, R. V., Kilbourne, E. M., & Blake, P. A. (1990). Lethal paralytic shellfish poisoning in Guatemala. *Am. J. Trop. Med. Hyg.* 42(3), 267-272 (89-262).

Rupp, G. S., De Oliveira Neto, F. M., & Guzenski, J. (2008). Estado actual del cultivo de moluscos bivalvos en la región sudeste-sur de Brasil. In: Lovatelli, A., Farías, A., Uriarte, I. (Eds.), *Estado actual del cultivo y manejo de moluscos bivalvos y su proyección futura: factores que afectan su sustentabilidad en América Laina, Organización de las naciones unidas para la agricultura y la alimentación,* Roma, pp. 77-90.

Seguel, M. (2008). Manejo de cultivos bivalvos contaminados con marea roja. In: Lovatelli, A., Farías, A., & Uriarte, I. (Eds.), *Estado actual del cultivo y manejo de moluscos bivalvos y su proyección futura: factores que afectan su sustentabilidad en América LainaOrganización de las naciones unidas para la agricultura y la alimentación,* Roma, pp. 309-316.

Sierra-Beltrán, A. P., Cruz, A., Nunez, E., Del Villar, L. M., Cerecero, J., & Ochoa, J. L. (1998). An overview of the marine food poisoning in Mexico. *Toxicon* 36(11), 1493-1502.

Smith, S. L. (1978). Nutrient regenration by zooplankton during a red tide off Peru, with notes on biomass and species composition of zooplankton. *Marine Biology* 49, 125-132.

Toyofuku, H. (2004). International standard setting on marine biotoxins in shellfish. In: Hall, S., Etheridge, S., Anderson, D., Kleindinst, J., Zhu, M., & Zou, Y. (Eds.), *Harmful Algae Management and Mitigation,* pp. 84-87.

Uriarte, I. (2008). Estado actual del cultivo de moluscos bivalvos en Chile. In: Lovatelli, A., Farías, A., Uriarte, I. (Eds.), Estado actual del cultivo y manejo de moluscos bivalvos y su proyección futura: factores que afectan su sustentabilidad en América Laina,

Organización de las naciones unidas para la agricultura y la alimentación, Roma, pp. 61-76.

van de Riet, J. M., Gibbs, R. S., Chou, F. W., Muggah, P. M., Rourke, W. A., Burns, G., Thomas, K., & Quilliam, M. A. (2009). Liquid chromatographic post-column oxidation method for analysis of paralytic shellfish toxins in mussels, clams, scallops, and oysters: single-laboratory validation. J. AOAC Intl. 92(6), 1690-1704.

Vargas, M. & Freer, E. (2003). Co-occurrence of different morphotypes of Pyrodinium bahamense during an extensive bloom in the gulf of Nicoya, Costa Rica. In: Villalba, A., Reguera, B., Romalde, L. J., & Beiras, R. (Eds.), *Molluscan Shellfish Safety, Xunta de Galicia*, IOC of UNESCO, Santiago de Compostela, Spain, pp. 211-217.

Vargas-Montero, M. & Freer, E. (2004). Paralytic shellfish poisoning outbreaks in Costa Rica. In: Steidinger, K. A., Landsberg, J. H., Tomas, C. R., & Vargo, G. A. (Eds.), *Harmful Algae 2002*. Florida Fish and Wildlife Conservation Commission, Florida Institute of Oceanography, and Intergovernmental Oceanographic Commission of UNESCO, pp. 482-484.

Velasco, L. A. & Barros, J. (2008). Cultivo de bivalvos en Colombia: utopía o apuesta de futuro? In Estado actual del cultivo y manejo de moluscos bivalvos y su proyección futura: factores que afectan su sustentabilidad en América Laina, Lovatelli, A., Farías, A., Uriarte, I., (Eds.). *Organización de las naciones unidas para la agricultura y la alimentación,* Roma, pp. 115-128.

Wright, J. L. C., Boyd, R. K., deFreitas, A. S. W., Falk, M., Foxall, R. A., Jamieson, W. D., Laycock, M. V., McCulloch, A. W., McInnes, A. G., Odense, P., Pathak, V., Quilliam, M. A., Ragan, M., Sim, P. G., Thibault, P., Walter, J. A., Gilgan, M., Richard, D. J. A., & Dewar, D. (1989). Identification of domoic acid, a neuroexcitatory amino acid, in toxic mussels from P.E.I. *Canada. Can. J. Chem.* 67, 481-490.

Wright, J. L. C. (1995). Dealing with seafood toxins: present approaches and future options. Food Research International 28(4), 347-358.

Yasumoto, T., Murata, M., Oshima, Y., Matsumoto, K., & Clardy, J. (1984). Diarrheic shellfish poisoning. In: Ragelis, E. P. (Ed.), Seafood toxins. Washington, D. C.: American Chemical Society. *ACS Symposium Series,* 262, pp. 207-214.

In: New Trends in Marine and Freshwater Toxins
Editors: A. G. Cabado and J. M. Vieites
ISBN: 978-1-62257-787-3

Chapter 12

Phytoplankton and Biotoxin Monitoring Programs for the Safe Exploitation of Shellfish in Europe

Philipp Hess
Ifremer, Unité de Recherche et de Service Environnement, Microbiologie & Phycotoxines, Rue de l'Ile d'Yeu, Nantes, Cedex, France

Abstract

This chapter examines the design of control programs for toxic phytoplankton and shellfish toxins in different types of seafood in Europe, with a focus on bivalve molluscs. A historic overview is given of the legislative changes over the last 20 years. Methodological developments and increasing toxicological understanding over this time period are also discussed. Subsequently, the different approaches to monitoring of harmful algae and shellfish toxins are reviewed considering the availability of resources. As European legislation has different statutory instruments for official control of harvesting areas, post-harvest control and end-product control, these aspects are discussed separately. Verification of the implementation of EU legislation by the European Union's Food and Veterinary Office is discussed to outline where harmonisation has not been achieved yet. Global change is considered a particularly delicate challenge in the design of monitoring programs and is examined prior to an outlook onto possible future evolutions of monitoring programs.

1. Introduction

Among several thousand species of microscopic algae known in the marine environment, around one hundred species have been associated with the production of toxins. Fish and shellfish may become contaminated with algal toxins through accumulation via marine food webs. In the case of bivalve molluscs, such contamination may occur directly through the uptake of toxic algae through filter-feeding, whereas more complex mechanisms may be involved in the uptake of algal toxins by fish, e.g. through their feeding on contaminated fish or shellfish which in turn have fed directly on algae. Whenever such contamination occurs, the consumption of this seafood poses a risk to humans, often resulting in gastro-intestinal disorders. Some toxins may, however, have more dramatic consequences, including the permanent loss of short-term memory (amnesic shellfish poisoning), neurological disorders (neurotoxic shellfish poisoning and ciguatera fish poisoning) or paralytic effects and even death (paralytic shellfish poisoning). At least 9 compound groups have been shown to lead to acute illness (fish or shellfish poisoning); the list below states the toxin group (its causative organism), and the illness associated, more detailed information can be found in a recent review (Rossini and Hess, 2010):

Azaspiracids, AZAs (*Azadinium spinosum*), gastro-intestinal disorder
Brevetoxins, BTXs (*Karenia brevis*), neurotoxic shellfish poisoning, skin & respiratory defects
Ciguatoxins, CTXs (*Gambierdiscus* spp.), ciguatera fish poisoning
Domoic acid, DA (*Pseudo-nitzschia* spp.), amnesic poison (gastro-intestinal and neurotoxic symptoms, potentially death)
Maitotoxin, MTXs (*Gambierdiscus* spp.), gastro-intestinal disorders
Okadaic acid and Dinophysistoxins, OA and DTXs (*Dinophysis* spp., *Prorocentrum* spp.), gastro-intestinal disorder
Palytoxins, PlTXs (*Ostreopsis* spp.), cupleotoxism (potentially death), skin & respiratory defects
Saxitoxins, STXs (*Alexandrium* spp., *Gymnodinium catenatum*, cyanobacteria), paralytic effects (potentially death)
Tetrodotoxins, TTXs (bacteria), paralytic effects (potentially death)

Most countries world-wide have developed monitoring systems to prevent and diminish the risk of such poisoning events. The systems for monitoring harmful algae and their toxins are influenced by legislation and available methodology, as well as the coastal marine environment itself, which appears heavily affected by global change. The legislative development is reviewed prior to the description of the monitoring systems currently practised across Europe. Subsequently, inspection of monitoring systems and failures of these systems are described. Finally, drivers for future evolutions are examined, including global change and the development of novel technologies.

There is a number of sources of information that can be used for the comprehension of the design and implementation of the various EU member states' (MS) monitoring systems for toxic phytoplankton and shellfish toxins:

a) legislation
b) reports of EU working groups of the network of reference laboratories
c) inspection reports of the EU Food and Veterinary Office (FVO)
d) description of the systems used in individual MS on websites or in publications

1.1. Evolution of Legislation for Biotoxins in Europe

Generally, monitoring systems are designed as a function of financial efforts by shellfish producers and governments, efforts that are directly related to the political will to deal with a public health issue. In contrary to the North American, Australian and New Zealand systems, Europe has, similar to Japan, developed a legal system with a strong emphasis on government involvement in the control of shellfish toxins. Only recently, legislation has looked to equilibrate these international discrepancies.

Despite the risk of poisoning, the production and consumption of fish and shellfish has risen very significantly since the middle of the last century. While fish stocks had been increasingly over-exploited, the production of farmed fish and shellfish continues to gain importance. At global level, the main reason for this increased consumption of seafood is of course the increase in global population. In addition to the growing number of consumers, there is also more and more awareness of the health benefits of seafood as a low fat diet rich in proteins. While seafood had always been a major constituent of the diet of coastal populations in Europe, in recent years, the consumption of such foods has expanded throughout Europe, favoured by the onset of mass tourism during the 1960s and continued integration of the European market from the 1950s to the abolition of intra-European borders in 1993. The rising desire to consume shellfish combined with the integration of European markets have led to a requirement to produce and market shellfish virtually year round throughout Europe. Additionally, harmful algae typically occur from late spring to early autumn, a period which coincides with the tourist season. Thus, a more pronounced risk of poisoning events has developed and hence, an increased need arose for harmonising the control of shellfish toxins within Europe.

Countering this new challenge, a major step of harmonisation of European legislation had been taken in Council Directives 91/492/EEC and 91/493/EEC (Anonymous, 1991a and b). This law prescribed method types and limits for two major poisoning types known at the time: diarrhetic and paralytic shellfish poisoning; in brief, the legislation stated:

- The total amount of paralytic shellfish poison (PSP) must not exceed 80 μg/100 g (in the whole body or any part edible separately), using the biological testing method (in association with the chemical method if necessary). In case of dispute the biological method is the reference method.
- The customary biological testing methods must not give a positive result to the presence of diarrhetic shellfish poison (DSP).

Looking closely, this legislation allowed for differing interpretation in several aspects. First of all, it is noted that no compound groups were specified. This lack of specification may not have been considered a problem for the PSP and DSP groups for which the reference test was a bioassay with no specificity other than the sample preparation procedure. However, if chemical methods were to be applied as stipulated possible for PSP, then this lack of specification is rather disturbing for the implementation of any method. Thus, *de facto* it was not possible to use chemical methods for the detection of PSP toxins. Another aspect of this legislation generating criticism by experts in the field was the lack of methodological description. While the biological method for PSP (a mouse bioassay) had been invented in

1937 by Sommer and Meyer, and had been validated in 1991 as AOAC-method, there were several chemical methods available at the time of this legislation (e.g. Bates & Rapaport, 1975; Lawrence & Ménard, 1991).

This absence of specified methods and of specified compounds has to be seen in a context of both a lack of information (e.g. on all compound groups and analogues involved) and a lack in proper control alternatives. A number of essentials had not been available at the time when the legislation had been written:

- Validated methods: the DSP bioassays (mouse and rat) had been neither standardised nor validated.
- Physico-chemical methods had only been validated for a limited range of PSP compounds (namely a small number of saxitoxin analogues).
- Reference materials were rather limited.
- There was no proficiency testing organised for official control laboratories in the shellfish toxin area.

It was during the 1990s and early 2000s that such issues had started to being addressed at international level. In particular, the National Research Council Canada's certified reference materials program had contributed to an increasing number of certified calibrants and tissue reference materials. For full details on available certified reference materials see the current NRCC website (NRCC, 2011). In 1993, the European Union's executive organ for food safety and consumer health (DG Sanco) introduced the system of national reference laboratories (NRL) and a community reference laboratory (EU-RL) to supervise national and European policies regards official testing of marine biotoxins (Anonymous, 1993). By 1995, the network was in place, and progress could be made in this area through annual meetings and working groups. However, the preparation and certification of calibrants and tissue reference materials is far from being trivial and much research remains to be done for a number of compound groups; a recent review of the process leading up to these materials outlines the difficulties (Hess et al., 2007). Open access to proficiency testing in the area of shellfish toxins has been made available through the proficiency test provider "Quality Assurance in Marine Environmental Matrices in Europe" (Quasimeme, 2011). Initially, this scheme had started with proficiency testing for domoic acid, in 2003. In 2005, the scheme had been widened to also comprise lipophilic toxins (OA and AZA-groups) and subsequently, since 2009, saxitoxins are also covered. A review had been carried out on the effectiveness of the scheme demonstrating the high quality of the materials and data assessment techniques (Hess and McCarron, 2007).

An interesting – although minor – change in legislation took place on request of Spain (Anonymous, 1996). This piece of legislation is mentioned here to demonstrate the possibility, in principle, to influence legislation through the provision of better understanding of the physico-chemical behaviour of shellfish toxins. This regulation allows for harvesting of a particular shellfish species, *Acanthocardia tuberculatum*, even when concentrations of PSP toxins exceed the legal level up to ca. four-fold, since subsequent processing was demonstrated to be efficient in reducing these levels of PSP toxins to below the safe threshold.

As a function of the introduction of a network of expert laboratories, further knowledge on biotoxins had also been acquired in the area of the amnesic shellfish poisoning toxin, domoic acid. Its occurrence had been found in shellfish harvested in Europe despite the complete absence of poisoning events. Consequently, a Council Directive amended the above mentioned legislation from 1991 to also include amnesic shellfish poisoning toxins (Anonymous, 1997): "*The total amount of amnesic shellfish poison (ASP) must not exceed 20 mg/kg (in the whole body or any part edible separately), using the HPLC method*". (More information related to these legislations is reported in chapter 5).

However, in the case of ASP where physico-chemical detection was to be applied, the specification of which compounds were concerned would have been an advantage. Particularly, since a further tightening of the performance duties of official control laboratories took effect in 1999 (Anonymous, 1999a), forcing official control laboratories for food stuffs to comply with European Normative Standards EN 45001, 45002 and 45003, and hence to comply with in-house validation rules, proficiency testing and formal accreditation. Council Decision 1999/312/EC (Anonymous, 1999a) also ensured that these conditions did not only apply for official control laboratories but also for the reference laboratories of the EU and individual member states. Still, the accreditation of non-validated methods such as the DSP bioassay posed significant problems in a number of member states, particularly since there had been legislation in place that prohibits the use of laboratory animals when alternatives are available (Anonymous, 1986).

Council Decision 1999/468/EC (Anonymous, 1999b) had paved the way for more flexible amendments in the area of food safety in giving legislative power to the Commission directly operating through committees such as the Standing Committee for Food Contaminants and Animal Health (SCoFCAH) which has also dealt with marine biotoxins since its introduction.

However, the lack of finances in this area had been very limiting in terms of producing calibrants and provision of contaminated samples. It had not been possible to culture many of the causative organisms and chemical synthesis of toxins is a very complex undertaking. Thus progress was heavily limited by naturally occurring toxic events. As a consequence, a vicious circle prevented progress in the area of proficiency testing: biological methods consume a lot of sample and samples were very precious due to sporadic occurrence of natural events of phycotoxins and high operational costs of organising sample collection, storage, treatment and postage. On the other hand, physico-chemical methods require standards and reference materials for validation which again depended on natural events. Hence, validation of novel methods that would have been consuming less of the precious samples did not take place due to lack of toxin standards. Without natural samples, there was no validation of novel methods, no validation of existing methods and virtually no proficiency testing despite the poor validation status of existing methods.

Still, continuous political pressure to work on alternatives to animal testing, and to introduce fully quantitative assays for the determination of lipophilic toxins led to a series of working groups on the toxicology of lipophilic toxin groups involved in May 2001 (CRL-MB, 1998, 1999b and 2001b). These working groups suggested that not only OA-group toxins but also AZAs, pectenotoxins (PTXs) and yessotoxins (YTXs) may be implicated in shellfish poisoning. The working groups recommended that the EU authorities should financially support the production of calibrants for toxins of these groups and that they should also fund research into detection technology and toxicology of these groups. In the meantime

Member States were encouraged to monitor for these compound groups. As a consequence, the legislation was amended in parallel to state which compounds were to be monitored (Anonymous, 2002a). This legislation details for the first time individual compounds to be monitored, and gives more specific guidance of the application of different bioassays in the area of lipophilic toxins. As such it consolidated the different bioassays carried out in different member states (mouse and rat bioassays) and continued to support a system that lacked in harmonisation. Alternative tests were specifically allowed as long as they would monitor for all the compounds covered by the legislation and as long as validation according to international standards would be carried out. This put more pressure on production of calibrants and reference materials, yet funding in this area had remained very limited until a new call for research projects in 2004. Unfortunately, an error was introduced in formulating this Commission Decision: while the above-mentioned working groups recommended to regulate PTXs at the same level as OA-group toxins, the legislation stated that the sum of OA-group toxins and PTXs must not exceed a common limit. Also, the concept of toxic equivalence factors (TEFs) has been introduced, and PTXs were regulated in terms of OA-equivalents of toxicity. This equivalence does not make sense as PTXs have a distinct mode of action (i.e. unlike OA they do not inhibit phospho-protein phosphatases PP1 and PP2a); as their OA-equivalent is zero, PTXs could *de facto* be considered not to be regulated which is clearly not the spirit of the decision.

In parallel, shellfish monitoring for amnesic shellfish poisoning toxins, introduced in the legislation in 1997, had led to substantial knowledge on the occurrence and distribution of domoic acid. While concentrations observed in the majority of bivalve molluscs at the time had been very low, the levels had been found particularly high in the king scallop, *Pecten maximus* (e.g. Hess et al., 2001). A series of working groups (CRL-MB, 1999a, 2000 and 2001a) led to the conclusion that fished scallops should be managed separately from aqua-cultured bivalve molluscs. In particular, the effects of processing of contaminated scallops may lead to contamination of edible scallop tissues (Hess et al., 2000; CRL-MB, 2001a). Also, in depth statistical analysis of the distribution and variability of domoic acid in scallops (data from Ireland, Spain and UK) led to the conclusion that monitoring of scallops should be carried out to manage scallop harvesting areas. Thus, scallops should not be found in monitoring to be contaminated at levels above 4.6 mg kg^{-1} to ensure that final product on the market does not contain more than the 20 mg kg^{-1} that are considered to be a safe level for consumption (Anonymous, 2002b).

In 2004, the EU Hygiene Package of regulations was published, bringing together and replacing the existing hygiene regulations for the food sector previously contained in numerous individual vertical Directives. In Annex II Section VII Chapter V (2) to Regulation 853/2004/EC, maximum levels for ASP, PSP and DSP toxins had been explicitly described as a simple transcription from previous pieces of legislation. 854/2004 also laid down rules for official control of these toxins (Anonymous, 2004a and b). The methodological specifications are given in a separate document (Anonymous, 2005). While the spirit of this legislation places more emphasis on the responsibility of shellfish producers themselves, in practice, there is very little change in terms of implementation of legislation by industry. This lack of implementation of legislation by shellfish producers is partly due to the fact that much of the European production is sold as live bivalve molluscs, a process so rapid that test results cannot be obtained in time and must therefore rely on official control results. Additionally, it is also due to a continued lack of easy-to-use tests that could be implemented in shellfish processing plants. Finally, there is also a lack of technical guidance by NRLs or competent authorities to food business operators (see section 1.2.1.3 of this chapter, Table 4).

In 2006, EU legislation has for the first time specified an alternative method to a bioassay. In particular, the Lawrence method for STXs by HPLC-FLD in its 2005.06 version of AOAC has been accepted as an alternative to the PSP mouse bioassay (Anonymous 2006). Significant progress had also been made in the area of domoic acid. An ELISA test had been developed and validated in compliance with AOAC (2006.02 ASP ELISA), an internationally accepted body for formal method validation. Initially, a single laboratory validation had been carried out (Kleivdal et al., 2007a) and subsequently, a full collaborative study (Kleivdal et al., 2007b) involving a number of international laboratories. This progress had also been rapidly translated into legislative acceptance in Europe. EU legislation has thus accepted the use of this ELISA for screening in official or end-product control (Anonymous, 2007). The test had been implemented in Scottish processing plants and further information is available from Biosense (Biosense website, 2011).

A number of working groups on the risk assessment and available methodology have been carried out at global level (FAO/IOC/WHO, 2005) and at EU level (EFSA, 2008a and b, 2009a to g, 2010a to d). More specifically, expert advice on regulated toxins (EFSA, 2009f) has been a driver contributing to the most recent change in legislation in 2011 with the change of the reference assay for lipophilic toxins from the mouse bioassay to LC-MS/MS testing (Anonymous, 2011). This change results in a suite of reference methods that are quantitative for all toxin groups regulated in Europe for the area of bivalve molluscs. This legislation is also in accordance with limits agreed by consensus in Codex Alimentarius (2008). However, some inconsistencies between EU and international regulation still remains in that the most recent EU regulation (Anonymous, 2011) still covers two toxin groups that are not regulated internationally: YTXs and PTXs. Both toxin groups had been discovered in the mid-1980s through use of the lipophilic mouse bioassay (Murata et al., 1986 and 1987). However, their involvement in acute shellfish poisoning could not be unequivocally substantiated over the last 25 years. The most recent risk assessment suggested that the EU-level in vigour for YTXs (1 mg kg^{-1} shellfish flesh; Anonymous, 2005) is more than sufficiently protective (EFSA, 2009a). Thus, DG Sanco are currently seeking advice from Member States on a potential increase of the regulatory limit for yessotoxins. The situation is somewhat more complex for pectenotoxins which are typically produced concomitantly with OA-group toxins by *Dinophysis* spp. This co-occurrence made it difficult so far to completely exclude a toxic effect of PTX-analogues.

Compounds causing ciguatera are produced by species belonging to the genus *Gambierdiscus* (benthic dinoflagellates), and may comprise CTXs, gambiertoxins and MTXs. These toxins mainly pose problems through the consumption of contaminated fish and had been reported until very recently only from tropical areas. Current EU legislation does give neither quantitative limits for ciguatoxins nor states which analogues are relevant. However, the relevant regulation (Anonymous, 2004b) simply states that "*Checks are to take place to ensure that the following fisheries products are not placed on the market: fisheries products containing biotoxins such as Ciguatera or other toxins dangerous to human health*". No specific details or requirements about methodology to be used are given. The most recent risk assessment of ciguatoxins in Europe estimates safe levels of ciguatoxins in fish to be around 0.01 μg kg^{-1}, and that the mouse bioassay is not an efficient tool to detect these levels (EFSA,

2010a). This topic, as well as other toxin groups such as tetrodotoxin, brevetoxins and cyclic imines are also reconsidered in section 1.4.2 (emerging toxins).

Finally, it is noted that EU legislation places relatively little emphasis on regulating the occurrence and monitoring of toxin-producing phytoplankton species. Current legislation (Anonymous, 2004b) simply states that "*Classified relaying and production areas must be periodically monitored to check for the presence of toxin-producing plankton in production and relaying waters*". This situation is somewhat of a paradox considering the fact that all the toxin groups involved in acute fish and shellfish poisoning are produced by unicellular micro-organisms, and 8 of these toxin groups are produced by marine phytoplankton (diatoms and dinoflagellates). Still, some further requirements are made in this regulation as to the use of phytoplankton monitoring in the context of the design of monitoring programs, in particular with regards to sampling representativeness and frequency:

"Sampling plans to check for the presence of toxin-producing plankton in production and relaying waters and for biotoxins in live bivalve molluscs must take particular account of possible variations in the presence of plankton containing marine biotoxins. Sampling must comprise:

(a) periodic sampling to detect changes in the composition of plankton containing toxins and their geographical distribution. Results suggesting an accumulation of toxins in mollusc flesh must be followed by intensive sampling;
(b) periodic toxicity tests using those molluscs from the affected area most susceptible to contamination."

Section 7 specifically states that "*With regard to the monitoring of plankton, the samples are to be representative of the water column* and to *provide information on the presence of toxic species as well as on population trends*. If any changes in toxic populations that may lead to toxin accumulation are detected, the sampling frequency of molluscs is to be increased or precautionary closures of the areas are to be established until results of toxin analysis are obtained."

The impact of these requirements on the design of monitoring programs and their evolution will be further discussed in section 1.2.2 and 1.4.

1.2. Biotoxin Monitoring Systems in Europe: A Harmonised Network?

The present section summarises several working group reports and FVO (Food and Veterinary Office) inspection reports.

1.2.1. Description of Official Control Systems and Guidance Given on End-Product Control

The EU-NRL working group on sampling in 2001 (3-4 October 2001, Brussels) clarified a number of aspects with regards to the representativeness of sampling points, sampling frequency, indicator species and the hierarchy of phytoplankton versus shellfish sampling.

Spatial representativeness of sampling points should be demonstrated through the study of the variability of toxins in any given production area as defined by legislation. Production areas may be further divided for management purposes if it can be demonstrated that significant spatial differences exist and that all subdivided areas are subsequently monitored during production. Typical field variability of the distribution of toxins across a production area is of the height of order of 25 to 30 %, as demonstrated by Hess et al., 2003, for AZA- and OA-group toxins, respectively, in mussels in Ireland. However, significantly larger variability can be observed for other toxin-shellfish couples, e.g. we found particularly high variability (71.3 to 106.2 %, depending on the body part) of domoic acid in scallops (*Pecten maximus*) in Irish waters (Bogan et al., 2007).

Due to possible rapid accumulation of toxic algae in filter feeding bivalves (hours or days), the frequency of sampling should be at least weekly. The frequency may be less if low risk can be demonstrated through periodically reviewed historic data. The frequency should not be higher than every 48 h as most shellfish species have detoxification rates (half-lives) significantly larger than this period, and resurgence of algal blooms could easily occur during short detoxification periods. Fished shellfish may be controlled either through monitoring of fishing areas or through control at landing. During fishing periods for *Pectinidae*, controls should be at least weekly.

Even though the accumulation of toxins in bivalve molluscs may depend in principle on a number of factors such as species-specific metabolism, filtration and detoxification rates, total phytoplankton community composition, and environmental parameters, trends on the difference of accumulation in different shellfish species have been obtained. For instance, the accumulation of domoic acid has been demonstrated to be notoriously high in the king scallop *P. maximus* (Hess et al., 2001; Blanco et al., 2002), even though it may also reach significant levels in blue mussels (Quilliam and Wright, 1989; McCarron and Hess, 2006). For lipophilic toxins, a consistent difference was noted between mussels and oysters, the latter species always showing less contamination when cultivated in the same area (Hess et al., 2003). These trends may be used to establish indicator species for specific harvesting areas. For instance in France, based on historical trends, king scallops are considered the indicator species for domoic acid, while mussels are considered sentinel species for all other toxin groups.

Generally shellfish sampling is the preferred indicator for the management of shellfish production areas due to the fact that phytoplankton occurrence is typically "patchy" and the representativeness of sampling is more difficult to obtain. Thus, most countries use phytoplankton only as early warning or for temporary closure orders. However, if historical trends can be established phytoplankton monitoring can be carried out during low risk periods to allow decentralisation of monitoring activities (e.g. practised in France and Ireland) and only when potentially toxic phytoplankton cells are detected, shellfish sampling is triggered or the frequency of shellfish sampling can be increased. For the sampling of phytoplankton, the representativeness should be increased wherever possible through integrated sampling of the water column.

Following recommendations of the CRL/NRL meeting in Cyprus, October 2007, the Irish-NRL prepared a questionnaire on the "Management Practices for Monitoring and Surveillance of Marine Biotoxins in Bivalve Molluscs". It should be noted that this section describes the situation as reported at the time, and it is likely that further changes have taken place since then, in particular in relation to the most recent legislative changes (Anonymous,

2011). The aim was to gather information on how the surveillance for marine biotoxins in bivalve molluscs is organised and carried out in different EU Member States. The questionnaire contained 21 questions related to the tools and the management practises used for the opening and closing of production areas, the pre- and post-harvest official control, the status of the shellfish industry's own checks and official agency involvement in the organisation of own checks. The questionnaire was sent to all EU-NRLs (with the exception of Luxemburg and Bulgaria) and to the Norwegian NRL. Of the 26 countries contacted, 20 replied to the questionnaire. In the EU and Norway, there are 14 states that produce bivalve molluscs (SPC = shellfish producing country). All SPCs replied to the questionnaire; among the non-shellfish producing countries, the response to the questionnaire was lower (6 of 13 replied). An overview of the involvement of NRLs in pre- and post-harvest monitoring is given in Table 1. A total of 8 NRLs are involved in pre-harvest monitoring, 9 are involved in post-harvest monitoring; 7 are involved in both pre- and post-harvest monitoring while 8 NRLs have no involvement in either stage.

1.2.1.1. Pre-Harvest Control

In the 14 SPCs, 10 have developed a best practise guide (BPG) for the official control of bivalve molluscs. In Germany, the "Länder" (counties) have responsibility for food control; one of the two shellfish producing counties has also produced a BPG for official control. Half of the NRLs in SPCs provide data used in the management of production areas (7), while 9 provide advice for the management of shellfish production areas. For country-by-country details see Table 2 below.

The numbers of monitoring sites varies greatly between MSs, ranging from 3 in Slovenia to 314 in Spain (reflecting the production volumes in the different countries). Generally, it is noted that there is a good correlation between the number of production areas (as an indicator for overall production) and the number of monitoring points, which means that countries with a large production also have many monitoring points.

Overall, it is noted that pre-harvest control systems differed mostly in two main domains between different MSs: (i) the use of phytoplankton monitoring as alert and (ii) the methods applied for analysis of shellfish in the areas of saxitoxins and lipophilic toxins.

Pre-harvest monitoring practises are not related to numbers of monitored areas, as management differs greatly between countries with large number of monitoring points (e.g. ES, FR, UK) or intermediate numbers of sites (e.g. NL, DE). Almost half the SPCs use the results of flesh tests as the primary decision-making tool for all toxin groups, irrespective of phytoplankton monitoring (DE, ES, IT, NO, PT, UK). Some Member States use phytoplankton monitoring to trigger flesh testing specifically for ASP and PSP toxins (FR, IE, SE, SI); in one production area in Ireland where PSP toxins are observed occasionally flesh testing is used during high risk periods in parallel with phytoplankton monitoring. Similarly, in Greece, the standard frequency for ASP and PSP flesh tests is every two weeks and interim phytoplankton results are used to trigger more frequent flesh tests. Two countries (DK and NL) require dual compliance of both phytoplankton results and flesh test results for opening of shellfish areas. France has clearly defined high risk periods for lipophilic toxins. These periods have been developed over a number of years and the monitoring system relies on phytoplankton testing as a primary tool for monitoring harvesting areas outside these periods. Outside the high risk periods, flesh testing is also used at sentinel sites at monthly frequency

Table 1. Overview of member states, national reference laboratories and shellfish production

Code	Country	NRL abbreviation	Questio-nnaire returned	Shellfish Producer	NRL in Official Control	NRL in Official Control
					Preharvest	Postharvest
AT	Austria	AAHFS	+	-	-	+
BE	Belgium	IPH	+	+	+	+
CY	Cyprus	LFCAO	+	-	-	-
CZ	Czech	SVI-Jihlava	+	-	-	
DE	Germany	BFR	+	+	-	-
DK	Denmark	NFI	+	+	+	+
EE	Estonia		-	-	-	-
ES	Spain	EURLMB	+	+	-	-
FI	Finland		+	-	-	-
FR	France	AFSSA	+	+	-	+
GR	Greece	IFH-Thessaloniki	+	+	+	+
HU	Hungary		+	-	-	-
IE	Ireland	MI	+	+	+	-
IT	Italy	CRM	+	+	+	+
LT	Lithuania		-	-	-	-
LV	Latvia		-	-	-	-
LU	Luxembourg		-	-	-	-
NL	Netherlands	RIVM	+	+	-	-
NO	Norway	NSVS	+	+	+	+
PL	Poland	NVRI-Pulawy	+	-	-	-
PT	Portugal	IPIMAR	+	+	+	+
RO	Roumania		-	-	-	-
SE	Sweden	NFA	+	+	-	-
SI	Slovenia	Vet Fac - IFH	+	+	+	+
SK	Slovakia		-	-	-	-
UK	UK	FRS-MLA/AFBNI	+	+	-/+	-
SUM	26		20	14	8	9

Notes: *nota bene*: even when NRLs are not involved in pre- or post-harvest control, other agencies do carry out this control but these could not be referenced here.

or following phytoplankton triggers to monitor and re-open bays that were closed. It was not possible in the frame of this questionnaire to establish whether all countries using phytoplankton results to trigger flesh testing are doing so based on a risk assessment following historical data.

The second major point of lacking harmonisation was related to the methods used in official control. A number of countries had used LC-MS for lipophilic toxins in official control as primary tool for shellfish production area management (DE, PT, SE), despite the reference method being the bioassay. This discrepancy outlines the general conflict existing between two legislative areas, i.e. shellfish safety (1991, 2002, 2004b and 2005) and animal protection (Anon. 1986). Ireland had used LC-MS in parallel with mouse bioassay testing since 2001 and had made best use of this kind of information in the management of shellfish harvesting areas. The Netherlands were the only country still using the rat bioassay for lipophilic toxins, despite its clearly lacking capability of detecting YTXs and PTXs. Five countries were in non-compliance with legislation in the area of PSP toxins: Germany, The Netherlands, Norway, Portugal and the UK had alternative testing in place for saxitoxins. While the most recently introduced legislation for lipophilic toxins is likely to change this lack of harmonisation over the next few years, the path for harmonisation in the area of STX remains to be decided.

1.2.1.2. Post-Harvest Control

Post-harvest official control was carried out in 13 Member States. The number of samples analysed however, varied greatly, ranging from 10/y (AU) to 400/y (DE, ES).

Six countries used the reference tests for the post-harvest OC (DK, ES, FR, GR, IT, SE) for all toxin groups. Poland used the Jellett-kit for control of ASP and PSP, and ELISA for PSP in their post-harvest control. Only Netherlands used LC-MS for all toxin groups in post-harvest control. Five countries used LC-FLD for post-harvest control of PSP toxins (AU, DE, NO, PT, UK). Ireland did not test for PSP in post-harvest control. Five countries used LC-MS in the post-harvest control of lipophilic toxins (DE, IE, NL, NO, PT), however, not all toxin groups were necessarily monitored for, and emphasis was generally placed on the OA group of toxins amongst the lipophilic toxins.

Table 3 shows the summary for all MS that replied for the post-harvest control section of the questionnaire.

1.2.1.3. HACCP and End-Product Testing

Fourteen NRLs replied to this part of the questionnaire. The UK was the only Member State that had a best practise guide for Hazard Assessment at Critical Control Points (BPG-HACCP) procedures. This however, is only applied in Scotland and not in England, Wales and Northern Ireland. Five countries had a BPG in preparation, however, Ireland was the only Member State where the NRL was involved in the design of this BPG-HACCP. In Denmark, Food Business Operators (FBOs) paid for most of the official control, and therefore, a large part of the official control may be considered a form of "own checks". No NRL is involved in inspections, training and proficiency testing for FBOs. Four countries had reported that their FBOs used on-site testing (IE, FR, SE, UK). Two countries (NO, UK) had specifically stated that they do not consider this area as part of the NRL remit. Table 4 shows the summary of country-by-country involvement in HACCP best practise for shellfish toxins and status of best practise guides. Unfortunately, no systematic data on shellfish business operators' efforts were available to the author. Still, in 2010 Spain reported that 3443 end-product tests or "own checks" had been carried out in a single laboratory carrying out such tests for separate producers and processors.

Table 2. Monitoring basis (phytoplankton monitoring or shellfish flesh testing) and methods used in official control (OC) by 14 European countries. Numbers of production areas and monitoring points are given as well as the status of best practise guides for official control and NRL involvement in management of production areas

Code	Country	Best Practise guide for OC?	NRL provides data for closure ?	NRL provides advice for closure ?	Primary tool for closure	Primary tool for closure	Primary tool for closure	Primary Method	Primary Method	Primary Method	Second. tool for closure	2ndary tool used for	No of production areas	No of monitoring points
					ASP	PSP	Lipophilic	ASP	PSP	Lipophilic				
BE	Belgium													
DE	Germany	-/+	-	-	Flesh	Flesh	Flesh	LC-UV	LC-FLD	LC-MS			15	48
DK	Denmark	+	+	+	Flesh+Phyto	Flesh+Phyto	Flesh+Phyto[4)]	LC-UV	MBA	MBA	Phyto	YTX/PTX	142	142
ES	Spain	+	-	-	Flesh	Flesh	Flesh	LC-UV	MBA	MBA	Phyto		177	314
FR	France	+	-	+	Phyto[1)]	Phyto[1)]	Flesh+Phyto[1)]	LC-UV	MBA	MBA	Flesh		82	271
GR	Greece	+	+	+	Flesh/Phyto	Flesh/Phyto[5)]	Flesh	LC-UV	MBA	MBA	Flesh/Phyto		70	70F/60P
IE	Ireland	+	+	+	Phyto	Phyto/Flesh	Flesh	LC-UV	MBA	MBA/LC-MS	Phyto/Flesh		95	95F/100P
IT	Italy	+	+	+	Flesh	Flesh	Flesh	LC-UV	MBA	MBA	LC-MS		200	200
NL	Netherlands	-	-	-	Flesh+Phyto	Flesh+Phyto	Flesh+Phyto[6)]	LC-UV	LC-MS	RBA[2)]			16	23
NO	Norway	+	+	+	Flesh	Flesh	Flesh	LC-UV	LC-FLD	MBA			> 50	50
PT	Portugal	-	+	+	Flesh	Flesh	Flesh	LC-UV	LC-FLD	LC-MS[3)]			20	20
SE	Sweden	+	-	+	Phyto	Phyto	Flesh	LC-UV	MBA	LC-MS/MBA			33	33F/6P
SI	Slovenia	+	+	+	Phyto	Phyto	Flesh	LC-UV	MBA	MBA			3	3
UK	UK	+	-	-	Flesh	Flesh	Flesh	LC-UV	LC-FLD	MBA			296	296
		10	7	9										

1) In France phytoplankton is used for triggering flesh testing outside risk periods in addition to monthly flesh tests, during risk periods phytoplankton is used for ASP & PSP

2) the RBA is not tested for PTXs and YTXs

3) PT does not regularly test by LC-MS for PTXs, YTXs and AZAs, YTX by ELISA when tested

4) The danish OC for the OA group is based on dual phyto and flesh testing while for the other lipophilic groups it is based on flesh testing only.

5) The Greek OC tests for ASP and PSP are carried out every 2 weeks and interim positive phyto results will trigger more frequent flesh tests

6) The Dutch OC tests require dual compliance of phyto and flesh results

Table 3. Methods and samples numbers for post-harvest official control of shellfish toxins

Code	Country	Methods			No of samples
		ASP	*PSP*	*Lipophilics*	analysed p.a.
AT	Austria	Not given	LC-FLD	Not given	10
BE	Belgium	n/a[1)]	n/a	n/a	0
CY	Cyprus	Not given	Not given	Not given	Not given
DE	Germany	LC-UV	LC-FLD	LC-MS	300-400
DK	Denmark	LC-UV	MBA	MBA	125
ES	Spain	LC-UV	MBA	MBA	350-400
FI	Finland	n/a	n/a	n/a	0
FR	France	LC-UV	MBA	MBA	250
GR	Greece	LC-UV	MBA	MBA	20-30
HU	Hungary	n/a	n/a	n/a	0
IE	Ireland	LC-UV	n/a	LC-MS	100-200
IT	Italy	LC-UV	MBA	MBA	75-100
NL	Netherlands	LC-MS	LC-MS	LC-MS[1)]	100-200
NO	Norway	LC-UV	LC-FLD	LC-MS	occasional on imports
PL	Poland	Jellett	ELISA, Jellett	n/a	30
PT	Portugal	LC-UV	LC-FLD	LC-MS[3)]	occasional BIP[4)] samples
SE	Sweden	LC-UV	MBA	MBA	Not given
SI	Slovenia	n/a	n/a	n/a	0
UK	UK	LC-UV	HPLC-FLD ELISA	PP2A	Not given

N/a not applicable.

NL don't routinely monitor for PTX and YTX in postharvest control.

Portugal monitor using single quadrupole MS.

BIP = border inspection post.

Table 4. HACCP best practise guides and the involvement of NRLs in the design

Code	Country	Replied	BPG HACCP	BPG in prep.	FBO on-site tests	FBO commercial	FBO OCL	FBO No tests	NRL Inspect.	NRL Training	NRL PT	NRL BPG	NRL Other
DE	Germany	yes	-	-	-	yes	yes	-	-	-	-	-	-
DK	Denmark	yes	-	yes	-	-	yes	-	-	-	-	-	-
ES	Spain	yes	-	yes	-	-	yes	-	-	-	-	-	-
FR	France	yes	-	yes	some	-	-	-	-	-	-	-	-
GR	Greece	yes	-	yes	-	-	-	-	-	-	-	-	-
IE	Ireland	yes	-	yes	some	yes	yes	-	-	-	-	yes	-
IT	Italy	yes	-	-	-	-	yes	-	-	-	-	-	-
NL	Netherlands	yes	-	-	-	-	-	yes	-	-	-	-	-
NO	Norway	yes	1)	1)	1)	1)	1)	1)	1)	-	-	-	-
PL	Poland	yes	-	-	-	-	-	-	-	-	-	-	-
PT	Portugal	yes	-	-	-	-	-	-	-	-	-	-	-
SE	Sweden	yes	-	-	some	-	yes	yes	-	-	-	-	-
SI	Slovenia	yes	-	-	-	-	-	-	-	-	-	-	-
UK	UK	yes	yes 2)	-	some	-	-	-	-	-	-	-	-
		14	**0.5**										

1) Please consult Norwegian Food Control Authority

2) Existing but not necessarily specific to England, Wales and Northern Ireland

1.2.2. Verification of the Control Systems Implemented by EU Member States

The Food and Veterinary Office (FVO) is a branch of DG Sanco which inspects the compliance of European monitoring systems with legislation. The FVO had been founded in 1979 and has now an extensive website at which much information can be found describing legislation, inspection reports and others (http://ec.europa.eu/food/fvo/index_en.cfm, last accessed 20/02/2011). Of particular interest to the current review were the reports available at the following link: http://ec.europa.eu/food/fvo/ir_search_en.cfm, accessed 20/02/2011.

The inspection reports from 1998 to 2010 have been reviewed. Over this period, there had been three series of inspections: 1998/99, 2001/02 and 2004/05. Countries visited over this period included the major shellfish producing countries in the EU: Belgium (BE), Denmark (DK), France (FR), Germany (DE), Greece (GR), Ireland (IE), Italy (IT), the Netherlands (NL), Portugal (PO), Spain (ES), Sweden (SE) and the United Kingdom (UK). A total of 26 inspections had been carried out in these 12 countries for the area of live bivalve molluscs.

The study of these reports revealed at least 34 major non-compliances in the area of monitoring for potentially toxic phytoplankton and shellfish toxins. These non-compliances can be classified into several areas:

i. Altered procedures for bioassays (number of animals and observation periods), 10 non-compliances
ii. Non-use of bioassays where these were prescribed as reference tests, 7 non-compliances
iii. Sampling frequency not sufficient, 4 non-compliances
iv. Delays between sampling and reporting/closure not sufficient, 4 non-compliances
v. Phytoplankton surveillance not integrated into the official control for biotoxins, 3 non-compliances
vi. Regional differences within member states and traceability issues, 3 non-compliances
vii. Toxin-groups not monitored for, 2 non-compliances
viii. Different limit for OA-group toxins in national legislation, 1 non-compliance

Clearly, this number of non-compliances outlines that there was a definite lack of harmonisation in a number of areas of official control of bivalve molluscs for shellfish toxins. The highest number of non-compliances were observed in the area of bioassays, either through the lack of their use or through the use of alternative protocols (points (i) and (ii)). Germany, Sweden and the Netherlands have had serious problems with the conflicting legislation for food safety and animal health protection, typically resulting in non-use of bioassays. Similarly, the Home Office in UK had also imposed strict limitation resulting in numerous changes in the protocol of bioassays (both DSP and PSP). In addition, 5h or 12h observation had been practised in FR, UK and ES, i.e. major shellfish producing countries (on different grounds of justification), and the number of animals used in bioassays varied grossly (1 rat in NL, 1 mouse in UK for PSP initially, 2 mice in DSP in UK). These discrepancies in the area of bioassays were also consistent with the declared differences in monitoring practises in the 2007 survey of NRLs (Table 2). Ultimately, this situation has been one of the major drivers for the latest legislative change introducing chemical testing for the official control of lipophilic toxins.

Points (iii) and (iv). The shortcomings in terms of sampling frequency and the long delays observed between sampling and reporting or closing of areas are logistic issues which are directly related to the lack of funding by the appropriate authority. No pattern or trend for these situations could be established except the fact that small member states are typically more affected by this situation (DK, PT). However, also large member states occasionally suffer from this problem, in particular when monitoring is the affair of regional authorities (IT). It is anticipated that this issue will become more prominent with the introduction of LC-MS/MS testing for lipophilic toxins.

Point (v). The lack of implementation of phytoplankton monitoring as integral part of the surveillance system for shellfish toxins had also been noted in the 2007 NRL survey. While it may partly be related again to funding issues, there are also other problems such as lack of legislative pressure in this area and lack of EU guidance. The lack of EU guidance can be understood from the fact that the EU reference laboratory had itself suffered from a total lack of funding in this area from 1995 to 2004, and even when the funding for the EU-RL increased in 2005, emphasis was placed on the toxins rather than the potentially toxic phytoplankton. Due to this lack of funding there is also a certain lack of competence of EU authorities in this area, contrary to international bodies such as the international Council for the Exploration of the Seas (ICES) or the Intergovernmental Oceanographic Commission (IOC) and the Intergovernmental Panel on Harmful Algal Blooms (IPHAB). This problem could be overcome by making better use of the expertise in the NRLs and their official control laboratories through a standing working group on potentially toxic phytoplankton.

Among the other issues which can be dealt with through normal inspection and sanctioning practises, the lower limit for OA-group toxins in German legislation was noted as an unusual anomaly, which is now considered rare thanks to the conscientious risk evaluation exercises carried out by EFSA over the period from 2006 to 2010 (EFSA, 2008 a and b, 2009 a to g, 2010 a to d).

1.3. Failures of Monitoring Systems

The European Community's Rapid Alert System for Food and Feed (RASFF, website 2011) is a web-based system allowing for rapid distribution of information pertaining to non-compliance of food- and feed-stuffs. This tool allows for an independent assessment of the functioning of monitoring systems from the consumer point of view. In the field of bivalve mollusc consumption there have been 297 alerts or notifications of irregularities over the 5-year period from 2006 to 2010. Biotoxins have been implicated in 71 of these notifications, i.e. ca. 25% of all non-compliances; other infringements were due to spoilage or contamination with bacteria, viruses or trace metals. Even though the post-harvest control in most EU member states examines an even spread of products from national, European and non-European sources, there were only 4 alerts related to imports from non-EU countries (Peru, Korea and Vietnam). Within the EU, product from Spain caused most of the alerts (23), followed by UK (11), Italy (10) and France (8), which is partially in relation to the overall relatively large shellfish production in these countries. The high incidence of toxins in shellfish from Spain is probably related to two facts: (i) Spain is the largest shellfish producer and has a high percentage of export compared to France and Italy which mainly produce for local consumption (ii) in 2005, several toxic events involving different toxin groups resulted in 9 months closure of shellfish production areas in Galicia (the main production area in Spain).

With respect to shellfish processing, legislation has failed to transpose important new knowledge on the behaviour of shellfish toxins during processing procedures into legislative requirements for shellfish producers. A number of studies have outlined the importance of shellfish processing, in particular in the area of lipophilic toxins. For instance, an increase of toxin concentrations had been noted in mussel processing. While some of the effects can be simply attributed to changes in water content of shellfish during processing (Hess et al., 2005; McCarron and Hess, 2006; McCarron et al., 2008), other changes are related to chemical transformation of analogues (McCarron et al., 2009). For lipophilic toxins, the concentrations may increase two-fold. Formal evaluation of these phenomena in an EFSA working group had led to the recommendation to harmonise sample pre-treatment in the analysis of lipophilic toxins (EFSA, 2009b). However, such harmonisation has not yet been achieved. As a result of this phenomenon, shellfish processors may legally obtain live mussels on the market, and create mussel products which are, after processing, above the regulatory limit. It is most likely such flaws that would have led to certain failures of the system detected for instance in RASFF alerts 2006.0175 and 2006.0309 where German post-market control detected OA-group toxins above the regulatory limit in cooked mussels originating from Norway and imported via Sweden (190 and 212 $\mu g\, kg^{-1}$ in the first case and 233 $\mu g\, kg^{-1}$ in the second case). These phenomena involving shellfish processing could be addressed either in legislation or in best practise guides for end-product control, which unfortunately have not yet been put in place in most EU member states (see above).

1.4. OUTLOOK

1.4.1. Global Change Impacting on Harmful Algae and Biotoxin Occurrence

It has become widely accepted that the anthropogenic impact on the biosphere has reached global dimensions. Thus, global change can be seen as the sum of changes deriving from human and natural changes over the last few decades or centuries. Recent studies have clarified the extent to which the oceans have changed overall. Interesting examples of these dramatic changes are for instance the re-distribution of phytoplankton species between the Atlantic, Arctic and Pacific oceans due to the disappearance of ice barriers in the arctic (Beaugrand and Reid, 2003; Reid et al., 2007). Similarly, the oceans have started to change significantly towards becoming more acidic (Lenton et al., 2009), which may be a limiting factor to some phytoplankton species.

The most heavily debated of the indicators of global change is global warming, which should be referred to as climate change since warming observed in some areas may also be associated with cooling in other areas. Due to very small changes in climate compared to changes in weather, it is generally necessary to examine data sets covering several decades before any trends can be established. The increased water temperature observed in some regions will obviously allow some species to develop, however, as the pelagic ecosystem is very little understood it remains a formidable challenge to predict any specific change or occurrence as a function of climate change (Hallegraeff G.M., 2010). Increase of the frequency of hurricanes in the US is an example that change in temperature may not be the sole effect of climate change. Increased turbidity and re-suspension of nutrients typically associated with such events may well also impact on the development of harmful algae. A recent review made it clear that occurrences of algal blooms have augmented both spatially and temporally for some toxin producers at global level, such as producers of paralytic shellfish poisoning toxins (Anderson D.M., 2009). Such growth in occurrence and spread is likely to be related to a number of factors including shipping (through ballast water) and relocation of shellfish themselves.

It has been considered impossible to model the occurrence of individual phytoplankton species due to a lack of models capable of dealing with the complexity of the species composition and diversity of unicellular micro-algae (Anderson T.R., 2005). Exceptions to this dogma have been given through simplified experimentally-fed models on major bloom species such as *Alexandrium* (Stock et al;, 2007), *Karenia* (Vanhoute-Brunier et al., 2008) or *Pseudo-nitzschia* (Pénard et al., 2007; Ifremer, 2011). However, most toxic algae are minor species in the total composition of plankton at any given time, even if they reach cell concentration sufficient for the accumulation of unsafe levels of toxins in shellfish. Allen et al., 2010 and Finkel et al., 2010, have attempted to develop functional models of pelagic ecosystems including compartments based on phytoplankton community structure. However, it could not yet be demonstrated that these models are capable of predicting phytoplankton community changes for toxic algal species in an operational context applicable to monitoring systems.

Due to the more and more rapidly changing ecosystems and the low capability to predict any changes in minor components such as toxic species occurrence in the overall composition of phytoplankton community, it remains necessary to observe these changes experimentally in real time. Traditionally, such measures were only possible through light microscopy and the determination of species composition and enumeration by taxonomic experts. More recently a wealth of studies have provided tools for detecting phytoplankton changes also in more automated formats (immuno-chemical tests for toxins, molecular probes for potentially toxic species, automated recognition of cells in microscopy), sometimes with devices that may be deployed *in situ* (Hess et al., 2010). A particularly interesting development is the environmental sampler allowing for *in situ* measurement of a growing array of immuno-chemical and bio-molecular parameters (Kudela et al., 2010). Finally, a convincing study by Campbell et al., 2010, demonstrated the applicability of automated microscopic and video techniques for morpho-taxonomic detection of phytoplankton community changes. It is expected that more intense developments in these areas will allow for more common application of such specialised techniques in operational contexts.

A further technique of monitoring for phycotoxins has gained importance in recent years: passive sampling of phycotoxins dissolved in seawater. This technique has the advantage of integrating water samples over time and thus not being prone to the patchiness of phytoplankton samples in the water column. While it could not be demonstrated that this technique necessarily bears the potential for early warning (Fux et al., 2009), it could always be applied to the exploration of new shellfish production areas prior to shellfish cultures being placed into an area.

1.4.2. Monitoring of Biological Effects and Emerging Toxins

Traditionally, shellfish toxins had been monitored using bioassays based on live animals. Even though animal assays have a lot of ethical issues associated with them, their application allowed to overcome a number of shortcomings in the area of phycotoxins:

- Lack of knowledge on the actual compounds causing health problems
- Lack of calibration standards for method development, validation and quality assurance
- Lack of shellfish tissue reference materials for validation and verification of alternative tests

As described above, these issues have been successfully addressed over the last decade, even though much work remains to be done to ensure the sustainable supply of such tools and knowledge.

In the area of emerging toxins, the following compound-groups appear of increasing interest to European monitoring agencies:

- Cyclic imines (gymnodimines, spirolides, pinnatoxins and pteriatoxins)
- *Ostreopsis*-toxins (palytoxins, ostreocins, ovata-toxins)
- Ciguatera-related toxins (ciguatoxins, maitotoxins, gambiertoxins)
- Tetrodotoxins

Cyclic imines have been identified in a number of EU countries, however, as they are not regulated monitoring is not systematic and the consequent lack of data often results in very preliminary risk evaluation (EFSA, 2010c). While it could be shown that for gymnodimines and spirolides there appears to be a great discrepancy between intra-peritoneal and oral toxicity in animals (Munday R., 2008), no such re-assuring data exists for pinnatoxins (EFSA, 2010c). Existing data suggest that low levels of these compounds are observed in shellfish globally and that there is no acute health effects associated with cyclic imines (Miles et al., 2009; Selwood et al., 2010), however, a lack of comprehensive surveillance in combination with notorious under-reporting of poisoning events hinders the establishment of safe levels. As there is a number of causative organisms and several compound groups involved, it is likely that concerted efforts are needed to gain accuracy in risk assessment. Similarly, the lack of toxicological studies on the long-term effects and the lack of knowledge on the synergistic effects of multiple co-occurring toxin groups contribute to the preliminary character of many risk evaluation exercices.

The genus *Ostreopsis* has been associated with palytoxins, ostreocins and ovata-toxins. Initial reports had outlined problems with consumption of contaminated fish and crabs (EFSA, 2010g). However, the European concern has been triggered by more recent events in

Mediterranean coastal areas (Brescianini et al., 2006), in which people have been affected while bathing or simply being present on beaches during harmful algal events. The toxicity of palytoxin is very elevated and compounds for this group may well have several relevant molecular mechanisms of action (EFSA, 2010g). Aligizaki et al., 2008, have shown that shellfish accumulate toxins from these compound groups, and thus systematic assessment of the extent of the problem is required through official control.

Ciguatera-related compounds (ciguatoxins, maitotoxins and gambiertoxins) are produced by species of the genus *Gambierdiscus*, and have effects that may last for several months after consumption of contaminated fish (Caillaud et al., 2010c). This review also describes the recent poisoning event in the North-east Atlantic (Spanish waters), and also the occurrence of the causative algal species in the Mediterranean, in association with accumulating fish species. A recent unpublished report from French Mediterranean waters also corroborates that ciguatera fish poisoning may play an increasing role in Mediterranean waters (de Haro L., personal communication, 2011). The greatest stumbling block about any surveillance in the domain of ciguatoxins is that fish carrying the toxins can swim large distances, and thus the official control monitoring program has very little stronghold in terms of spatial representativeness. A further major complication is the fact that none of the current methods have been validated; only LC-MS/MS appears to be sensitive enough to detect the extremely low levels necessary to protect the consumer effectively (0.01 μg pacific ciguatoxin-equivalent / kg flesh), however, a lack of pure toxin prevents further development and validation of methods (EFSA, 2010a).

Tetrodotoxins have also recently been reported from European waters (Rodriguez et al., 2008). Even though this event has remained a single incident it outlines a significant shortcoming of most existing European monitoring programs, i.e. the absence of regular surveillance of gastropods for a wide variety of shellfish toxins. This lack has also been outlined in a recent review (Marcaillou et al., 2010), and should be addressed as a matter systematic optimisation of official control systems.

Thus, the application of bioassays has resulted in the discovery of a number of bioactive compounds that had and still have to be examined through risk evaluation exercises, including among others: pectenotoxins, yessotoxins, azaspiracids, gymnodimines, spirolides, pinnatoxins and palytoxins. The process and the approaches involved in the initial elucidation of such bioactive compounds were elegantly described by Quilliam M.A., 2003. A recent review of French official control data also suggested that there were still a number of unknown bioactive compounds to be discovered in bivalve molluscs (Belin et a., 2009).

A major effort has been made in the detection of neurotoxic bioactive molecules (Ledreux et al., 2009). The authors have developed an approach with cellular assays to allow for the detection of a variety of known shellfish toxin groups and for the identification of large groups of biological targets of neurotoxic compounds. However, the shellfish matrix still poses problems of interference with the assay and further efforts will be necessary to overcome this issue. A test applicable to both marine and freshwater toxins was optimised using recombinant enzymes (PP2A) (Ikehara et al., 2008), and quantitative structure/activity relationships of a number of analogues could be established, thus allowing for more effective monitoring of microcystins (Ikehara et al., 2009). In the area of lipophilic toxins, significant progress has also been achieved using a neuro-2A cell bioassay in one laboratory (Canete and Diogene 2008; Canete et al., 2010; Canete and Diogene 2010), however, it will be interesting to see whether this approach can be reproduced in a larger number of laboratories. The same authors also proved that their model can be used in combination with LC-MS-MS and PP2a for detection of OA in *P. rathymum* (Caillaud et al., 2010a) or for previously unknown maitotoxin-like compounds (Caillaud et al., 2010b). Finally, these authors also demonstrated the effectiveness of the cellular assay when coupled with fractionation HPLC for both shellfish and micro-algal matrices, although the assay format requires 48 h exposure to reach the best level of sensitivity (Caillaud et al., 2009).

Recent approaches involving passive sampling and proteomics have been outlined by Hess and Nicolau, 2010. In parallel, emerging anthropogenic contaminants have to be monitored for in a comprehensive manner (Kantiani et al., 2010) and novel approaches for this surveillance also need to be investigated. A recent study also outlined the overlapping effects of natural phycotoxins and anthropogenic contaminants, possibly leading to the need to comprehensive and distinctive surveillance methods (Tresch et al., 2011). Bearing in mind the various ethical and scientific disadvantages of bioassays using small mammals, such as high consumption of sample and high detection limits (EFSA, 2009f; Hess et al., 2006), it becomes apparent that novel techniques and approaches must be developed to screen for bioactive substances in the marine environment.

CONCLUSION

On one hand, the review of the evolution of legislation in the area of shellfish toxin and phytoplankton surveillance over a 20-year period has shown that legislation can change rapidly when political consensus exists for such change. On the other hand, despite strong political pressure from conflicting legislation on animal handling and protection, the regulations for shellfish toxins had remained largely the same over this period in terms of methodology due to lacking funds to develop alternative methods using specific quality assurance tools. It was finally mostly national or global efforts that have allowed such tools to be developed and the system to evolve. It would be desirable that the legislator take a more active role at EU level allowing for improvement and harmonisation of the legal framework in the area of shellfish toxins.

Due to conflicting legislation, the implementation of the existing framework - as assessed through examining FVO reports over a 10-year period - was relatively poor, leading to consistent repetitive non-implementation by a number of EU countries. These discrepancies were worsened by a lack of guidance documents for the shellfish industry. Analysis of the RASFF reports and risk evaluations also showed that the non-implementation or poor harmonisation of the system brought failures of monitoring systems with it, resulting in a number of cases of illnesses.

Global change is an additional challenge in the effective management of harmful algae and shellfish toxins; hence, it is anticipated that more emphasis will need to be placed on techniques allowing for the rapid detection of changes in the phytoplankton community structure. Recent progress in the area of in situ-buoys equipped with immuno-chemical sensors and biomolecular probes demonstrates the applicability of such techniques at operational level and thus suggests further efforts into this direction should be pursued.

Much research remains to be conducted in the area of emerging toxins including cyclic imines (spirolides, pinnatoxins etc.), and spreading toxins such as ciguatoxins, toxins produced by *Ostreopsis* species and tetrodotoxins. Efforts similar to the STX-, OA- and AZA-groups of toxins need to be made for these spreading / emerging toxins groups, i.e. standards and reference materials must be obtained to allow for systematic monitoring and effective risk evaluation.

At more generic level, biological effects monitoring for bioactive compounds in the marine environment needs to be developed for natural compounds in parallel with man-made contaminants, and recent efforts show that an array of miniaturised bioassays are available for further development and validation of high-throughput bio-monitoring.

ACKNOWLEDGMENTS

The author has received funding for this publication by French Ministry of Education, Research and Technology through Programme 187 of the National Finance Law. Additional funds have been obtained through the ASTOX2 project (Grant-Aid Agreement No. PBA-AF-08-001) carried out under the Sea Change strategy with the support of the Marine Institute and the Marine Research Sub-Programme of the Irish National Development Plan 2007–2013, co-financed under the European Regional Development Fund, and the COLNACOQ project, financed by the French Region "Pays de la Loire" (2009 - 2013).

REFERENCES

Aligizaki K., Katikou P., Nikolaidis G., Panou A. (2008). First episode of shellfish contamination by palytoxin-like compounds from *Ostreopsis* species (Aegean Sea, Greece). *Toxicon* 51 (3), 418-427.

Allen J. I., Aiken J., Anderson T. R., Buitenhuis E., Cornell S., Geider R. J., Haines K., Hirata T., Holt J., Le Quéré C., Hardman-Mountford N., Ross O. N., Sinha B. and While J. (2010). Marine ecosystem models for earth systems applications: The MarQUEST experience. *J. Marine Systems* 81(1-2): 19-33.

Anderson D. M. (2009). Approaches to monitoring, control and management of harmful algal blooms (HABs). *Ocean & Coastal Management* 52(7): 342-347.

Anderson T. R. (2005). Plankton functional type modelling: running before we can walk? *J. Plankton Res.* 27(11): 1073-1081.

Anonymous (1986) Council Directive 86/609/EEC of 24 November 1986 on the approximation of laws, regulations and administrative provisions of the Member States regarding the protection of animals used for experimental and other scientific purposes. *Official Journal* L358, 1–29.

Anonymous (1991a) Council Directive 91/492/EEC of 15 July 1991 laying down the health conditions for the production and the placing on the market of live bivalve molluscs. *Official Journal* L 268, Volume 34, 24/09/1991, pp. 1-14.

Anonymous (1991b) Council Directive 91/493/EEC of 22 July 1991 laying down the health conditions for the production and the placing on the market of fishery products. *Official Journal* L 268 Volume 34, 24/09/1991 pp. 15 - 34

Anonymous (1993) Council Decision 93/383/EEC of 14 June 1993 on reference laboratories for the monitoring of marine biotoxins. *Official Journal* L 166, 08/07/1993 pp. 31 – 33.

Anonymous (1996) Commission Decision 96/77/EC of 18 January 1996 establishing the conditions for the harvesting and processing of certain bivalve molluscs coming from areas where the paralytic shellfish poison level exceeds the limit laid down by Council Directive 91/492/EEC. *Official Journal* L 15 , 20/01/1996 pp. 46 – 47

Anonymous (1997) Council Directive 97/61/EC of 20 October 1997 amending the Annex to Directive 91/492/EEC laying down the health conditions for the production and placing on the market of live bivalve molluscs. *Official Journal* L 295 , 29/10/1997 pp. 35 – 36.

Anonymous (1999a) Council Decision 1999/312/EC of 29 April 1999 amending Decision 93/383/EEC on reference laboratories for the monitoring of marine biotoxins. Official Journal L120, 8/5/1999 pp. 37 – 39.

Anonymous (1999b) Council Decision 1999/468/EC of 28 June 1999 laying down the procedures for the exercise of implementing powers conferred on the Commission. *Official Journal* L184, 17/7/1999 pp. 23 – 26.

Anonymous (2002a) Commission Decision 2002/225/EC of 15 March 2002 laying down detailed rules for the implementation of Council Directive 91/492/EEC as regards the maximum levels and the methods of analysis of certain marine biotoxins in bivalve molluscs, echinoderms, tunicates and marine gastropods. *Official Journal* L75, 16/3/2002 pp. 62-64.

Anonymous (2002b) Commission Decision 2002/226/EC of 15 March 2002 establishing special health checks for the harvesting and processing of certain bivalve molluscs with a level of amnesic shellfish poison (ASP) exceeding the limit laid down by Council Directive 91/492/EEC. *Official Journal* L75, 16/3/2002 P. 65-66.

Anonymous (2004a) Regulation (EC) No 853/2004 of the European Parliament and of the Council of 29 April 2004 laying down specific hygiene rules for food of animal origin. *Official Journal* L 139, 30.4.2004, pp. 55–205.

Anonymous (2004b) Regulation (EC) No 854/2004 of the European Parliament and of the Council of 29 April 2004 laying down specific rules for the organisation of official controls on products of animal origin intended for human consumption. *Official Journal* L 139, 30/4/2004, pp. 206–320.

Anonymous (2005) Commission Regulation (EC) No 2074/2005 of 5 December 2005 laying down implementing measures for certain products under Regulation (EC) No 853/2004 of the European Parliament and of the Council and for the organisation of official controls under Regulation (EC) No 854/2004 of the European Parliament and of the Council and Regulation (EC) No 882/2004 of the European Parliament and of the Council, derogating from Regulation (EC) No 852/2004 of the European Parliament and of the Council and amending Regulations (EC) No 853/2004 and (EC) No 854/2004 *Official Journal* L 338, 22/12/2005, pp. 27–59.

Anonymous (2006) Commission Regulation (EC) No 1664/2006 of 6 November 2006 amending Regulation (EC) No 2074/2005 as regards implementing measures for certain products of animal origin intended for human consumption and repealing certain implementing measures. *Official Journal* L 320, 18/11/2006, pp. 13–45.

Anonymous (2007) Commission Regulation (EC) No 1244/2007 of 24 October 2007 amending Regulation (EC) No 2074/2005 as regards implementing measures for certain products of animal origin intended for human consumption and laying down specific rules on official controls for the inspection of meat. *Official Journal* L 281, 25/10/2007, pp. 12–18.

Anonymous (2011) Commission Regulation (EU) No 15/2011 of 10 January 2011 amending Regulation (EC) No 2074/2005 as regards recognised testing methods for detecting marine biotoxins in live bivalve molluscs. *Official Journal* L6, 11/1/2011, pp. 3–6.

Bates, H. A. and Rapoport, H. (1975). Chemical assay for saxitoxin, the paralytic shellfish poison. *J. Agric. Food Chem.* 23(2): 237-239.

Beaugrand G. and Reid P.C. 2003. Long-term changes in phytoplankton, zooplankton and salmon related to climate. *Global Change Biol.*, 9, 801-817.

Belin C., Soudant D. and Amzil Z. (2009) Surveillance des toxines lipophiles dans les coquillages. Analyses statistique et comparaison des résultats obtenus par deux méthodes d'analyse : les bio-essais sur souris et les analyses chimiques par CL-SM/SM. Rapport exécuté dans le cadre de la Convention Etudes DGAL / Ifremer, correspondant à la Subvention pour charges de service public, Programme 206, 97 p. http://wwz.ifremer.fr/envlit/documents/publications (rubrique Phytoplancton et Phycotoxines, last accessed : 11/03/2011).

Biosense (2011) website of the Biosense company, supplier for ASP-ELISA test kit, last accessed on 2011-03-06: www.biosense.com/docs/A31300401-sheet.pdf.

Blanco J., Acosta C.P., Bermùdez de la Puente M., Salgado C. (2002) Depuration and anatomical distribution of the amnesic shellfish poisoning (ASP) toxin domoic acid in the king scallop *Pecten maximus*. *Aquat. Toxicol.* 60 (1-2) 111-121.

Bogan, Y.M., Bender, K., Hervas, A., Kennedy, D.J., Slater, J.W., Hess, P. (2007) Spatial variability of domoic acid concentration in king scallops *Pecten maximus* off the southeast coast of Ireland. *Harmful Algae*, 6, 1-14.

Brescianini C., Grillo C., Melchiorre N., Bertolotto R., Ferrari A., Vivaldi B., Icardi G., Gramaccioni L., Funari E., Scardala S. (2006) *Ostreopsis ovata* algal blooms affecting human health in Genoa, Italy, 2005 and 2006. *Eurosurveillance* 11 (36) 3040. Available online (last accessed: 23/03/2011:
http://www.eurosurveillance.org/ViewArticle.aspx?ArticleId=3040.

Caillaud A., Cañete E., de la Iglesia P., Giménez G. and Diogène J. (2009). Cell-based assay coupled with chromatographic fractioning: A strategy for marine toxins detection in natural samples. *Toxicology in Vitro* 23(8): 1591-1596.

Caillaud A., de la Iglesia P., Campàs M., Elandaloussi L., Fernández M., Mohammad-Noor N., Andree K. and Diogène J. (2010a) Evidence of okadaic acid production in a cultured strain of the marine dinoflagellate Prorocentrum rhathymum from Malaysia. *Toxicon* 55(2-3): 633-637.

Caillaud A., Yasumoto T. and Diogène J. (2010b) Detection and quantification of maitotoxin-like compounds using a neuroblastoma (Neuro-2a) cell based assay. Application to the screening of maitotoxin-like compounds in Gambierdiscus spp. *Toxicon* 56(1): 36-44.

Caillaud A., de la Iglesia P., Darius H.T., Pauillac S., Aligizaki K., Fraga S., Chinain M. and Diogène J. (2010c) Update on Methodologies Available for Ciguatoxin Determination: Perspectives to Confront the Onset of Ciguatera Fish Poisoning in Europe. *Marine Drugs* 8, 1838-1907

Campbell L., Olson R.J., Sosik H.M., Abraham A., Henrichs D.W., Hyatt C.J., Buskey E.J. (2010) First harmful Dinophysis (*Dinophyceae, Dinophysiales*) bloom in the U.S is revealed by automatic imaging flow cytometry. *J. Phycol.* 46, 66-75.

Canete E., Campas M., de la Iglesia P. and Diogene J. (2010) NG108-15 cell-based and protein phosphatase inhibition assays as alternative semiquantitative tools for the screening of lipophilic toxins in mussels. Okadaic acid detection. *Toxicology in Vitro* 24(2): 611-619.

Canete E. and Diogene J. (2008) Comparative study of the use of neuroblastoma cells (Neuro-2a) and neuroblastorna x glioma hybrid cells (NG108-15) for the toxic effect quantification of marine toxins. *Toxicon* 52(4): 541-550.

Canete, E. and Diogene, J. (2010) Improvements in the use of neuroblastoma x glioma hybrid cells (NG108-15) for the toxic effect quantification of marine toxins. *Toxicon* 55(2-3): 381-389.

Codex Alimentarius (2008) *Codex Standard* 292/2008. Standard for live and raw bivalve molluscs. Pp 1-7.

CRL-MB / DG Sanco (1998) Report of the meeting of the Working Group of the CRL and NRLs on DSP toxins Brussels - 26 November 1998. Available (after registration) at URL: https://www.aesa.msc.es/crlmb/web/CRLMB.jsp.

CRL-MB / DG Sanco (1999a) Report of the meeting of the Working group of the CRL and NRL on Marine Biotoxins - ASP determination Wednesday, 3 Feb 1999. Available (after registration) at URL: https://www.aesa.msc.es/crlmb/web/CRLMB.jsp.

CRL-MB / DG Sanco (1999b) Report of the meeting of the Working Group on Toxicology of YTXs, PTXs and the OA group toxins. Brussels, 24 February 1999. Available (after registration) at URL: https://www.aesa.msc.es/crlmb/web/CRLMB.jsp.

CRL-MB / DG Sanco (2000) Report of the meeting of the Working groups of the CRL and NRL on ASP, Brussels 17 February 2000 and 25 October 2000 (Doc SANCO/4024/2000). Available (after registration) at URL: https://www.aesa.msc.es/crlmb/web/CRLMB.jsp.

CRL-MB / DG Sanco (2001a) Report of the meeting of the Working group of the CRL and NRL on ASP toxins, Brussels 3-4 April 2001. Available (after registration) at URL: https://www.aesa.msc.es/crlmb/web/CRLMB.jsp.

CRL-MB / DG Sanco (2001b) Report of the meeting of the Working Group on Toxicology of DSP and AZP 21 - 23 May 2001, Brussels. Available (after registration) at URL: https://www.aesa.msc.es/crlmb/web/CRLMB.jsp.

CRL-MB (2007) Standard Operating Procedure for detection of Okadaic acid, Dinophysistoxins and Pectenotoxins by Mouse Bioassay. Version 4.0 (April 2007). Available (after registration) at URL: https://www.aesa.msc.es/crlmb/web/CRLMB.jsp.

De Haro L. (2011) Personal communication, 12/01/2011: two patients have reported at the poison centre of Marseille, France, to suffer illness after consumption of fish that were later identified to be Barracuda, caught off the South coast of Corsica. The symptoms described by medical experts appear to fit the description of ciguatera fish poisoning. Initial analysis by the French national Reference Laboratory (Anses) using animal assays did not preclude the possibility of ciguatera as the causative agent for the poisoning.

De Schrijver K., Maes I., De Man L. and Michelet J. (2002) An outbreak of diarrhoeic shellfish poisoning in Antwerp, Belgium. *Eurosurveillance* 7, 139-141.

EFSA (European Food Safety Authority) (2008a), Marine biotoxins in shellfish – Okadaic Acid and analogues, Scientific Opinion of the Panel on Contaminants in the Food chain, adopted on 27 November 2007, *EFSA Journal* 589, 1-62

EFSA (European Food Safety Authority) (2008b), Marine biotoxins in shellfish – Azaspiracid group, Scientific Opinion of the Panel on Contaminants in the Food chain, adopted on 9 June 2008, *EFSA Journal* 723, 1-52.

EFSA (European Food Safety Authority) (2009a), Marine biotoxins in shellfish – Yessotoxin group, Scientific Opinion of the Panel on Contaminants in the Food chain, adopted on the 2 December 2008, *EFSA Journal* 907, 1-62.

EFSA (European Food Safety Authority) (2009b), Influence of processing on the levels of lipophilic marine biotoxins in bivalve molluscs, Statement of the Panel on Contaminants in the Food Chain (Question No EFSA-Q-2009-00203), Adopted on 25 March 2009, *EFSA Journal* 1016, 1-10.

EFSA (European Food Safety Authority) (2009c), Marine biotoxins in shellfish – Saxitoxin group; Scientific Opinion of the Panel on Contaminants in the Food Chain, Adopted on 25 March 2009, *EFSA Journal* 1019, 1-76.

EFSA (European Food Safety Authority) (2009d), Marine biotoxins in shellfish – Pectenotoxin group, Scientific Opinion of the Panel on Contaminants in the Food chain; adopted on 27 May 2009, *EFSA Journal* 1109, 1-47.

EFSA (European Food Safety Authority) (2009e), Marine biotoxins in shellfish – Domoic acid group, Scientific Opinion of the Panel on Contaminants in the Food chain; adopted on 2 July 2009, *EFSA Journal* 1181, 1-61.

EFSA (European Food Safety Authority) (2009f), Marine biotoxins in shellfish – Summary Opinion, Scientific Opinion of the Panel on Contaminants in the Food chain; adopted on 13 August 2009, *EFSA Journal* 1306, 1 - 23.

EFSA (European Food Safety Authority) (2009g), Marine biotoxins in shellfish – Palytoxin Group, Scientific Opinion of the Panel on Contaminants in the Food chain; adopted on 13 August 2009, *EFSA Journal* 7 (12) 1393, 1 - 38.

EFSA (European Food Safety Authority) (2010a), Marine biotoxins in shellfish – Emerging Toxins: Ciguatoxin-group, Scientific Opinion of the Panel on Contaminants in the Food chain; adopted on 18 May 2010, *EFSA Journal* 8 (6) 1627, 1 - 38.

EFSA (European Food Safety Authority) (2010b), Marine biotoxins in shellfish – Emerging Toxins: Brevetoxin-group, Scientific Opinion of the Panel on Contaminants in the Food chain; adopted on 5 July 2010, *EFSA Journal* 8 (7) 1677, 1 - 29.

EFSA (European Food Safety Authority) (2010c), Marine biotoxins in shellfish – Cyclic Imines (spirolides, gymnodimines, pinnatoxins and pteriatoxins), Scientific Opinion of the Panel on Contaminants in the Food chain; adopted on 5 July 2010. *EFSA Journal* 8 (6) 1628, 1-39.

EFSA (European Food Safety Authority) (2010d), Statement on further elaboration on consumption figure of 400 g shellfish meat on the basis of new consumption data, EFSA Panel on contaminants in the food chain.; adopted on 31 July 2010, *EFSA Journal* 8 (8) 1706, 1 - 20.

FAO/IOC/WHO (2005) Report of the Joint FAO/IOC/WHO ad hoc Expert Consultation on Biotoxins in Bivalve Molluscs, Oslo, Norway, Sept. 26-30, 2004. Available at: ftp://ftp.fao.org/es/esn/food/biotoxin_report_en.pdf (accessed 5 June 2009)

Finkel Z. V., Beardall J., Flynn K. J., Quigg A., Rees T. A. V. and Raven J. A. (2010). Phytoplankton in a changing world: cell size and elemental stoichiometry. *J. Plankton Res.* 32(1): 119-137.

Fux E., Biré R., Hess P. (2009) Comparative accumulation and composition of lipophilic marine biotoxins in passive samplers and in mussels (M. edulis) on the West Coast of Ireland. *Harmful Algae*, 8, 523-537.

Hallegraeff G.M. (2010) Ocean Climate Change and Harmful Algal Blooms: a formidable predictive challenge. Keynote lecture at the 14th International Conference on Harmful Algae, Crete, 1 – 5 November 2010, Hersonissos, Crete, Greece.

Hess P., Brown N.A., Bates L.A., Turriff J.J., Howard F.G. and Moffat C.F. (2000) A comparison of domoic acid concentrations in scallop (*Pecten maximus*) gonads processed by FRS MLA and those prepared by a commercial processor. Fisheries Research Services - Report No 03/00. 12pp.

Hess P., Gallacher S., Bates L.A., Brown N., Quilliam M.A. (2001) Determination and confirmation of the amnesic shellfish poisoning toxin, domoic acid, in shellfish from Scotland by liquid chromatography and mass spectrometry. *J. Ass. Off. Anal. Chem.* 84 (5) 1657-1667.

Hess P., Grune B., Anderson D.B., Aune T., Botana L.M., Caricato P., van Egmond H.P., Halder M., Hall S., Lawrence J.F., Moffat C., Poletti R., Richmond J., Rossini G.P., Seamer C. and Serratosa Vilageliu J., 2006. Three Rs Approaches in Marine Biotoxin Testing, The Report and Recommendations of a joint ECVAM/DG SANCO Workshop (ECVAM Workshop 55). *Altern. Labor. Anim. (ATLA)* 34, 193-224. Available at URL: http://altweb.jhsph.edu/publications/journals/atla/34_2/ECVAM%20Workshop%2055%2034.2.pdf.

Hess P. and McCarron P. (2007) Review of Proficiency Testing for Shellfish Toxins. Oral presentation at 6th Intl. Conf. Molluscan Shellfish Safety (ICMSS), March 18 – 23, 2007 Blenheim, New Zealand.

Hess P., McCarron P., Quilliam M.A. (2007) Fit-for-purpose Shellfish Reference Materials for Phycotoxins in Internal and External Quality Control. *Anal. Bioanal. Chem.* 387, 2463-2474.

Hess P., McMahon T., Slattery D., Swords D., Dowling G., McCarron M., Clarke D., Gibbons W., Silke J., O'Cinneide M. (2003) Use of LC-MS testing to identify lipophilic toxins, to establish local trends and interspecies differences and to test the comparability of LC-MS testing with the mouse bioassay: an example from the Irish biotoxin monitoring programme 2001". in : *Molluscan Shellfish Safety*; editors: A. Villalba, B. Reguera, J.L. Romalde, R. Beiras; publishers: Consellería de Pesca e Asuntos Marítimos da Xunta de Galicia and Intergovernmental Oceanographic Commission of UNESCO, Santiago de Compostela, Spain. ISBN: 84-453-3638-X.

Hess P., Nguyen L., Aasen J., Keogh M., Kilcoyne J., McCarron P., Aune T. (2005) Tissue distribution, effects of cooking and parameters affecting the extraction of azaspiracids from mussels, Mytilus edulis, prior to analysis by liquid chromatography coupled to mass spectrometry. *Toxicon* 46, 62-71.

Hess P. and Nicolau E. (2010) Recent developments in the detection of phycotoxins. Advances and new technologies in Toxinology, Meeting on Toxinology, 2010, SFET Editions, pp. 35 – 41. Free on-line access on the website : http://www.sfet.asso.fr

Hess P., Amzil Z., Belin C., Compere C., Lassus P., Menesguen A. (2010) Evolution of European monitoring systems for shellfish toxins – an increased need for the surveillance of phytoplankton. Oral presentation at the 14th Intl. Conference on Harmful Algal Blooms, 1 – 5 November 2010, Hersonissos, Crete.

Ifremer (2011) PREVIMER-website (last accessed: 22/03/2011): http://www.previmer.org/previsions/production_primaire/modele_eco_mars3d_bretagne

Ikehara T., Imamura S., Oshiro N., Ikehara S., Shinjo F. and Yasumoto T. (2008) A protein phosphatase 2A (PP2A) inhibition assay using a recombinant enzyme for rapid detection of microcystins. *Toxicon* 51(8): 1368-1373.

Ikehara T., Imamura S., Sano T., Nakashima J., Kuniyoshi K., Oshiro N., Yoshimoto M. and Yasumoto T. (2009) The effect of structural variation in 21 microcystins on their inhibition of PP2A and the effect of replacing cys269 with glycine. *Toxicon* 54(4): 539-544.

Kantiani L., Llorca M., Sanchís J., Farré M., Barceló D. (2010) Emerging food contaminants: a review. *Anal. Bioanal. Chem.* 398, 2413–2427

Kleivdal H., Kristiansen S. I., Nilsen M. V. and Briggs L. (2007a). Single-laboratory validation of the biosense direct competitive enzyme-linked immunosorbent assay (ELISA) for determination of domoic acid toxins in shellfish. J. AOAC Intl. 90(4): 1000-1010.

Kleivdal H., Kristiansen S. I., Nilsen M. V., Goksoyr A., Briggs L., Holland P. and McNabb P. (2007b). Determination of domoic acid toxins in shellfish by Biosense ASP ELISA - A direct competitive enzyme-linked immunosorbent assay: Collaborative study. *J. AOAC Intl.* 90(4): 1011-1027.

Kudela R. M., Howard M. D.A., Jenkins B. D., Miller P. E. and Smith G. J. (2010). Using the molecular toolbox to compare harmful algal blooms in upwelling systems. *Progr. Oceanogr.* 85 (1-2) 108-121.

Lawrence J. F. and Menard C. (1991) Liquid chromatographic determination of paralytic shellfish poisons in shellfish after pre-chromatographic oxidation. *J. AOAC Int.*: 1006-1012.

Ledreux, A., Krys, S. and Bernard, C. (2009) Suitability of the Neuro-2a cell line for the detection of palytoxin and analogues (neurotoxic phycotoxins). *Toxicon* 53(2): 300-308.

Lenton A., Codron F., Bopp L., Metzl N., Cadule P., Tagliabue A. and Le Sommer J. (2009). Stratospheric ozone depletion reduces ocean carbon uptake and enhances ocean acidification. *Geophys. Res. Lett.* 36 (12) L12606.

Marcaillou-Le Baut C., Vernoux J.P., Arnich N., Frémy J.M. (2010) Occurrence of phycotoxins in marine shellfish other than bivalve molluscs: an update. Pp. 220 – 228 in Proc. 7th Intl. Conf. Molluskan Shellfish Safety June 14th – 19th 2009, Nantes, France. Editor: P. Lassus, Ifremer edition, ISBN: 978-2-84433-161-8.

McCarron P. and Hess P. (2006) Tissue distribution and effects of heat treatments on the content of domoic acid in blue mussels, Mytilus edulis. *Toxicon* 47, 473-479.

McCarron P., Kilcoyne J., Hess P. (2008) Effects of cooking and heat treatment on concentration and tissue distribution of okadaic acid and dinophysistoxin-2 in mussels (*Mytilus edulis*). *Toxicon* 51, 1081-1089.

McCarron P., Kilcoyne J., Miles C.O., Hess P. (2009) Formation of Azaspiracids-3, -4, -6, and -9 via decarboxylation of carboxyazaspiracid metabolites from shellfish. *J. Agric. Food Chem.* 57, 160-169.

Miles C.O., Rundberget T., Sandvik M., Aasen J.A.B., Selwood A. (2009) The presence of pinnatoxins in norwegian mussels. *National Veterinary Institute`s Report series* 7b-2010. Oslo: National Veterinary Institute.

Munday R., 2008. Toxicology of cyclic imines: gymnodimine, spirolides, pinnatoxins, pteriatoxins, prorocentrolide, spiro-prorocentrimine, and symbioimines. In: Seafood and Freshwater toxins: Pharmacology, Physiology and Detection. 2nd edition. Ed LM Botana. CRC Press (Taylor and Francys Group), Boca Raton, FL, 581-594.

Murata M., Kumagai M., Lee J. S. and Yasumoto T. (1987). Isolation and structure of yessotoxin, a novel polyether compound implicated in diarrhetic shellfish poisoning. *Tetrahedron Lett.* 28: 5869-72.

Murata M., Sano M., Iwashita T., Naoki H. and Yasumoto T. (1986). The structure of pectenotoxin-3, a new constituent of diarrhetic shellfish toxins. *Agric. Biol. Chem* 50(10): 2693-5.

NRCC (2011) National Research Council Canada, web-site of the Certified Reference Materials Programme : http://www.nrc-cnrc.gc.ca/eng/programs/imb/crmp.html (last accessed on 23/3/2011).

Penard, C., Menesguen, A., Dumas, F. and Guillaud, J. F. (2007) Towards operational modelling of the fate of nutrients in the coastal zone off Brittany (France). Houille Blanche - Revue Internationale de l'Eau: 62-67.

QUASIMEME (2011) web-site of proficiency test provider for shellfish toxins (last accessed: 23/03/2011): http://www.quasimeme.org/nl/25222726-Home.html

Quilliam M.A. (2003) The role of chromatography in the hunt for red tide toxins. *J. Chromatogr. A* 1000, 527 – 548.

Quilliam M.A. and Wright J.L.C. (1989) The amnesic shellfish poisoning mystery. *Anal. Chem.* 61 (18) 1053A – 1060A.

RASFF (2011) EU rapid alert web-sit (last accessed: 22/03/2011) http://ec.europa.eu/food/food/rapidalert/rasff_portal_database_en.htm

Reid P.C., Johns D.G., Edwards M., Starr M., Poulin M., Snoeijs P. (2007) A biological consequence of reducing Arctic ice cover: arrival of the Pacific diatom *Neodenticula seminae* in the North Atlantic for the first time in 800 000 years. *Global Change Biol.* 13, 1910-1921.

Rodriguez P., Alfonso A., Vale C., Alfonso C., Vale P., Tellez A. and Botana L. M. (2008) First toxicity report of tetrodotoxin and 5,6,11-trideoxyTTX in the trumpet shell *Charonia lampas lampas* in Europe. *Anal. Chem.* 80 (14): 5622-5629.

Rossini G.P. and Hess P. (2010) Phycotoxins: chemistry, mechanism of action and shellfish poisoning. Chapter 2.3 in "Clinical Toxicology", Vol. 2, pages: 65-122. Andreas Luch (ed.) *Experientia Supplementa* (EXS) *book series*, Birkhäuser Publishing, Basel.

Selwood A. I., Miles C. O., Wilkins A. L., van Ginkel R., Munday R., Rise F. and McNabb P. (2010). Isolation, Structural Determination and Acute Toxicity of Pinnatoxins E, F and G. *J. Agric. Food Chem.* 58 (10): 6532-6542.

Stock C. A., McGillicuddy D. J., Anderson D. M., Solow A. R. and Signell R. P. (2007). "Blooms of the toxic dinoflagellate *Alexandrium fundyense* in the western Gulf of Maine in 1993 and 1994: A comparative modeling study." *Continental Shelf Research* 27(19): 2486-2512.

Tresch S., Schmotz J. and Grossmann K. (2011). Probing mode of action in plant cell cycle by the herbicide endothall, a protein phosphatase inhibitor. *Pestic. Biochem. Physiol.* 99(1): 86-95.

UK-COT (The Committee on Toxicity of Chemicals in Food, Consumer Products and the Environment), 2006. Statement on risk assessment of marine biotoxins of the OA, PTX,

AZA and YTX groups in support of human health. Available at URL: http://www.food.gov.uk/multimedia/pdfs/cotstatementlipophilic200616.pdf .

Utermöhl, H., 1958. Zur Vervollkommnung der quantitativen Phytoplankton-Methodik. *Mitteilungen-Internationale Vereinigung für Limnlogie* 9, 1–38.

Vanhoutte-Brunier A., Fernand L., Ménesguen A., Lyons S., Gohin F. and Cugier P. (2008). Modelling the Karenia mikimotoi bloom that occurred in the western English Channel during summer 2003. *Ecological Modelling* 210 (4): 351-376.

List of Contributors

Amparo Alfonso,
Departamento de Farmacología,
Facultad de Veterinaria, USC,
27002 Lugo, Spain

Miroslava Atanassova,
(ANFACO-CECOPESCA) Microbiology and Biotoxins Area,
Spanish National Association of Sea Food Producers-Technological Centre,
Crta. Col. Univ. 16, 36310 Vigo, Spain

Juan Blanco,
CIMA, Pedras de Corón s/n. Apdo. 13,
36620 Vilanova de Arousa, Pontevedra, Spain

Ana M. Botana,
Departamento de Química Analítica,
Facultad de Ciencias, USC,
27002 Lugo, Spain

Luis M. Botana,
Departamento de Farmacología,
Facultad de Veterinaria, USC,
27002 Lugo, Spain

Larry E. Brand,
Marine Biology and Fisheries,
Rosenstiel School of Marine and Atmospheric Science,
University of Miami,
Miami, Florida 33149, US

Ana G. Cabado,
(ANFACO-CECOPESCA) Microbiology and Biotoxins Area,
Spanish National Association of Sea Food Producers-Technological Centre,
Crta. Col. Univ. 16, 36310 Vigo, Spain

Katrina Campbell,
Institute of Agri-food and Land Use,
Queen's University, University Road Belfast,
BT7 1NN, Northern Ireland, UK

María-José Chapela,
(ANFACO-CECOPESCA) Microbiology and Biotoxins Area,
Spanish National Association of Sea Food Producers-Technological Centre,
Crta. Col. Univ. 16, 36310 Vigo, Spain

Martiña Ferreira,
(ANFACO-CECOPESCA) Microbiology and Biotoxins Area,
Spanish National Association of Sea Food Producers-Technological Centre,
Crta. Col. Univ. 16, 36310 Vigo, Spain

Mª Elina Garet Fernández,
Manager Nanoimmunotech S.L.,
Plaza de Fernando Conde Montero-Ríos, 9,
36201 Vigo, Spain

África González-Fernández,
Director of Biomedical Research Centre,
University of Vigo, 36310 Vigo, Spain

Stacey L. DeGrasse,
U.S. FDA Center for Food Safety and Applied Nutrition,
Office of Regulatory Science,
Division of Analytical Chemistry,
College Park, Maryland, 20740 US

Christopher T. Elliott,
Institute of Agri-food and Land Use,
Queen's University, University Road Belfast,
BT7 1NN, Northern Ireland, UK

José María Franco,
Instituto de Investigaciones Marinas,
Unidad Asociada de Fitoplancton Tóxico (CSIC-IEO),
Ap. 1552, 36080 Vigo, Spain

Philipp Hess,
Ifremer, Unité de Recherche et de Service,
Environnement, Microbiologie & Phycotoxines,
Nantes, Cedex 03, BP21105, France

Hans Kleivdal,
Uni Research, Department of Environment,
Postboks 7800, NO-5020, Bergen, Norway

Jean-Pierre Lacaze,
Marine Scotland Science,
Aberdeen, AB11 9DB, Scotland

Jorge Lago,
(ANFACO-CECOPESCA) Microbiology and Biotoxins Area,
Spanish National Association of Sea Food Producers-Technological Centre,
Crta. Col. Univ. 16, 36310 Vigo, Spain

Elena López,
Laboratorio de Sanidad Exterior,
Ministerio de Política Territorial y Administración Pública,
Vigo, Spain

Mari Carmen Louzao,
Departamento de Farmacología,
Facultad de Veterinaria, USC,
27002 Lugo, Spain

Karen Martinez-Diaz,
U.S. FDA Center for Food Safety and Applied Nutrition,
Office of Regulatory Science,
Division of Analytical Chemistry,
College Park, Maryland 20740 US

Alberto Otero,
(ANFACO-CECOPESCA) Microbiology and Biotoxins Area,
Spanish National Association of Sea Food Producers-Technological Centre,
Crta. Col. Univ. 16, 36310 Vigo, Spain

Beatriz Reguera,
IEO, Centro Oceanográfico de Vigo,
Cabo Estay, Canido, Aptdo 1552,
36200 Vigo, Spain

Pilar Riobó,
Instituto de Investigaciones Marinas,
Unidad Asociada de Fitoplancton Tóxico (CSIC-IEO),
Ap. 1552, 36080 Vigo, Spain

Francisco Rodríguez,
IEO, Centro Oceanográfico de Vigo,
Cabo Estay, Canido, Aptdo 1552,
36200 Vigo, Spain

Toshiyuki Suzuki,
National Research Institute of Fisheries Science,
2-12-4 Fukuura, Kanazawa, Yokohama,
Kanagawa 236-8648, Japan

Carmen Vale,
Departamento de Farmacología,
Facultad de Veterinaria, USC,
27002 Lugo, Spain

V. Vasconcelos,
Laboratory of Ecotoxicology, Genomics and Evolution,
Centro Interdisciplinar de Investigação Marinha e Ambiental (CIIMAR),
Rua dos Bragas, 289, University of Porto,
4050-123 Porto, Portugal

Mercedes R. Vieytes,
Departamento de Fisiología,
Facultad de Veterinaria USC,
27002 Lugo, Spain

Natalia Vilarino,
Departamento de Farmacología,
Facultad de Veterinaria, USC,
27002 Lugo, Spain

Ryuichi Watanabe,
National Research Institute of Fisheries Science,
2-12-4 Fukuura, Kanazawa, Yokohama,
Kanagawa 236-8648, Japan

INDEX

B

C

D

E

F

G

H

I

N

O

P

T

U